Hamiltonian Mechanics

Integrability and Chaotic Behavior

NATO ASI Series

Advanced Science Institutes Series

A series presenting the results of activities sponsored by the NATO Science Committee, which aims at the dissemination of advanced scientific and technological knowledge, with a view to strengthening links between scientific communities.

The series is published by an international board of publishers in conjunction with the NATO Scientific Affairs Division

A	**Life Sciences**	Plenum Publishing Corporation
B	**Physics**	New York and London
C	**Mathematical and Physical Sciences**	Kluwer Academic Publishers
D	**Behavioral and Social Sciences**	Dordrecht, Boston, and London
E	**Applied Sciences**	
F	**Computer and Systems Sciences**	Springer-Verlag
G	**Ecological Sciences**	Berlin, Heidelberg, New York, London,
H	**Cell Biology**	Paris, Tokyo, Hong Kong, and Barcelona
I	**Global Environmental Change**	

Recent Volumes in this Series

Volume 328 —Quantum Field Theory and String Theory
edited by Laurent Baulieu, Vladimir Dotsenko, Vladimir Kazakov, and Paul Windey

Volume 329 —Nonlinear Coherent Structures in Physics and Biology
edited by K. H. Spatschek and F. G. Mertens

Volume 330 —Coherent Optical Interactions in Semiconductors
edited by R. T. Phillips

Volume 331 —Hamiltonian Mechanics: Integrability and Chaotic Behavior
edited by John Seimenis

Volume 332 —Deterministic Chaos in General Relativity
edited by David Hobill, Adrian Burd, and Alan Coley

Volume 333 —Perspectives in the Structure of Hadronic Systems
edited by M. N. Harakeh, J. H. Koch, and O. Scholten

Volume 334 —Frontier Topics in Nuclear Physics
edited by Werner Scheid and Aurel Sandulescu

Volume 335 —Hot and Dense Nuclear Matter
edited by Walter Greiner, Horst Stöcker, and André Gallmann

Series B: Physics

Hamiltonian Mechanics

Integrability and Chaotic Behavior

Edited by

John Seimenis

University of the Aegean
Samos, Greece
and N. Copernicus University
Torun, Poland

Plenum Press
New York and London
Published in cooperation with NATO Scientific Affairs Division

Proceedings of a NATO Advanced Research Workshop on
Hamiltonian Mechanics,
held June 28–July 2, 1993, in Torun, Poland

NATO-PCO-DATA BASE

The electronic index to the NATO ASI Series provides full bibliographical references (with keywords and/or abstracts) to more than 30,000 contributions from international scientists published in all sections of the NATO ASI Series. Access to the NATO-PCO-DATA BASE is possible in two ways:

—via online FILE 128 (NATO-PCO-DATA BASE) hosted by ESRIN, Via Galileo Galilei, I-00044 Frascati, Italy

—via CD-ROM "NATO Science and Technology Disk" with user-friendly retrieval software in English, French, and German (©WTV GmbH and DATAWARE Technologies, Inc. 1989). The CD-ROM also contains the AGARD Aerospace Database.

The CD-ROM can be ordered through any member of the Board of Publishers or through NATO-PCO, Overijse, Belgium.

Library of Congress Cataloging in Publication Information

Hamiltonian mechanics : integrability and chaotic behavior / edited by John Seimenis.
p. cm. -- (NATO ASI series. Series B, Physics ; v. 331)
xPublished in cooperation with NATO Scientific Affairs Division."
"Proceedings of a NATO Advanced Research Workshop on Hamiltonian Mechanics, held June 28-July 2, 1993, in Toruń, Poland"--T.p. verso.
Includes bibliographical references and index.
ISBN 0-306-44808-4
1. Mathematical physics--Congresses. 2. Hamiltonian systems--Congresses. 3. Chaotic behavior in systems--Congresses. 4. Astrophysics--Mathematics--Congresses. I. Seimenis, John. II. North Atlantic Treaty Organization. Scientific Affairs Division. III. NATO Advanced Research Workshop on Hamiltonian Mechanics (1993 : Toruń, Poland) IV. Series.
QC19..H36 1994
530.1'55352--dc20 94-40008
CIP

ISBN 0-306-44808-4

A Division of Plenum Publishing Corporation
233 Spring Street, New York, N.Y. 10013

Printed in the United States of America

SPECIAL PROGRAM ON CHAOS, ORDER, AND PATTERNS

This book contains the proceedings of a NATO Advanced Research Workshop held within the program of activities of the NATO Special Program on Chaos, Order, and Patterns.

Volume 208 — MEASURES OF COMPLEXITY AND CHAOS
edited by Neal B. Abraham, Alfonso M. Albano, Anthony Passamante, and Paul E. Rapp

Volume 225 — NONLINEAR EVOLUTION OF SPATIO-TEMPORAL STRUCTURES IN DISSIPATIVE CONTINUOUS SYSTEMS
edited by F. H. Busse and L. Kramer

Volume 235 — DISORDER AND FRACTURE
edited by J. C. Charmet, S. Roux, and E. Guyon

Volume 236 — MICROSCOPIC SIMULATIONS OF COMPLEX FLOWS
edited by Michel Mareschal

Volume 240 — GLOBAL CLIMATE AND ECOSYSTEM CHANGE
edited by Gordon J. MacDonald and Luigi Sertorio

Volume 243 — DAVYDOV'S SOLITON REVISITED: Self-Trapping of Vibrational Energy in Protein
edited by Peter L. Christiansen and Alwyn C. Scott

Volume 244 — NONLINEAR WAVE PROCESSES IN EXCITABLE MEDIA
edited by Arun V. Holden, Mario Markus, and Hans G. Othmer

Volume 245 — DIFFERENTIAL GEOMETRIC METHODS IN THEORETICAL PHYSICS: Physics and Geometry
edited by Ling-Lie Chau and Werner Nahm

Volume 256 — INFORMATION DYNAMICS
edited by Harald Atmanspacher and Herbert Scheingraber

Volume 260 — SELF-ORGANIZATION, EMERGING PROPERTIES, AND LEARNING
edited by Agnessa Babloyantz

Volume 263 — BIOLOGICALLY INSPIRED PHYSICS
edited by L. Peliti

Volume 264 — MICROSCOPIC ASPECTS OF NONLINEARITY IN CONDENSED MATTER
edited by A. R. Bishop, V. L. Pokrovsky, and V. Tognetti

Volume 268 — THE GLOBAL GEOMETRY OF TURBULENCE: Impact of Nonlinear Dynamics
edited by Javier Jiménez

SPECIAL PROGRAM ON CHAOS, ORDER, AND PATTERNS

Volume 270 — COMPLEXITY, CHAOS, AND BIOLOGICAL EVOLUTION
edited by Erik Mosekilde and Lis Mosekilde

Volume 272 — PREDICTABILITY, STABILITY, AND CHAOS IN N-BODY DYNAMICAL SYSTEMS
edited by Archie E. Roy

Volume 276 — GROWTH AND FORM: Nonlinear Aspects
edited by M. Ben Amar, P. Pelcé, and P. Tabeling

Volume 278 — PAINLEVÉ TRANSCENDENTS: Their Asymptotics and Physical Applications
edited by Decio Levi and Pavel Winternitz

Volume 280 — CHAOS, ORDER, AND PATTERNS
edited by Roberto Artuso, Predrag Cvitanović, and Giulio Casati

Volume 284 — ASYMPTOTICS BEYOND ALL ORDERS
edited by Harvey Segur, Saleh Tanveer, and Herbert Levine

Volume 291 — PROTEIN TRANSFER IN HYDROGEN-BONDED SYSTEMS
edited by T. Bountis

Volume 292 — MICROSCOPIC SIMULATIONS OF COMPLEX HYDRODYNAMIC PHENOMENA
edited by Michel Mareschal and Brad Lee Holian

Volume 298 — CHAOTIC DYNAMICS: Theory and Practice
edited by T. Bountis

Volume 312 — FUTURE DIRECTIONS OF NONLINEAR DYNAMICS IN PHYSICAL AND BIOLOGICAL SYSTEMS
edited by P. L. Christiansen, J. C. Eilbeck, and R. D. Parmentier

Volume 320 — SINGULAR LIMITS OF DISPERSIVE WAVES
edited by N. M. Ercolani, I. R. Gabitov, C. D. Levermore, and D. Serre

Volume 323 — SOFT ORDER IN PHYSICAL SYSTEMS
edited by Y. Rabin and R. Bruinsma

Volume 329 — NONLINEAR COHERENT STRUCTURES IN PHYSICS AND BIOLOGY
edited by K. H. Spatschek and F. G. Mertens

Volume 331 — HAMILTONIAN MECHANICS: Integrability and Chaotic Behavior
edited by John Seimenis

PREFACE

This volume contains invited papers and contributions delivered at the International Conference on Hamiltonian Mechanics: *Integrability and Chaotic Behaviour*, held in Torun, Poland during the summer of 1993. The conference was supported by the NATO Scientific and Environmental Affairs Division as an Advanced Research Workshop. In fact, it was the *first* scientific conference in all Eastern Europe supported by NATO.

The meeting was expected to establish contacts between East and West experts as well as to study the current state of the art in the area of Hamiltonian Mechanics and its applications. I am sure that the informal atmosphere of the city of Torun, the birthplace of Nicolaus Copernicus, stimulated many valuable scientific exchanges.

The first idea for this cnference was carried out by Prof Andrzej J. Maciejewski and myself, more than two years ago, during his visit in Greece. It was planned for about forty well-known scientists from East and West. At that time participation of a scientist from Eastern Europe in an Organising Committee of a NATO Conference was not allowed. But *always there is the first time.* Our plans for such a "small" conference, as a first attempt in the new European situation - the Europe without borders - quickly passed away. The names of our invited speakers, authorities in their field, were a magnet for many colleagues from all over the world. More than one hundred and forty colleagues from twenty-five countries and all continents, expressed their interest to participate in our conference. Among them, distinguished scientists from the United States and the former Soviet Union. We were very glad to have the majority of them among us, in Torun.

The conference was funded by a major grant from NATO. Additional support was provided by the N. Copernicus University. I am indebted to Professor Andrzej Jamiolkowski, Rector of the N. Copernicus University and Chairman of the Local Organising Committee of our conference, for his continued help and support during the organisational period. I am sure that without his help, we would not have been able to finish this meeting succesfully.

I would like to express my thanks to my colleague and friend Andrzej J. Maciejewski for his efforts during the time he was a member of the Scientific Organising Committee. Many thanks are due to Professors Carles Simo and Tassos Bountis for their advice as members of the Scientific Organising Committee. I would like to also thank, the members of the Local Organising Committee, Krzysztof Gozdziewski and Ewa Derkowska as well as the members of the staff of the Faculty of Physics and Astronomy of the N. Copernicus University for their help, during the period of the conference.

All papers of the volume are refereed. I am indebted to the referees - especially to the external experts - for their kind assistance and their efforts to be fast, in order to give the volume for publication, in proper time.

In the end, I would like to apologise to my son Kimon for the time I spent far from him, in order to finish successfully my work as Director of the NATO Advanced Research Workshop and editor of the proceedings. The part of the work that was carried out by me during the last two years, is dedicated to him.

John Seimenis, Editor

CONTENTS

Invited papers

Non-integrability criterion of Hamiltonian systems based on Ziglin's Theorem and its relation to the singular point analysis 1
H. Yoshida

Averaging under fast quasiperiodic forcing 13
C. Simo

Natural boundaries of normalizing transformations 35
L. Billi, M. Malavasi and G. Turchetti

Singular perturbation in Hamiltonian Mechanics 43
A.D. Bruno

The structure of Chaos 51
G. Contopoulos

From Integrability to Chaos: Examples of interrelations between physics and dynamics for minor bodies in the solar system 73
C. Froeschlé, Ph. Bendjoya and A. Cellino

Successive elimination of harmonics: A way to explore the resonant structure of a Hamiltonian system 89
A. Morbidelli

Periodic solutions of nonlinear Schrodinger equations and the Nash-Moser method 103
W. Craig and C. Eugene Wayne

On the tendency toward ergodicity with increasing number of degrees of freedom in Hamiltonian systems 123
L. Hurd, C. Grebogi and E.Ott

Gibbsian check of the validity of Gibbsian calculation through dynamical observables 131
D.F. Escande, H. Kantz, R. Livi and S. Ruffo

Adiabatic invariants and time scales for energy sharing in models of classical gases 139
G. Benettin

Numerical integration of Hamiltonian systems in the presence of additional integrals: Application of the observer method 153
A.J. Maciejewski and J-M Strelcyn

Symmetries and topology of dynamical systems with two degrees of freedom 167
V. Kozlov

Contributed papers

Variational criteria for nonintegrability and chaos in Hamiltonian systems 173
S. Bolotin

Exponentially small splitting in Hamiltonian systems 181
A. Delshams and T.M. Seara

Integrable and chaotic behaviour in the Paul trap and the hydrogen atom in a generalized van der Waals potential 189
D. Farrelly, J.E. Howard and T. Uzer

Recent applications of Hamiltonian dynamics to accelerator physics 197
E. Todesco

Singularity analysis of 2D complexified Hamiltonian systems 205
S. Abenda and A. Bazzani

Perturbation theory for systems without global action-angle coordinates 213
F. Fassò

A non-integrability test for perturbed Hamiltonian systems of two degrees of freedom 221
E. Meletlidou and S. Ichtiaroglou

Librational invariant surfaces in the spin-orbit problem 229
A. Celletti

Normalization of resonant Hamiltonians 237
D. Farrelly, J. Humpherys and T. Uzer

Effective stability for periodically perturbed Hamiltonian systems 245
À. Jorba and C. Simó

Bihamiltonian systems and Lax representation 253
J.J. Morales and R. Ramirez

An efficient method for computing periodic orbits of conservative dynamical systems 261
M.N. Vrahatis and T.C. Bountis

The dynamics of trace maps 275
J.A.G. Roberts and M. Baake

Scars in groups of eigenfunctions 287
F. Borondo, G.G. de Polavieja and R.M. Benito

A model of Poincaré and rigorous proof of the second element of thermodynamics from mechanics 295
L.D. Pustyl'nikov

Nekhoroshev and KAM theorems revisited via a unified approach 299
A. Delshams and P. Gutierrez

Dynamics and k-symmetries 307
G.R.W. Quispel and J.S.W. Lamb

Quantal-classical mixed mode dynamics of coupled oscillators 315
E. Yurtsever

The three-wave interaction of four waves revisited: A Lax pair and possibly general solution 321
F.J. Romeiras

Chaotic Friedman-Robertson-Walker Cosmology coupled to a real free massive scalar field in Maupertuis picture 329
M. Szydlowski

The method of modular smoothing 335
N. Buric and I.C. Percival

Soliton chaos in elastic chains and turbulence 341
T. Kapitaniak

Stochastic webs with fourfold rotation symmetry 345
J.S.W. Lamb

Antibrackets and supersymmetric Mechanics 353
A. Nersessian

Integrable systems and confocal quadrics 361
Y. Fedorov

An elementary approach to integrability condition for the Euler equations on Lie algebra so(4) 371
S.I. Popov and J-M. Strelcyn

Integrable Hamiltonian systems and Poisson actions with simple singular points 377
L.M. Lerman and Y.L. Umanskii

Dynamics and bifurcations in two-parameter unfolding of a Hamiltonian system with a homoclinic orbit to a saddle-center 385
O.Y. Koltsova and L.M. Lerman

Self-similar isomonodromy solutions of nonlinear Schrödinger equation 391
M. Can

A method for visualizing the 4-dimensional space of section in 3-D Hamiltonian systems 399
P.A. Patsis and L. Zachilas

Non-adiabatic aspects of time-dependent Hamiltonian systems 405
F. Casas, J.A. Oteo and J. Ros

A study of a finite-dimensional dynamical system approximating the evolution of quantum averages 413
S.Y. Sadov

Index 421

NON-INTEGRABILITY CRITERION OF HAMILTONIAN SYSTEMS BASED ON ZIGLIN'S THEOREM AND ITS RELATION TO THE SINGULAR POINT ANALYSIS

Haruo Yoshida

National Astronomical Observatory
Mitaka, Tokyo 181, Japan
e-mail:yoshida@C1.mtk.nao.ac.jp

ABSTRACT After two examples of the singular point analysis, a sufficient condition (criterion) for the non-existence of an additional analytic integral is given for n-degree-of-freedom Hamiltonian systems with a homogeneous potential, which justifies the validity of the singular point analysis. This criterion is based on Ziglin's theorem, which will be reviewed extensively from the basic ideas.

1 Introduction

For a given Newtonian equation of motion

$$\dot{q} = p, \quad \dot{p} = -\frac{\partial V}{\partial q}, \tag{1}$$

one can always obtain a solution, in the vicinity of (regular) initial condition (q_0, p_0) at $t = 0$, in the form of Taylor series expansion

$$q(t) = q_0 + tp_0 - (t^2/2)V_q + \dots, \quad p(t) = p_0 - tV_q + \dots. \tag{2}$$

All Newtonian equations of motion, or all Hamiltonian equations of motion look like the same at this level. They are locally integrable. On the other hand, we know that completely integrable systems are rare, and most systems are non-integrable. The distinction between integrable and non-integrable systems can been seen easily at least numerically. *Chaotic* phenomena is usually related with the global or long-time behavior of a dynamical system, and for short time interval all systems are, in some sense, integrable. On the other hand, if one looks for a computable criterion of integrability

Hamiltonian Mechanics, Edited by J. Seimenis,
Plenum Press, New York, 1994

and non-integrability of a given system, one must compute something local, since global property is not computable by hand in general. In early 80's, several people began to discuss the behavior of the singularity of solution in connection with integrability of the system. The history of this singular point analysis goes back to Kowalevski(1890) in her discovery of a new integrable case of the rigid body motion. Although integrability does not affect a local property near a regular point of solution, there is a hope that integrability and the local behavior near a singularity of solution are strongly related. In this paper, after two examples of the singular point analysis, the underlying theorem will be presented. The proof is, for the moment, based on Ziglin's theorem.

2 Examples of the singular point analysis

2.1 A 2-dimensional quartic potential system

The model Hamiltonian is given by

$$H = \frac{1}{2}(p_1^2 + p_2^2) + \frac{1}{4}(q_1^4 + q_2^4) + \frac{\epsilon}{4}q_1^2 q_2^2, \tag{3}$$

where ϵ is a constant parameter. The equations of motion for q are

$$\ddot{q}_1 = -q_1(q_1^2 + \epsilon q_2^2), \quad \ddot{q}_2 = -q_2(q_2^2 + \epsilon q_1^2). \tag{4}$$

This system is known to be integrable when $\epsilon = 0, 1, 3$. Suppose the solution $q(t)$ has a singularity at $t = t_0$. By redefining $t := t - t_0$, we can always assume that $t = 0$ is a singularity. One particular (singular) solution is given by

$$q_1 = d_1 t^{-1}, \quad q_2 = d_2 t^{-1}, \tag{5}$$

where d_1 and d_2 are solution of the algebraic equations

$$2d_1 = -d_1(d_1^2 + \epsilon d_2^2), \quad 2d_2 = -d_2(d_2^2 + \epsilon d_1^2). \tag{6}$$

There are essentially two distinct solutions of these equations;

$$(i) \quad d_1 = 0, \quad d_2 = \sqrt{-2}, \tag{7}$$

$$(ii) \quad d_1 = \pm\sqrt{\frac{-2}{1+\epsilon}}, \quad d_2 = \pm\sqrt{\frac{-2}{1+\epsilon}}. \tag{8}$$

The fact that these coefficients are imaginary number implies that singularities of the solution do not appear on the real axis of t. On the other hand, at this level, how the integrability depends on the value of ϵ is not clear at all. All values of ϵ look like the same. Moreover solution (5) is a particular solution because there is no arbitrary integration constants in the expression.

In order to obtain a singular solution with some integration constants, which has (5) as its leading order, put

$$q_1 = d_1 t^{-1} + e_1 t^{-1+\rho}, \quad q_2 = d_1 t^{-1} + e_2 t^{-1+\rho}, \tag{9}$$

with some constants e_1 and e_2. The exponent ρ is known to be a solution of the following quartic equation

$$(i) \quad (\rho+1)(\rho-4)\left(\rho^2 - 3\rho + 2(1-\epsilon)\right) = 0, \tag{10}$$

in the case of leading order behavior (7), and

$$(ii)\ \ (\rho+1)(\rho-4)\left(\rho^2-3\rho-4\frac{1-\epsilon}{1+\epsilon}\right)=0, \tag{11}$$

in the case of leading order behavior (8). By denoting these four roots as ρ_i (with $\rho_3 = 4$ and $\rho_4 = -1$), we have the following expansion of solution

$$q_i = t^{-1}\left[d_i + f_i\left(I_1 t^{\rho_1}, I_2 t^{\rho_2}, I_3 t^4\right)\right], \tag{12}$$

with three arbitrary constants I_1, I_2, I_3, where $f_i(x, y, z)$ denotes a (convergent) Taylor series in its arguments without constant term. This expansion is not necessarily a Laurent series expansion because the exponent ρ_i can be arbitrary complex number in general. In order that this solution to be a Laurent series (i.e., the solution is single-valued around the singularity), the exponents ρ_1 and ρ_2 must be integer in both cases of leading order behavior. This is possible only when $\epsilon = 0, 1, 3$, as can be checked easily.

Thus the known integrable cases are identified with the cases having the property of single-valuedness. This property is often called the *Painlevé property* in the literatures. See Ramani et al.(1989) for a review. By now a lot of integrable Hamiltonian systems are shown to have this Painlevé property and even some integrable systems were discovered by postulating this property. There remained a rigorous relation between the integrability and having the Painlevé property. On the other hand, the next example of classical two body problem shows that the Painlevé property is not a necessary condition for integrability.

2.2 Classical 2-body v.s. 3-body problems

Consider, in general, the classical n-body problem with Newton's low of gravitation, which is described by the Hamiltonian

$$H = \sum_{i-1}^{n} \frac{p_i^2}{2m_i} - \sum_{i,j} \frac{m_i m_j}{r_{ij}}. \tag{13}$$

The case of $n = 2$ is the well known Kepler problem and is integrable. On the other hand, if $n \geq 3$, the system becomes non-integrable although the proof is not so popular at all. As mentioned in the introduction, 2-body and 3-body problems look like the same in the neighborhood of a regular initial condition. It has been known, on the other hand, that there is a sharp distinction between 2-body and 3-body problems when one looks at the solution near a collision singularity.

The 2-body problem can be easily reduced to the 1-body problem in a central field, for which

$$H = \frac{1}{2}p^2 - \frac{1}{r}. \tag{14}$$

As in the previous example, this system has a singular solution of the form

$$q_i = d_i t^{2/3}, \tag{15}$$

with real constants d_i. This solution represents a collision of 2 bodies, since $q_i \to 0$ as $t \to 0$. The energy of this solution is 0. Other collision solution for non-zero energy can be obtained, in the form

$$q_i = t^{2/3}\left[d_i + f_i\left(I_1 t^{2/3}\right)\right]. \tag{16}$$

This expression shows that real solution defined before collision (i.e. $t < 0$), can be analytically continued after the collision ($t > 0$), because the function $t^{2/3}$ has a real branch for both $t < 0$ and $t > 0$. Indeed this collision solution can be obtained from the elliptic, parabolic, and hyperbolic orbits of the 2-body problem by taking the limit $e \to 1$ (eccentricity).

The situation is quite different for the 3-body problem. As shown by Block(1908) and by Siegel(1941), the triple collision solution has the form

$$q_i = t^{2/3}\left[d_i + f_i\left(I_1 t^{2/3}, I_2 t^{\alpha}, I_3 t^{\beta}, \ldots\right)\right]. \tag{17}$$

where exponents $\alpha, \beta, \ldots$ are irrational or imaginary number, in general, and the continuation of real solution across the collision is no more possible. This example suggests that although the Painlevé property (single-valuedness) is not necessary for integrability, its weaker version (i.e., no irrational or imaginary exponents in the expansion) may be a necessary condition. Indeed this is justified in the form of the following theorem.

3 A non-integrability criterion

Consider a Hamiltonian system with n-degrees of freedom of the form

$$H = \frac{1}{2}\mathbf{p}^2 + V(\mathbf{q}), \tag{18}$$

where $V(\mathbf{q})$ is assumed to be a homogeneous function of integer degree k, but $k \neq 0, \pm 2$. As in the previous examples, a singular solution of (18) is obtained in the form

$$\mathbf{q}(t) = t^{-g}\left[\mathbf{d} + \mathbf{f}\left(I_1 t^{\rho_1}, I_2 t^{\rho_2}, \ldots\right)\right]. \tag{19}$$

Here $g = 2/(k-2)$, $\mathbf{d}$ is a constant vector, I_j's are integration constants, and $\mathbf{f}(x, y, z, ...)$ denotes a convergent Taylor series. For (18), the exponents ρ_i, called the Kowalevski Exponents (KE), have a pairing property, $\rho_i + \rho_{i+n} = 2g + 1, (i = 1, 2, ..., n)$. Because of this pairing property, ρ_i and ρ_{i+n} can be rational numbers if and only if the difference of KE $\Delta\rho_i := \rho_{i+n} - \rho_i$ is a rational number. Further, one can always assume that $\Delta\rho_n$ is rational, $\Delta\rho_n = (3k-2)/(k-2)$. Then,

Theorem 1 (Yoshida, 1989) *If the n numbers* $(\Delta\rho_1, \Delta\rho_2, \ldots, \Delta\rho_n)$ *are rationally independent, then the Hamiltonian system* (18) *has no additional analytic integral* $\Phi(\mathbf{q}, \mathbf{p}) =$ *const. besides the Hamiltonian itself.*

When $n = 2$, there is only one non-trivial difference of KE ($\Delta\rho_1$), and if $\Delta\rho_1$ is not a rational number, the system has no additional analytic integral. In order to be integrable, all KE must be rational number. Some applications of this theorem are found in Yoshida(1989). For a given homogeneous potential, $\Delta\rho_i$ can be computed by the following algorithm;

Step 1 Solve and fix a solution $\mathbf{c} = (c_1, c_2, ..., c_n)$ of the algebraic equation

$$gradV(\mathbf{c}) = \mathbf{c}. \tag{20}$$

This constant $\mathbf{c}$ is related with the constant $\mathbf{d}$ in (19) by $\mathbf{d} = [-g(g+1)]^{g/2}\mathbf{c}$.

Step 2 Let $D^2V(\mathbf{c})$ be the Hessian matrix of $V(\mathbf{q})$ evaluated at $\mathbf{q} = \mathbf{c}$. Compute the eigenvalues $(\lambda_1, \lambda_2, ..., \lambda_n)$ of $D^2V(\mathbf{c})$.

Step 3 The pair of the KE (ρ_i, ρ_{i+n}) are the two roots of the quadratic equation

$$\rho^2 - (2g+1)\rho + g(g+1)(1-\lambda_i) = 0, \tag{21}$$

so that

$$\Delta\rho_i = \sqrt{1 + \frac{8k\lambda_i}{(k-2)^2}} \quad (i = 1, 2, ..., n). \tag{22}$$

4 Ziglin analysis and sketch of the proof

4.1 Basics of Ziglin's theorem

Since the time of Poincaré, it has been known that the integrability of a system is related with the stability of periodic solutions. Suppose that for a given N-dimensional autonomous system,

$$\frac{dx}{dt} = F(x), \tag{23}$$

there exists a periodic solution, $x = \psi(t)$, having a period T. By linearizing (23) around this periodic solution, one gets the (linear) variational equations

$$\frac{d\xi}{dt} = \left(\frac{\partial F}{\partial x}\right)\xi, \tag{24}$$

where the Jacobian matrix $\left(\frac{\partial F}{\partial x}\right)$ is evaluated on the periodic solution $x = \psi(t)$. The monodromy matrix $M(T)$ is defined by the time evolution of the solution of (24) from $t = 0$ to $t = T$ as

$$\xi(T) = M(T)\xi(0), \tag{25}$$

which is a linear symplectic mapping. The eigenvalues σ_i of the monodromy matrix $M(T)$ are called the characteristic multipliers, and quantities α_i defined by

$$\sigma_i = \exp(\alpha_i T) \tag{26}$$

are called the characteristic exponents. The presence of characteristic exponent with its modulus greater than one ($|\alpha| > 1$) implies that the periodic solution $x = \psi(t)$ is exponentially unstable. For a Hamiltonian system, there is a pairing property of characteristic multipliers and of characteristic exponents. Furthermore for an autonomous Hamiltonian system, two of the characteristic multipliers are unity and two of the characteristic exponents are zero. Therefore for a 2-degrees-of-freedom autonomous Hamiltonian, the multipliers and the exponents have the spectrum $(1, 1, \sigma, 1/\sigma)$ and $(0, 0, \alpha, -\alpha)$.

Poincaré has shown that if a 2-degrees-of-freedom Hamiltonian system is integrable (i.e., an additional analytic integral $\Phi(q,p) = const.$ exists), and if the two vectors $gradH$ and $grad\Phi$ are linearly independent on the periodic solution $x = \psi(t)$, then all the characteristic multipliers must be unity and all the characteristic exponents must be zero. Therefore exponentially unstable periodic solutions are exceptional in integrable systems. However, because of the necessity of the assumption on $grad\Phi$, this statement cannot be used as integrability/non-integrability criterion. In fact there is an integrable Hamiltonian system which has an exponentially unstable periodic solution, as seen in the next example.

4.1.1 Example of an integrable system which has an exponentially unstable periodic solution

$$H = \frac{1}{2}(p_1^2 + p_2^2) - \frac{1}{2}(q_1^2 + q_2^2) + \frac{1}{4}(q_1^4 + q_2^4). \tag{27}$$

This is a trivial integrable (separable) system. A choice of the additional integral is

$$\Phi = \frac{1}{2}p_1^2 - \frac{1}{2}q_1^2 + \frac{1}{4}q_1^4. \tag{28}$$

One can easily confirm (at least, numerically) that a straight-line periodic solution $q_2 = p_2 = 0$ is exponentially unstable. This instability is possible because on this periodic solution,

$$gradH = grad\Phi = (-q_1 + q_1^3, 0, p_1, 0), \tag{29}$$

and $gradH$ and $grad\Phi$ are not linearly independent. Remark that if we choose

$$\Phi' := H - \Phi = \frac{1}{2}p_2^2 - \frac{1}{2}q_2^2 + \frac{1}{4}q_2^4 \tag{30}$$

as the additional integral, then $grad\Phi'$ vanishes identically on the periodic solution.

4.1.2 First step of Ziglin

In order to obtain a criterion of integrability we must find a property which is *always* a consequence of the existence of an additional integral, without any strong assumptions on the integral. The previous statement of Poincaré on the characteristic exponents of integrable systems is derived from the fact that, if Φ is an integral, then $\xi \cdot (\partial\Phi/\partial x)$ becomes an integral for the linear variational equation (24). But if $(\partial\Phi/\partial x)$ vanishes identically on the periodic solution, this does not give any useful information. The first step of Ziglin's theorem is to confirm the following fact. Even if $\xi \cdot (\partial\Phi/\partial x)$ vanishes identically, the second order derivative,

$$D^2\Phi := \left(\xi \cdot \frac{\partial}{\partial x}\right)^2 \Phi, \tag{31}$$

may not vanish identically. Then, $D^2\Phi$ becomes an integral for the variational equation (24). In general, by the assumption of analyticity of the integral Φ, there exists always a first no vanishing derivative $D^k\Phi$ and this *always* becomes an integral for the variational equation. So the problem is, what is the consequence of the existence of this integral.

4.1.3 Second step of Ziglin

So far we have only assumed that $x = \psi(t)$ is a periodic solution. Suppose now that our Hamiltonian is a complex analytic function (e.g., polynomial, rational function, etc.) so that we can define the solution as a complex analytic function of t. Furthermore, assume that there exists a particular solution $x = \psi(t)$ which has two independent periods T_1 and T_2, like a solution expressed in terms of the Jacobi elliptic functions. For each period, we can define monodromy matrices $M(T_1)$ and $M(T_2)$ by the variational equations. These monodromy matrices do not commute in general. The fact is that the presence of the previous integral $D^k\Phi$ for the variational equation yields a common invariant function of the two symplectic mappings described by the monodromy matrices $M(T_1)$ and $M(T_2)$. This gives a big constraint on these monodromy matrices and leads to an useful and computable criterion. In order to present the explicit statement of the theorem, let us come back to the example of the quartic potential system (3).

4.2 Ziglin analysis for the 2-dimensional quartic potential system

From the equations of motion (4), it is evident that the straight-line $q_1 = 0$ is a solution. For this straight-line solution, q_2 satisfies

$$\ddot{q}_2 = -q_2^3, \tag{32}$$

and the solution is expressed in terms of the Jacobi elliptic function

$$q_2(t) = cn(t, 1/\sqrt{2}). \tag{33}$$

This Jacobi elliptic function has two independent periods

$$T_1 = 4K(1/\sqrt{2}), \quad T_2 = i4K(1/\sqrt{2}), \tag{34}$$

in the complex t plane, so that the straight-line solution $q_1 = 0$ is a doubly periodic solution. The variational equations around this straight-line solution $q_1 = 0$ take the form

$$\ddot{\xi}_1 + \epsilon[q_2(t)]^2\xi_1 = 0, \tag{35}$$
$$\ddot{\xi}_2 + 3[q_2(t)]^2\xi_2 = 0. \tag{36}$$

The second equation (36) for ξ_2 describes the variation along the straight-line solution and gives the trivial characteristic multipliers $(1,1)$ and exponents $(0,0)$ and have nothing to do with the integrability of the system. The first equation (35) for ξ_1, on the other hand, describes the variation normal (perpendicular) to the straight-line solution, and the stability depends on the value of ϵ. For later use, only the equation for ξ_1 is necessary and will be called the Normal Variational Equation (NVE). A monodromy matrix for the NVE is a 2×2 matrices with the determinant unity, $\det M(T) = 1$. We now define a 2×2 monodromy matrix to be *resonant* if the eigenvalue (characteristic multiplier) σ is a root of unity. *Non-resonant* monodromy matrix is defined as the negative of resonant one. For the NVE, monodromy matrices M_1, M_2 are defined for each period T_1 and T_2. As explained before, if the system (3) has an additional integral, then there must be always a common invariant function for these two matrices (area-preserving mappings), and we end up with the following statement;

Theorem 2 (Ziglin, 1983) *Suppose that the Hamiltonian* (3) *has an additional analytic integral* $\Phi = const.$ *and that one of the monodromy matrices* M_1 *is non-resonant. Then either* (i) M_1 *commutes with* M_2, *or* (ii) $traceM_2 = 0$, *must follow.*

For the proof, see Yoshida (1987, 1988) in addition to the Ziglin's original paper. If $traceM_2 = 0$, the eigenvalues are $\pm i$ so that M_2 becomes a resonant matrix. Therefore we have the following statement.

Corollary 1 *If both of two monodromy matrices* M_1 *and* M_2 *are non-resonant, and if they do not commute, then the system has no additional analytic integral* $\Phi = const.$ *besides the Hamiltonian itself.*

For a given variational equation, there is no algorithmic way to express its monodromy matrix. For the variational equation (35), however, monodromy matrices permit the

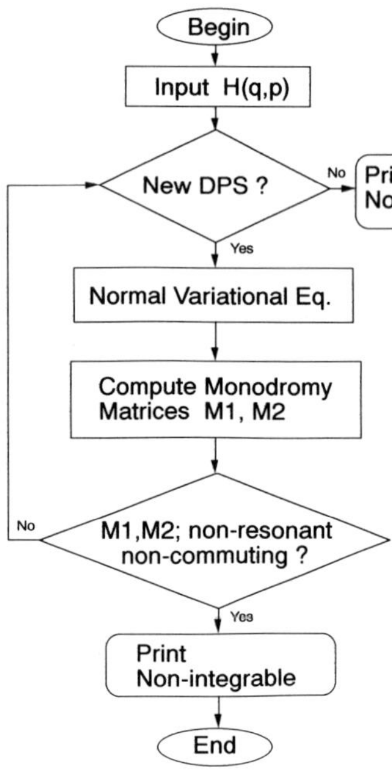

Figure 1. Flow chart of Ziglin analysis based on Corollary 1. DPS means Doubly Periodic Solution.

explicit expression, because (35) can be transformed to the Gauss hypergeometric equation by the change of the independent variable. As the result, we have the explicit expression of the monodromy matrices M_1 for the real period $T_1 = 2K$ and M_2 for the imaginary period $T_2 = i2K$. Note that the minimum period of $[q_2(t)]^2$ in the NVE (35) is the half of (34). For these matrices it is shown (Yoshida, 1986) that

$$trace M_1 = trace M_2 = 2\sqrt{2}\cos\left(\frac{\pi}{4}\sqrt{1+8\epsilon}\right), \tag{37}$$

and, M_1 and M_2 commute only when the value of trace is ± 2. Therefore if ϵ has a value in the region,

$$\epsilon < 0,\ 1 < \epsilon < 3,\ 6 < \epsilon < 10, \ldots \tag{38}$$

then these two monodromy matrices are both non-resonant and do not commute each other. This implies non-integrability of (3) by Corollary 1.

If we repeat the same analysis with another trivial periodic solution, $q_1 = q_2$, we can prove the non-integrability for the remaining region $0 < \epsilon < 1, 3 < \epsilon$, and only three values $\epsilon = 0, 1, 3$ can survive. For these values of ϵ, the system (3) is indeed integrable.

4.2.1 Another set of monodromy matrices

For the NVE (35), the choice of two independent periods is not unique. The Jacobi elliptic function $cn(t, 1/\sqrt{2})$ has poles located doubly periodically in the complex t plane. The nearest one from the origin $t = 0$ are located at $t = \pm i2K(1/\sqrt{2})$. A loop which starts from and comes back to $t = 0$ after a counterclockwise circuit around the singularity at $t = i2K$ is called C_1. Another loop C_2 makes a circuit around $t = -i2K$. These two loops can be considered the periods of the function $cn(t)$, and Ziglin's theorem may be applied. Monodromy matrices for these periods are not necessarily trivial identity matrices, since the solution of NVE (35) is not single-valued around the poles in general.

Around the singularity $t = \pm i2K(1/\sqrt{2})$, the function $[q_2(t)]^2$ is expanded in the form ($\tau := t \mp i2K$)

$$[q_2(t)]^2 = \frac{-2}{\tau^2}\left(1 + O(\tau^4)\right). \tag{39}$$

Thus, two independent solutions of NVE (35) around these singularities have the form

$$\xi^{(1)} = \tau^{\rho_1 - 1} f_1(\tau), \quad \xi^{(2)} = \tau^{\rho_2 - 1} f_2(\tau), \tag{40}$$

where ρ_1 and ρ_2 are roots of the quadratic equation

$$\rho^2 - 3\rho + 2(1-\epsilon) = 0, \tag{41}$$

and $f_i(\tau)$ are convergent Taylor series. Remark that these two exponents are exactly the same one as appeared in (10). These two independent solutions of NVE (35) change after the periods (C_1 or C_2) as

$$\xi^{(1)} \to e^{2\pi i\rho_1}\xi^{(1)}, \quad \xi^{(2)} \to e^{2\pi i\rho_2}\xi^{(2)}, \tag{42}$$

so that the common value of the trace of monodromy matrices is

$$trace M_{1,2} = e^{2\pi i\rho_1} + e^{2\pi i\rho_2} = -2\cos\left(\pi\sqrt{1+8\epsilon}\right) = -2\cos\left(\pi(\rho_1 - \rho_2)\right). \tag{43}$$

From this expression it is clear that the monodromy matrices are non-resonant if $(\rho_1-\rho_2)$ is not a rational number. Furthermore, one non-trivial fact is that these two matrices M_1 and M_2 do not commute if they are non-resonant. This fact can be proved only from the explicit expression of monodromy matrices, at this moment. Anyway by Corollary 1, we conclude that the system is non-integrable if the roots of (41) are not rational numbers. For integrability, ρ_1 and ρ_2 must be rational number.

This statement is, in some sense, weaker than one derived from the previous choice of two monodromy matrices. Nevertheless, if one try to apply Ziglin's theorem to higher dimensional systems, only this choice turns out to be effective.

4.3 Ziglin analysis for N-dimensional homogeneous potential systems

For N-dimensional homogeneous potential system (18), there exists always a straight-line solution of the form

$$\mathbf{q} = \mathbf{c}\phi(t), \tag{44}$$

with a constant vector $\mathbf{c}$ given by (20), and $\phi(t)$ satisfying

$$\ddot{\phi} + \phi^{k-1} = 0. \tag{45}$$

In fact $\phi(t)$ is fixed to be a solution of

$$(1/2)\dot{\phi}^2 + (1/k)\phi^k = const. = 1/k. \tag{46}$$

This function $\phi(t)$ has multi-periods in general, and in this sense the straight-line solution (44) is a multi-periodic solution. The linear variational equations around this straight-line solution become

$$\ddot{\xi} = -\phi(t)^{k-2}D^2V(\mathbf{c})\xi, \tag{47}$$

Thus after a proper rotation of coordinates, (47) is decoupled into the form

$$\ddot{\xi}_i + \lambda_i\phi(t)^{k-2}\xi_i = 0, \ (i = 1, 2, ..., n) \tag{48}$$

where λ_i are eigenvalues of the Hessian matrix $D^2V(\mathbf{c})$, and one can always assume that $\lambda_n = (k-1)$. Therefore the $(2n-2) \times (2n-2)$ monodromy matrix for the Normal Variational Equations (NVE) has the form

$$M = diag[M(\lambda_1), M(\lambda_2), ..., M(\lambda_{n-1})], \tag{49}$$

with 2×2 matrices $M(\lambda_i)$ such that $\det M(\lambda_i) = 1$. A monodromy matrix of the form (49) is called *resonant* when its eigenvalues $(\sigma_i, 1/\sigma_i)$, $(i = 1, 2, \ldots, n-1)$ have a relation, with integers m_i, of the form

$$\sigma_1^{m_1}\sigma_2^{m_2}....\sigma_{n-1}^{m_{n-1}} = 1. \tag{50}$$

Note that when $n = 2$, this means that σ_1 is a root of unity. Let M_1 and M_2 be two monodromy matrices of the form (49) associated with two different periods of $\phi(t)$. As a natural extension of corollary 1, it is shown (Yoshida, 1987) that if both M_1 and M_2 are non-resonant and if none of the pairs, $M_1(\lambda_i)$ and $M_2(\lambda_i)$ commute, then the system (18) has no additional analytic integral besides the Hamiltonian itself.

As mentioned before, (48) is transformed, by the change of the independent variable $z = \{\phi(t)\}^k$, to the Gauss hypergeometric equation

$$z(1-z)\frac{d^2\xi_i}{dz^2} + [c - (a+b+1)z]\frac{d\xi_i}{dz} - ab\xi_i = 0, \tag{51}$$

with

$$a + b = 1/2 - 1/k, \; ab = -\lambda_i/(2k), \; c = 1 - 1/k. \tag{52}$$

Thus, using the explicit expression for the two fundamental monodromy matrices $M(\gamma_0) := M(\gamma_0; k, \lambda_i)$ and $M(\gamma_1) := M(\gamma_1; k, \lambda_i)$ of the hypergeometric equation,

$$M(\gamma_0) = \begin{pmatrix} 1 & B - C \\ 0 & C \end{pmatrix}, \quad M(\gamma_1) = \begin{pmatrix} AB/C & 0 \\ 1 - A/C & 1 \end{pmatrix}, \tag{53}$$

with $A = e^{-2\pi i a}, B = e^{-2\pi i b}, C = e^{-2\pi i c}$, one can compute arbitrary monodromy matrices of (48) as a proper product of $M(\gamma_0)$ and $M(\gamma_1)$. We choose two periods of function $\phi(t)$, such that

$$M_1(\lambda_i) = \{M(\gamma_0)M(\gamma_1)\}^{4k}, \quad M_2(\lambda_i) = \{M(\gamma_1)M(\gamma_0)\}^{4k}. \tag{54}$$

This choice of two periods is essentially the same as the previous one (loops around different singularities) which leads to (43). With this expression (54), one can confirm that $M_1(\lambda_i)$ and $M_2(\lambda_i)$ have the common value of trace

$$trace\, M_{1,2}(\lambda_i) = 2\cos[2\pi\theta_i], \tag{55}$$

with

$$\theta_i = \sqrt{(k-2)^2 + 8k\lambda_i}, \tag{56}$$

and that $M_1(\lambda_i)$ and $M_2(\lambda_i)$ commute only when $trace\, M_{1,2}(\lambda_i) = \pm 2$ or $k = \pm 2$. Let us check the non-resonance condition. Since the eigenvalue of $M_{1,2}(\lambda_j)$ is $\sigma_j = \exp(2\pi i\theta_j)$ and $\theta_n = 3k - 2$, an integer, M_1 and M_2 are non-resonant whenever the n numbers θ_i $(i = 1, 2, ..., n)$ are rationally independent. Furthermore, with the same assumption on θ_i, none of the pair $M_1(\lambda_i)$ and $M_2(\lambda_i)$ can commute unless $k = \pm 2$. This implies the non-existence of an additional analytic integral.

Let us compute the Kowalevski Exponents (Yoshida, 1983) for (18). Consider an autonomous system

$$\frac{dx_i}{dt} = F_i(x_1, x_2, ..., x_N), \tag{57}$$

which is invariant by a scaling transformation,

$$t \to \alpha^{-1}t, x_i \to \alpha^{g_i}x_i \;.(i = 1, 2, ..., N) \tag{58}$$

Fix a solution $\mathbf{d} = (d_1, d_2, ..., d_N)$ of the algebraic equations

$$F_i(d_1, d_2, ..., d_N) = -g_i d_i\;. \tag{59}$$

Then the eigenvalues of the $N \times N$ matrix $K = (K_{ij})$ with

$$K_{ij} = \left(\frac{\partial F_i}{\partial x_j}\right)_{x=d} + \delta_{ij}g_i, \tag{60}$$

are called the Kowalevski Exponents (KE). For the system (18), the characteristic polynomial $K(\rho) = \det|\rho I - K|$ whose roots are the KE, reduces to

$$K(\rho) = \det|(\rho - g)(\rho - g - 1)I + D^2V(\mathbf{d})|, \tag{61}$$

where **d** is a solution of

$$gradV(\mathbf{d}) = -g(g+1)\mathbf{d}. \tag{62}$$

The solution of (62) and that of (20) are related by $\mathbf{d} = [-g(g+1)]^{g/2}\mathbf{c}$, and since the Hessian matrix $D^2V(\mathbf{q})$ is a homogeneous function of degree $k-2$ $(= 2/g)$, we have

$$D^2V(\mathbf{d}) = -g(g+1)D^2V(\mathbf{c}). \tag{63}$$

Thus the roots of $K(\rho) = 0$ are given by (21), and since $\theta_i = |k-2|\Delta\rho_i$, the statement of the Theorem 1 follows.

References

[1] Block, H., 1908, *Arkiv för Mat. Astron. Fys.* **5**, No. 9, 1.

[2] Kowalevski, S., 1890, *Acta Math.* **14**, 81.

[3] Ramani, A., Grammaticos, B. and Bountis, T., 1989, *Phys. Rep.* **180**, 159.

[4] Siegel, C.L., 1941, *Ann. Math.* **42**, 127.

[5] Yoshida, H., 1983, *Celestial Mech.* **31**, 363.

[6] Yoshida, H., 1986, *Physica D* **21**, 163.

[7] Yoshida, H., 1987, *Physica D* **29**, 128.

[8] Yoshida, H., 1988, 'Ziglin analysis for proving non-integrability of Hamiltonian system', in *Finite Dimensional Integrable Nonlinear Dynamical Systems*, P.G.L. Leach and W.H. Steeb, eds. (World Scientific, Singapore) 74.

[9] Yoshida, H., 1989, *Phys. Lett. A* **141**, 108

[10] Ziglin, S.L., 1983, *Functional Anal. Appl.* **16**, 181

AVERAGING UNDER FAST QUASIPERIODIC FORCING

Carles Simó

Departament de Matemàtica Aplicada i Anàlisi
Universitat de Barcelona
Gran Via 585, 08007 Barcelona (Spain)

Abstract

We consider a non autonomous system of ordinary differential equations. Assume that the time dependence is quasiperiodic with large basic frequencies, ω/ε and that the ω vector satisfies a diophantine condition. Under suitable hypothesis of analyticity, there exists an analytic (time depending) change of coordinates, such that the new vector field is the sum of an autonomous part and a time dependent remainder. The remainder has an exponentially small bound of the form $\exp(-c\varepsilon^{-a})$, where c and a are positive constants. The proof is obtained by iteration of an averaging process. An application is made to the splitting of the separatrices of a two-dimensional normally hyperbolic torus, including several aspects: formal approximation of the torus and their invariant manifolds, numerical computations of the splitting, a first order analysis using a Melnikov approach and the bifurcations of the set of homoclinic orbits.

1 Introduction and statement of results

In many problems in Mechanics appear differential equations with fast periodic forcing of the form $z' = f(z, \tau/\varepsilon)$, $z \in D \subset \mathbb{R}^n$, D compact, $t \in \mathbb{R}$ and f being $2\pi\varepsilon$-periodic in τ. The well known result of Neishtadt,[12] by doing a suitable number of averaging steps, tell us that, if f is analytic in D and continuous with respect to t, then there is an analytic change of variables, $w = z + \varphi(z, \tau/\varepsilon)$, such that the new differential equation is of the form $w' = g(w) + r(w, \tau/\varepsilon)$ and there exists $c > 0$ such that $|r| < \exp(-c/\varepsilon)$.

It is natural to ask for for an analogous result when the dependence of f with respect to t is fast quasiperiodic. The main objective of this paper is to study this problem and to give some application. Some more restrictive conditions have to be imposed then on f: analycity with respect to the angles varying with time has also to be required and, furthermore, the frequencies have to satisfy some diophantine condition.

By introducing the new time $t = \tau/\varepsilon$, the initial equation can be written as $\dot{z} = \varepsilon f(z, \tau)$, where $\dot{}$ denotes $\frac{d}{dt}$.

First we state the main result.

Theorem A. *(Averaging theorem) Let*

$$\dot{z} = \varepsilon f(z, \theta, \varepsilon), \tag{1}$$

where f is analytic in (z, θ) for $z \in \Omega \subset \mathbf{C}^n$, $\Omega = D + \Delta$, a Δ-neighbourhood of D in $\mathbf{C}^n$, D a compact in $\mathbf{R}^n$ and $\theta \in \mathbf{T}^p + \Delta$, $p \geq 2$, where $\mathbf{T}^p$ is a p-dimensional torus. Assume f is bounded in ε for $|\varepsilon| \leq \varepsilon_0$ and $\theta = \omega t$, where $\omega \in \mathbf{R}^n$ is a vector of frequencies satisfying

$$|(k, \omega)| \geq b|k|^{-\tau}, \tag{2}$$

where $b > 0$, $\tau > p - 1$ and $|k| = \sum_{i=1}^{p} |k_i|$. Then, if ε_0 is small enough, for a fixed ε with $|\varepsilon| \leq \varepsilon_0$, there exists a change of variables $z = h(w, \theta, \varepsilon)$, analytic in (w, θ) for $w \in D + \Delta/2$, $\theta \in \mathbf{T}^p + \Delta/2$, such that the new equation is $\dot{w} = \varepsilon(g(w, \varepsilon) + r(w, \theta, \varepsilon))$ and the remainder satisfies an exponentially small bound

$$|r|_{\Delta/2} < c_1 \exp(-c_2/\varepsilon^{c_3}), \tag{3}$$

where c_1, $c_2 > 0$, $c_3 = \frac{1}{\tau+1}$. The constants c_1, c_2 depend only on $|f|_\Delta$, the dimensions and the constants in (2). Furthermore $|g|_{\Delta/2} < 2|f|_\Delta$. Here $|f|_\Delta$ denotes the supremum norm of f in $D + \Delta$, $\mathbf{T}^p + \Delta$ for the fixed value of ε.

Remark 1. If $\tau = p - 1$ (a non generic situation) one can replace $\varepsilon^{\frac{1}{\tau+1}}$ by $(\varepsilon|\ln\varepsilon|)^{1/p}$.

Remark 2. In the particular case of (1) when f has the form

$$f = f_0(z, \varepsilon) + \varepsilon f_1(z, \varepsilon) + \cdots + \varepsilon^{k-1} f_{k-1}(z, \varepsilon) + \varepsilon^k f_k(z, \theta, \varepsilon),$$

then $|f - g|_{\Delta/2} = O(\varepsilon^k)$ and c_1, c_2 depend only on $|f_j|_\Delta$, $j = 0, \ldots, k$, the dimensions and the constants in (2). An application of Remark 2 to the periodic case can be found in Broer et al.[3] Note that in the periodic case the diophantine condition uses $b = |\omega|$, $\tau = 0$. The periodic case is a particular case of Theorem A, but the hypothesis can be weakened, as said before.

Theorem B. *In the hypothesis of Theorem A, if f is the vector field associated to a Hamiltonian H, then the quasiperiodically time depending change of variables h can be chosen in a canonical way. Hence the final vector field, $g + r$, is also Hamiltonian.*

Remark 3. Theorem B has many consequences on upper bounds for the splitting of separatrices, remainders of normal forms, etc.

The proof of Theorem A is given in Section 2. The proofs of Remarks 1 and 2 and Theorem B will appear elsewhere.[13]

In order to see how good are the upper bounds given in (3) and to illustrate Remark 3, Section 3 studies the following example: Consider the equation

$$x'' = \sin(x + \sin(\tau/\varepsilon) + \sin(\gamma\tau/\varepsilon)), \tag{4}$$

where γ denotes the golden number ($\gamma = (1 + \sqrt{5})/2$). The first problem is to show that (4) has a normally hyperbolic invariant two-dimensional torus, $\mathbf{T}^2$. This is easy after one step in the averaging process. The invariant $\mathbf{T}^2$ is located at an $O(\varepsilon^2)$ distance of $x = 0 \pmod{2\pi}$.

Approximations of the invariant $\mathbf{T}^2$ and their stable and unstable manifolds, $W^S_{\mathbf{T}^2}$, $W^U_{\mathbf{T}^2}$, are obtained in Section 4. The algorithm to obtain those approximations is given in detail. By symbolic manipulation those approximations are obtained to high order on a computer implementation of the algorithm. Also in Section 4 we show how to globalize $W^U_{\mathbf{T}^2}$ by starting at points given by the previous approximation and integrating (4) numerically. This is done until the points reach $\{x = \pi\}$ for the first time. By symmetry the intersection of $W^S_{\mathbf{T}^2}$ with $\{x = \pi\}$ is also obtained. As intersections two tori are obtained. They can be described by giving the values of x' as a function of two angles, φ_1 and φ_2. Let $x'_{W^{U,S}_{\mathbf{T}^2}\cap\{x=\pi\}} = m^{U,S}(\varphi_1, \varphi_2, \varepsilon)$ and let $m = m^U - m^S$. A convenient measure of the splitting, in this first intersection, is

$$d(\varepsilon) = \sup_{(\varphi_1,\varphi_2)\in\mathbf{T}^2} |m(\varphi_1, \varphi_2, \varepsilon)|. \tag{5}$$

The function (5) has been computed for several values of ε.

In Section 5 we carry out a "first order" analysis of the splitting, that is, we use formally Melnikov's method after one step of averaging. The computations give a theoretical value for the function m described before. We do not claim this approximation is rigourously justified, but the results are in good agreement with the numerical ones.

Assuming the Melnikov approach gives a good approximation to the splitting, we analyze the function m in the case of small values of ε. The splitting is given by an infinite sum, each term being, essentially, of the form

$$\alpha_k \exp\left(-\frac{|(k,\omega)|}{b\varepsilon}\right) \sin((k,\varphi)), \quad k \in \mathbf{Z}^2, \tag{6}$$

where $\omega = (1, \gamma)^T$ and b is independent of ε and k. However, the coefficients α_k in (6) depend strongly on k and the function $d(\varepsilon)$ defined by (5) behaves as

$$\exp\left(-\frac{c(\varepsilon)|\ln\varepsilon|^{1/2}}{\varepsilon^{1/2}}\right), \tag{7}$$

when ε goes to zero, where $c(\varepsilon)$ is a bounded function: $0 < c(\varepsilon) < \hat{c}$. The values $\overline{c} = \limsup_{\varepsilon\to 0} c(\varepsilon)$ and $\underline{c} = \liminf_{\varepsilon\to 0} c(\varepsilon)$ are computed and they satisfy $0 < \underline{c} < \overline{c}$. This shows, under the assumption above, that the exponent of ε in (3) can not be improved, in general. The extra $|\ln\varepsilon|$ factor in the exponent in (7) (note that in our case $\tau = 1$ in (2) and use Remark 1) is due to the fact that *sin* is not only analytic but entire. Some extensions and improvements are studied.

Section 6 is devoted to the set of points in $W^U_{\mathbf{T}^2} \cap W^S_{\mathbf{T}^2} \cap \{x = \pi\}$ giving rise to "primary" homoclinic orbits. This is a curve in a two-torus, defined by $m(\varphi_1, \varphi_2, \varepsilon) = 0$. The topology of the curve changes when ε does. This is a very natural phenomenon. The behaviour of the set of bifurcations is described. Let ε_k, $k \in \mathbf{N}$, $\varepsilon_1 > \varepsilon_2 > \ldots$, the set of values for which a bifurcation of the set of primary homoclinic orbits is produced. Then $\lim_{k\to\infty} \varepsilon_{k+1}/\varepsilon_k = \gamma^2$. The curves close after n_1 revolutions in the φ_1 direction and n_2 in the φ_2 direction. The succesive values of the pairs (n_1, n_2) are (1,2), (2,3), (3,5), (5,8),..., i.e., given by two successive Fibonacci numbers (F_j, F_{j+1}), where $F_1 = 1$, $F_2 = 2$, $F_j = F_{j-1} + F_{j-2}$.

2 Proof of the Averaging Theorem

To prove Theorem A we have to introduce some notation. We denote by ${}^j f$, $j = 1, \ldots, n$, the components of f. Let $D + \Delta = \{(z_1, \ldots, z_n) \in \mathbf{C}^n$ such that it exists $(x_1, \ldots, x_n) \in$

D, a compact in $\mathbf{R}^n$, with $\max_j |z_j - x_j| \le \Delta\}$, and $\mathbf{T}^p + \Delta = \{(\theta_1, \ldots, \theta_n)$ such that $\theta_j = \rho_j + i\sigma_j$, $\rho_j \in S^1$, $\sigma_j \in \mathbf{R}$, $|\sigma_j| \le \Delta\}$. In the previous expressions $|\ \ |$ denote the modulus of a complex or the absolute value of a real. We remark that the values of Δ (in $D + \Delta$, $\mathbf{T}^p + \Delta$) could be different, but this is not essential.

For a vector f we use the supremum norm

$$|f|_\Delta = \max_{j=1,\ldots,n} \max_{z \in D+\Delta} \max_{\theta \in \mathbf{T}^p+\Delta} \max_{|\varepsilon| \le \varepsilon_0} |{}^j f(z, \theta, \varepsilon)|,$$

while for a matrix $M(z, \theta, \varepsilon) = ({}^{ij}M(z, \theta, \varepsilon))_{i,j=1,\ldots,n}$, we use the induced norm

$$|M|_\Delta = \max_{i=1,\ldots,n} \sum_{j=1}^{n} |{}^{ij}M|_\Delta. \tag{8}$$

Let us define the average of a vector, h, as $\overline{h}(z, \varepsilon) = \int_{\mathbf{T}^p} h(z, \theta, \varepsilon)\, d\theta$, and the purely quasiperiodic part as $\tilde{h}(z, \theta, \varepsilon) = h(z, \theta, \varepsilon) - \overline{h}(z, \varepsilon)$.

For convenience we relabel our initial vector, f, as f_0. So we have $f_0 = \overline{f}_0 + \tilde{f}_0$. We also relabel Δ as Δ_0. Let $\alpha = |f_0|_{\Delta_0}$. Hence $|\overline{f}_0|_{\Delta_0} = \overline{\alpha}_0$, $|\tilde{f}_0|_{\Delta_0} = \tilde{\alpha}_0$ satisfy

$$\overline{\alpha}_0 \le \alpha, \quad \tilde{\alpha}_0 \le 2\alpha. \tag{9}$$

For any real positive r let

$$\Delta_r = \Delta_0 - r\delta, \tag{10}$$

where δ is a suitable constant, depending on ε, to be determined later.

We shall perform a finite number of steps of some averaging process. The number of steps depends on ε. Assume that, after j steps, we have a current variable, z_j, that we rename as z for shortness, and a current equation

$$\dot{z} = \varepsilon(\overline{f}_j(z, \varepsilon) + \tilde{f}_j(z, \theta, \varepsilon)), \tag{11}$$

where $(z, \theta) \in (D + \Delta_j) \times (\mathbf{T}^p + \Delta_j)$. We denote this domain by S_j. We assume that the bounds

$$|\overline{f}_j|_{\Delta_j} \le \overline{\alpha}_j, \quad |\tilde{f}_j|_{\Delta_j} \le \tilde{\alpha}_j \tag{12}$$

hold. Next we do one more step of the averaging process, changing to a new variable, z_{j+1}, replacing $\overline{f}_j$, $\tilde{f}_j$ by $\overline{f}_{j+1}$, $\tilde{f}_{j+1}$ and reducing a little the domain (from Δ_j to Δ_{j+1}).

In fact we shall still introduce intermediate domains $D + \Delta_{j+1/2}$. Assume $(z, \theta) \in S_j$ and let

$$z = w + \varepsilon v(w, \theta, \varepsilon), \tag{13}$$

for $(w, \theta) \in R_{j+1}$, where $R_{j+1} = (D + \Delta_{j+1/2}) \times (\mathbf{T}^p + \Delta_{j+1})$. The norm in this domain will be denoted by $|\ \ |_{R_{j+1}}$. By substitution of (13) in (11) we get

$$\begin{aligned}(Id + M_j(w, \theta, \varepsilon))\dot{w} + \varepsilon \frac{\partial v}{\partial \theta}(w, \theta, \varepsilon)(w) &= \\ = \varepsilon(A_j(w, \theta, \varepsilon) + \overline{f}_j(w, \varepsilon) + B_j(w, \theta, \varepsilon) + \tilde{f}_j(w, \theta, \varepsilon)).\end{aligned} \tag{14}$$

In the expression (14) Id denotes the n-dimensional identity matrix, and M_j, A_j, B_j are defined by

$$M_j(w, \theta, \varepsilon) = \varepsilon \frac{\partial v}{\partial w}(w, \theta, \varepsilon), \tag{15}$$

$$A_j(w, \theta, \varepsilon) = \overline{f}_j(w + \varepsilon v((w, \theta, \varepsilon), \varepsilon) - \overline{f}_j(w, \varepsilon), \tag{16}$$

$$B_j(w, \theta, \varepsilon) = \tilde{f}_j(w + \varepsilon v((w, \theta, \varepsilon), \theta, \varepsilon) - \tilde{f}_j(w, \theta, \varepsilon). \tag{17}$$

Let us choose v such that

$$\frac{\partial v}{\partial \theta}(w,\theta,\varepsilon)(\omega) = \tilde{f}_j(w,\theta,\varepsilon) \text{ and } \overline{v}(w,\varepsilon) = 0. \tag{18}$$

This defines v uniquely and can be obtained by equating Fourier coefficients. Let

$$\tilde{f}_j(w,\theta,\varepsilon) = \sum_{k\in\mathbb{Z}^p\setminus\{0\}} \tilde{f}_{j,k}(w,\varepsilon)e^{2\pi i(k,\theta)}$$

and

$$v(w,\theta,\varepsilon) = \sum_{k\in\mathbb{Z}^p\setminus\{0\}} v_k(w,\varepsilon)e^{2\pi i(k,\theta)}.$$

From (18) we get

$$v_k(w,\varepsilon) = \tilde{f}_{j,k}(w,\varepsilon)/(2\pi i(k,\omega)).$$

We have to bound v in R_{j+1}. By analyticity of $\tilde{f}_j$ in S_j we have $|\tilde{f}_{j,k}(w,\varepsilon)|_{\Delta_{j+1/2}} \leq \frac{\tilde{\alpha}_j}{2\pi}e^{-\Delta_j|k|}$. Then

$$|v|_{R_{j+1}} \leq \frac{\tilde{\alpha}_j}{2\pi}\sum_{k\in\mathbb{Z}^p\setminus\{0\}} \frac{e^{-\Delta_j|k|}}{|(k,\omega)|}e^{-\Delta_{j+1}|k|} = \frac{\tilde{\alpha}_j}{2\pi}\sum_{k\in\mathbb{Z}^p\setminus\{0\}} \frac{e^{-\delta|k|}}{|(k,\omega)|}. \tag{19}$$

At this point we need a technical Lemma on small divisors.

Lemma 1. *Let $\omega \in \mathbb{R}^p$ satisfy the diophantine condition (2) with $\tau > p-1$. Let $s_m = \sum_{k\in\mathbb{Z}^p\setminus\{0\},|k|\leq m} |(k,\omega)|^{-1}$. Then*

$$s_m \leq \Gamma m^\tau, \tag{20}$$

where Γ is a positive constant depending on ω but independent of m. If $\tau = p-1$ the previous inequality is replaced by $s_m \leq \Gamma' m^\tau \ln m$, where Γ' is also independent of m.

This result is rather well known. A complete proof can be found in Simó.[13]

Let us define $c_m = s_m - s_{m-1}$, with $s_0 = 0$. Then we can finish the bound in (19) by using (20):

$$\begin{aligned} |v|_{R_{j+1}} &\leq \frac{\tilde{\alpha}_j}{2\pi}\sum_{m\geq 1} e^{-\delta m}c_m = \frac{\tilde{\alpha}_j}{2\pi}\sum_{m\geq 1} s_m(e^{-\delta m} - e^{\delta(m+1)}) \leq \\ &\leq \frac{\tilde{\alpha}_j}{2\pi}(1-e^{-\delta})\Gamma\sum_{m\geq 1} e^{-\delta m}m^\tau \leq \tilde{\alpha}_j\frac{K}{\delta^\tau}, \end{aligned} \tag{21}$$

where K is a constant independent of $\tilde{\alpha}_j$ and δ.

If $w \in D + \Delta_{j+1/2}$, we must be sure that $z \in D + \Delta_j$. It is enough to have

$$\varepsilon\tilde{\alpha}_j\frac{K}{\delta^\tau} < \frac{\delta}{2}, \tag{22}$$

a requirement to be checked later by a suitable election of δ.

The equation (14) can be written as

$$\dot{w} = \varepsilon(\overline{f}_{j+1}(w,\varepsilon) + \tilde{f}_{j+1}(w,\theta,\varepsilon)) \tag{23}$$

for $(w,\theta) \in S_{j+1}$. Then (23) is like (11) after one more step of averaging, and we take w as the new variable z_{j+1}. We should relate the bounds of $|\overline{f}_{j+1}|_{\Delta_{j+1}}$, $|\tilde{f}_{j+1}|_{\Delta_{j+1}}$

to the ones of $|\overline{f}_j|_{\Delta_j}$, $|\widetilde{f}_j|_{\Delta_j}$ as given by (12). We use the identity $(Id + M_j)^{-1} = Id - (Id + M_j)^{-1}M_j$ to write

$$\dot{w} = \varepsilon[(Id + M_j)^{-1}(A_j + B_j - M_j\overline{f}_j) + \overline{f}_j], \tag{24}$$

where we have used the cancellation due to (18). By comparing (23) and (24) we see that all the terms in the square brackets in (24) can give contributions to $\overline{f}_{j+1}$, and all but the last one to $\widetilde{f}_{j+1}$.

First we need some preliminary estimates:

a) From the Cauchy inequality we have

$$\left|\frac{\partial({}^k v)}{\partial({}^j w)}\right|_{\Delta_{j+1}} \leq \tilde{\alpha}_j \frac{2K}{\delta^{\tau+1}}, \tag{25}$$

where we have reduced the strip by $\delta/2$ and we have used the bound (21). Then, from (15) and because of (8) and (25), we have

$$|M_j|_{\Delta_{j+1}} \leq \frac{2Kn\varepsilon\tilde{\alpha}_j}{\delta^{\tau+1}}, \tag{26}$$

b) Again by using Cauchy inequality and the mean value Theorem, we have from (16) and (17), and using again (21), the bounds

$$|A_j|_{\Delta_{j+1}} \leq \overline{\alpha}_j \varepsilon \frac{2Kn\tilde{\alpha}_j}{\delta^{\tau+1}}, \quad |B_j|_{\Delta_{j+1}} \leq \tilde{\alpha}_j \varepsilon \frac{2Kn\tilde{\alpha}_j}{\delta^{\tau+1}}. \tag{27}$$

c) Assume $|M_j|_{\Delta_{j+1}} < 1$. For this it is enough to have

$$\varepsilon\tilde{\alpha}_j \frac{K}{\delta^{\tau+1}} < \frac{1}{2n}. \tag{28}$$

This is stronger than (22). Hence we had to select δ later on to satisfy (28) and then (22) will hold. Under this assumption we have

$$|(Id + M_j)^{-1}|_{\Delta_{j+1}} \leq \left(1 - \frac{\varepsilon\tilde{\alpha}_j 2Kn}{\delta^{\tau+1}}\right)^{-1}. \tag{29}$$

From (24) we get

$$\overline{f}_{j+1}(w,\varepsilon) = \overline{f}_j(w,\varepsilon) + \overline{g(w,\theta,\varepsilon)}, \quad \widetilde{f}_{j+1}(w,\theta,\varepsilon) = \widetilde{g(w,\theta,\varepsilon)}, \tag{30}$$

where

$$g(w,\theta,\varepsilon) = (Id + M_j(w,\theta,\varepsilon))^{-1}(A_j(w,\theta,\varepsilon) + B_j(w,\theta,\varepsilon) - M_j(w,\theta,\varepsilon)\overline{f}_j(w,\varepsilon)). \tag{31}$$

By using (27), (29), (26) and (12) we bound the norm in (31). To simplify the notation we introduce

$$\psi = \frac{2Kn\varepsilon}{\delta^{\tau+1}}, \tag{32}$$

and then

$$|g|_{\Delta_{j+1}} \leq (1 - \psi\tilde{\alpha}_j)^{-1}((\overline{\alpha}_j + \tilde{\alpha}_j)\psi + \overline{\alpha}_j\psi)\tilde{\alpha}_j. \tag{33}$$

We note that, now, the requirement (28) is just

$$\tilde{\alpha}_j \psi < 1. \tag{34}$$

From (30) and (33) we get

$$\begin{aligned} |\overline{f}_{j+1}|_{\Delta_{j+1}} &\leq \overline{\alpha}_j + (1-\psi\tilde{\alpha}_j)^{-1}\psi(\tilde{\alpha}_j + 2\overline{\alpha}_j)\tilde{\alpha}_j, \\ |\tilde{f}_{j+1}|_{\Delta_{j+1}} &\leq 2(1-\psi\tilde{\alpha}_j)^{-1}\psi(\tilde{\alpha}_j + 2\overline{\alpha}_j)\tilde{\alpha}_j. \end{aligned} \tag{35}$$

Let us denote by $\overline{\alpha}_{j+1}$ and $\tilde{\alpha}_{j+1}$, respectively, the right-hand terms in the inequality of (35).

Claim *By a suitable selection of ψ we have*

$$2\alpha > \cdots > \overline{\alpha}_j > \overline{\alpha}_{j-1} > \cdots > \overline{\alpha}_0, \quad 2\alpha \geq \tilde{\alpha}_0 > \tilde{\alpha}_1 > \cdots > \tilde{\alpha}_j > \cdots. \tag{36}$$

Let $0 < \eta < 1/2$ and choose ψ such that

$$\psi\alpha \leq \frac{\eta}{12+2\eta}. \tag{37}$$

Then, if (36) holds it does (34).

We shall prove the claim by induction. If $j = 0$ it follows from (35), (37) and (9) that

$$\overline{\alpha}_0 < \overline{\alpha}_1 < \overline{\alpha}_0 + \frac{\psi 4\alpha}{1-\psi 2\alpha}2\alpha < \alpha\left(1+\frac{2\eta}{3}\right) < 2\alpha,$$

$$\tilde{\alpha}_1 < \frac{2\psi 4\alpha}{1-\psi 2\alpha}\tilde{\alpha}_0 < \frac{2\eta}{3}\tilde{\alpha}_0 < \tilde{\alpha}_0.$$

Assume (36) holds till some value of j. Then

$$\overline{\alpha}_j < \overline{\alpha}_{j+1} < \overline{\alpha}_j + \frac{\psi 6\alpha}{1-\psi 2\alpha}\tilde{\alpha}_j, \tag{38}$$

and

$$\tilde{\alpha}_{j+1} < \frac{2\psi 6\alpha}{1-\psi 2\alpha}\tilde{\alpha}_j < \eta\tilde{\alpha}_j. \tag{39}$$

The relation (39) shows the second part in (36) and, furthermore, it gives

$$\tilde{\alpha}_m \leq \eta^m 2\alpha \text{ for } m = 0, \ldots, j+1. \tag{40}$$

Going back to (38) we have

$$\overline{\alpha}_{j+1} < \overline{\alpha}_j + \frac{\eta}{2}\tilde{\alpha}_j < \cdots < \overline{\alpha}_0 + \frac{\eta}{2}(\tilde{\alpha}_0 + \cdots + \tilde{\alpha}_j) < \alpha + \frac{\eta}{2}\sum_{m=0}^{j}\eta^m 2\alpha < \alpha\left(1+\frac{\eta}{1-\eta}\right) < 2\alpha.$$

This ends the proof of the Claim.

For convenience we select $\eta = e^{-1} = \exp(-1)$. By going to (32) we see that one can take $\delta = \overline{K}\varepsilon^{c_3}$, where $c_3 = (1+\tau)^{-1}$, $\overline{K} = (2K(12e+2))^{c_3}$. As we want our final domain to contain $(D + \Delta/2) \times (\mathrm{T}^p + \Delta/2)$, it follows from (10) that the number of averaging steps is $N = N(\varepsilon) = E\left(\frac{\Delta/2}{\delta}\right) > \frac{\Delta/2}{\overline{K}\varepsilon^{c_3}} - 1$, where $E(\ )$ denotes the integer part. The final vector field, after these transformations, is $\varepsilon(\overline{f}_N + \tilde{f}_N)$. The formulae (36) and (40) give $|\overline{f}_N|_{\Delta/2} < 2|f|_\Delta$ and $|\tilde{f}_N|_{\Delta/2} < 2\alpha e^{-N} < c_1\exp(-c_2\varepsilon^{-c3})$, where $c_1 = 2\alpha\varepsilon$, $c_2 = \frac{\Delta/2}{\overline{K}}$. This proves the formula (3) and ends the proof of Theorem A.

3 An example: first step of averaging and existence of a quasiperiodic solution

As an illustration of the averaging theorem and to gain insight in the splitting under fast quasiperiodic forcing we consider the example (4). The change of the time variable $t = \tau/\varepsilon$ leads to the equation

$$\ddot{x} = \varepsilon^2 \sin(x + \sin t + \sin(\gamma t)), \quad \gamma = (1+\sqrt{5})/2. \tag{41}$$

Equivalently this can be written as

$$\dot{x} = \varepsilon y, \quad \dot{y} = \varepsilon \sin(x + \sin t + \sin(\gamma t)), \tag{42}$$

which is of the form (1) and, furthermore, is a Hamiltonian vector field with Hamiltonian function

$$H(x, y, t, \varepsilon) = \varepsilon \left(\frac{y^2}{2} + \cos(x + \sin t + \sin(\gamma t)) \right). \tag{43}$$

As stated in Theorem B, the averaging process can be carried out, in this case, by using canonical transformations. It is useful, to discuss the behaviour of (41), to do first one step of averaging. To this end we split (43) in the averaged and the purely quasiperiodic part. Let us use the angles θ_1 and θ_2 to denote t and γt. Then (43) reads

$$\begin{aligned} H(x, y, \theta, \varepsilon) &= \varepsilon \left(\frac{y^2}{2} + \cos x \cos(\sin\theta_1 + \sin\theta_2) - \sin x \sin(\sin\theta_1 + \sin\theta_2) \right) = \\ &= \varepsilon(\overline{H}_0 + \widetilde{H}_0), \end{aligned} \tag{44}$$

where $\overline{H}_0 = \frac{y^2}{2} + \cos x \overline{\cos(\sin\theta_1 + \sin\theta_2)}$. Note that $\widetilde{H}_0$ does not depend on y. This makes the first average easier. More explicitely (44) is of the form $\varepsilon(\frac{y^2}{2} + A\cos x + m_1(t)\sin x + m_2(t)\cos x)$, where $\overline{m_1} = \overline{m_2} = 0$.

We change from (x, y) to a new set of variables (ξ, η) by means of the canonical transformation $x = \frac{\partial S}{\partial y}$, $\eta = \frac{\partial S}{\partial \xi}$, where $S = S(\xi, y)$ is a generating function of the form $S = \xi y + \varepsilon S_1(\xi, \eta, t)$ and S_1 is determined by $\frac{\partial S_1}{\partial t} = \widetilde{H}_0$, $\overline{S_1} = 0$. Hence $S_1 = M_1(t)\sin\xi + M_2(t)\cos\xi$, where M_j, $j = 1, 2$, is the primitive of m_j with zero average.

Hence, the canonical change is given by

$$x = \xi, \quad \eta = y + \varepsilon(\cos\xi M_1(t) - \sin\xi M_2(t)),$$

leading to the Hamiltonian after the first average, which can be obtained in full explicit form

$$\begin{aligned} &\varepsilon\left(\frac{\eta^2}{2} + A\cos\xi\right) + \varepsilon^2\eta\left(-\cos\xi M_1(t) + \sin\xi M_2(t)\right) + \\ &\quad + \varepsilon^3\left(-\sin(2\xi)\frac{1}{2}M_1 M_2 + \frac{1}{4}\cos(2\xi)(M_1^2 - M_2^2)\right) = \varepsilon H_1 + \varepsilon^2 H_2 + \varepsilon^3 H_3. \end{aligned} \tag{45}$$

To determine the values of A, $m_1(t)$ and $m_2(t)$ we use the expansions (see Abramowitz and Stegun[1])

$$\begin{aligned} \cos(z\sin\alpha) &= J_0(z) + 2\sum_{k\geq 1} J_{2k}(z)\cos(2k\alpha), \\ \sin(z\sin\alpha) &= 2\sum_{k\geq 0} J_{2k+1}(z)\sin(2(k+1)\alpha), \end{aligned} \tag{46}$$

where $J_n(z)$ stands for the Bessel function of first kind, order n and argument z. Therefore $A = (J_0(1))^2$ and M_1, M_2 can be easely obtained from (44), (45) and (46) as

$$
\begin{aligned}
-M_1(t) &= 2J_0 \sum_{k\geq 0} J_{2k+1} \left(\frac{\cos((2k+1)\theta_1)}{2k+1} + \frac{\cos((2k+1)\theta_2)}{(2k+1)\gamma} \right) + \\
&+2 \sum_{k\geq 0} \sum_{l\geq 1} J_{2k+1} J_{2l} \left(\frac{\cos((2k+1)\theta_1 + 2l\theta_2)}{2k+1+2l\gamma} + \frac{\cos((2k+1)\theta_1 - 2l\theta_2)}{2k+1-2l\gamma} + \right. \\
&\left. + \frac{\cos(2l\theta_1 + (2k+1)\theta_2)}{2l+(2k+1)\gamma} + \frac{\cos(-2l\theta_1 + (2k+1)\theta_2)}{-2l+(2k+1)\gamma} \right),
\end{aligned} \tag{47}
$$

$$
\begin{aligned}
M_2(t) &= 2J_0 \sum_{k\geq 1} J_{2k} \left(\frac{\sin(2k\theta_1)}{2k} + \frac{\sin(2k\theta_2)}{2k\gamma} \right) + \\
&+2 \sum_{k,l\geq 1} J_{2k} J_{2l} \left(\frac{\sin(2k\theta_1 + 2l\theta_2)}{2k+2l\gamma} + \frac{\sin(2k\theta_1 - 2l\theta_2)}{2k-2l\gamma} \right) + \\
&+2 \sum_{k,l\geq 0} J_{2k+1} J_{2l+1} \left(\frac{\sin((2k+1)\theta_1 + (2l+1)\theta_2)}{2k+1+(2l+1)\gamma} - \right. \\
&\left. - \frac{\sin((2k+1)\theta_1 - (2l+1)\theta_2)}{2k+1-(2l+1)\gamma} \right),
\end{aligned}
$$

where all the Bessel functions are evaluated at 1.

At this point we want to do some remarks.

a) The example (4) can be generalized to

$$x'' = \sin(x + \varepsilon^p(a\sin(\tau/\varepsilon) + b\sin(\gamma\tau/\varepsilon))), \quad a, b \in \mathbb{R}, \quad p \geq 0. \tag{48}$$

Then H_1 is replaced by $\frac{\eta^2}{2} + J_0(a\varepsilon^p)J_0(b\varepsilon^p)\cos\xi$. If ε goes to zero and $p > 0$ the numerical coefficient tends to 1. If $p = 0$ we find $J_0(a) \cdot J_0(b)$. The character of the fixed points of H_1, $\xi = 0 \pmod{\pi}$, $\eta = 0$, depends on $J_0(a) \cdot J_0(b)$. If the product is positive then $(0, 0)$ is hyperbolic and $(\pi, 0)$ elliptic. They exchange the stability for negative values of the product.

b) A lot of papers in the last years deal with equations like

$$x'' = \sin x + \varepsilon^q \sin(\tau/\varepsilon), \quad q \in \mathbb{R}, \tag{49}$$

(see Holmes et al.[8], Gelfreich[7], Delshams and M. Seara[4]). The difficulties seem to appear for $q \leq 0$. However, consider again (48). By introducing $z = x + \varepsilon^p(a\sin(\tau/\varepsilon) + b\sin(\gamma\tau/\varepsilon))$ we have

$$z'' = \sin z - \varepsilon^{p-2}(a\sin(\tau/\varepsilon) + b\gamma^2\sin(\gamma\tau/\varepsilon)),) \tag{50}$$

which reduces to the form (49) if $q = p - 2$, $a = -1$, $b = 0$. But (50) has a Hamiltonian like (45) with arguments of the Bessel functions equal to $a\varepsilon^p$ or to $b\varepsilon^p$. Instead of (49) we have, introducing $z = x + \varepsilon^{q+2}\sin(\tau/\varepsilon)$, the equation $z'' = \sin(z - \varepsilon^{q+2}\sin(\tau/\varepsilon))$. Proceeding as in the passage from (4) to (45) we get a Hamiltonian like (45) but now

$$
\begin{aligned}
&A = J_0(\varepsilon^{q+2}), \quad M_1(t) = -2\sum_{k\geq 0} J_{2k+1}(\varepsilon^{q+2}) \frac{\cos((2k+1)t)}{2k+1}, \\
&M_2(t) = 2\sum_{k\geq 1} J_{2k}(\varepsilon^{q+2}) \frac{\sin(2kt)}{2k}.
\end{aligned} \tag{51}
$$

From (51) it follows that, for any $q \geq -2$, there is no difficulty in the study of (49), the Hamiltonian (45) being a relatively small perturbation of a Hamiltonian, εH_1, with H_1 autonomous and with well defined non degenerated limit when ε goes to zero. On the other hand, if $q < -2$ then $J_0(\varepsilon^{q+2})$ has infinitely many oscillations when ε goes to zero.

We return to the study of (45). The first thing is to prove the existence of an invariant two-dimensional torus which replaces the hyperbolic fixed point $(0,0)$ appearing in H_1. To this end we state the following result.

Theorem C. *Let us consider*

$$\dot{x} = \varepsilon[(A(\varepsilon) + \varepsilon Q(\theta, \varepsilon))x + \varepsilon g(\theta, \varepsilon) + h(x, \theta, \varepsilon)] \tag{52}$$

where $|\varepsilon| \leq \varepsilon_0$, *$x$ belongs to a closed ball of radius r around 0 in* $\mathbf{R}^n$, $\theta \in \mathbf{T}^p + \Delta$, $\theta = \omega t$. *Assume $A(\varepsilon)$ is a uniformly hyperbolic matrix (i.e.* $|\mathrm{Re}\,\lambda| > a > 0$, $\forall \lambda \in \mathrm{Spec}\, A(\varepsilon)$, $|\varepsilon| \leq \varepsilon_0$*) and any two eigenvalues of $A(\varepsilon)$,* $\lambda_i(\varepsilon)$, $\lambda_j(\varepsilon)$, *satisfy* $\lambda_i(\varepsilon) \neq \lambda_j(\varepsilon)$ *if* $i \neq j$ $\forall |\varepsilon| \leq \varepsilon_0$. *Furthermore assume ω satisfies a diophantine condition* $|(k, \omega)| \geq \frac{c}{|k|^\tau}$ $\forall k \in \mathbf{Z}^p \setminus \{0\}$ *for some* $c, \tau > 0$ *and A, Q, g, h are bounded with respect to ε, Q, g and h are analytic with respect to θ in* $\mathbf{T}^p + \Delta$, *h is analytic with respect to x and* $h(0, \theta, \varepsilon) = 0$, $D_x h(0, \theta, \varepsilon) = 0$ $\forall \theta \in \mathbf{T}^p + \Delta$, $|\varepsilon| \leq \varepsilon_0$.

Then there exists $\varepsilon_1 \leq \varepsilon_0$ *such that for any ε with* $0 < |\varepsilon| < \varepsilon_1$, *there exists an affine change of coordinates* $y = R(\theta)x + v(\theta)$, *depending analytically on θ for* $\theta \in \mathbf{T}^p + \Delta/2$, *transforming (52) to*

$$\dot{y} = \varepsilon(A_\infty(\varepsilon)y + h_\infty(y, \theta, \varepsilon)), \tag{53}$$

where $h_\infty = O(|y|^2)$ *uniformly in θ, ϵ. The matrix A_∞ is hyperbolic. Furthermore the equation (52) has a quasiperiodic solution,* $x_\varepsilon(t)$, *with basic frequencies* $\omega_1, \ldots, \omega_p$ *and* $|x_\varepsilon| = O(\varepsilon^2)$ *for* $\varepsilon \to 0$.

The proof is close to the one of similar results that can be found in Bogoljubov et al.[2] or in Jorba and Simó.[9] The main idea is to do a sequence of transformations. After each one the equation (52) is of the form

$$\dot{x}_j = \varepsilon[(A_j(\varepsilon) + \varepsilon^{2^j} Q_j(\theta, \varepsilon))x_j + \varepsilon^{2^j} g_j(\theta, \varepsilon) + h_j(x_j, \theta, \varepsilon)]. \tag{54}$$

In (54) x_j is the current variable and the variables (x_j, θ) range on a somewhat smaller domain, $B_{r_j}(0) \times (\mathbf{T}^p + \Delta_j)$, where r_j and Δ_j have positive limits when j goes to ∞. In this case the difference $\Delta_j - \Delta_{j+1}$ can be chosen of the form $\frac{d}{j^2}$, for a suitable $d > 0$, for instance. Despite the norms of Q_j and g_j increase without bound for $j \to \infty$, the functions $\varepsilon^{2^j}|Q_j|$ and $\varepsilon^{2^j}|g_j|$ behave like η^{2^j} for some $0 < \eta < 1$ provided ε is small enough. The existence of the normally hyperbolic quasiperiodic solution follows from the fact that (53) has 0 as solution, by going back through the sequence of transformations. Full details on the proof and on the properties of the invariant manifold of the solution will appear elsewhere.[13]

Returning to our example (41), the quasiperiodic solution fills up densely a two-dimensional torus, $\mathbf{T}^2$. We shall talk loosely of $\mathbf{T}^2$ as the solution replacing the fixed point. The related manifolds, $W^{U,S}_{\mathbf{T}^2}$, are close to the separatrices of H_1 in (45). From the averaging theorem it follows that the splitting must be bounded by some expression like (3) (see Fontich and Simó[5,6] and references therein for the proximity argument and simpler cases). To check the behaviour of (41) we shall carry out some symbolic and numerical computations in the next section.

4 Approximations of the invariant torus and their invariant manifolds - numerical results

To obtain approximations of the invariant torus and their invariant manifolds we can proceed as explained before. That is, we can do some steps of the averaging process and then some ones along the lines of the proof of Theorem C. However we shall proceed, formally, in a simpler and more direct way. In contrast to the approach following the proofs of those Theorems, we do not claim the method proposed here is convergent.

We start directly with equation (41): $\ddot{x} = \varepsilon^2 \sin(x + \sin t + \sin(\gamma t))$. We expand it as

$$\ddot{x} = \varepsilon^2 \sum_{k \geq 0} x^k f_k(\overline{\theta}_1, \overline{\theta}_2), \tag{55}$$

where now $\overline{\theta}_1 = \exp(it)$, $\overline{\theta}_2 = \exp(i\gamma t)$, i.e. $\overline{\theta}_j = \exp(i\theta_j)$. The symbolic computation of the quasiperiodic solution is done by keeping $\overline{\theta}_1$, $\overline{\theta}_2$ as symbols and using complex numerical coefficients, for instance in FORTRAN double precision. If we keep terms up to 10^{-16}, as $\sin t + \sin(\gamma t)$ can reach any value close to 2 and $|t|$, $|\gamma t|$ can be reduced to something bounded by π, we use powers up to order 28. Let $C = \cos(\sin t + \sin(\gamma t))$, $S = \sin(\sin t + \sin(\gamma t))$ and $\sin t = \frac{1}{2i}(\overline{\theta}_1 - \overline{\theta}_1^{-1})$, $\sin(\gamma t) = \frac{1}{2i}(\overline{\theta}_2 - \overline{\theta}_2^{-1})$. Then $f_k = (-1)^{E(k/2)} \frac{1}{k!} \binom{C}{S}$, where C is used for k odd and S for k even.

We look for a formal solution expanded in even powers of ε:

$$x = \sum_{j \geq 1} \varepsilon^{2j} x_j(\overline{\theta}_1, \overline{\theta}_2). \tag{56}$$

By inserting (56) in (55) we get $\sum_{j\geq 1} \varepsilon^{2j} \ddot{x}_j = \varepsilon^2 \sum_{k \geq 0} f_k (\sum_{l \geq 1} \varepsilon^{2l} x_l)^k$ and, in particular, $\ddot{x}_1 = f_0$. When x_{n-1} has been obtained we compute the required products $x_{l_1}, x_{l_2}, \ldots, x_{l_s}$ with $l_1 + \cdots + l_s = n - 1$, to obtain x_n by a double integration. If we have memory enough in the computer it is better to keep for x^k the terms of order $s = k, \ldots, n-1$ in ε^2 and actualize them to order n (computing also the terms of order n in ε^2 in the expansion of x^n) when x_n is available.

To obtain the invariant manifolds we split (55) as two first order equations

$$\dot{x} = \varepsilon y, \quad \dot{y} = \varepsilon \sin(x + \sin t + \sin(\gamma t)). \tag{57}$$

Then, from (56) we obtain $y = \varepsilon^{-1} \sum_{j \geq 1} \varepsilon^{2j} \dot{x}_j$. Let

$$q^{(0)} = \sum_{j \geq 1} \varepsilon^{2j} x_j, \quad p^{(0)} = \sum_{j \geq 1} \varepsilon^{2j-1} \dot{x}_j, \tag{58}$$

the approximation of the quasiperiodic solution. We introduce

$$x = q^{(0)} + \xi, \quad y = p^{(0)} + \eta. \tag{59}$$

Then, as $q^{(0)}$, $p^{(0)}$, as defined by (58), satisfy the system (57), we get

$$\begin{aligned}
\dot{\xi} &= \varepsilon \eta, \\
\dot{\eta} &= \varepsilon[\sin(q^{(0)} + \sin t + \sin(\gamma t) + \xi) - \sin(q^{(0)} + \sin t + \sin(\gamma t))] = \\
&= \varepsilon(\widehat{C}\xi - \widehat{S}\frac{\xi^2}{2!} - \widehat{C}\frac{\xi^3}{3!} + \ldots) = \varepsilon \sum_{k \geq 1} \xi^k m_k(\overline{\theta}_1, \overline{\theta}_2, \varepsilon),
\end{aligned} \tag{60}$$

where

$$\widehat{C} = \cos(q^{(0)} + \sin t + \sin(\gamma t)), \quad \widehat{S} = \sin(q^{(0)} + \sin t + \sin(\gamma t)). \tag{61}$$

Those are available in powers (positive and negative, of course) of $\overline{\theta}_1$, $\overline{\theta}_2$ when (58) is available.

We search for the invariant manifold (either the unstable or the stable one) writting η as a power series in ε, ξ and coefficients sums of powers of $\overline{\theta}_1$, $\overline{\theta}_2$:

$$\eta = \sum_{k\geq 1} \xi^k p^{(k)}(\overline{\theta}_1, \overline{\theta}_2, \varepsilon) = \sum_{k\geq 1} \xi^k \sum_{l\geq 0} \varepsilon^l p_{k,l}(\overline{\theta}_1, \overline{\theta}_2) = \sum_{k\geq 1} \xi^k \sum_{l\geq 0} \varepsilon^l \sum_{r,s} p_{k,l,r,s} \overline{\theta}_1^r \overline{\theta}_2^s. \tag{62}$$

The invariance of (62) by the equation (60) gives

$$\begin{aligned} \varepsilon \sum_{k\geq 1} \xi^k \sum_{s\geq 0} \varepsilon^s m_{ks}(\overline{\theta}_1, \overline{\theta}_2) &= \sum_{k\geq 1} \xi^k \sum_{l\geq 0} \varepsilon^l i \left(\frac{\partial p_{kl}}{\partial \overline{\theta}_1} \overline{\theta}_1 + \gamma \frac{\partial p_{kl}}{\partial \overline{\theta}_2} \overline{\theta}_2 \right) + \\ &+ \sum_{k\geq 1} k \xi^{k-1} \sum_{l\geq 0} \varepsilon^l p_{kl}(\overline{\theta}_1, \overline{\theta}_2) \cdot \varepsilon \sum_{r\geq 1} \xi^r \sum_{m\geq 0} \varepsilon^m p_{rm}(\overline{\theta}_1, \overline{\theta}_2). \end{aligned} \tag{63}$$

If $\overline{p}_{k,l}$ denotes the average of $p_{k,l}$ then we write $p_{k,l} = \overline{p}_{k,l} + \tilde{p}_{k,l}$. By looking at the terms in ε^0 in (63) we have $\tilde{p}_{k,0} = 0$. Equating the terms in ξ^u for $u \geq 1$ and among them the terms in ε^v, $v \geq 1$, in (63) we obtain, recurrently, the functions $\tilde{p}_{u,v}$ and $\overline{p}_{u,v-1}$. The only algebraic condition to be satisfied to make the recurrence possible is $\overline{p}_{1,0} \neq 0$. For $u = v = 1$ we obtain $\overline{p}_{1,0}^2 = \overline{m}_{1,1}$. The value of $\overline{m}_{1,1}$ is the average of $\widehat{C}$ in (60) when $q^{(0)}$ is set equal to zero, that is, $(J_0(1))^2$. By taking $\overline{p}_{1,0} = J_0(1)$ we obtain the unstable manifold. If $\overline{p}_{1,0} = -J_0(1)$ we get the stable one. Then all the successive terms are uniquely determined. Furthermore, if $p_{k,l}^U$ and $p_{k,l}^S$ denote the functions $p_{k,l}$ appearing in the expansion of the unstable and stable manifolds, respectively, from the symmetries of the functions in (61) we have $p_{kl}^S = p_{kl}^U(-1)^{1+l}$. The function p_{kl}, as a function of $\overline{\theta}_1$ and $\overline{\theta}_2$, has the parity of $1+k+l$. Ax η must be real for ξ, ε and t real, it follows from (62) that $p_{k,l,-r,-s}$is the complex conjugate of $p_{k,l,r,s}$. This, together with the symmetries mentioned above save computer memory: It is enough to store terms with $r \geq 0$ and if $r = 0$, with $s \geq 0$.

The previous algorithms have been implmented to obtain the coefficients $p_{k,l,r,s}$ with $1 \leq k \leq 12$, $0 \leq l \leq 24$, $|r|, |s| \leq 28$. In this way we obtain a formal representation of the invariant torus and the related manifolds.

To extend numerically the local manifolds obtained so far, given ε, ξ and t, we substitute in (56) and (62). As $(t, \gamma t) \pmod{2\pi}$ fill $\mathbf{T}^2$ densely when t ranges in $\mathbf{R}$, for practical purposes we have taken a different approach. Initial values of the angles (θ_1, θ_2), say (ψ_1, ψ_2), are taken in $\mathbf{T}^2$ and t is set to zero. Then the system to be integrated is

$$\dot{x} = \varepsilon y, \quad \dot{y} = \varepsilon \sin(x + \sin\theta_1 + \sin\theta_2), \quad \dot{\theta}_1 = 1, \quad \dot{\theta}_2 = \gamma, \tag{64}$$

with the initial conditions

$$t = 0, \quad \theta_1 = \psi_1, \quad \theta_2 = \psi_2, \quad x = q^{(0)}(\overline{\theta}_1, \overline{\theta}_2, \varepsilon) + \xi, \quad y = \sum_{k\geq 0} \xi^k p^{(k)}(\overline{\theta}_1, \overline{\theta}_2, \varepsilon), \tag{65}$$

as follows from (59) and (62). Before globalizing the manifolds the following tests have been performed:

a) If $\xi = 0$ in (64) we have the invariant torus. Given ε, ψ_1, ψ_2 we can compute x, y and also, by formal derivation, the values of $\dot{x}$, $\dot{y}$. These values have been checked against direct evaluation in (42). The agreement is uniformly very good in $\mathbf{T}^2$

until $\varepsilon = 0.07$ (say, errors less than 10^{-12} and much less if ε is very small) and acceptable till $\varepsilon = 0.1$ (errors less than 10^{-8}). No direct numerical integration of (64) with the initial conditions (65) has been done because of the hyperbolic character of $\mathbf{T}^2$.

b) If $\xi \neq 0$ is small (say $\xi = 0.02 \div 0.04$) and ε is fixed, we can integrate (64) with the conditions (65) for a moderate time interval (we have selected $t = t_f = 2\pi$ as final time). Then the final values x_f, y_f, $\theta_{1,f}$, $\theta_{2,f}$ have been obtained. From the third equation in (65) the current value of ξ, ξ_f, is obtained. Then the value, y_w, of y from the fourth equation in (65), evaluated at ε, ξ_f, $\theta_{1,f}$, $\theta_{2,f}$ is obtained. This value is compared with y_f. The agreement is similar to the one described for $\mathbf{T}^2$ or even better.

After the tests have been passed we have globalized the manifold $W^U_{\mathbf{T}^2}$ till x reaches the value π. Starting with values of ψ_1, ψ_2 in an equispaced lattice in $\mathbf{T}^2$ (typically 100×100 points are used) we obtain as final values $y = y^U$, $\theta_1 = \varphi_1$, $\theta_2 = \varphi_2$. Of course, the values of (φ_1, φ_2) are not equispaced in $W^U_{\mathbf{T}^2} \cap \{x = \pi\}$, because the travelling times are different. However, the map $(\psi_1, \psi_2) \rightarrow (\varphi_1, \varphi_2)$ (for a fixed initial value of ξ) is a diffeomorphism. It is not difficult to set up an interpolation procedure to obtain y^U as a function of (φ_1, φ_2) in $\mathbf{T}^2$ with enough accuracy: $y^U = m^U(\varphi_1, \varphi_2, \varepsilon)$. If we do the same process for $W^S_{\mathbf{T}^2}$, we obtain y^S in $W^S_{\mathbf{T}^2} \cap \{x = \pi\}$ (by backwards integration in time taking the correct "branch" of $W^S_{\mathbf{T}^2}$), given by $y^S = m^S(\varphi_1, \varphi_2, \varepsilon)$. By symmetry one has $m^S(\varphi_1, \varphi_2, \varepsilon) = m^U(2\pi - \varphi_1, 2\pi - \varphi_2, \varepsilon)$.

The splitting between the two manifolds is measured by the function

$$m(\varphi_1, \varphi_2, \varepsilon) = m^U(\varphi_1, \varphi_2, \varepsilon) - m^S(\varphi_1, \varphi_2, \varepsilon). \tag{66}$$

A measure of the "size" of the splitting can be the function

$$d(\varepsilon) = \max_{(\varphi_1, \varphi_2) \in \mathbf{T}^2} |m(\varphi_1, \varphi_2, \varepsilon)|.$$

It is enough to let φ_1 range in $[0, 2\pi]$ and φ_2 in $[0, \pi]$. The Table I gives the values of $d(\varepsilon)$ obtained numerically for several values of ε. With the accuracy used on the computations we can only be confident on the results for the range of values of ε displayed here.

We note also that the values of φ_1, φ_2 such that $m(\varphi_1, \varphi_2, \varepsilon) = 0$ correspond to homoclinic orbits to the invariant $\mathbf{T}^2$. The figure 1 shows some plots of the sets $C_h(\varepsilon) = \{(\varphi_1, \varphi_2) \in \mathbf{T}^2 | m(\varphi_1, \varphi_2, \varepsilon) = 0\}$ for different values of ε. We remark that the topology of C_h changes with ε. This is quite natural. In the limit when ε is very small the dynamics of (x, y) in (64) is quite slow. If a point belongs to $C_h(\varepsilon)$ then the transport of that point for the irrational flow on the torus must be rather close to points belonging to $C_h(\varepsilon)$. We shall return to this point in Section 6.

5 The first order splitting

In this section we shall compute formally, the "first order" terms of the splitting by using Melnikov's method after one step of averaging, i.e., using (45). So we obtain a theoretical value for the function m defined in (66). We do not claim this approximation is rigourously justified. However in figure 2 we plot the data of Table I against

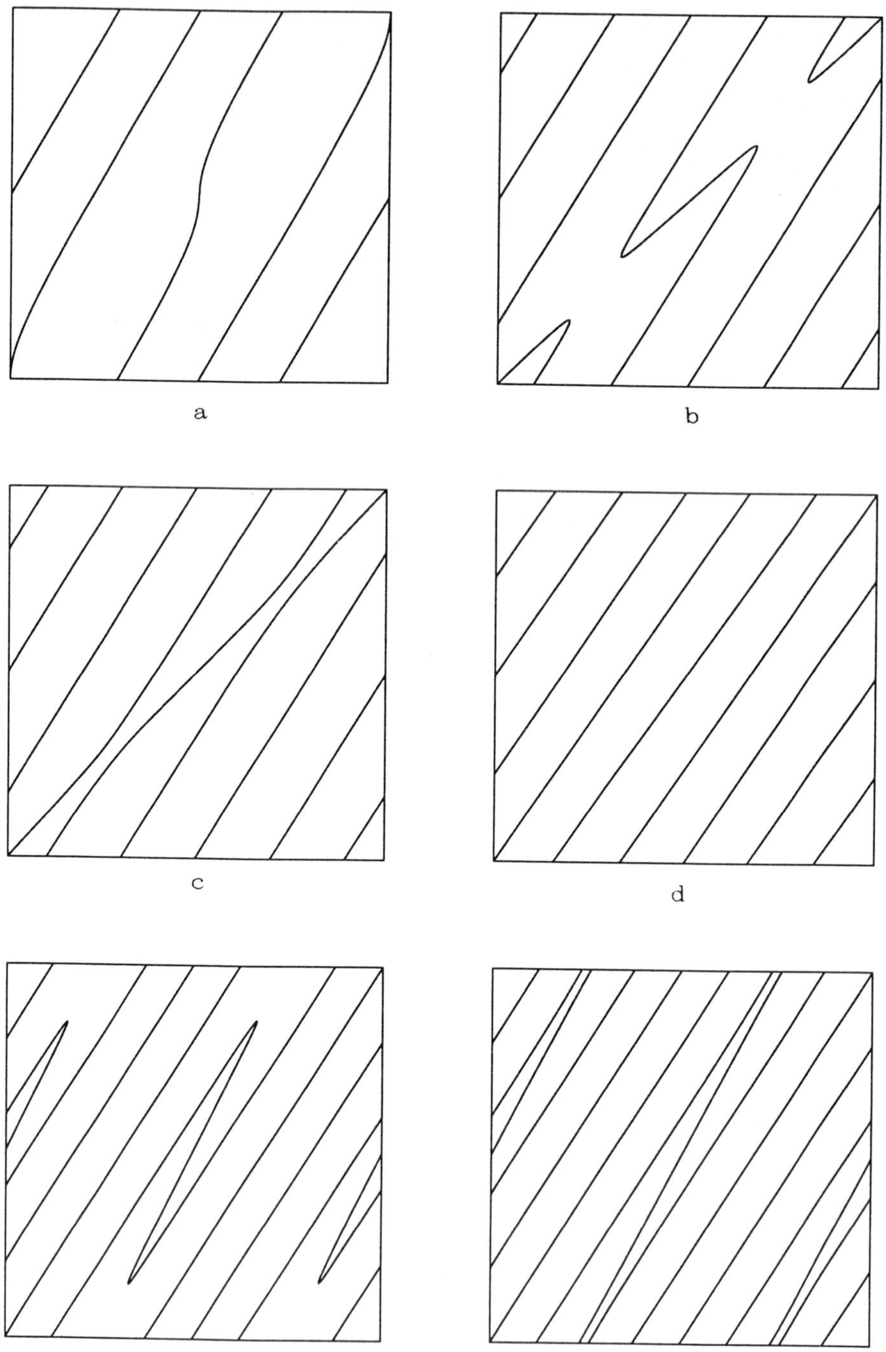

Figure 1. The set $\mathcal{C}_h(\varepsilon)$ of primary homoclinic points in the section $x = \pi$ for different values of ε. a) $\varepsilon = 0.098$; b) $\varepsilon = 0.08$; c) $\varepsilon = 0.078$; d) $\varepsilon = 0.05$; e) $\varepsilon = 0.0308$; f) $\varepsilon = 0.0304$

Table I

ε	$10^6 \times d(\varepsilon)$	ε	$10^6 \times d(\varepsilon)$	ε	$10^6 \times d(\varepsilon)$
0.098	12309.4645	0.072	2148.8343	0.046	88.72111
0.096	11097.3677	0.070	1795.1467	0.044	57.03980
0.094	9963.8012	0.068	1486.7868	0.042	38.12663
0.092	8907.2905	0.066	1220.0508	0.040	24.57281
0.090	7926.2890	0.064	991.2397	0.038	15.19353
0.088	7019.1189	0.062	796.7054	0.036	8.97125
0.086	6183.9246	0.060	632.8989	0.034	5.03753
0.084	5418.6365	0.058	496.4097	0.032	2.67983
0.082	4720.9480	0.056	383.9985	0.030	1.34597
0.080	4088.3052	0.054	292.6163	0.028	0.63507
0.078	3517.9080	0.052	219.4063	0.026	0.27932
0.076	3006.7227	0.050	161.6829	0.024	0.11257
0.074	2551.5056	0.048	116.9086	0.022	0.04077

the predictions of the Melnikov's approach observing a good agreement. So, we will pursue in this direction, assuming that the Melnikov function gives a good description of the behaviour, and we shall derive several consequences. The ones concerning the bifurcation of the set of primary homoclinic points in $x = \pi$ are presented in Section 6.

From (45) the variation of H_1 due to the perturbations H_2 and H_3 is

$$\Delta H_1(\varphi_1, \varphi_2, \varepsilon) = \int_{\mathbb{R}} (\varepsilon^2 [H_1, H_2] + \varepsilon^3 [H_1, H_2])|_{\sigma(t,\varphi_1,\varphi_2,\varepsilon)}\, dt, \tag{67}$$

where the integration is carried out along the separatrix of H_1 and, in (67), the angles θ_1, θ_2 appearing in the functions M_1 and M_2 have phases φ_1, φ_2 at $\xi = \pi$. On the other hand the separatrix of H_1 is given by

$$\frac{\eta^2}{2} + J_0^2 \cos\xi = J_0^2. \tag{68}$$

By differentiation with respect to η at $\xi = \pi$ we have

$$m(\varphi_1, \varphi_2, \varepsilon) = \Delta\eta \simeq \frac{\Delta H_1}{\eta_0}, \tag{69}$$

where $\Delta H_1$is given by (67) and η_0 by $2J_0$. The result from (69) has to be compared with the values in (66). Here, and in what follows, the Bessel functions are evaluated at the point 1 unless explicitely stated.

The brackets in (67) are easely computed and the integrand is given by

$$\begin{aligned} &\varepsilon^2 J_0^2 (M_1(-2\sin\xi + 3\sin\xi\cos\xi) + M_2(-2\cos\xi + 2\cos^2\xi - \sin^2\xi)) + \\ &+ \varepsilon^2 2J_0 \sin\frac{\xi}{2}\left(\cos 2\xi M_1 M_2 + \sin 2\xi \frac{M_1^2 - M_2^2)}{2}\right). \end{aligned} \tag{70}$$

The separatrix is parametrized by

$$\xi = 4\arctan(e^{\varepsilon J_0 t}), \tag{71}$$

and η is obtained from (68). By substitution of (71) in (70) the formula (67) is expressed as

$$\int_{\mathbb{R}} \varepsilon^2 J_0^2 \left[M_1(t)\left(-\frac{\sinh(u)}{\cosh^2(u)}+\frac{12\sinh(u)}{\cosh^4(u)}\right)+M_2(t)\left(-\frac{8}{\cosh^2(u)}+\frac{12}{\cosh^4(u)}\right)\right]+$$

$$+2J_0\varepsilon^3\left[M_1(t)M_2(t)\left(\frac{1}{\cosh(u)}-\frac{8}{\cos^3(u)}+\frac{8}{\cos^5(u)}\right)+\right. \tag{72}$$

$$\left.+\frac{1}{2}(M_1(t)^2-M_2(t)^2)\left(-\frac{4\sinh(u)}{\cosh^3(u)}+\frac{8\sinh(u)}{\cosh^5(u)}\right)\right],$$

where $u=\varepsilon J_0 t$. The functions $M_1(t)$ and $M_2(t)$ in (72) are of the form

$$\begin{aligned} M_1(t) &= \sum_s \mathrm{coef}_s \cos(\beta_s t+p_s\varphi_1+q_s\varphi_2)= \\ &= \sum_s \mathrm{coef}_s(\cos(\beta_s t)\cos(p_s\varphi_1+q_s\varphi_2)-\sin(\beta_s t)\sin(p_s\varphi_1+q_s\varphi_2)), \\ M_2(t) &= \sum_s \mathrm{coef}_s \sin(\beta_s t+p_s\varphi_1+q_s\varphi_2)= \\ &= \sum_s \mathrm{coef}_s(\sin(\beta_s t)\cos(p_s\varphi_1+q_s\varphi_2)-\cos(\beta_s t)\sin(p_s\varphi_1+q_s\varphi_2)). \end{aligned} \tag{73}$$

Due to the parities of the terms in (72) and (73), the only terms giving a non zero contribution to (72) are the ones which appear in (73) multiplying $\sin(p_s\varphi_1+q_s\varphi_2)$. The terms M_1, M_2 behave like M_2 and $M_1^2-M_2^2$ like M_1. We summarize the values of the integrals appearing in (72):

$$\begin{aligned} \int_{\mathbb{R}}\sin(\beta t)\left(-\frac{\sinh(u)}{\cosh^2(u)}+\frac{12\sinh(u)}{\cosh^4(u)}\right)dt &= \frac{2\pi\beta^3}{\varepsilon^4 J_0^4\cosh(\frac{\beta\pi}{2J_0\varepsilon})}, \\ \int_{\mathbb{R}}\cos(\beta t)\left(-\frac{8}{\cosh^2(u)}+\frac{12}{\cosh^4(u)}\right)dt &= \frac{2\pi\beta^3}{\varepsilon^4 J_0^4\sinh(\frac{\beta\pi}{2J_0\varepsilon})}, \\ \int_{\mathbb{R}}\cos(\beta t)\left(\frac{1}{\cosh(u)}-\frac{8}{\cos^3(u)}+\frac{8}{\cos^5(u)}\right)dt &= \left(\frac{\beta^4}{3J_0^5\varepsilon^5}-\frac{2\beta^2}{3J_0^3\varepsilon^3}\right)\frac{\pi}{\cosh(\frac{\beta\pi}{2J_0\varepsilon})}, \\ \int_{\mathbb{R}}\sin(\beta t)\left(-\frac{4\sinh(u)}{\cosh^3(u)}+\frac{8\sinh(u)}{\cosh^5(u)}\right)dt &= \left(\frac{\beta^4}{3J_0^5\varepsilon^5}-\frac{2\beta^2}{3J_0^3\varepsilon^3}\right)\frac{\pi}{\sinh(\frac{\beta\pi}{2J_0\varepsilon})}, \end{aligned} \tag{74}$$

where, as before, $u=\varepsilon J_0 t$. As we shall see later, for a fixed value of ε, the dominant terms in (67), and hence in (69), have values of β which are roughly of the orfder $\varepsilon^{1/2}$, while the ones coming from H_3 have a factor ε^0. However the coefficients in front of those terms (like coef_s in (73)) having H_3 as source are larger, and they become much more important when ε decreases. So we have to keep both the contribution of H_2 and H_3 in (67).

Finally we obtain for ΔH_1 an expression of the form

$$\sum_s c_s(p_s,q_s,\beta_s)\varepsilon^{r_s}\frac{1}{F(\frac{\beta_s\pi}{2J_0\varepsilon})}\sin(p_s\varphi_1+q_s\varphi_2), \tag{75}$$

where $p_s, q_s \in \mathbb{Z}$, $\beta_s=p_s+\gamma q_s$, the coefficients c_s do not depend on ε and F denotes one of the hyperbolic functions cosh or sinh. Looking at (75) we see that each one of

Table II

ε	p	q	coef	ε	p	q	coef	ε	p	q	coef
0.03	2	-1	-.185166E-8	0.01	3	-2	.805440E-19	0.003	8	-5	-.107506E-30
	3	-2	.953812E-6		5	-3	-.205424E-13		13	-8	.728265E-29
	5	-3	-.103376E-5		8	-5	-.546092E-13	0.001	13	-8	.496686E-61
	8	-5	-.122753E-8		13	-8	.247705E-18		21	-13	.682913E-58

the terms is of the form $\exp(-c/\varepsilon)$, where c is a positive constant. However, for a fixed ε, if β_s is big the contribution of this term is negligible. If β_s is very small, coming from a small divisor associated to suitable values of p_s and q_s, then $1/F$ is big, but the constant c_s turns out to be very small. The dominant terms in (75) shall be studied later.

The figure 2 shows the data of Table I against the maximum values of the function given by (69) by using (72), (74) and the explicit expressions given by (47). Despite the complexity of the formulae only a few of the harmonics present in (75) have relevant coefficients, specially when ε becomes small. Table II gives the dominant harmonics and the final coefficients in (75) for a few values of ε.

We see that the dominant contributions come from values (p_s, q_s) of the form $(F_j, -F_{j-1})$, where F_j denote the Fibonacci numbers. This is natural: this is the way to have β_s small, and then $1/F$ relatively big without having a coefficient c_s too small. To be explicit we define $F_1 = 1$, $F_2 = 2$, $F_j = F_{j-1} + F_{j-2}$. A short analysis shows that the terms coming from $M_1 M_2$ and $M_1^2 - M_2^2$ are the most important. As those harmonics, of the form

$$\sin(F_j\varphi_1 - F_{j-1}\varphi_2), \tag{76}$$

come from products of elements of M_1 and M_2, we have to examine the contributions of pairs (p_{s_1}, q_{s_1}), (p_{s_2}, q_{s_2}) such that

$$p_{s_1} + p_{s_2} = F_j, \quad q_{s_1} + q_{s_2} = -F_{j-1}. \tag{77}$$

Again an examination of the values of the coefficients $c_{s_i}(p_{s_i}, q_{s_i}, \beta_{s_i})$, $i = 1, 2$, with the constraints (77), shows that a large contribution comes from the choice $p_{s_1} = F_{j-1}$, $p_{s_2} = -q_{s_1} = F_{j-2}$, $-q_{s_2} = F_{j-3}$. In fact, there are several decompositions of (77) giving similar or even larger contributions. Using just this one, we underestimate, at most, by a factor of the form γ^{2j}, which turns out to be like ε^{-1}.

We recall here, for future use the explicit expression of F_j:

$$F_j = (\gamma^{j+1} + \gamma^j + (-1)^j\gamma^{-j})/(\gamma + 2), \tag{78}$$

and then $|F_j - F_{j-1}\gamma| = \gamma^{-j}$.

At this moment, to see the effect of powers of ε in the "perturbation" we consider a more general case. We shall see (48) with $a = b = 1$ instead of (4). This amounts to replace the argument 1 in the Bessel functions by ε^p. We also use the asymptotic value

$$J_n(z) \simeq \left(\frac{z}{2}\right)^n \frac{1}{n!}, \tag{79}$$

for z fixed and n large. Even for $z = 1$ this gives a relative error less than $\frac{1}{8(n+1)}$. Furthermore, for j large we can approximate (78) by

$$F_j \simeq \frac{\gamma + 1}{\gamma + 2}\gamma^j. \tag{80}$$

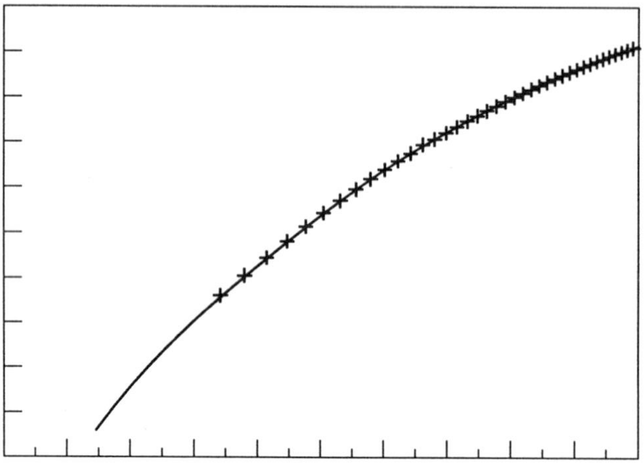

Figure 2. Comparison of the numerical values of the measure of the splitting, $d(\varepsilon)$, and the theoretical ones computed by using the Melnikov function. The figure displays $d(\varepsilon)$ versus ε, using $\log_{10}$ scales for both variables. The window in the horizontal variable is $[-2, -1]$, and in the vertical one is $[-11, -1]$.

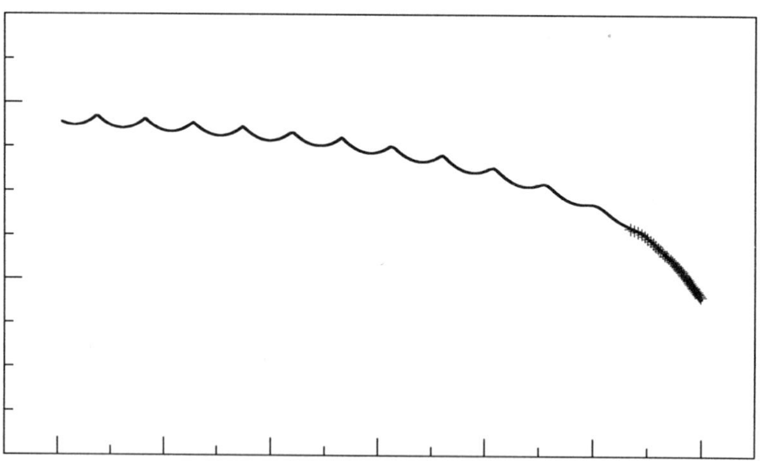

Figure 3. Behaviour of the function $c(\varepsilon)$. The function $c(\varepsilon)$ is plotted against ε. The $\log_{10}$ scale is used for ε. Window in the horizontal variable: $[-7.5, -0.5]$. In the vertical one $[0, 2.5]$

By using (79) and (80) and assuming β_s/ε not too small (to be checked later), the dominant part, ρ_j, of the coefficient of the harmonic (76) in (67) is given by

$$\rho_j = \frac{(\varepsilon^p)^{F_{j+1}}}{2^{F_{j+1}}(F_{j-1})!((F_{j-2})!)^2(F_{j-3})!} \frac{\gamma^{-2j-3}}{\exp(\frac{\pi\gamma^{-j}}{2\varepsilon J_0(\varepsilon^p)})}. \tag{81}$$

By taking logarithms and skipping some terms simply linear in j (where the contribution of the other decompositions in (77) is included), we have

$$\begin{aligned} \alpha_j &= \ln\rho_j \simeq \frac{\gamma+1}{\gamma+2}\left[\gamma\left(1-\ln 2-\ln\frac{\gamma+1}{\gamma+2}\right)+\left(1+\frac{3}{\gamma}\right)\ln\gamma\right]\gamma^j - \\ &-\frac{\gamma+1}{\gamma+2}\gamma\ln\gamma\cdot\gamma^j\left(j-p\frac{\ln\varepsilon}{\ln\gamma}\right)-\frac{\pi\gamma^{-j}}{2\varepsilon J_0(\varepsilon^p)}, \end{aligned} \tag{82}$$

where in (81) we have used Stirling's formula for the factorials and explicit use of (80) and the recurrence of the F_j has been made. Given ε we have to select j in (82) so that α_j is maximized. To this end we rewrite (82) as

$$R\gamma^j - S\gamma^j\left(j-p\frac{\ln\varepsilon}{\ln\gamma}\right)-T\gamma^{-j}\varepsilon^{-1}. \tag{83}$$

In (83) R and S are independent of ε and j. Also T can be considered as independent of ε. Indeed, if $p=0$ then $T=\frac{\pi}{2J_0(1)}$, and if $p>0$ T tends to $\frac{\pi}{2}$ when ε goes to zero. Let $\widetilde{S}=\frac{S}{T}$. Then the value of j maximizing (83) is

$$j = -\frac{\ln\varepsilon+\ln[\widetilde{S}(-\frac{1}{2}-p)\frac{\ln\varepsilon}{\ln\gamma}](1+o(1))}{2\ln\gamma}, \quad \text{as } \varepsilon\to 0. \tag{84}$$

Of course, in (84) one has to take the closest integer to obtain the right value of j, and for some values of ε there are two consecutive values of j giving the same contribution.

By inserting the value of j as given by (84) in (82) we get the maximum value of ρ as a function of ε. This is given by

$$\rho_{\max} = -\left(\frac{|\ln\varepsilon|}{\varepsilon}\right)^{1/2}\left(\frac{2\gamma+1}{\gamma+2}\frac{\pi}{J_0}\right)^{1/2}(1+2p)^{1/2}. \tag{85}$$

As j must be taken integer there is a small difference in the correct value when j, as given by (84), is not an integer. Let us introduce the function $c(\varepsilon)$ defined by

$$c(\varepsilon) = -\ln d(\varepsilon)\left|\frac{\varepsilon}{\ln\varepsilon}\right|^{1/2}. \tag{86}$$

The minimum value of $c(\varepsilon)$ is obtained when the value of j given by (84) is integer. The maximum (and then two harmonics have equal contribution to $d(\varepsilon)$) when (84) produces (roughly) a value in $\mathbb{Z}+1/2$. The behaviour of $c(\varepsilon)$ is shown in figure 3. The values corresponding to the data in Table I are also shown. When ε goes to zero we have

$$\begin{aligned} L &= \liminf_{\varepsilon\to 0} c(\varepsilon) = \left(\frac{2\gamma+1}{\gamma+2}\frac{\pi}{J_0}\right)^{1/2}(1+2p)^{1/2}, \\ U &= \limsup_{\varepsilon\to 0} c(\varepsilon) = \left(\frac{2\gamma+5/4}{\gamma+2}\frac{\pi}{J_0}\right)^{1/2}(1+2p)^{1/2}. \end{aligned} \tag{87}$$

The value of L in (87) follows immediately of (85) and the one of U is obtained by a small modification of the argument (asking for equal coefficients for two consecutive values of j). For completeness we give the numerical values of L and U:

$$\begin{aligned} L(p) &\simeq 1.917874016(1+2p)^{1/2}, \text{ if } p>0, \quad L(0)=2.192467937, \\ U(p) &\simeq 1.973656366(1+2p)^{1/2}, \text{ if } p>0, \quad U(0)=2.256236993. \end{aligned} \tag{88}$$

From (86) and (88), and always under the assumption of a good prediction of the splitting when using Melnikov's function, we have

$$d(\varepsilon) > \exp\left(-\frac{c|\ln\varepsilon|^{1/2}}{\varepsilon^{1/2}}\right) \tag{89}$$

for $\varepsilon \to 0$, where c is some positive constant. The relation (89) shows the optimality of the exponent in (3). We cannot change c_3 by anything else, at least in the current example. By comparing more carefully with (3) including the Remark 1, we notice in (89) an extra factor $|\ln\varepsilon|$. This is due to the special form of the equation (4) (or (48)). Indeed those are entire functions and in (47) the coefficient of the k-th harmonic decrases like $1/k!$. For analytic functions in some strip we would have a behaviour like $\exp(-\sigma|k|)$, with some positive σ, for the k-th harmonic. Furthermore, due to the special form of (47) in our case, in (77) only a reduced set of decompositions of $(F_j - F_{j-1})$ plays a significant role. A behaviour like $\exp(-\sigma|k|)$ produces that many terms contribute in a significant way. This implies that in the general case, even the estimates using Melnikov's approach give larger estimates, closer to the ones predicted by (3) and Remark 1.

As a final comment we point out that the averaging process leading first to (45), can be continuated for further steps. However it is not possible then to express it as a sum $\sum_j \varepsilon^j H_j$ with a finite number of terms. The second and successive averages introduce successive integrals of the functions M_1 and M_2 and their products. This gives rise to additional terms in the Melnikov function. The contribution of those terms is more relevant for smaller values of ε, because then some powers of small divisors play a significant role. This can explain the better agreement observed in figure 2 for larger rather than for small values of ε.

6 Bifurcations of the set of homoclinic orbits

From the discussion in the preceding section we have seen that for some values of ε, two different consecutive values of j give the same contribution. Let us consider in general the contribution to $m(\varphi_1, \varphi_2, \varepsilon)$ given by the two dominant harmonics of the form (76) with coefficients a and b:

$$m(\varphi_1,\varphi_2,\varepsilon) = a\sin(F_j\varphi_1 - F_{j-1}\varphi_2) + b\sin(F_{j-1}\varphi_1 - F_{j-2}\varphi_2), \quad a=a(\varepsilon), \quad b=b(\varepsilon). \tag{90}$$

Primary homoclinic orbits are given by $m(\varphi_1,\varphi_2,\varepsilon)=0$. This gives a curve $\mathcal{C}_h(\varepsilon)$ in $\mathbb{T}^2$. If, say, $\left|\frac{b}{a}\right|$ is small, then $\mathcal{C}_h(\varepsilon)$ is close to the line in $\mathbb{T}^2$ with slope F_j/F_{j-1}. Changes in the topology of $\mathcal{C}_h(\varepsilon)$ are produced when m, $\frac{\partial m}{\partial\varphi_1}$ and $\frac{\partial m}{\partial\varphi_2}$ are simultaneously zero. From (90) and the properties of F_j it follows immediately that this happens if and only if $|a|=|b|$. Then the values of the arguments, $F_j\varphi_1 - F_{j-1}\varphi_2$, $F_{j-1}\varphi_1 - F_{j-2}\varphi_2$, must belong to $\{\pi/2, -\pi/2\}$. It follows also that the corresponding values of φ_1, φ_2 are multiples of $\pi/2$.

Going back to the numerical results, it is seen that a bifurcation is produced for some ε in $[0.078, 0.080]$, where the average slope of $\mathcal{C}_h(\varepsilon)$ changes from 2 to 3/2 and, more accurately, for some ε in $[0.03065, 0.03066]$, where it changes from 3/2 to 5/3 (see also figure 1). The predicted values of the bifurcations, looking only at the two dominant terms in (67) are close to 0.079199 and 0.030389, respectively.

From (84) it follows that to advance j in one unit, when ε goes to zero, one should decrease $\ln \varepsilon$ by $2 \ln \gamma$. Hence, if the successive bifurcations are produced for values of ε equal to $\varepsilon_1, \varepsilon_2, \varepsilon_3, \ldots$ one has $\lim_{j\to\infty} \frac{\varepsilon_{j+1}}{\varepsilon_j} = \gamma^2$. At the bifurcation some of the orbits associated to $\mathcal{C}_h(\varepsilon)$ are no longer transversal. It is precisely at those orbits where the change of topology of $\mathcal{C}_h(\varepsilon)$ is produced. Of course, other bifurcations appear when the invariant manifolds are continuated and the successive folds give rise to homoclinic tangencies. But these are related to "secondary" homoclinic orbits, crossing several times the plane $\xi = \pi$.

Changes in the topology of the set of primary homoclinic orbits have been found in the context of the elliptic planar restricted three-body problem by Martínez and Pinyol[10] and in the general planar three-body problem by Martínez and Simó,[11] but the behaviour with respect to the parameters (for instance, the eccentricity of the orbit of the two dominant bodies) is different from the one displayed here. These bifurcations seem generically unavoidable in 3 degrees of freedom Hamiltonian systems depeding on parameters or in similar problems.

If a higher number of frequancies appears in the problem then the geometry of the bifurcations depends on the diophantine properties of the frequency vector and the current dominant harmonics for each value of ε.

7 Conclusions

A general Theorem on averaging under fast quasiperiodic forcing has been proved. Certainly this is not the most general case to be considered. Specially, to understand the dynamics in the case of Hamiltonian systems, it is necessary to take into account that the "forcing" frequencies are also varying. So the crossing of resonances is unavoidable. Even without having the frequencies locked at resonance, this problem presents major difficulties.

However the results obtained are useful when the forcing frequencies can be assumed constant and satisfying a diophantine condition. An example can be the study of the motion of asteroids not influencing the major planets, whose orbit can be considered as quasiperiodic for a moderate time interval.

In particular we have learned about the behaviour of the splitting of separatrices under quasiperiodic forcing. Furthermore, just one step of the averaging procedure can make easier the analysis of some problems.

A formal approach using the Melnikov function followed by a detailed analysis, seems to give some light on the amazing fact that each contribution to the splitting is of the type $\exp(-c/\varepsilon)$, but all together give much larger contributions, say $\exp(-c/\varepsilon^a)$, $a < 1$, as happens in the example presented here. The global behaviour is a kind of envelope of the splittings due to the individual harmonics associated to small divisors. As usual the small divisors become relevant when they are of the order of $\varepsilon^{1/2}$.

Finally a new kind of bifurcation of the set of primary homoclinic orbits shows up in a natural way and this fact makes a big difference with respect to the splitting under fast periodic forcing.

Acknowledgements. The author is indebted to A. Neishtadt, A. Delshams, T.M. Seara, A. Jorba and E. Fontich for very useful comments, suggestions and remarks. The author has been supported by CICYT Grants PB90-0580 and ESP91-0403.

References

1. Abramowitz M. and Stegun I.: "Handbook of Mathematical Functions", Dover, 1972.

2. Bogoljubov N.N, Mitropolski Ju. A. and Samoilenko A.M.: "Methods of Accelerated Convergence in Nonlinear Mechanics", Springer, 1976.

3. Broer H., Roussarie R. and Simó C.: Invariant circles in the Bogdanov-Takens bifurcation for diffeomorphisms, in preparation.

4. Delshams A. and Seara T.M.: An asymptotic expansion for the splitting of separatrices of the rapidly forced pendulum, *Comm. Math. Phys.* 150, 433-463, 1992.

5. Fontich E. and Simó C.: Invariant manifolds for near identity differentiable maps and splitting of separatrices, *Erg. Th. and Dyn. Systems*, 10, 319-346, 1990.

6. Fontich E. and Simó C.: The splitting of separatrices for analytic diffeomorphisms, *Erg. Th. and Dyn. Systems*, 10, 295-318, 1990.

7. Gelfreich V.G.: Splitting of separatrices for the rapidly forced pendulum, preprint, 1990.

8. Holmes P., Marsden J. and Scheurle J.: Exponentially small splittings of separatrices with applications to KAM Theory and degenerate bifurcations, *Contemporary Mathematics* 81, 213-244, 1988.

9. Jorba A. and Simó C.: On quasiperiodic perturbations of elliptic equilibrium points, preprint, 1993.

10. Martínez R. and Pinyol C.: Parabolic orbits in the elliptic restricted three-body problem, to appear in J. of Diff. Eq.

11. Martínez R. and Simó C.: A note on the existence of heteroclinic orbits in the planar three-body problem, in V.F. Lazutkin, editor, Proceed. S. Petersburg Workshop on Hamiltonian Dynamical Systems, to appear.

12. Neishtadt A.I.: The separation of motions in systems with rapidly rotating phase, *Prikladnaja Matematika i Mekhanika* 48, 133-139, 1984.

13. Simó C.: Averaging and splitting for Hamiltonians with fast quasiperiodic forcing, in preparation.

NATURAL BOUNDARIES OF NORMALIZING TRANSFORMATIONS

Luca Billi, Monica Malavasi, Giorgio Turchetti

Department of Physics,University of Bologna
INFN, Sezione di Bologna
Via Irnerio 46, 40126, Bologna, Italy

INTRODUCTION

The complex time singularities of Hamilton's equations have been investigated by Chang et al.[1] in order to distinguish integrable from nonintegrable systems. The analysis of the singularities for the transformation which brings a complexified hamiltonian in normal form [2] extends to hamiltonian maps; bounds are provided by KAM techniques, but the exact location of the singularities and their nature remain unknown. Some heuristic information is provided by perturbation theory[3,4], but at best only the main singularities can be identified. Given an area preserving map near an elliptic point, the singularities in the action plane are poles, at any finite order; the Padé Approximants (P.A.) are adequate to detect them, as shown in [5] and to provide the analytic continuation of the perturbation series beyond the resonances.
The angle analyticity is the object of this note, where we show how the natural boundaries, first observed by Percival[6,7], appear when a sequence of complex frequencies with resonant real part approach a real nonresonant frequency. The P.A. have been used to investigate the singularities in the coupling for the standard map [8,9,10]. The Fourier amplitudes have been considered in[11] filtering the dynamical poles from noise. A systematic investigation carried out in the case of linearizable quadratic maps of the complex plane[12,13] has shown that the poles of P.A. reproduce the singularity patterns of conjugation functions only at large enough orders provided that a sufficiently accuracy is used.
We examine here the singularities of the normalizing transformation of a torus, in the standard map. For a complex frequency with resonant real part of order q the P.A. poles in the angle θ form two families of real lines parallel to the imaginary axis (or rays in $z = e^{i\theta}$); in the limit of a real irrational frequency these lines become

Hamiltonian Mechanics, Edited by J. Seimenis,
Plenum Press, New York, 1994

dense creating a natural boundary in the analyticity strip. This picture is explained analytically at the level of homologic equation when the non linear term has itself a singularity.
The Siegel problem exhibits the same features when the frequency is complexified and the same correspondence between the homologic level and the complete solution exists. The rays of singularities in $z = e^{i\theta}$ are explained as the cuts connecting the sheets of the Reimann surface on which the mapping is injective. The natural boundaries in the angle plane seem to be quite generally related to a uniformization process.

PADÉ APPROXIMANTS AND NOISE

Given an analytic function $f(z)$ with Taylor expansion $f(z) = f_0 + zf_1 + \ldots + z^n f_n + \ldots$ (convergent or asymptotic), its $[M/N]$ Padé Approximants[14] are defined by

$$[M/N]_f(z) \equiv \frac{P_M(z)}{Q_N(z)} = f(z) + O(z^{N+M+1}) \tag{1}$$

and the coefficients of the polynomial $Q_N(x) = 1 + q_1 x + \ldots + q_N\, x^N$ are determined by solving the system of N linear equations. We shall consider the diagonal sequences $[N-1/N]_f$ which converge for the class of Stjeltjes functions and have a partial fractions expansions according to $[N-1/N]_f(z) = \sum_{k=1}^{N} \gamma_k^{(N)}/(z - z_k^{(N)})$. The poles and zeroes cluster on the singularities of the function; this can be proved for Stjeltles functions since $x^N\, Q_N(x^{-1})$ are orthogonal polynomials with respect to a positive measure. The accuracy of q_n depends on the number d of significant digits used in the calculations, which were implemented in FORTRAN using a package[17] which allows to choose d arbitraryly large; typically [99/100] P.A. were computed with $d \geq 200$ and the poles were evaluated with comparable accuracy. The effect of finite precision is simulated by adding a noise function $r(z)$ to $f(z)$

$$f_\epsilon(z) \equiv f(z) + \epsilon\nu(z) = \sum_{n=0}^{\infty} f_n(1 + \epsilon\nu_n)z^n, \qquad \epsilon = 10^{-d} \tag{2}$$

If $|f_n|^{-1/n} \to R$ and ν_n are random variables belonging to $[0,1]$ with probability 1, then the radius of convergence of $f_\epsilon(z)$ is R with probability 1 as proved in[15]. The poles of $[N-1/N]$ are with probability 1 on the circle as $N \to \infty$, while at finite order they can be inside; the residues are proportional to ϵ as confirmed by numerical experiments, which show that by increasing N, with d fixed, the number of poles corresponding to $f(z)$ remains constant while the number of noise poles increases.

THE STANDARD MAP

We examine first the conjugation Φ of the standard map M with its normal form U defined by

$$M : \begin{cases} \jmath' = \jmath + \epsilon f(\theta) \\ \theta' = \theta + \jmath' \end{cases}, \quad U : \begin{cases} \Omega' = \Omega \\ \Theta = \Theta + \Omega \end{cases}, \quad \Phi : \begin{cases} \jmath = \Omega + v(\Theta, \Omega; \epsilon) \\ \theta = \Theta + u(\Theta, \Omega; \epsilon) \end{cases} \tag{3}$$

where $f(\theta) = f(\theta + 2\pi)$. The functional equation $M \circ \Phi = \Phi \circ U$ implies the following difference equations for u, v

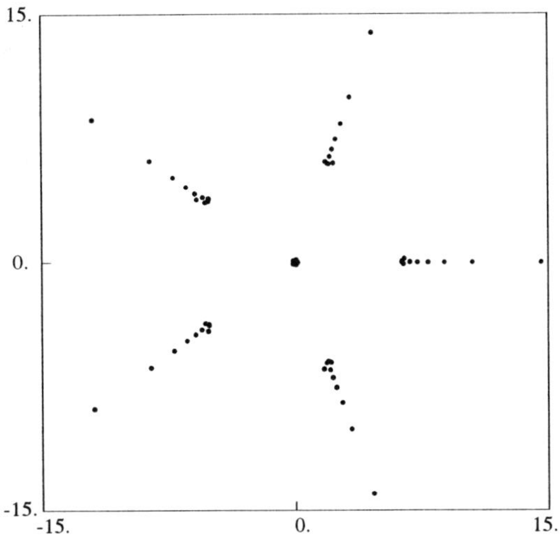

FIG 1. [59/60] for S.M. $\Omega = 2\pi\,(3/5\text{-i }0.001)$ ε=0.1

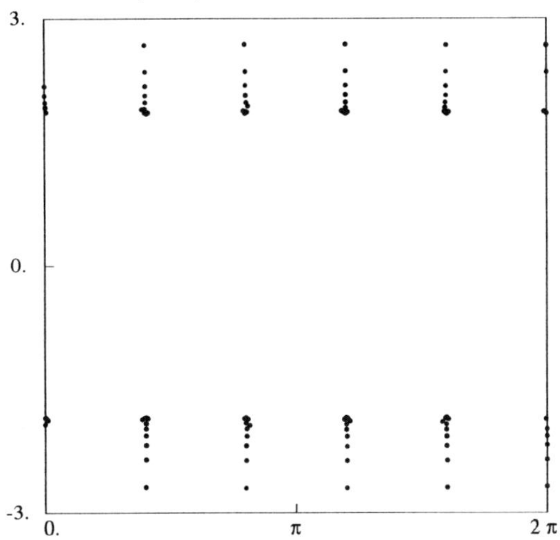

FIG. 2 The same as Fig. 1 in Θ plane

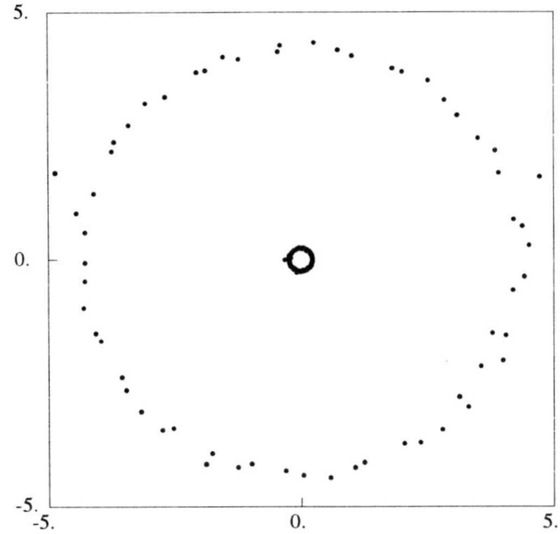

FIG 3. [59/60] for S.M. $\Omega = 2\pi\,(0.61803\text{-i }0.001)$ ε=0.3

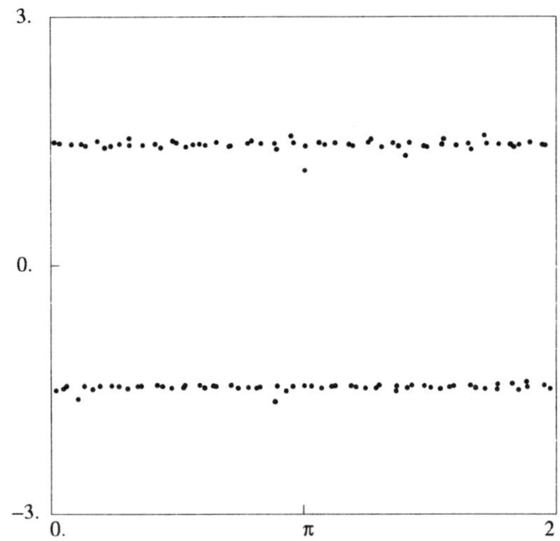

FIG. 4 The same as Fig. 3 in Θ plane

$$\begin{cases} u(\Theta+\Omega) - 2u(\Theta) + u(\Theta-\Omega) = f(\Theta + \epsilon u(\Theta)) \\ v(\Theta) = u(\Theta) - u(\Theta - \Omega) \end{cases} \tag{4}$$

which are recurrently solved after expanding in ϵ the Fourier coefficients according to

$$u(\Theta) \equiv u_+ + u_- \qquad u_+ = \sum_{k=1}^{+\infty} u_k(\epsilon) z^k, \qquad u_- = \sum_{k=1}^{+\infty} u_{-k}(\epsilon) z^{-k} \tag{5}$$

where $z = e^{i\Theta}$ and

$$u(\Theta;\epsilon) = \sum_{n=1}^{\infty} u_n(\Theta)\epsilon^n = \sum_{n=1}^{+\infty} \epsilon^n \sum_{k=-n}^{n} u_{n,k} e^{ik\Theta}, \quad u_k(\epsilon) = \epsilon^{|k|} \sum_{m=0}^{\infty} \epsilon^{2m} u_{|k|+2m,k} \tag{6}$$

We have considered the extension to a complex frequency Ω for the usual choice $f(\theta) = \sin\theta$. The series $u_\pm(z)$ where computed up to order 120 with $d \leq 200$ choosing ϵ small enough to insure a truncation error below 10^{-d}. The picture that emerges is simple: when Ω is resonant with a small imaginary part $\Omega/2\pi = p/q + i\alpha$ then we have q symmetric rays in the z plane issued at about the same distance from the origin, see Fig 1. For u_- the picture is quite similar and the rays in z terminate at the origin. In the Θ plane we observe two families of straight lines starting from horizontal lines, see Fig 2. By increasing q and decreasing α one approaches the case of a nonresonant real frequency and the the singularities become dense on the border of the strip to form two circles, Fig 3., or on the border of the analytic strip Fig. 4 in the Θ plane to form a natural boundary.

THE HOMOLOGIC EQUATION

An explanation of the analyticity structure described above can be found at the level of the homologic equation for a hamiltonian system such as a nonlinear oscillator submitted to a time periodic perturbation[3]. We show here that, at the homologic level, the complex quadratic map, the circle map and the standard map exhibit the same behaviour. We recall that locally the standard map is close to a circle map with the same frequency (real or complex)

$$\theta' = \theta + \omega + \epsilon f(\theta) \tag{7}$$

which is conjugated with a rotation $\Theta' = \Theta + \Omega$ where $\Omega(\epsilon, \omega) = \omega + O(\epsilon^2)$. When $\mathrm{Im}\,\theta \ll 0$ it is well approximated by the map

$$z' = e^{i\omega} z \exp\left(ze^{-i\omega}\right) \tag{8}$$

by setting $z = \epsilon e^{i(\theta+\omega)}/2$. We remark that the exponential map has the same qualitatively behaviour as the quadratic map

$$z' = F(z) = \lambda z + z^2, \qquad \lambda = e^{i\omega} \tag{9}$$

When $\mathrm{Im}\,\theta \gg 0$ then the circle map is well described by (8) with $\omega \to -\omega$ by setting $z = \epsilon e^{-i(\theta+\omega)}/2$.
We shall consider here the conjugation Φ of the quadratic map with its linear part and the inverse function $\Psi = \Phi^{-1}$ satisfying

$$F(\Phi(z)) = \Phi(\lambda z), \qquad \Psi(F(z)) = \lambda\Psi(z) \tag{10}$$

for complex frequencies $\omega + i\alpha$ with $\alpha > 0$ so that $|\lambda| < 1$. For $\alpha \neq 0$ the real part ω is allowed to be resonant.
Before discussing the results on the analyticity of Φ which are numerically the same as for $u_\pm(z)$, we consider the homologic equation for the map

$$z' = \lambda z + \epsilon \frac{z^2}{1 - \frac{z}{R}} \tag{11}$$

whose solution Φ_1 explicitlty reads

$$\Phi_1(z) = z - \epsilon\, z^2 \sum_{k=0}^{\infty} \frac{\lambda^{k-1}}{1 - \frac{\lambda^k z}{R}} \tag{12}$$

The map (11) has a pole at $z = R$ and critical points at $z_c = R[1 \pm i(\epsilon R/\lambda)^{1/2} + O((\epsilon R)^{3/2})]$; we see that $z_c \to R$ as $\epsilon \to 0$. The function Φ has a sequence of poles lying on the arms of a spiral $z_k = \lambda^{-k} R$. If $\omega/(2\pi) = p/q$ and $\alpha > 0$ is resonant the poles are on q half lines. When $\alpha \to 0$ and $\omega/(2\pi)$ is irrational the poles accumulate on a circle to form natural boundary. For $\epsilon \to 0$ and $R \to 0$ the trajectory of z_c, which is inside the circle $|z| = R$, tends to the circle itself so that the singularities location of Φ and Φ_1 is the same.

For the homologic equation the results for map of the circle and for the standard map are very similar. Indeed let

$$f(\theta) = r\left(R^{-1}\sin\theta\right) - \left\langle r\left(R^{-1}\sin\theta\right)\right\rangle, \qquad r(x) = \frac{x}{1-x} \tag{13}$$

where $z = e^{i\theta}$, we can write

$$f(\theta) \equiv \hat{f}(z) = \sum_{n=1}^{\infty}[z^n C_n(R) + z^{-n}C_{-n}(R)], \qquad C_{\pm n}(R) = \frac{(\pm 1)^n}{(2iR)^n}\left(1 + O(R^{-2})\right) \tag{14}$$

If we consider then the circle map $\theta' = \theta + \omega + \epsilon f(\theta)$ this is conjugated with $\Theta = \Theta + \Omega$ by $\theta = \Theta + \epsilon u_1(\Theta) + O(\epsilon^2)$ where $\Omega = \omega + O(\epsilon^2)$. Letting $z = e^{i\Theta}$ replacing ω with a complex frequency $\omega + i\alpha$ and setting $\lambda = \exp(i\omega - \alpha)$ the solution reads

$$u_1(\Theta) \equiv \hat{u}_1(z) = \sum_{k=0}^{\infty}\left[-r\left(\frac{z\lambda^k}{2iR}\right) + r\left(\frac{-z^{-1}\lambda^{1+k}}{2iR}\right)\right]\left(1 + O(R^{-2})\right) \tag{15}$$

The terms of higher order in R^{-2} change the residues of the first order poles and introduce higher order poles at the same locations. Here it is evident that there are two families of poles on two spirals accumulating at zero and infinity according to

$$z = -\lambda^{1+k}(2iR)^{-1}, \qquad z = \lambda^{-k}(2iR) \tag{16}$$

For the standard map the conclusions are the same even though the computations are slightly more involved. Indeed the homologic equation now reads $\hat{u}_1(\lambda z) - 2\hat{u}_1(z) + \hat{u}_1(\lambda^{-1}z) = \hat{f}(z)$ and expanding for $\alpha > 0$ the divisors

$$D_n = \frac{1}{\lambda^n + \lambda^{-n} - 2} = \frac{1}{2[\cos n(\omega + i\alpha) - 1]} = \sum_{k=0}^{\infty} b_k \lambda^{n(k+1)} \tag{17}$$

we obtain

$$\hat{u}_1(z) = \sum_{k=0}^{\infty} b_k\left[r\left(\frac{z\lambda^{k+1}}{2iR}\right) + r\left(\frac{-z^{-1}\lambda^{1+k}}{2iR}\right)\right]\left(1 + O(R^{-2})\right) \tag{18}$$

To conclude we observe that the singularity pattern of $u(z)$ for the standard map, suggested by the P.A. when $f(\theta) = \sin(\theta)$ and ϵ is small is basically the same as for $u_1(z)$, solution of the homologic equation, for a map where $f(\theta)$ has a singularity. In the former case ϵ small implies the analyticity annulus is large: indeed the radii are proportional to ϵ and ϵ^{-1} respectively; this corresponds to the approximation of large R we have considered.

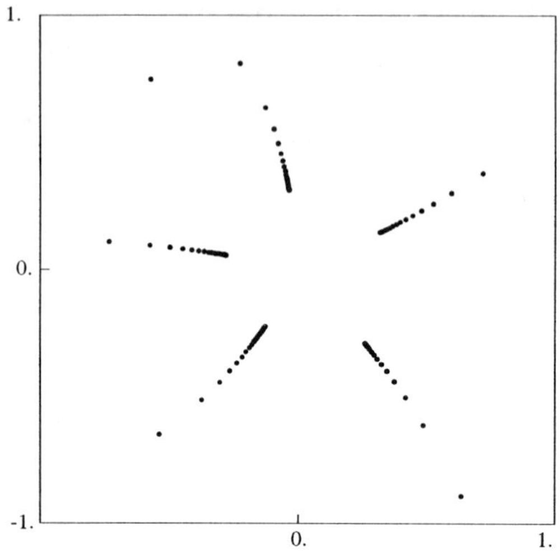

FIG. 5 [100/100] Poles $\omega = 2\pi\,(3/5+i\,.001)$ d= 200

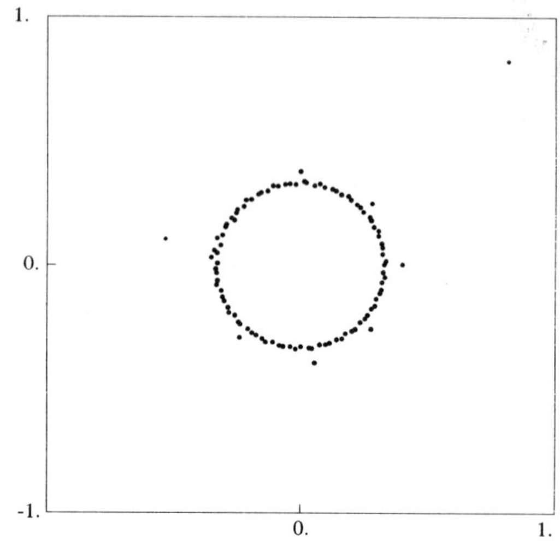

FIG. 7 [100/100] Poles $\omega = 2\pi\,0.61803$ d= 200

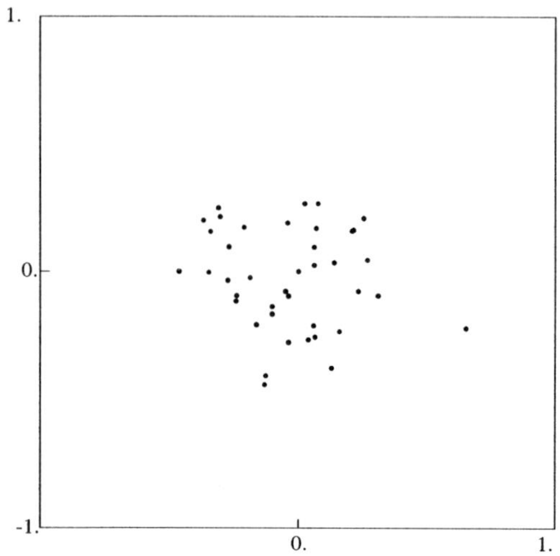

FIG. 6 the same as Fig 5. with d= 40

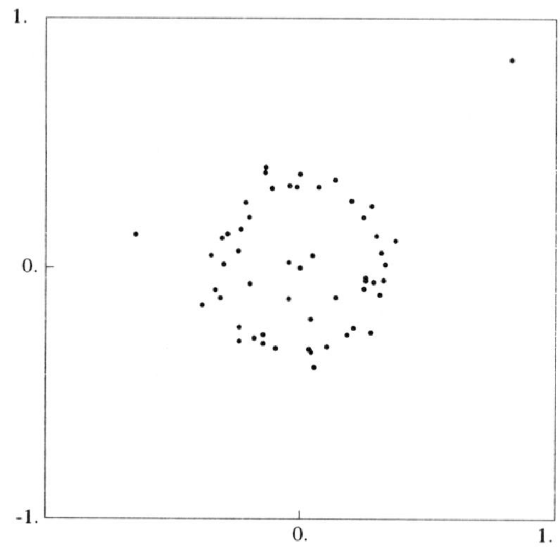

FIG. 8 the same as Fig. 7 with d= 40

ANALYTICITY OF Φ FOR THE SIEGEL PROBLEM

In the Siegel problem for the quadratic map the singularity pattern of Φ is the same, see Fig. 5,7: q rays for a resonant frequency with a positive imaginary part and a circle for nonresonant real frequency as first observed by Xie[18]. Fig. 6,8 show that when the accuracy is insufficient very poor information is extracted from the series.

The singularity pattern for Φ and Ψ is quite different: Ψ is analytic within the Fatou set $\mathcal{F}$ when $|\lambda| < 1$, while the singularities of Φ belong to symmetric rays which cluster on a circle for real nonresonant frequencies. The preimages of the critical point $z_c = -\lambda/2$

$$z_0 = z_c, \qquad z_k = F_k^{-1}(z_c), \qquad \ldots, \qquad z_{k_1,\ldots,k_n} = F_{k_n}^{-1} \circ \ldots \circ F_{k_1}^{-1}(z_c) \qquad k_i = \pm 1 \quad (19)$$

where $F_{\pm}^{-1} = \frac{-\lambda \pm \sqrt{\lambda^2+4z}}{2}$ are mapped into

$$\zeta_n = \Psi(z_{k_1,\ldots,k_n}) = \lambda^{-n}\zeta_c \qquad (20)$$

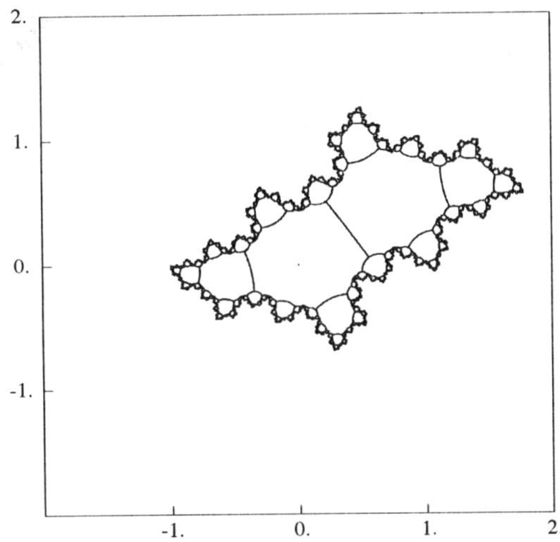

FIG. 9 λ=0.96 exp[2 π 3/5]

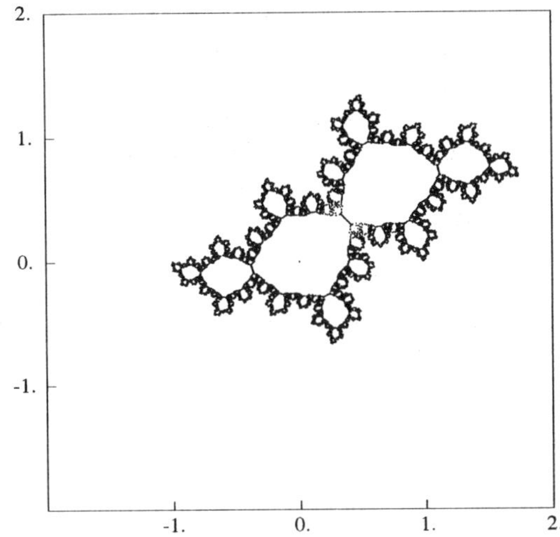

FIG. 10 λ=0.9999 exp[2π 21/34]

and the following results hold[12,13,19].

Proposition If z_c is a critical point of F then z_c and its preimages $z_{k_1,\ldots,k_n}$ are critical points of $\Psi(z)$.

As a consequence the singularities at ζ_n are branch points of square root type. At the first point ζ_0 a cut is drawn up to ζ_1 joining the first two sheets Σ_1, Σ_2 at ζ_1 there is a square root branch point in both sheets. Therefore two other sheets Σ_3, Σ_4 are needed for a bijective correspondence with the z plane; the sheets Σ_3, Σ_4 have boundaries in common with both Σ_1, Σ_2. We may deform these sheets bringing all of them in a unique plane Π. At order n we have 2^n sheets and at the point ζ_n we have to introduce 2^n new sheets connected pairwise to the previous ones. On the plane Π the infinity is reached only once by a continuous family of straight lines in each domain. The image of the same domain in $\mathcal{F}$ has only one point belonging to $\mathcal{J}$. The construction of the Riemann surface of Φ gives a simple explanation of the presence of the cut in the ζ plane. When $\lambda = \exp(2\pi i\, p/q)$ the construction is similar but in the ζ plane we have q cuts: indeed the points $\zeta_n = \lambda^{-n}\zeta_c$ are on a spiral but aligned along q half lines. The genesis of the natural boundary is evident: consider the sequence p_j/q_j of best rational approximations to an irrational frequency $\omega/(2\pi)$ and choose correspondingly the sequence α_j such that the convergence radius r_j is maximum (recall that $r_j = 0$ at $\alpha = 0$ and $\alpha = +\infty$).

q_j	$e^{-\alpha_j}$	$\log(1/\alpha_j)$	$r_j(\alpha_j)$
5	.96	3.2	.277
8	.98	3.9	.2961
13	.9915	4.76	.3081
21	.9965	5.65	.3184
34	.9985	6.50	.3190
55	.9995	7.60	.3214
89	.9998	8.51	.3229

The values r_j reported in the table have to be compared with the exact value for the real golden mean frequency $r_{\text{exact}} = 0.326$. In the Fig. 9,10 we show the domains tasselating $\mathcal{F}$ for the Fibonacci numbers $q_j = 5, 34$; their boundary belongs to $\mathcal{J}$ as $j \to \infty$ and each of them becomes a simply connected component of $\mathcal{J}$. In the domain containing the origin the map is conjugated to a rotation and its boundary is the trajectory of the critical point; the interior is the Siegel disk as proved by Herman[16].

Acknowledgements. One of us (GT) wishes to thank S.Abenda, A. Bazzani for helpful discussions.

REFERENCES

1 Y. F. Chang, M. Tabor, J. Weiss "Analytic structure of Henon Heiles hamiltonian in integrable and nonintegrable regimes" *J. Math. Phys.* **23**, 531 (1982)

2 A. Bazzani, G. Turchetti "Singularities of normal forms and topology of orbits in area preserving maps" *J. Phys.* **A 25**, L427 (1992)

3 A. Bazzani, S. Abenda contribution to this volume

4 A. Bazzani, G. Turchetti "Perturbation Theory and Analyticity of Normalizing Transformation" *Chaos: theory and practice* ed. T. Bountis, Springer Verlag (1993)

5 G. Servizi, G. Turchetti "Normal forms singularities in area preserving maps and analytic continuation by Padé Approximants" *Physics Letters* **151**, 485 (1990)

6 I. C. Percival, J. M. Greene "Hamiltonian maps in the complex plane" *Physica* **3D**, 530, (1981)

7 I. C. Percival "Chaotic boundary of a hamiltonian map" *Physica* **6D**, 67, (1982)

8 A. Berretti, L. Chierchia "On the complex analytic structure of the golden invariant curve for the standard map" *Nonlinearity* **3**, 39 (1990)

9 C. Falcolini, R. de La Llave "Numerical calculation of domains of analyticity for perturbation theories in presence of small divisors" *J. Stat. Phys.* **67**, 645 (1992)

10 A. Berretti, S. Marmi "Standard map at complex rotation numbers: creation of natural boundaries" *Phys. Rev. Lett.* **68**, 1443 (1992)

11 M. Malavasi "Analysis of singularities of the Standard Map conjugation function" *Chaos: theory and practice* Ed. T. Bountis, Springer Verlag (1992)

12 L. Billi, E. Todesco, G. Turchetti " Singularity analysis by Padé Approximants of some holomorphic maps" Dep. of Phys. Bologna preprint (1993)

13 L. Billi, G. Turchetti, R. Xie "Natural boundaries for hamiltonian maps and the genesis of the Siegel disc" Phys. Rev. Lett. **71**, 2513 (1993)

14 J. Baker, P. Graves Morris, "Padé Approximants" in *Encyclopedia of Mathematics*, vol 13, 14 Add. Wesley (1981)

15 J. Gilewicz, B. Truong-Van "Froissart doublets in the Padé approximation and noise" in *Constructive theory of functions* Sophia 1988

16 M. Herman Comm. Math. Phys. **99**, 593 (1985)

17 Brent "MPA: multiple precision floating point arithmetic" CERN Library A105

18 R. Xie "Padé Approximants applied to the analysis of the singularity structure for conjugacy problem" *Chaos: theory and practice* Ed. T. Bountis, Springer Verlag (1992)

19 J. Yoccoz "Théorème de Seigel, nombres de Bruno et polynomes Quadratiques" *Preprint-Université Paris Sud* (1993)

SINGULAR PERTURBATION IN HAMILTONIAN MECHANICS

A. D. Bruno

Institute of Applied Mathematics
Russian Academy of Sciences
Miusskaja sq. 4, Moscow 125047 Russia

INTRODUCTION

Recently I have developed an algorithm of the local analysis of singularities. It is equally applied in the systems of algebraic equations and in the systems of ordinary equations and in the partial differential equations [1,2]. Here I shall show its effectivity by two examples of Hamiltonian systems. Indeed I prefer to study properties of arbitrary systems and to apply them to the Hamiltonian systems [3].

THE RESTRICTED THREE–BODY PROBLEM

Let the body J of mass μ move on the unit circular orbit around the body S of mass $1-\mu$. In the synodical (rotating) coordinates the motion of the body A of zero mass is described by a Hamiltonian system with two degrees of freedom and with one parameter μ. The Hamiltonian function is [4]:

$$H = \frac{1}{2}(y_1^2 + y_2^2) + x_2 y_1 - x_1 y_2 - \frac{1-\mu}{\sqrt{x_1^2 + x_2^2}} - \frac{\mu}{\sqrt{(x_1^2 - 1) + x_2^2}} + \mu x_1. \qquad (1)$$

We consider small values of the massratio μ. For $\mu = 0$ the restricted problem is the problem of two bodies S and A in the synodical coordinates. But in its four-dimensional phase space we must delete the two-dimensional plane $J^* = \{x, y : x_1 = 1, x_2 = 0\}$ corresponding to the body J. Due to the integrability of the two-body problem, all its solutions were described and from them the solutions with consecutive collisions of bodies A and J were selected [4]. The set of symmetrical such solutions was described in [5]. It consists of infinitely many one parameter families (families S). Their intersection with the symmetry plane $\Pi = \{x, y : x_2 = y_1 = 0\}$ are shown in

Hamiltonian Mechanics, Edited by J. Seimenis,
Plenum Press, New York, 1994

the Fig. 1 in coordinates $\tilde{a}, \tilde{e}$ which are simply connected with the major semiaxes a and the excentricity e of the elliptic orbits of the problem of two bodies S and A. The dot–dash curve represents J^*.

For small $\mu > 0$ near the plane J^* we have singular perturbation. To find all first approximations of the restricted three body problem we do as follows [6]. Near the plane J^* we introduce the local coordinates

$$\xi_1 = x_1 - 1, \quad \xi_2 = x_2, \quad \eta_1 = y_1, \quad \eta_2 = y_1 - 1$$

and develop the Hamiltonian function in these coordinates. After expanding $1/\sqrt{(\xi_1+1)^2+\xi_2^2}$ in a Taylor series, the Hamiltonian function (1) takes the form

$$\begin{aligned} H &= const = \frac{1}{2}(\eta_1^2+\eta_2^2) + \xi_2\eta_1 - \xi_1\eta_2 - \xi_1^2 + \frac{1}{2}\xi_2^2 \\ &+ f(\xi_1,\xi_2) + \mu\{\xi_1^2 - \frac{1}{2}\xi_2^2 - \frac{1}{\sqrt{\xi_1^2+\xi_2^2}}\} - f(\xi_1,\xi_2), \end{aligned} \tag{2}$$

where f is a convergent power series not containing terms lower than the third power. We will say that the quantities ξ_1 and ξ_2 are of the same order of smallness and similarly for the quantities η_1 and η_2. We will consider the exponents of terms in expression (2) as as points in a three-dimensional space with coordinates

$$q_1 = ord(\xi_i), \quad q_2 = ord(\eta_i), \quad q_3 = ord(\mu).$$

Let D be the set for those points (q_1, q_2, q_3), excluding zero, and Γ be the convex hull of D. Then Γ is the Newton polyhedron of the series (2). The surface $\partial\Gamma$ of the polyhedron Γ consists of faces $\Gamma_j^{(2)}$, edges $\Gamma_j^{(1)}$ and vertices $\Gamma_j^{(0)}$. To each such element $\Gamma_j^{(d)}$ there corresponds the truncated Hamiltonian function $\hat{H}_j^{(d)}$, which is the sum of those terms of the series (2) whose points (q_1, q_2, q_3) are on the $\Gamma_j^{(d)}$. The truncated Hamiltonian functions $\hat{H}_j^{(d)}$ are different first approximations of the H valid in different domains of the space (ξ, η, μ). Fig. 2 shows the Newton polyhedron Γ of H, which is the truncated prism with the skew basis [6,7].

Let us consider one edge ξ and two faces:

1. The edge $\Gamma_2^{(1)}$. It contains the points $(0,2,0)$ and $(-1,0,1)$ of the set D. The corresponding truncated Hamiltonian function is

$$\hat{H}_2^{(1)} = \frac{1}{2}(\eta_1^2+\eta_2^2) - \frac{\mu}{\sqrt{\xi_1^2+\xi_2^2}}$$

   It describes the problem of two bodies J and A.

2. The face $\Gamma_2^{(2)}$. It contains the points $(2,0,0)$, $(1,1,0)$, $(0,2,0)$ and $(-1,0,1)$ of the set D. The corresponding truncated Hamiltonian function is

$$\hat{H}_2^{(2)} = \frac{1}{2}(\eta_1^2+\eta_2^2) + \xi_2\eta_1 - \xi_1\eta_2 - \xi_1^2 + \frac{1}{2}\xi_2^2 - \frac{\mu}{\sqrt{\xi_1^2+\xi_2^2}}$$

   It describes the Hill's problem.

3. The face $\Gamma_1^{(2)}$ has $q_3 = 0$, i.e., $\mu = 0$. The corresponding truncated Hamiltonian function is (2) with $\mu = 0$. It describes the problem of two bodies S and A.

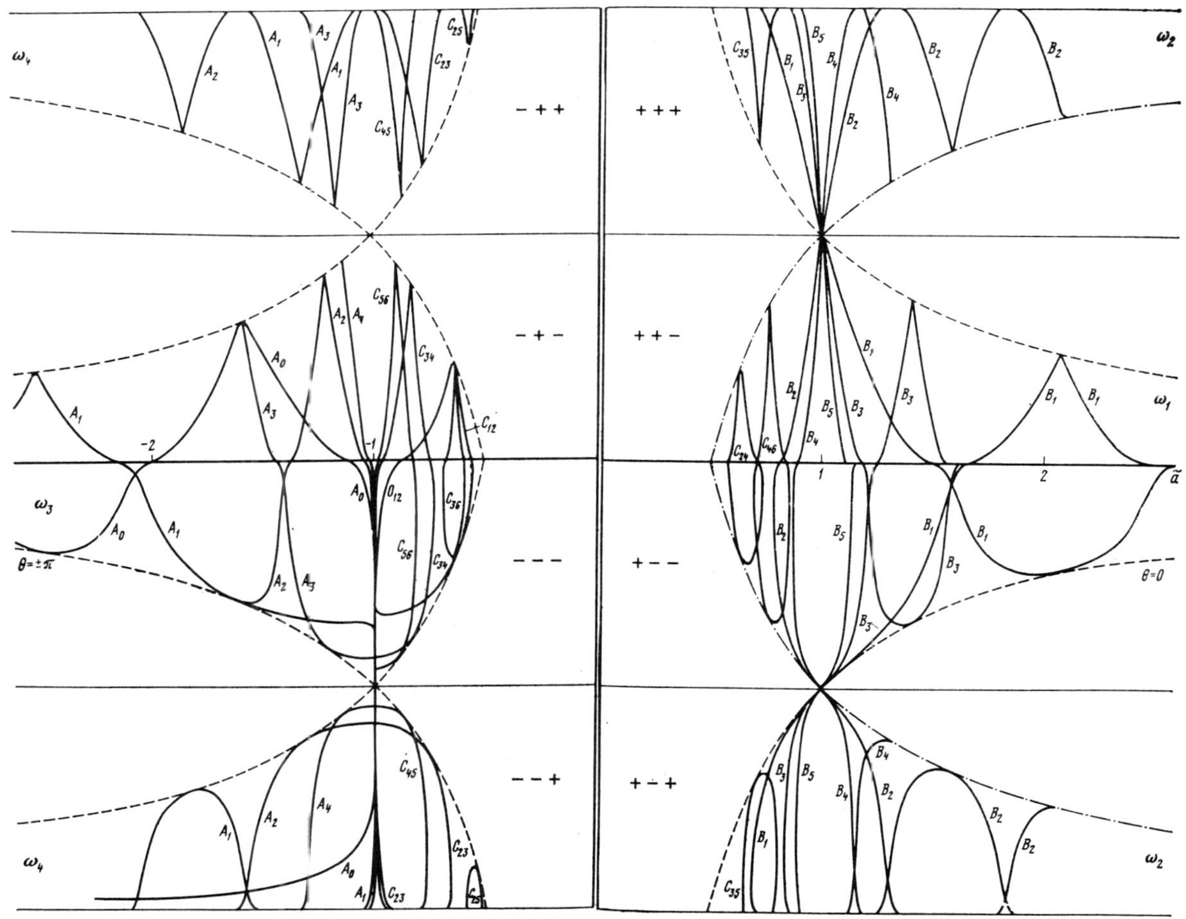

Figure 1

So very close to the body J the first approximation is the problem 1, not very close to J it is the problem 2 and far from J it is problem 3.

Periodic solutions passing near J are pasted from the axes with consecutive collisions of the problem 3 and hyperbolic orbits of the problem 1. In addition we have periodic solutions of the Hill's problem.

For $\mu = \mu_J = 0.00095388$ corresponding to Sun(S)–Jupiter(J) case I have computed several families of periodic solutions corresponding to the first quadrant of Fig. 1 [8]. Results are shown in Fig. 3, where the dot–dash curve represents J^* and L_1 and L_2 are the Lagrangian points. Big pieces of the computed families correspond to families S. But not all of them. Besides other parts are very near to the periodic solutions of the Hill's problem, shown in Fig. 4, where the vertical axis represents J^*. Fig. 1,3 and 4 show closeness of the computed families to the generating ones. Indeed the more detailed analysis of the first approximations $\hat{H}_j^{(d)}$ gives more information on the solutions of the restricted problem with small μ.

OSCILLATIONS OF A SATELLITE ON AN ELLIPTIC ORBIT

They are described by the differential equation

$$(1 + e\cos\nu)\frac{d^2\delta}{d\nu^2} - 2e\sin\nu\frac{d\delta}{d\nu} + \mu\sin\delta = 4e\sin\nu \tag{3}$$

where δ is the doubled angle of the position of the satellite's axis, e is the excentricity of the orbit, ν is the true anomaly on the orbit [9,Ch. V]. For $e < 1$ the equation (3) is regular. For $e = 1$ and $\nu = \pi$ the equation (3) is singular because the coefficient before main derivative vanishes. To consider a vicinity of the singularity let us introduce the local coordinates and parameter near the singularity:

$$\theta = \pi - \nu, \quad \epsilon = 1 - e.$$

Then the equation (3) takes the form

$$\{\epsilon + \frac{1}{2}\theta^2 + o(\epsilon, \theta^2)\}\frac{d^2\delta}{d\theta^2} + 2\{\theta + o(\epsilon, \theta^2)\}\frac{d\delta}{d\theta} + \mu\sin\delta = -4\{\theta + o(\epsilon, \theta^2)\} \tag{4}$$

For each term $\theta^\alpha\epsilon^\beta\frac{d^\gamma\delta}{d\theta^\gamma}$ in (4) we put

$$\begin{aligned} q_1 &= \alpha - \gamma \quad \text{that is the order of } \theta, \\ q_2 &= \beta \quad \text{that is the order of } \epsilon. \end{aligned}$$

Let D be the set of points (q_1, q_2) corresponding to all terms in the equation (4). The convex hull Γ of the set D contains two edges (Fig. 5). On the edge $\Gamma_1^{(1)}$ the unique vector $R_1(1,0)$. The corresponding variable is $\theta^1\epsilon^0 = \theta$. On the edge $\Gamma_2^{(1)}$ the unique vector $R_2(1, -\frac{1}{2})$. The corresponding variable is $\theta^1\epsilon^{-1/2}$. But that variable is not convenient because the equation (3) is not periodic in it. So we want to find such a variable which near the point $\theta = \epsilon = 0$ behaves as $\theta/\sqrt{\epsilon}$ and in which the equation (3) is periodic one. Such a variable is the excentric anomaly E.

Now we can regularize the singularity. Let us determinate the number τ by the condition

$$\pi - \nu = -\theta = E = \tau.$$

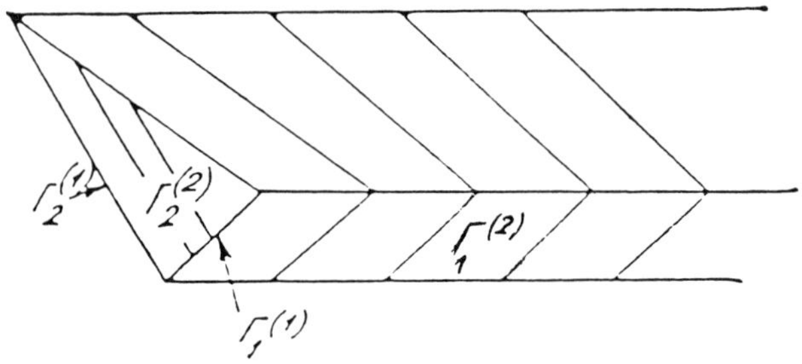

Figure 2

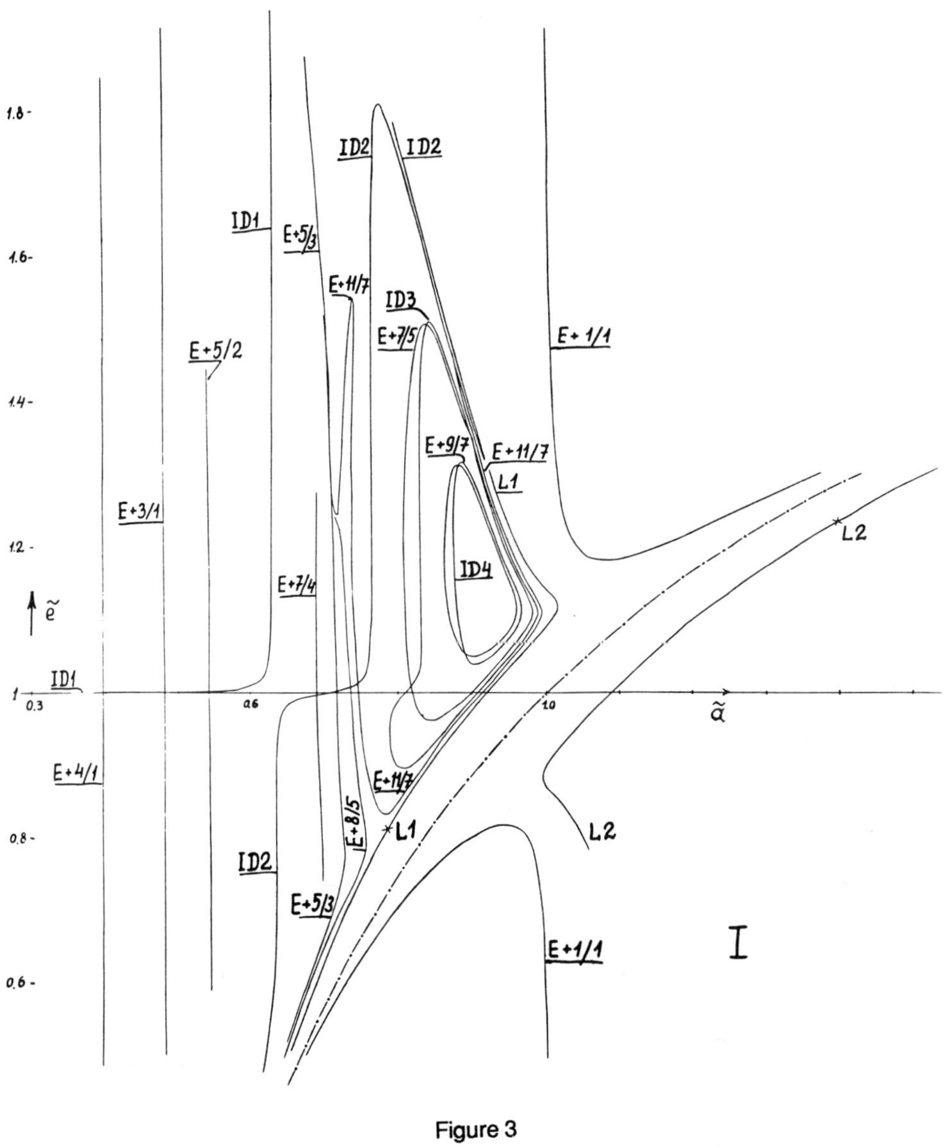

Figure 3

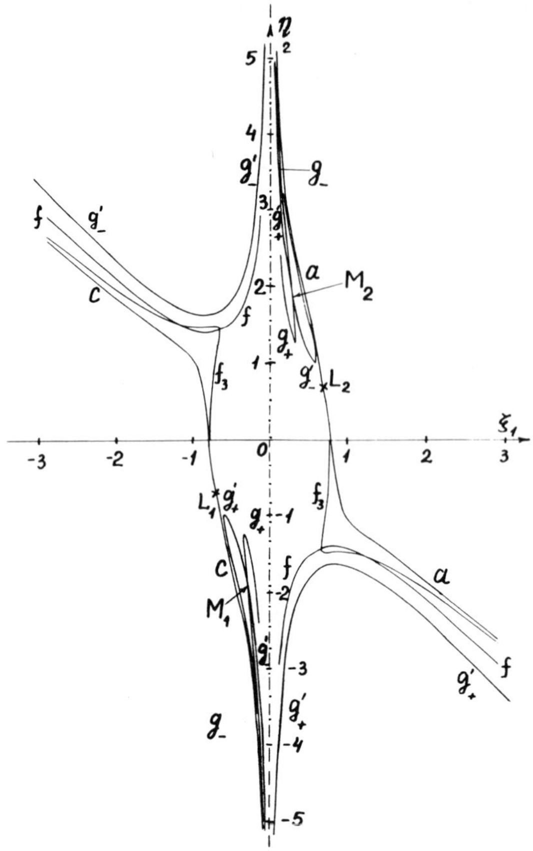

Figure 4

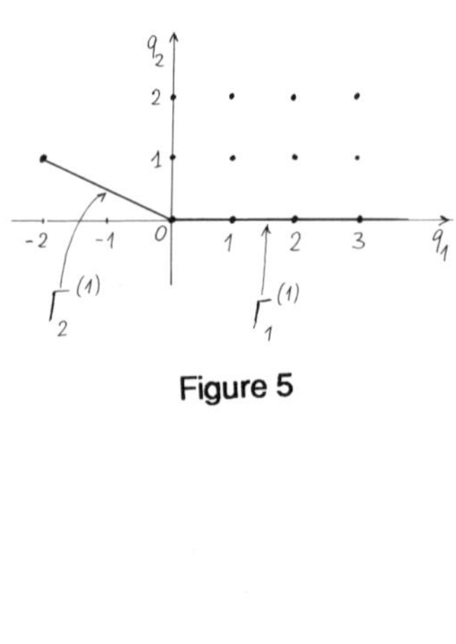

Figure 5

On the interval $0 \leq \nu \leq \pi - \tau$ we integrate the equation (3) along the variable ν and on the interval $\pi - \tau \leq \nu \leq \pi$ — along the variable E. But for E the second interval is $\tau \leq E \leq \pi$. It gives the regularization of the equation (3) near the singularity and allows to compute periodic solutions for $e = 0.9999\ldots$.

By that we have found new family of critical symmetrical periodic solutions for which Trace equals to 2 [10]. It has $-2 < \mu < 0$ and lies very closely to $e = 1$. When $e \to 1$ we have $\tau \to 0$ and the first interval tends to $0 \leq \nu \leq \pi$ and the second interval tends to $0 \leq E \leq \pi$. So our regularization blows up the point $\nu = \pi$ into interval $0 \leq E \leq \pi$.

REFERENCES

1. A. D. Bruno and A. Soleev, *First approximations of algebraic equations*, Doklady Akad. Nauk, 230 (to appear), in Russian. Soviet Mathematics Doklady in English.
2. A. D. Bruno, *First approximations of differential equations*, Ibid.
3. A. D. Bruno, 1990, *A local analysis of Hamiltonian systems*, Preprint I.H.E.S./M/1990/33.
4. A. D. Bruno, 1990,*The Restricted Three Body Problem*, Nauka, Moscow (in Russsian). English translation: Walter de Gruyter, Berlin, 1993.
5. M. Hénon, 1968, *Sur les orbites interplanetaires qui rencontrent deux fois la terre*, Bull. Astron., Ser. 3, t. 3:3, pp. 377-402.
6. A. D. Bruno, 1978, *On periodic flybys of the moon*, Preprint No. 91 of Inst. Applied Math., Moscow (in Russian). English version: Celest. Mech, 1981, 24, pp. 255–268.
7. A. D. Bruno, 1992, *General approach to a study of complicated bifurcations*, Prikladnaya Mekhanika, 28:12, pp. 83–86 (in Russian). English translation in Soviet Applied Mathematics.
8. A. D. Bruno, 1993, *Multiple periodic solutions of the restricted three-body problem in the Sun–Jupiter case*, Preprint of Inst. Applied Math., Moscow (in Russian), to appear.
9. A. D. Bruno, 1979, *Local Metchods in Nonlinear Differential Equations*, Nauka, Moscow (in Russian). English translation: Springer–Verlag, Berlin etc., 1989.
10. A. D. Bruno and V. Yu. Petrovitch, *Regularization of oscillations of a satellite on a very stretched orbit*, 1993, Preprint of Inst. Appl. Math., Moscow (in Russian), to appear.

THE STRUCTURE OF CHAOS

G. Contopoulos

Department of Astronomy
University of Athens
Panepistimiopolis
GR 157 83 - Athens
Greece
and
Department of Astronomy
University of Florida
Gainesville, Fl. 32611
USA

ABSTRACT

We describe the structure of chaotic regions in 2 and 3 degrees of freedom. The lobes formed by the asymptotic curves of unstable periodic orbits follow certain rules that allow predictions of their structure. When the energy becomes larger than the escape energy there are some "limiting asymptotic curves" defining the regions of escape. When there are no escapes the asymptotic curves become longer as the energy increases and produce an infinity of new tangencies, followed by the appearance of new stable periodic orbits.

In systems of 3 degrees of freedom the asymptotic curves of complex unstable orbits are spirals, described by a linear theory. In their vicinity there is an infinity of periodic orbits arranged along certain spirals. The asymptotic curves of doubly unstable orbits oscillate in a way different from that of unstable 2-D systems.

INTRODUCTION

Chaos appears near the unstable periodic orbits of nonintegrable systems. If the deviation from an integrable system is small the chaotic regions are small, and increase with increasing deviation from integrability. But the chaotic regions are not completely disordered. There is some order in chaos that is best described by studying the asymptotic curves of the unstable periodic orbits.

We describe this order for system of 2 and 3 degrees of freedom.

The rotation number (rot) near an unstable periodic orbit of an integrable system (Fig. 1a) is represented by a smooth curve which has an inflection point at the unstable periodic orbit and a plateau around the stable periodic orbit. At the inflection point and at the boundaries of the plateau the tangent of the curve is perpendicular.

In the case of a slightly nonintegrable system the rotation number increases abruptly close to the unstable periodic orbit. An example is provided by the Hamiltonian

$$H \equiv \frac{1}{2}\,(x^2+y^2+Ax^2+By^2) - \varepsilon x^2 y = h \qquad (1)$$

with A=0.9, B=1.6, h=0.00765 and ε=4 (Fig. 1b; Contopoulos 1966). In general the rotation number is not uniquely defined in a region close to this orbit (intervals between the vertical lines in Fig. 1b). If one defines the rotation number over a given finite time (see, e.g. Laskar et al. 1992) the rotation curve is varying irregularly close to the unstable orbit. The chaotic behaviour appears in the region where the rotation number is not well defined. This is the region occupied by the asymptotic curves of the unstable periodic orbit.

When the system is strongly nonintegrable we have the following phenomena (ε=4.4; Fig. 1c). (1) The increase of rot near the periodic orbit is larger. (2) the irregularities are larger, and (3) we may have certain small regions where rot is constant. These regions correspond to stable periodic orbits in the neighbourhood of the initial unstable orbit, or orbits that have bifurcated from such a stable orbit for a smaller value of the perturbation. In general, we have an infinity of higher order periodic orbits near an unstable periodic orbit, that correspond to rational values of rot. Thus we have the interaction of several resonances in this region (Rosenbluth et al. 1966, Contopoulos 1966, Chirikov 1969), that generate chaos, through their homoclinic and heteroclinic intersections.

The appearance of heteroclinic intersections between the asymptotic curves of different multiplicities is the most obvious signature of chaos. These intersections represent doubly asymptotic orbits, whose rotation number tends to different values as $t \to \infty$ and as $t \to -\infty$.

In the integrable case the secondary resonant periodic orbits are at the boundary of instability, and do not form islands of stability. For small perturbations these orbits form pairs of stable and unstable orbits. The stable islands give constant rotation numbers, and this is how we have straight line segments in Fig. 1. The higher order unstable periodic orbits in the chaotic zone have asymptotic curves that intersect the asymptotic curves of the main unstable periodic orbit. Thus higher order heteroclinic points are generated.

The characteristics of various families of periodic orbits bifurcating from the "central family" (composed of simple periodic orbits intersecting perpendicularly the y-axis), are shown in Fig. 2. These families are called regular, because they are formed by a continuous deformation of the orbits of the central family. They undergo further

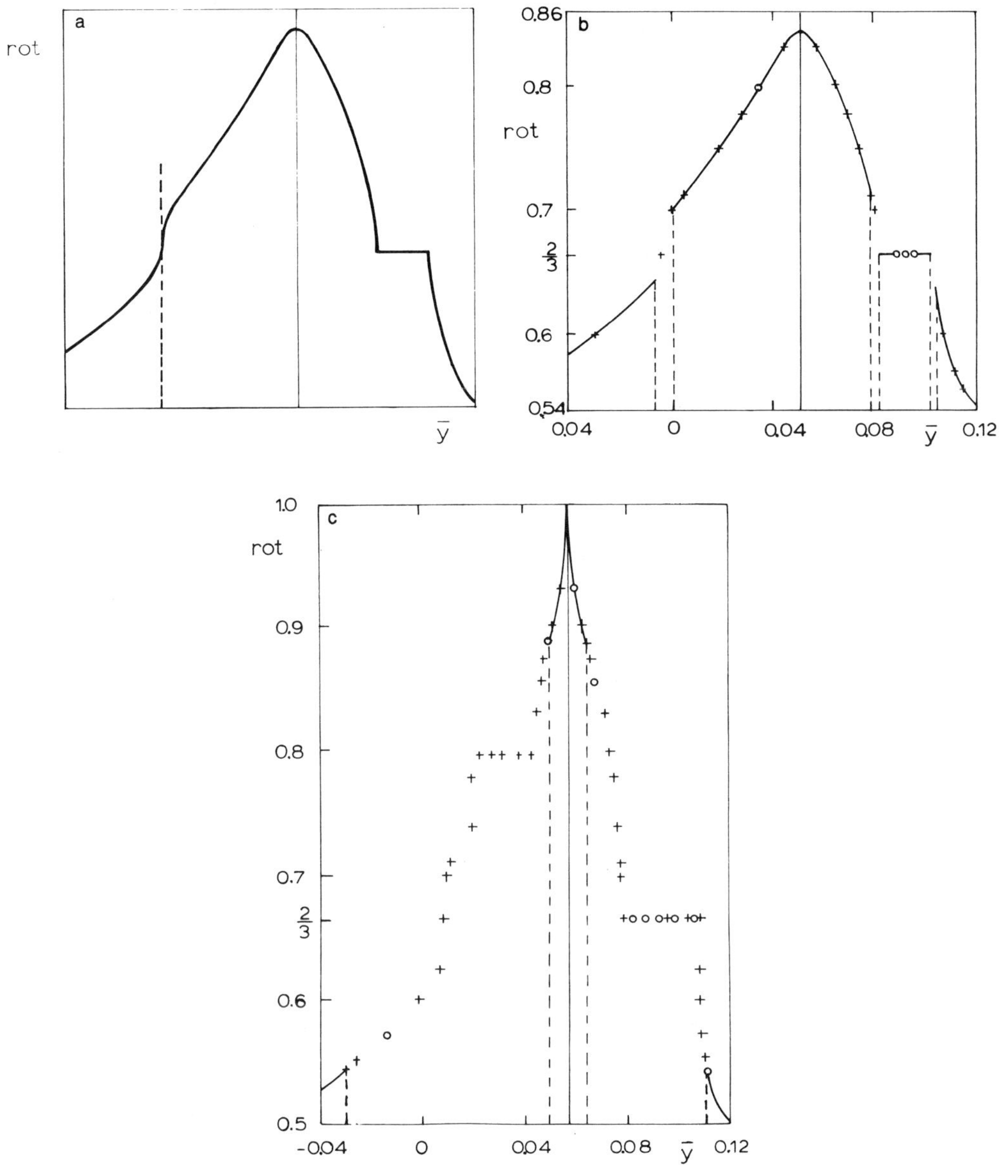

Figure 1. The rotation number as a function of x close to an unstable orbit located at the solid vertical line in an integrable system (a), in a slightly nonintegrable system (b), and in a strongly nonintegrable system (c); (o) stable and (+) unstable periodic orbits.

bifurcations of equal or double period (e.g the family 8/10 bifurcating from 4/5 in Fig. 2). However besides the regular families there are other irregular families that cannot be reduced to the orbit of the central family by a continuous change. Such is the family 6 (of multiplicity 6) in Fig. 2, which has one branch in the region 4/5 (of multiplicity 5) and another branch in the region 6/7 (of multiplicity 7). Such irregular families are characteristic of chaotic regions.

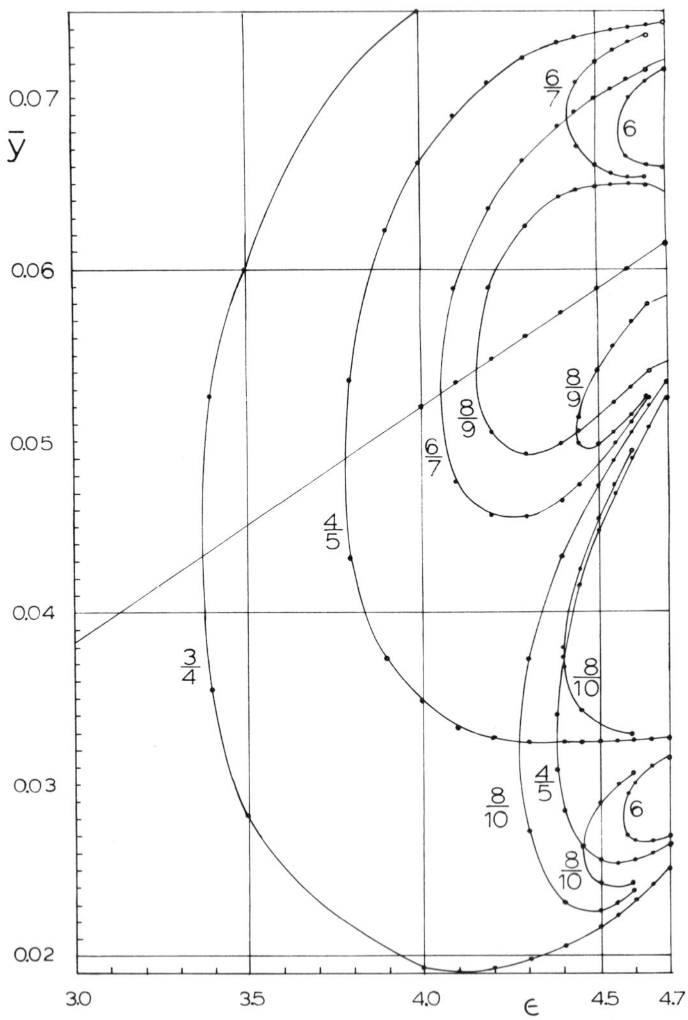

Figure 2. The characteristics of some families of periodic orbits in the Hamiltonian (1) ($\bar{y}=y\sqrt{B}$ as a function of ε for $A=0.9$, $B=1.6$, $h=0.00765$).

We consider now non-periodic orbits. If we start an orbit close to an unstable periodic orbit on a Poincaré surface of section its consequents are scattered in an irregular way over the whole chaotic region, giving the impression of chaos. Two nearby orbits may give a very different distribution of their consequents (Fig. 3). Therefore one may consider this distribution as random and unpredictable. However the distribution of the consequents can be derived by considering the form of the asymptotic curves and their homoclinic intersections.

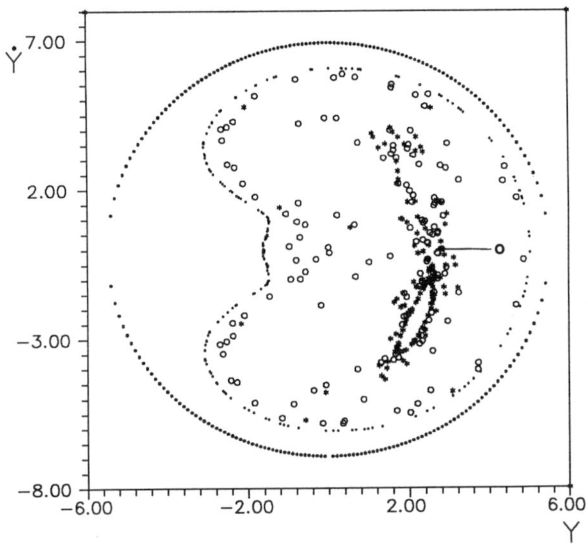

Figure 3. The distributions of 140 consequents of two nearby initial points (o) and (*) are very different although they cover the same stochastic region. The outer circle is defined by the energy integral. The other curve (dotted) is an invariant curve surrounding the consequents.

We study the asymptotic curves from the point O that represents a simply unstable periodic orbit in the Hamiltonian (1) with A=0.9, B=1.6, ε=0.08 and h=24 (Contopoulos and Polymilis 1993).

The asymptotic curves U(unstable) and S(stable) form oscillations and intersect at infinite homoclinic points like P_0, P_1', P_1 etc (Fig.4). The arcs of the curves U and S up to the point P_0, which is almost at equal distances from O along U and S, define the resonant region O_1. In a similar way the resonant region O_2 is defined by the curves UU, SS. The arcs of the curves U and S between two successive homoclinic points form lobes. Thus between P_o and P_1' we have the (outer) lobe U_1', between P_1' and P_1, we have the (inner) lobe U_1, etc. In a similar way we have inner lobes S_{-1}', outer lobes S_{-2}, etc, and lobes UU,UU', SS,SS'.

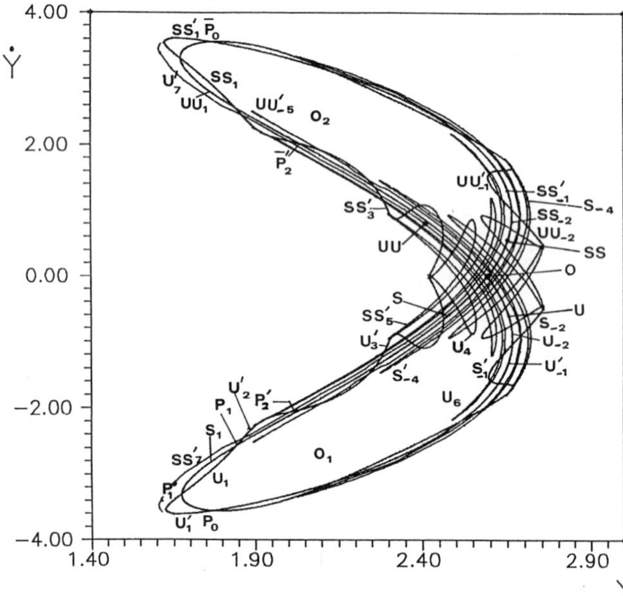

Figure 4. The asymptotic curves from the unstable periodic orbit O. There are two unstable curves (U and UU) and two stable curves (S and SS). The initial U and S up to the point P_o, roughly at equal distances from O along U and S, define the resonant region O_1. The curves U and S intersect at the homoclinic points P_o, P_1, etc, where U crosses S outwards,and P_1', P_2',etc, where U crosses S inwards. Analogous definitions apply to the resonant region O_2 and to the homoclinic points $\bar{P}_o$, etc.

Because of the area conservation the intersections of various lobes are mapped to other intersections of equal area. E.g. the intersection of the lobes U_6 and S_{-1}' is mapped into the intersection U_5 and S_{-2}' and, in general, to the intersection of U_m and S_{m-7}'.

The intersections of the various lobes follow two rules:

<u>Rule I</u>: If a lobe U_n intersects a lobe S_{n+q}' ($=U_{n+q}'$), or a lobe S_{n+q}($=U_{n+q}$), then the lobe U_{n+1} intersects the lobe S'_{n+1+q}, or S_{n+1+q} respectively.

<u>Rule II</u>: If a lobe U_n' intersects a lobe UU_{-n+q} ($=SS_{-n+q}$), or a lobe UU_{-n+q}'($=SS_{-n+q}'$), the lobe U_{n+1}' intersects the lobe UU_{-n-1+q}, or UU_{-n-1+q}' respectively.

Following these simple rules, we can explain the form of the higher order lobes from the form of the lower order lobes, although the higher order lobes are so complicated that their description seems hopeless.

In Figs. 5 and 6 we have drawn the lobes U_{12}' and U_{13}', respectively, together with some S,S',SS, SS', and UU,UU' lobes.

The lobe U_{12}' is a very long arc of the asymptotic curve U, starting at the intersection P_{11} with the curve S (Fig.5) and terminating at the intersection P_{12}', which is very close to P_{11}. (The thickness of the lobe is so small that it cannot be seen in Fig. 5). The most remote point of the lobe from P_{11} (and P_{12}') along the curve U is the "end point". Thus we describe the lobe as one thick curve from its origin (the points P_{11} and P_{12}') to its end point.

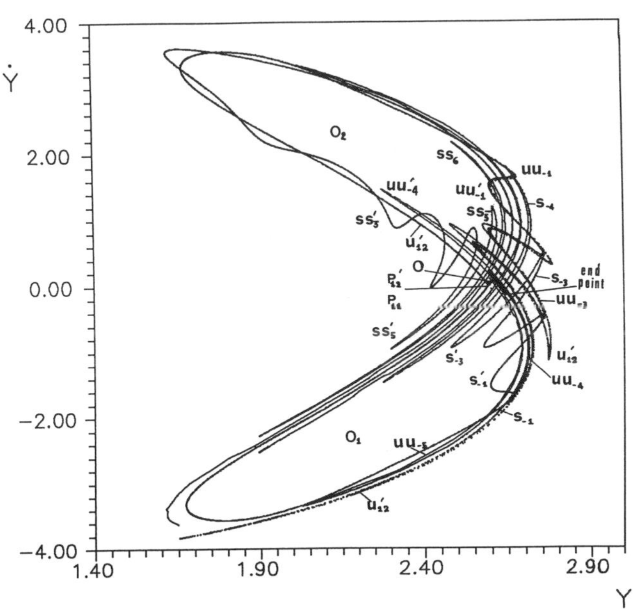

Figure 5. The outer lobe U_{12}'together with some S,S',SS,SS', UU and UU' lobes. The lobe starts at the point P_{11} and terminates at the nearby point P_{12}'. The "end point" is the most remote point from P_{11} and P_{12}' along the lobe.

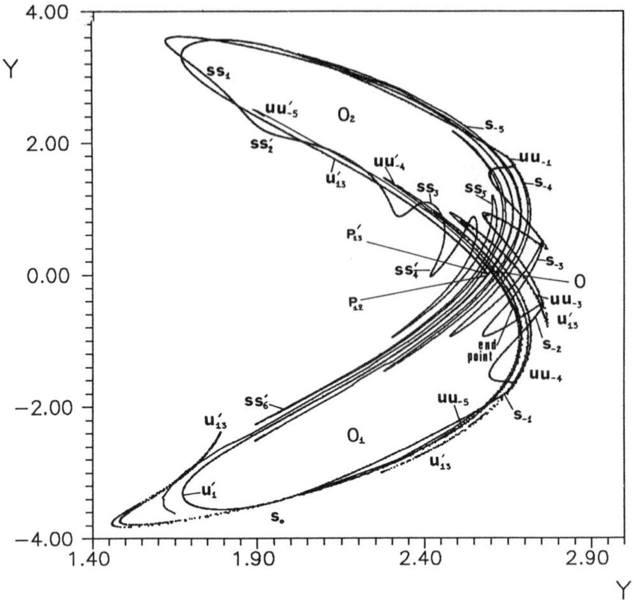

Figure 6. The same as in Fig. 5 for the lobe U_{13}'.

The lobe U_{12}' surrounds the region O_2 and then it follows closely, from the outside, the lobes UU_{-1}, UU_{-2}, UU_{-3}, and UU_{-4}. In order to compare these results with those of higher order lobes we write these lobes as $UU_{-n+q}(=SS_{-n+q})$, where n=12, so that the corresponding lobes intersected by $U_{n+1}'(=U_{13}')$ are $UU_{-n-1+q}(=SS_{-n-1+q})$ according to Rule II. E.g. the lobe UU_{-4} is written UU_{-n+8}, where n=12. Similarly U_{12}' enters into the lobes UU_{-1}', UU_{-2}', UU_{-3}', UU_{-4}'.

The last inner oscillation of U'_{12} is inside the lobe UU_{-4}', but it is well outside the tip of UU_{-4}', and a little beyond this last oscillation is the end point of U_{12}'.

The oscillation of U_{12}' in the lobe UU_{-4}' reaches the lobe SS_6 (i.e. U'_n reaches SS_{-n+18}) and the end point is a little outside the lobe S_{-4}. We write this lobe S_{n-16}, with n=12, so that we should be able to apply the Rule II when n becomes 13, etc. Furthermore U_{12}' has 2 intersections with $S_{-1}(S_{n-13})$, 4 intersections with S_{-2} (S_{n-14}), and 6 intersections with $S_{-3}(S_{n-15})$.

As a comparison U_{13}' starts at P_{12} and terminates at P_{13}' nearby (Fig. 6). It has oscillations around the same UU_{-n} and UU_{-n}' lobes as U'_{12}, plus around one outer lobe, $UU_{-n+8}(=UU_{-5})$ and one inner lobe, $UU_{-n+8}'(=UU_{-5}')$.

The outer oscillations of U_{13}' are shorter than the corresponding oscillations of U_{12}'. E.g. the outer oscillation of U_{13}' outside UU_{-4} is much shorter than the "last" outer oscillation of U_{12}' (outside UU_{-4}). On the other hand the inner oscillations of U_{13}' in the lobes UU_{-n}' are longer than the corresponding oscillations of U_{12}' (compare the oscillations of U_{12}' and U_{13}' in the lobe UU_{-4}').

The last inner oscillation of U_{13}' is in the lobe UU_{-5}', but much shorter than UU_{-5}'. This oscillation of U_{13}' reaches the lobe $SS_{-n+18}(=SS_5)$ according to the Rule II. The end point is a little outside $S_{n-16}(=S_{-3})$ according to the Rule I. Finally U_{13}' has 2 intersections with $S_{n-13}(=S_o)$, 4 intersections with $S_{n-14}(=S_{-1})$, etc.

We remark that the outer lobes enter into the resonant regions 0_1 and 0_2 and make a number of rotations around 0_1 and 0_2 before going far away from 0_1 and 0_2 (Fig. 7). Thus we have a stickiness phenomenon of the asymptotic curves close to the stable regions 0_1 and 0_2 (Contopoulos 1971, Shirts and Reinhardt 1982). In Fig. 7 we see that the lobes up to about U_{20}' stay close to the region $0_1, 0_2$, but they move gradually away from them. The lobes U_{25}', U_{26}' reach almost the maximum allowed distance from 0. Further out we have closed invariant curves (Fig. 3) that surround the whole chaotic region around 0.

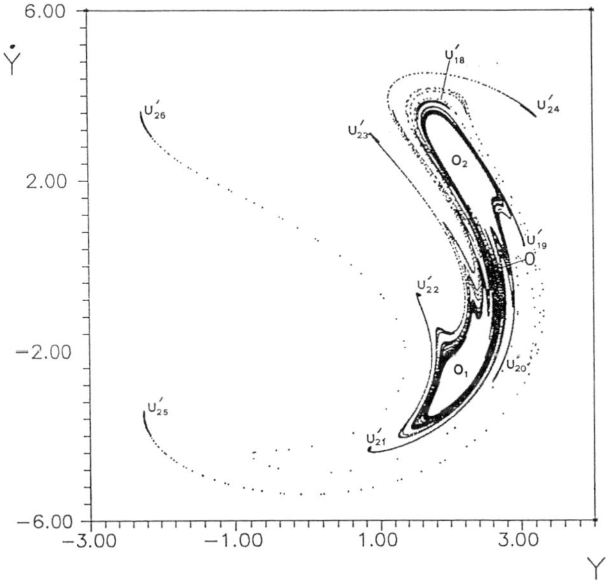

Figure 7. The asymptotic curve U up to the lobe U_{26}' for h=24. All the lobes have an important part near the resonant regions 0_1 and 0_2 and produce the "stickiness" phenomenon. But the high order lobes have also a part far from 0. In particular the last two lobes (U_{25}' and U_{26}') reach almost the maximum possible distance from 0.

The deviation from 0 occurs faster for larger values of the energy. E.g. when h=25 the stickiness phenomenon lasts less and the deviation of the lobes to large distance appears already with the lobes U_5' and U_6' (Fig. 8).

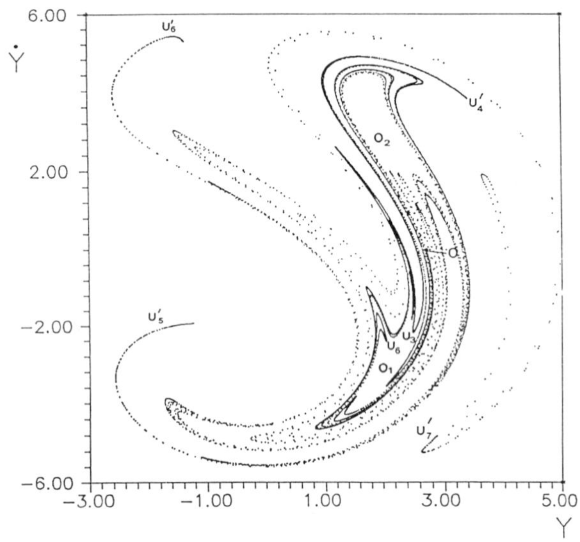

Figure 8. The asymptotic curve U, for h=25, up to the lobes U_6 and U_7'.

If the energy goes beyond the escape energy then some orbits escape to infinity. In fact then the curve of zero velocity (CZV) is open, with one or more channels of escape.

Across every opening of the curve of zero velocity there is an unstable periodic orbit, called Lyapunov orbit, which is the limit of the escaping orbits (Rod 1973, Churchill et al 1975, 1979). In the case of the Hamiltonian (1) we have two Lyapunov orbits, O_1, O_2 (Fig. 9). Every orbit crossing a Lyapunov orbit outwards escapes from the system.

Thus the limit of the escaping orbits consists of the orbits approaching asymptotically a Lyapunov orbit from the inner side. This asymptotic set intersects the Poincaré surface of section $(y,\dot{y})$ along a closed curve, called "limiting asymptotic curve" (Contopoulos 1990a). In the case of the Hamiltonian (1) for h>25.31 there are two escape channels and correspondingly two "limiting asymptotic curves" that contain orbits escaping to infinity as $t \to \infty$ through each of these channels. One such curve is called L1 in Fig. 10. The unstable asymptotic curve U from 0 tends asymptotically to this curve after infinite spiral rotations around it (Fig. 10). Similarly the stable asymptotic curve S tends asymptocically to the limiting asymptotic curve L2. Orbits inside this curve escape to infinity as $t \to -\infty$.

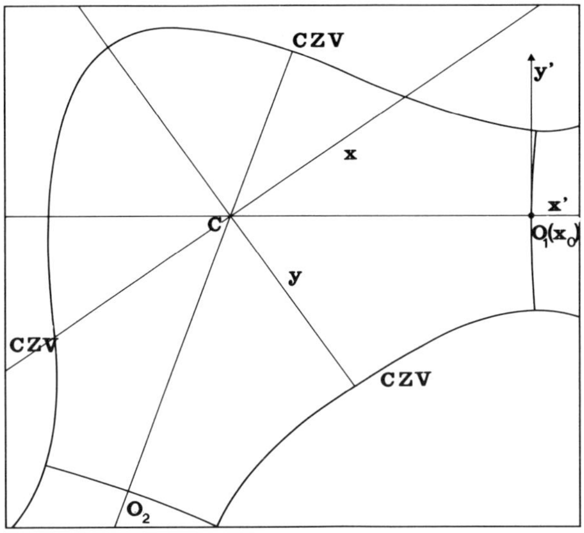

Figure 9. Escape channels in the case of the Hamiltonian (1) for an energy larger than the escape energy. Every channel is crossed by an unstable (Lyapunov) periodic orbit.

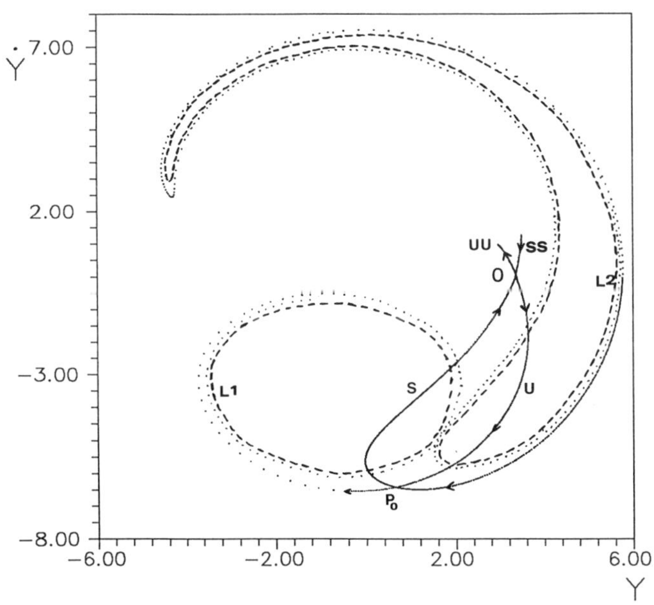

Figure 10. The limiting asymptotic curves L1 and L2 in the case h=31.30. The curve U makes infinite rotations around the limiting asymptotic curve L1. Similarly the curve S makes infinite rotations around the limiting asymptotic curve L2.

There are two more "limiting asymptotic curves" for orbits escaping when $t \to \infty$, or $t \to -\infty$. These curves are called L3 (escape for $t \to -\infty$) and L4 (escape for $t \to \infty$), and they are symmetric to L1 and L2 with respect to the axis $\dot{y}=0$.

When h is smaller (but larger than the escape energy h=25.31) the asymptotic curve U from the unstable orbit O makes some turns close to L1, but returns close to its original path to form further homoclinic points. But after approaching successively L1 and L4 a number of times, it finally makes infinite spiral rotations around either L1, or L4. A similar behaviour is shown by the asymptotic curve S.

A similar behaviour is shown by all the asymptotic curves of other unstable periodic orbits, provided they contain escape orbits. It is necessary to state this proviso, because there may be unstable periodic orbits whose asymptotic curves do not lead to escapes. Such is the case of unstable orbits inside closed invariant curves, which cannot ever communicate with the escape region. On the other hand if along the asymptotic curve of an unstable periodic orbit there are escape orbits for $t \to \infty$, these can escape only through one of the two escape channels above. The boundary of these orbits cannot be anything else, but the curve L1, or L2.

The problem of escapes is related to the problem of chaotic scattering. In chaotic scattering we consider orbits coming from infinity, interacting with the system and eventually escaping to infinity again through the same or a different channel. The limits of the escaping sets in different directions form Cantor sets (Jung and Scholz, 1987, 1988, Hénon 1988). In many cases, however, there are regions of initial conditions of orbits that do not escape to infinity (Contopoulos 1990a, Contopoulos and Kaufmann 1992).

A particular problem of this kind is the case of photons approaching a system of two fixed black holes (non-rotating). In this case all the periodic orbits are unstable (Contopoulos 1990b) and do not trap any nonperiodic orbits around them. Thus all orbits, except for a Cantor set of measure zero, "escape" either to infinity, or to one of the two black holes (in this case 2 of the 3 escape routes are at a finite distance). Orbits coming from infinity may fall into either one of the two black holes, or escape again to infinity. In Fig. 11 we show how a beam of photons coming from infinity is split into orbits escaping to the black holes M1 or M2, or to infinity, after one (not close) approach to the black hole M1. We notice that between an orbit reaching M1 and an orbit reaching M2 there is a set of orbits reaching infinity. Similarly between an orbit reaching M1 and an orbit reaching infinity there is a set of orbits reaching M2, etc. This property is the basic characteristic of chaotic scattering, that leads to the Cantor set of the separating orbits.

These phenomena are common to all Hamiltonians that have an escape energy. However there are also Hamiltonians, like

$$H \equiv \frac{1}{2}\,(\dot{x}^2+\dot{y}^2+x^2+y^2) + \varepsilon x^2 y^2 = h \qquad (2)$$

for $\varepsilon>0$, that do not have any escape energy. In such a case the asymptotic curves of an unstable periodic orbit become longer and longer as the energy increases.

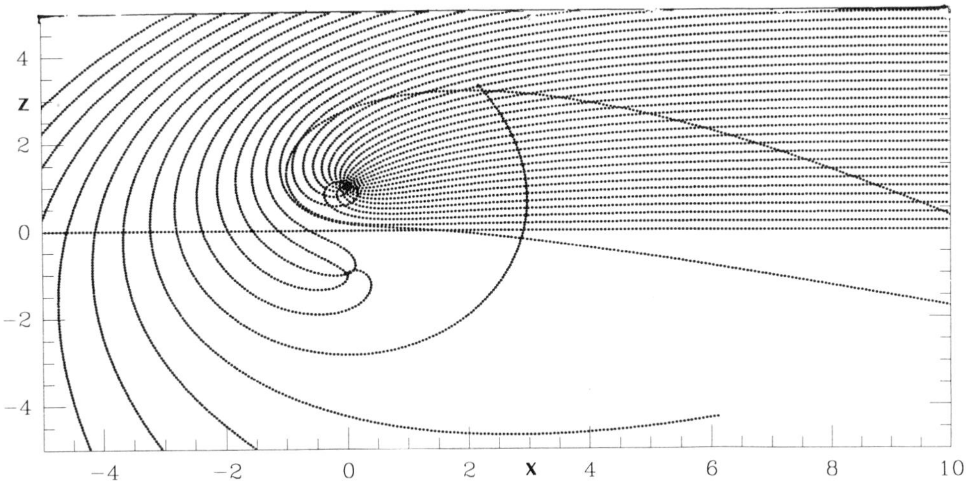

Figure 11. A thin beam of photons from infinity is split into 3 sets of orbits, falling into the black holes M_1, or M_2, or going to infinity.

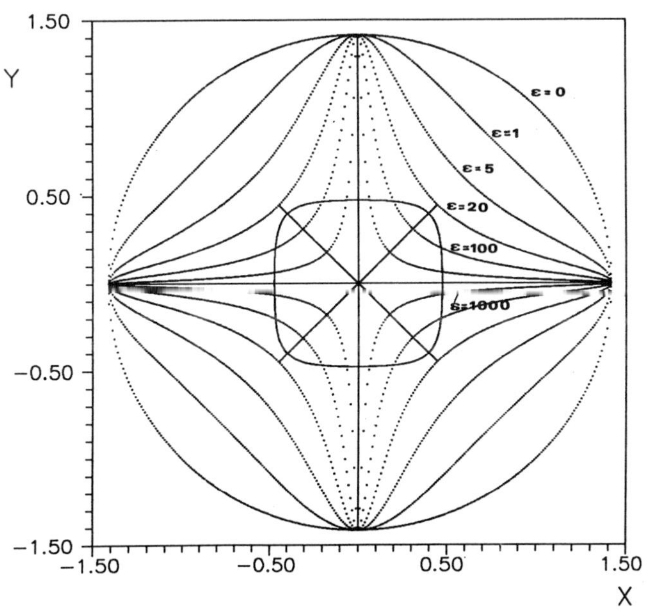

Figure 12. Equipotentials and the main periodic orbits in the Hamiltonian (2). The main periodic orbits are the axes, the diagonals and an almost square orbit.

We have studied this system by rescaling it in such a way that h remains constant (h=1) and ε is a parameter that increases to infinity.

One important property of this system is that the periodic orbits along the axes (Fig.12), which are represented on a Poincare surface of section by the central point $y=\dot{y}=0$, or by the boundary $y^2+\dot{y}^2=2$, have inifite intervals of stability and instability in ε. The transitions from stability to instability are given by the approximate formulae

$$\varepsilon=0.6174k^2+0.9013k-0.191 \quad , \tag{3}$$

and the transition from instability to stability by

$$\varepsilon=0.61874k^2+1.5189k-0.031 \quad . \tag{4}$$

These values are found by solving the variational equation, which is a Mathieu equation (Sohos et al. 1989). Therefore there is no value of ε beyond which the system is completely chaotic.

In such a system the asymptotic curves become longer and longer as ε increases. In Fig. 13 we show that the asymptotic curves of the periodic orbit B_1 (the diagonal orbit in Fig. 12) form spirals, going both clockwise and counterclockwise. As ε increases these spirals become longer, and have more and more intersections.

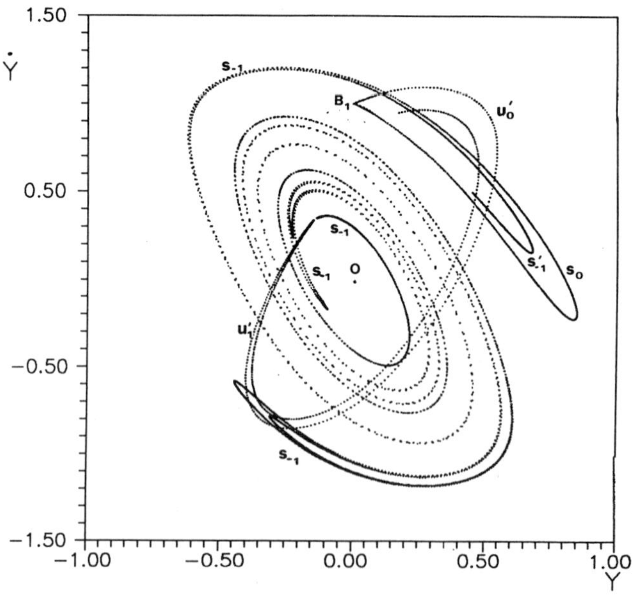

Figure 13. Asymptotic curves from the invariant point B_1 that spiral around the center, for $\varepsilon=25$.

Every new intersection follows a tangency of two spirals. But it is known that near a tangency of two asymptotic curves there is a stable periodic orbit (Newhouse 1977, 1983). Thus we have an infinity of tangencies that appear for arbitrarily large ε and are followed by an infinity of stable periodic orbits and sets of islands around them. These periodic orbits become eventually unstable after an infinite cascade of period doubling bifurcations. But there are infinite intervals Δε of values of ε that contain islands of stability. Such intervals appear for arbitrarily large ε; therefore chaos is never complete.

CHAOS IN 3 DEGREES OF FREEDOM

In systems of 3 degrees of freedom we have some new phenomena that do not appear in systems of 2 degrees of freedom. These are

a) Complex instability

b) Collisions of bifurcations

and c) Arnold diffusion.

Complex instability appears when two eigenvalues collide on the unit circle, but not on the real axis, and the eigenvalues leave the unit circle moving outwards and inwards. Then the successive consequents on a Poincaré surface of section (which is now 4-dimensional) form spirals. The projection of such a spiral is shown in Fig. 14.

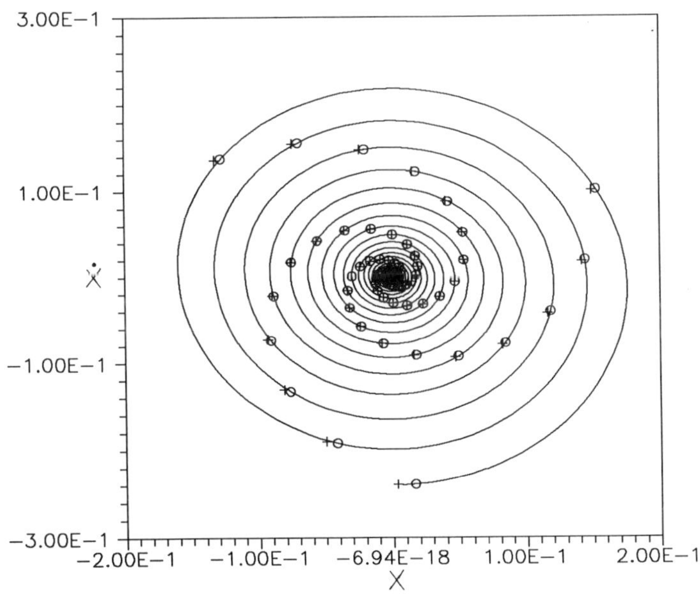

Figure 14. The projection of a spiral invariant curve from a complex unstable periodic orbit (with α>0 and θ≈2π/3). The theoretical points (+) are close to those found numerically (o).

A linear theory can describe not only qualitatively, but also quantitatively these spirals, close to the periodic orbit (Papadaki et al. 1994).

The eigenvalues are

$$\lambda_{\frac{1}{2}} = \rho(\cos d \pm i\sin\theta), \quad \lambda_{\frac{3}{4}} = \frac{1}{\rho}(\cos\theta \pm i\sin\theta), \tag{5}$$

where $\rho>1$.

The corresponding eigenvectors are complex

$$\underline{z}_{\frac{1}{2}} = \underline{x}_o \pm i\underline{y}_o, \quad \underline{z}_{\frac{3}{4}} = \underline{u}_o \pm i\underline{w}_o. \tag{6}$$

After one period T of the periodic orbit the eigenvectors become

$$\underline{z}_{\frac{1}{2}}' = \lambda_{\frac{1}{2}}\underline{z}_{\frac{1}{2}} = \rho[(\cos\theta\underline{x}_o - \sin\theta\underline{y}_o) \pm i(\sin\theta\underline{x}_o + \cos\theta\underline{y}_o)]$$

$$\underline{z}_{\frac{3}{4}}' = \lambda_{\frac{3}{4}}\underline{z}_{\frac{3}{4}} = \frac{1}{\rho}[(\cos\theta\underline{u}_o - \sin\theta\underline{w}_o) \pm i(\sin\theta\underline{u}_o + \cos\theta\underline{w}_o)]. \tag{7}$$

Orbits with initial conditions $\underline{x}_o, \underline{y}_o, \underline{u}_o, \underline{w}_o$ after n periods are respectively

$$\begin{aligned} \underline{x}(nT) &= \rho^n(\cos n\theta\underline{x}_o - \sin n\theta\underline{y}_o) \\ \underline{y}(nT) &= \rho^n(\sin n\theta\underline{x}_o + \cos n\theta\underline{y}_o) \\ \underline{u}(nT) &= \rho^{-n}(\cos n\theta\underline{u}_o - \sin n\theta\underline{w}_o) \\ \underline{w}(nT) &= \rho^{-n}(\sin n\theta\underline{u}_o + \cos n\theta\underline{w}_o) \end{aligned} \tag{8}$$

Thus an orbit with initial conditions

$$\underline{\xi}(0) = c_1\underline{x}_o + c_2\underline{y}_o + c_3\underline{u}_o + c_4\underline{w}_o \tag{9}$$

after n periods is

$$\begin{aligned} \underline{\xi}(nT) = {} & \rho^n[(c_1\cos n\theta + c_2\sin n\theta)\underline{x}_o + (c_2\cos n\theta - c_1\sin n\theta)\underline{y}_o] \\ & + \rho^{-n}[(c_3\cos n\theta + c_4\sin n\theta)\underline{u}_o + (c_4\cos n\theta - c_3\sin n\theta)\underline{w}_o] \end{aligned} \tag{10}$$

If we take $c_1=1$, $c_2=c_3=c_4=0$ we find the projections of $\underline{\xi}(nT)$ along the axes $i=1,2$

$$\xi_i = \rho(\cos n\theta\, x_{oi} - \sin n\theta\, y_{oi}), \tag{11}$$

Solving these equations for $\cos n\theta$, $\sin n\theta$ and eliminating $n\theta$ we find

$$\rho^{2n}(x_{o2}y_{o1} - x_{o1}y_{o2})^2 = (x_{o1}\xi_2 - x_{o2}\xi_1)^2 + (y_{o1}\xi_2 - y_{o2}\xi_1)^2 \tag{12}$$

This equation is of the form

$$\alpha_1\xi_1^2 + \beta_1\xi_1\xi_2 + \gamma_1\xi_2^2 = \delta\rho^{2n} \tag{13}$$

with $\alpha_1>0$, $\gamma_1>0$ and $\delta>0$. Thus the successive points with $n=1,2,3...$, spiral outwards on the plane (ξ_1, ξ_2).

In the particular case of Fig. 14 the theoretical positions are very close to the empirical positions for over 100 iterations.

In a similar way an orbit with $c_3=1, c_1=c_2=c_4=0$ forms a spiral inwards.

In the more general case when all c_i are different from zero the terms in $\underline{\xi}(nT)$ with ρ^n dominate, in comparison with the terms with ρ^{-n}, and the consequents spiral outwards for large n. Even in cases where c_1, c_2 are very small and we have initially a spiral inwards, after some order n the successive consequents start to move <u>outwards</u>.

This phenomenon is similar to the 2-D case of an unstable orbit with its asymptotic curves on a Poincare surface of section. If a initial point is exactly on a stable asymptotic curve, its consequents approach the invariant point that represents the periodic orbit. But if the initial point deviates, even slightly, from the stable asymptotic curve, its consequents reach a minimum distance from the periodic orbit and then move outwards, close to an unstable asymptotic curve.

If we write the quantities c_i in Eq. (9) in the form

$$c_1=R\cos\Phi,\ c_2=R\sin\Phi,\ c_3=\bar{R}\cos\bar{\Phi},\ c_4=\bar{R}\sin\bar{\Phi}, \tag{14}$$

we find

$$\underline{\xi}(nT) = R^{(n)}[\cos(\Phi-n\theta)\underline{x}_o+\sin(\Phi-n\theta)\underline{y}_o] + \bar{R}^{(n)}[\cos(\bar{\Phi}-n\theta)\underline{u}_o+\sin(\bar{\Phi}-n\theta)\underline{w}_o, \tag{15}$$

where

$$R^{(n)} = R\rho^n,\ \bar{R}^{(n)}=\bar{R}\rho^{-n}\ . \tag{16}$$

Thus we have

$$R^{(n)}\bar{R}^{(n)} = R\bar{R} = \text{const} \tag{17}$$

and if we consider the consequents on the 3-D space $(x_o, y_o, \bar{R})$ we have a spiral helix with the consequents moving outwards on a surface generated by the rotation of the hyperbola $R\bar{R}$ = const around the $\bar{R}$ axis (Fig. 15).

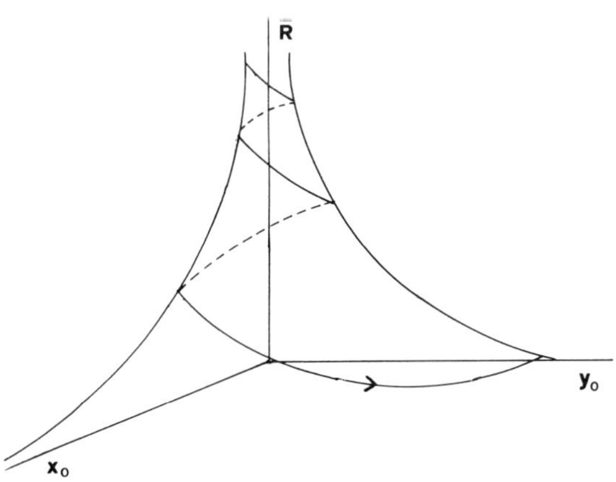

Figure 15. An invariant curve (in the linearized approximation) is a helix in the space $(x_o, y_o, \bar{R})$.

The above results are best seen if ρ is slightly larger than 1 and the spiral (13) is tight. In particular in the case of Fig. 14 the value of θ is close to $2\pi/3$ and this is why the successive consequents are approximately $2\pi/3$ apart in angle.

When $R^{(n)}=R\rho^n$ becomes large the linear theory is no more applicable. In fact the successive consequents reach a maximum distance from the center (the periodic orbit) and then move again inwards.

A nonlinear theory for the mapping generated by the Hamiltonian close to a complex unstable periodic orbit was developed by Heggie (1985). Heggie considered an integrable Hamiltonian that coincides with a given Hamiltonian up to 4th (or 5th) order terms, and studied the transition from stability to complex instability. A stable orbit is surrounded by quasiperiodic tori. When the orbit becomes complex unstable the tori continue to exist, provided the most important 4th order term (a_3r^4) is positive. However these tori do not exist close to the periodic orbit.

On the other hand if this 4th order term (a_3r^4) is negative the region of invariant tori decreases as the energy approaches the transition value to complex instability and shrinks to zero at the transition value itself.

This behaviour corresponds to our results (Contopoulos and Magnenat 1985, Contopoulos and Barbanis 1989) in the Hamiltonian system

$$H \equiv \frac{1}{2}(\dot{x}^2+\dot{y}^2+\dot{z}^2+Ax^2+By^2+Cz^2)-\varepsilon xz^2-\eta yz^2= h. \qquad (18)$$

If we keep η=fixed (small) and increase ε we have a transition from stability to complex instability. Then a small chaotic region is formed near the unstable periodic orbit. In this region the Lyapunov characteristic number is positive, but small. Further away from the periodic orbit the Lyapunov number is zero, and the motion is ordered (quasiperiodic).

As ε increases further, the chaotic region increases abruptly and the Lyapunov number also increases abruptly and reaches the value appropriate for the large stochastic sea covering most of the phase space.

When ε increases even further, the periodic orbit becomes again stable. However the region of stability (where we have zero Lyapunov number and a set of invariant tori) is small, close to the transition value, and further away we have a large chaotic zone, and the Lyapunov number is large.

The first transition corresponds to the case of Heggie where $a_3r^4>0$, while the second case corresponds to $a_3r^4<0$.

Near a given complex unstable orbit there are infinite invariant orbits of higher multiplicity. The perpendicular intersections of these orbits by the plane $(\bar{x},\bar{y})$ in the wide neighbourhood of the complex unstable periodic orbit have been studied by Barbanis (1993). These form regular spirals with a well defined order in the multiplicities of the periodic orbits. E.g. starting from an orbit called 8a that has two perpendicular intersections (Fig. 16) we have successive orbits 10a,12a,.... that approach along a spiral the periodic orbit 1a.

The second intersections of these orbits approach two other points on the plane $(\bar{x},\bar{y})$ called d and u. Namely the second points 8a,12a etc approach, along a spiral, the limiting point d, while the second points 10a,14a,... approach in the same way the point u. There seems to be an infinity of sequences of invariant points approaching 1a and d or u. Such are the sequences (2) 8b,10b,..., (3) 17a, 19a,...., (4) 17b, 19b,... and others.

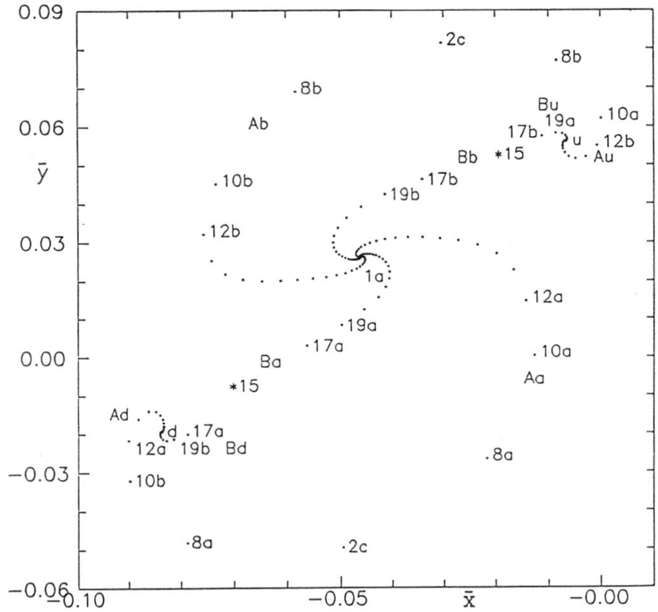

Figure 16. Perpendicular intersections of orbits of various multiplicities in the larger neighbourhood of the complex unstable orbit 1a ($\bar{x}=x\sqrt{A}$, $\bar{y}=y\sqrt{B}$).

These orbits may be stable or unstable. They show a rich structure of phase space that is difficult to explore, because each orbit has a different, and rather high, multiplicity. However the fact that the perpendicular intersections show such a nice order indicates that the structure of phase space is well organized.

The complex unstable orbits, that we have just described, are an important feature of 3-D systems, that does not appear in the case of 2 degrees of freedom. In this case we have two unstable and two stable asymptotic curves. The unstable asymptotic curves, unlike in the case of two degrees of freedom, do not make infinite oscillations close to the invariant point representing the periodic orbit, but go away from the plane defined by the unstable (U) and stable (S) directions of the corresponding 2-D problem, along the other unstable direction ($\bar{U}$) (Fig. 17). Thus after a finite number of oscillations close to the curve S the asymptotic curve U continues by making oscillations close to the asymptotic curve $\bar{U}$.

The curve U does not, in general, intersect the curve S, thus it does not have homoclinic points. However there is a whole 2-D unstable *surface* tangent to the 2-D plane def-

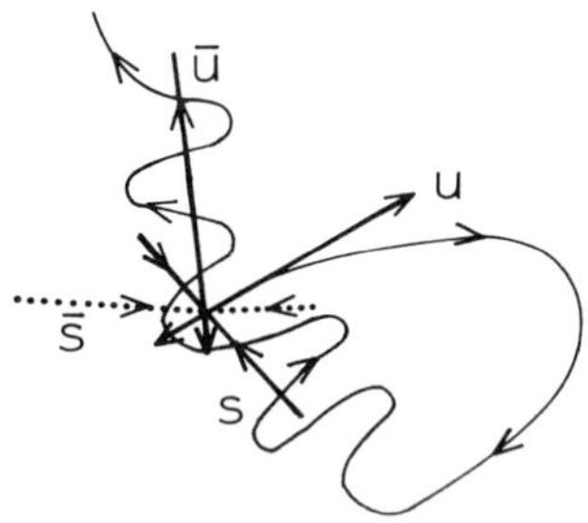

Figure 17. Oscillations of the (unstable) asymptotic curve U. While in the case of 2 degrees of freedom such an asymptotic curve makes infinite oscillations, intersecting the stable asymptotic curve S at infinite homoclinic points, in the case of 3 degrees of freedom the curve U makes a finite number of oscillations close to S, but then continues with oscillations close to the other unstable direction $\bar{U}$.

ined by U and $\bar{U}$. The unstable multiplicity $(U,\bar{U})$ intersects the stable multiplicity $(S,\bar{S})$ along isolated homoclinic *points* in general and not along homoclinic *lines*. This is due to the fact that the Poincaré surface of section is then 4-dimensional.

A further study of this problem is in progress.

ACKNOWLEDGEMENT

This research was supported in part by the Greek Ministry of Industry, Energy and Technology, under contract No. 9013502.

REFERENCES

Barbanis, B., 1993, Celest.Mech.Dyn.Astron.55:87.

Contopoulos, G., 1966, in "Les Nouvelles Methodes de la Dynamique Stellaire", F. Nahon and M. Henon eds., CNRS, Paris ≡ Bull.Astron. Ser.3, 2: 223.

Contopoulos, G.,1971, Astron. J. 76:147.

Contopoulos, G., 1990a, Astron.Astrophys. 231:41.

Contopoulos, G., 1990b, Proc.R.Soc.London A 431:183.

Contopoulos, G and Barbanis, B., 1989, Astron. Astrophys. 222:329.

Contopoulos, G. and Magnenat, P., 1985, Celest.Mech. 37:387.

Contopoulos, G. and Kaufmann, D., 1992, Astron. Astrophys. 253:379.

Contopoulos, G. and Polymilis, C.,1993, Phys.Rev.E 46:1546.

Contopoulos, G., Kandrup, H.E. and Kaufmann, D.,1993, Physica D 64:310.

Chirikov, B.V., 1969, Phys.Rep. 52:263.

Churchill, R.C. Pecelli, G. and Rod, D.L.,1975, J.Diff.Equ. 17:329.

Churchill, R.C., Pecelli, G. and Rod, D.L., 1979 in "Stochastic Behavior in Classical and Quantum Quantum Hamiltonian Systems", G. Casati, and J. Ford, eds., Springer Verlag, Heidelberg, p. 76.

Heggie, D.C.,1985, Celest.Mech. 35:357.

Henon, M., 1988, Physica D 33:132.

Jung, C. and Scholz, H.Z., 1987, J.Phys.A 20:3607.

Jung, C. and Scholz, H.Z., 1988, J.Phys.A 21:2301.

Laskar, J., Froeschle, C. and Celetti, A.,1992, Physica D 56:253.

Newhouse, S.E., 1977, Am.J.Math. 99:1061.

Newhouse, S.E.,1983, in "Chaotic Behaviour of Deterministic Systems", G.Iooss, R.H.G. Helleman and R. Stora, eds. North Holland, Amsterdam, p. 443.

Papadaki, H., Contopoulos, G. and Polymilis, C., 1994, in Roy, A.E. (ed). "From Newton to Chaos: Modern Techniques for Understanding and Coping with Chaos in N-Body Dynamical Systems", Plenum Press, N. York (in press).

Rod, D.L.,1973, J.Diff.Equ. 14:129.

Rosenbluth, M.N., Sagdeev,R.A., Taylor, J.B. and Zaslavskii, M.,1966, Nucl.Fusion 6:297.

Shirts, R.B. and Reinhardt, W.P.,1982, J.Chem.Phys. 77: 5204.

Sohos, G., Bountis, T. and Polymilis, H.,1989, Nuovo Cim. 104B:339.

FROM INTEGRABILITY TO CHAOS : EXAMPLES OF INTERRELATIONS BETWEEN PHYSICS AND DYNAMICS FOR MINOR BODIES IN THE SOLAR SYSTEM

C. Froeschlé [1], Ph. Bendjoya[1], and A. Cellino[2]

[1] Observatoire de la Côte d'Azur - B.P. 229
06304 Nice Cedex 4 - France
[2] Osservatorio Astronomico di Torino - Italy

ABSTRACT

We present some examples of dynamical problems in the solar system with strong physical implications and for which regular and chaotic behaviours are fundamental.

Due to their relatively small size, asteroids are mostly primordial (i.e. they have not been affected by geological processes) and therefore can give us essential information on the physical properties of the primitive matter in the solar system. Among asteroids, those belonging to the so called families (i.e. fragments of a parent body, with low enough ejection velocities to produce initially swarms of objects having similar orbital elements) give us a unique opportunity to get information about the interior of asteroidal bodies. The problem has two faces. First to obtain, under the assumption of quasi regularity, quasi first uniform integrals of the motion : the so called proper elements, and second, to use objective statistical methods of cluster analysis. Recently, decisive improvements came from the application of high-order perturbation theories in computing the proper elements (Williams, Yuasa, Knežević and Milani) and from the use of reliable cluster analysis methods (Zappalà et al. 1990, Bendjoya et al. 1991).

On the other hand, chaotic routes are essential to bring either asteroidal material or comets in the vicinity of Earth. In order to study the chaotic behaviour of asteroidal and cometary orbits, modelling becomes an aim and a tool. Monte Carlo mappings and Markov processes are the main tools which will be discussed in this paper as well as the strength and weakness of these methods versus the origin of chaos.

1. ORDER AND MEMORY : AN ASTEROID FAMILY STORY

1.1. INTRODUCTION

The existence of the asteroid dynamical families has been known since 1918. They were discovered by Hirayama, who first noted that some groupings are obvious in the space of the asteroid orbital elements a, e, i. He called these clusterings "*families*", and dedicated to this subject a series of papers from 1918 to 1933. The best known examples of families discovered by Hyrayama are the famous Eos, Themis and Koronis families.

The interest about asteroid families is a consequence of the improvement of our knowledge of the overall process of collisional evolution of the asteroid belt. Indeed, the most natural explanation of the existence of the families is that they are the products of catastrophic impacts between pairs of asteroidal bodies. In other words, the present members of asteroid families should be fragments of single asteroids destroyed by impacts. They were ejected with comparatively low relative velocities, so that their orbital elements did not change too much with respect to each other and to the original parent body. This gave origin to the presently observed clusterings in the space of orbital elements. It is obvious that, in this scenario, asteroid families have a crucial importance: their number can be considered as an important constraint for our models of asteroid collisional evolution, while the overall physical properties of their members can give us important information on the processes of catastrophic breakups of asteroidal sized bodies, as well as on the internal composition of asteroids.

For these reasons, it is not surprising that various attempts to unambiguously identify families have been carried out for several years. As was already recognized by Hirayama, the gravitational perturbations due to the planets, mainly Jupiter, cause time variations of the orbital osculating elements of asteroids, hence these elements are not well suited for the purposes of families identification. However, the motions look quasi periodic, reflecting "order" and therefore the search for quasi first integrals, called proper elements (a', e', i'), is natural. Obviously, these quantities which possibly change but extremely slowly with time (in the spirit of Nekhoroshev theory) provide a set of elements which, in principle, could be considered as constant and keep memory of initial conditions. For this reason proper elements have been computed and analyzed by the authors who have studied the problem of asteroid family identification during the last decades. Among them, and after the work of Hirayama, the most important studies have been published by Brouwer (1951), Arnold (1969), Lindblad and Southworth (1971), Carusi and Massaro (1978), Williams (1979), Kozai (1979).

In spite of the noticeable efforts produced by these authors, the situation concerning asteroid families has been far from satisfactory until very recently. This is due to the fact that a general agreement between the various proposed family lists has not been achieved neither for what concerns the number nor the membership of these families. Only in a few cases, namely the three major families of Eos, Themis and Koronis a good agreement has been found, but this is not surprising, since these clusterings can be very easily identified in both the $a' - i'$ and $a' - e'$ planes.

The discrepancies existing among the quoted classifications are due to various reasons. In particular, they are a consequence of: (1) the different data bases considered by the different authors; (2) the different methods used to compute the orbital proper elements; (3) the different identification procedures used.

In this paper we do not deal in detail with issues like the theory of asteroid proper elements, the properties of the several clustering identification methods, or the precise implications of asteroid family studies to the problem of asteroid collisional evolution. Such arguments are treated in detail by Milani et al. (1991), to which the reader will be often addressed to in the following. In the present chapter, after an overview of the different methods used to obtain proper elements, we mostly describe the two methods of asteroid families identification which have been most recently applied, namely the Hierarchical Clustering Method (HCM) and the Wavelet Analysis Method (WAM); we will show their respective results, in order to stress their general agreement as well as the problem which are presently open, and deserve further investigation. In particular, Section 1.3 will be devoted to a general description of the HCM, and its results, while Section 1.4 will present the WAM and the families identified in this way. Finally, Section 1.5 will be devoted to a comparison of the results obtained by means of these two methods, and to a general discussion of the open problems and future work.

1.2 THE PROBLEMATIC OF PROPER ELEMENTS

Even looking at the distribution of osculating elements of asteroidal orbits, some obvious groupings which could result from the fragmentation of single parent bodies are evident. To remove the time variations, due to the planets, of the osculating elements, i.e. to search invariants of the motion, called proper elements, becomes obviously one of the important parts of the dynamical analysis concerning the asteroidal families. Let us assume that we are able to calculate for each asteroid three invariants of the motion, depending only on the initial conditions, and staying constant for very long time. We can then imagine that planetesimals with very close proper elements could have been, in the past, physically very close to each other and perhaps fragments of the same parent body.

However, besides the calculation of the proper elements, there remains the problem of the interpretation of the results (when are proper elements "close enough" to be candidates to a same parent ?) and of the understanding of the fragmentation mechanism (are the fragments coming from a unique catastrophic event or are they secondary fragments of successive impacts ?)

Using a linear perturbation theory Hirayama in 1918 has computed the first proper elements. Then Yuasa (1973) developed by hand a theory of secular motion up to degree 4 in eccentricity and inclination and order 2 in the mass of any perturbing planet. Later Knežević (1988, 1989) checked and completed (addition of the indirect part) the Yuasa's work. Then Milani and Knežević (1990) revisited the problem using a Lie formalism and adding an iterative procedure on the final calculation. On the other hand, in a different spirit Williams (1969) developed an alternative theory of asteroid secular motion, and in 1979 he published a list of asteroid proper elements; these were computed without expansion in eccentricity and inclination of the asteroid, and therefore should be particularly accurate for highly inclined or eccentric asteroidal orbits.

In conclusion, nowadays two different main methods for the computation of proper elements exist : Milani and Knežević method and Williams' method. Actually, a third one mainly in the spirit of Williams' method, but more elaborated, has been carried out by A. Lemaître and A. Morbidelli (1993).

In the last part of this paragraph we will closely follow the paper of Lemaître and Morbidelli (1993) to introduce the different methods for computing proper elements.

Let us analyze these methods to find out their limits and advantages.

First of all, we recall that the proper elements are the results of two averaging processes, the first one over the short periodic terms (to get the mean elements), the second one over the long periodic terms (to get the proper elements), on an initial three or many body problem. In between those two operations, before the elimination of the long periods, there is an important rearrangement of the terms. Theories may differ on either the adopted Hamiltonian for the model (i.e the number of terms taken into account) or on the perturbation method used to find out the approximate constants of motion. In other words, one has to decide, first, what must be considered as an "integrable approximation" and what as a "perturbation", and, later, which perturbation technique to adopt in order to take into account the effects produced by these "perturbations".

Milani and Knežević's method is a very accurate one for asteroids with small inclination and eccentricity. This is due to two different reasons. The first one is obvious and concerns the model : their program works on a Hamiltonian which is truncated at degree 4 in eccentricity, e, and inclination, i, which is a good approximation provided only that e and i stay small during all the secular evolution of the asteroid orbit. The second reason is that they consider e and i to be of the same order of magnitude as the eccentricities and inclinations of the main planets for what concerns the order of magnitude (which is perfectly compatible with the truncation hypothesis); so, what they choose as an integrable approximation is a quadratic form in e, i, e_p, i_p, the "$_p$" variables referring to the perturbing planet(s) (i.e. the Hamiltonian of the linear theory of Laplace). Therefore, the term that depends on the argument of perihelion, namely $e^2 i^2 \cos(2\omega)$ is considered as a "perturbation".

This choice is important because, as pointed out first by Kozai (1962), at high inclination, the dynamic associated with the motion of the argument of perihelion is the most relevant one, and a resonance, namely a region where ω librates, exists. By consequence, any method that pretends to consider the term $e^2 i^2 \cos(2\omega)$ as a perturbation, must fail at high inclinations.

Let us see now the assumptions of Williams' theory, which are very different. First, no expansion in eccentricity or inclination of the asteroid is used; conversely, Williams uses a linear expansion in e_p and i_p. Therefore, one implicitly considers the asteroidal eccentricity or inclination to be much larger than those of the planets. Secondly, always following the same idea, the division between integrable part and perturbation is made in a completely different way : any term proportional to e_p or i_p is considered as a perturbation. The consequence of this choice is that the integrable part is a function of the argument of the perihelion and it is precisely the one studied by Kozai. Unfortunately, Williams did not include the second order in the perturbing masses in his work, and he calculated the proper elements directly on Lagrange equations, without the help of the Hamiltonian formalism and canonical transformations.

The method of Lemaître and Morbidelli for the computation of proper elements is very close to that of Williams. They make the same assumptions about the non expanded Hamiltonian and they separate integrable and perturbation parts by the same parameter. However, there are differences in the following features:

- They have included the second order in the perturbing masses originated by the first averaging; the role of the quadratic terms (in mass) is relevant close to the main mean motion resonances; in particular, it may significantly change the proper frequencies of the asteroid and, consequently, its relative position with respect to secular resonances. In the paper by Morbidelli and Henrard (1991), for example, the authors showed that the secular resonance of second order which affects the motion of the Eos family members cannot be found in the quadratic term if the masses are neglected.

- They have used more accurate values for the amplitudes, the frequencies and the phases of the secular planetary motion. Indeed, Williams' work is based on the values given by Brouwer (1950), whereas their program uses the values computed numerically by Nobili et al. (1989) in the framework of the LONGSTOP project.
- Their perturbation method is written in a Hamiltonian formalism, using Arnold action angle canonical variables.
- They have averaged over the long periods by another semi numerical method.
- They have implemented, following Milani and Knežević, an iterative procedure on the calculation of the proper elements. Actually, the computation from mean elements to proper elements involves an approximation of the proper frequencies which are computed only at the end of the whole computation. Hence the process is iterated as for the search of fixed points of the transformation.

In the same way as Williams' method, their procedure is particularly suited for high eccentric and, still more important, for high inclined orbits. Unfortunately the consideration of the quadratic terms in e_p, i_p is beyond the limits of their method.

It is obvious that the advantages of the method vanish for small inclinations and eccentricities. For these orbits Milani and Knežević's method turns out to be preferable for two reasons : it is much faster, and it takes into account the perturbations of 4 planets up to degree 2 in e_p, i_p. At this time (June 93) Lemaître and Morbidelli have not yet provided proper elements, therefore no statistical analysis has been performed.

A great deal has been achieved during the last few years for what concerns the determination and the reliability of the proper elements. Let's now switch to questions beside the field of celestial mechanics : do the structures within the proper elements distribution reflect only a pure dynamical evolution or also some physical processes? To answer this question one has to quantify the level of significance of the groupings versus chance fluctuations. Recently two methods of cluster analysis have given very similar results although their approaches are different and complementary. The more classical one, known as the single linkage hierarchical clustering method and the more recent one, based on the wavelet transform, are described in Zappalà *et al.* 1990, Bendjoya *et al. 1991* and Bendjoya *et al.* 1993. We only recall in the following sections the main characteristics of these two methods.

1.3. THE HIERARCHICAL CLUSTERING METHOD

The so-called Hierarchical Clustering method for the identification of asteroid families has been recently applied by Zappalà et al. (1990). This method is based on well known techniques of Multivariate data Analysis, which are explained in detail, for example, in Murtagh and Heck (1987).

The basic idea of this method is to introduce a definition of distance in the space of orbital proper elements a', e', i'. When this is done, one can compute all the mutual distances $d(i,\ j)$ between the N objects of the considered data set. The two objects, i and j , having the smallest mutual distance are identified and an "agglomeration" is then performed, so that the two objects i and j are replaced by one new object $i+j$. All the mutual distances of the resulting $N-1$ elements are computed again, with the assumption that the distance between the new $(i,\ j)$ object and any other object, k, is taken as the minimum between $d(i, k)$ and $d(j, k)$; the couple of objects having the minimum mutual distance are again identified and agglomerated as in the previous step, and this process continues as long as only two objects remain, which are in fact two big agglomerations of the N elements of the original data set. In this way, a *dendrogram* is created, which contains all necessary information for family identification purposes, since it gives immediately, for any given value d' of the distance, all the existing agglomerations of objects, whose mutual distances are less than d'. If one knows a threshold value of the distance, $\overline{d}$, for which no clustering can exist having a significant probability to be due to chance, then all the clusterings found at the $\overline{d}$ level can be reasonably called "families" . One must take into account that an eventual true collisional origin of these clusterings can be assessed only on the basis of further investigations concerning their overall physical properties (mass and velocity distributions, taxonomic types, etc ...). A typical shape of a dendrogram is a succession of stalactites giving the number of groupings and their abundance as a function of the distance. Usually, at larger distances only a few big agglomerations exist, containing most of the objects of the sample. The deepest stalactites are, in general, families, since they correspond to intrinsically denser groupings composed of objects mutually closer in the space of proper elements.

As can be seen, the procedure described above is in principle very easy, but contains two critical points: the definition of the metric in the space of proper elements, and the derivation of the threshold value $\overline{d}$ of the distance. For what concerns the first point, if we accept the idea of a collisional origin for the families, it is straightforward to introduce a definition of distance d in the space of the proper elements, in such a way that this distance can be put in relation with the ejection velocities of the fragments in an event of catastrophic disruption. In particular, indicating with δv_1, δv_2, δv_3 the three components of the velocity differences between two fragments of a collisional event along the directions which are, respectively, parallel, normal and binormal with respect to the direction of the orbital motion at the instant of breakup, the Gauss equations give the relation of δv_1, δv_2, δv_3, with the differences in the orbital elements δa, δe, δi. Without entering into details which are extensively explained in both Milani et al.(1991), and Zappalà et al. (1990), a "natural" choice of the metric in the space of proper elements has been found to be given by

$$d = \sqrt{k_1(\frac{\delta a}{a})^2 + k_2(\delta e)^2 + k_3(\delta i)^2} \qquad (1)$$

in which the three coefficients k_1, k_2, k_3, are given by 5/4, 2, 2, respectively. It has to be noticed that the distance defined in this way is dimensionally a velocity, and can be expressed in m/sec. The papers quoted above contain also exhaustive description of the procedure adopted to derive the value of the critical level of distance, $\overline{d}$, for which the existing clusterings can in general be assumed to be families. The idea is to create a fictitious population of objects randomly generated, but with the constraint to globally fit the a', e', i' distributions of the real population to be analyzed. The Hierarchical Clustering algorithm is then applied to this quasi-random population, and the critical level $\overline{d}$ is assumed to be the deepest value of d for which it is still possible to find one or more groupings of the fictitious population composed by at least 5 members. Having determined $\overline{d}$ in this way , the "families" are assumed to be all the clusterings of the real population having at least 5 members at distance levels $d \leq \overline{d}$, or having more than twice the number of members of the largest existing clustering of the fictitious population at the $\overline{d}$ level. In this way, not only the probability to find clusterings at low levels of distance, but also the probability to find noticeably large clusterings at the critical distance level is taken into account.

The method for families identification described above is fast and simple in principle, but in computer form it requires the computation of quite large matrices, whose elements are the mutual distances between each pair of objects belonging to the sample to be analyzed. For this reason, and also on the basis of physical considerations involving possible variations of the properties of different parts of the asteroid belt, due to both the presence of mean motion and secular resonances, and to different completeness of our asteroid sample at different heliocentric distances, the Hierarchical Clustering method has been applied separately to asteroids belonging to six different zones of the belt. These zones are delimited by the main mean motion resonances with Jupiter.

The sample of asteroids analyzed by Zappalà et al. (1990) has been the 4.2 version of the list of asteroid proper elements provided by Milani and Knežević (1990). Such a data set, including about 4100 asteroids, was the largest list ever used for families identification purposes; the same data set has been later used by Bendjoya et al. (1991), in their application of the Wavelet Analysis Method which is described in Section 1.4.

1.4 THE WAVELET ANALYSIS METHOD

The wavelet cluster analysis is a new method, able to extract from a set of points, structures with a degree of significance versus chance. It is a hierarchical method in the sense that the wavelet transform allows to point out the structures the size of which corresponds to the studied scale (a characteristic given size). By using a set of increasing scales, this analyzing method permits to obtain significant structures from the smallest ones to the biggest ones fitted into each other just as the well-known russian dolls. This method of analysis lies on the wavelet transform, a mathematical tool which has already given very interesting results in different domains such as: the signal restauration and filtering, the fractal behavior of a signal, the compression of information and the cluster analysis, which is a peculiar application of the wavelet transform due to the peculiar studied signal, since in this case it can be modelled by a set of points or, more precisely, a set of Dirac functions. The main advantage of the wavelet transform is that of being able to give information not only on the location of an event in a signal but also on the size of this event. The wavelet transform gives information both in time and frequencies for a time-signal or both in space and size for a space-signal.

Let $s(x)$ represent a one-dimension space-signal in $L^2(\Re)$. The wavelet transform of this signal is a two dimension function $S(\sigma, b)$, where σ is a strictly positive scale-variable (size) and b a space variable (location). $S(\sigma, b)$ is the decomposition of $s(x)$ onto a base of functions obtained from the dilations and the translations of a unique function $\psi(x)$ of $L^2(\Re)$.

$$S(\sigma, b) = \sigma^{-\frac{1}{2}} \int_{-\infty}^{+\infty} s(x)\psi^*(\frac{x-b}{\sigma})dx \tag{2}$$

$\psi(x)$ is called the analyzing wavelet. $\sigma^{-\frac{1}{2}}$ is a $L^2(\Re)$ normalization factor and * denotes the complex conjugate.

The wavelet transform acts like a mathematical zoom and $S(\sigma, b)$ can be seen as the σ-sized details of $s(x)$ around the location b. By varying the scale variable σ, thinner and thinner details are extracted from the signal when σ decreases. In reality the wavelet transform allows to consider a signal from a new point of view consisting to consider the signal itself as the sum of all its own details at the different scales.

In order to be a suitable analyzing wavelet, the function $\psi(x)$ must satisfy the admissibility condition:

$$\text{i.e.} \qquad \int_{-\infty}^{+\infty} \psi(x)dx = 0 \tag{3}$$

Also, $\psi(x)$ must be smooth enough and compact supported in order to give informations on the location.

These criteria make of the mexican hat function a good candidate for being an analyzing wavelet. This function comes from the second derivative function of a Gaussian function:

$$\psi(r) = (n - \frac{r^2}{\sigma^2}) \exp(-\frac{r^2}{\sigma^2}) \tag{4}$$

n is the dimension of the space in which the signal is defined and r is the distance defined from the metric chosen in the considered space.

This function fulfills condition (4) and can be considered practically equal to zero for $|r| > 5\sigma$.
The wavelet Analysis method has been applied to the issue of asteroid families identification by Bendjoya et al. (1991). In order to identify asteroid families, the wavelet analysis is carried out in the proper elements space, for the reasons explained in Section 1. Each asteroid is represented by a point in the (a', e', i') space (proper semi major axis, proper eccentricity, proper inclination). The signal to be studied is hence a set of Dirac functions $\delta(\vec{x} - \vec{\eta}_j)$, $\vec{\eta}_j$ is the position of the jth point. In computer form, the wavelet transform is derived in a discrete way by computing the wavelet coefficients of the signal on each node of a frame superimposed to it.

Let us consider a plane of data points over which a network is superimposed. The wavelet coefficient for the node $\vec{b}_i$ is obtained by substituting in eq. 2 $s(x)$ by the set of Dirac functions $\delta(\vec{x} - \vec{\eta}_j)$:

$$C(\sigma, \vec{b}_i) = \sum_{j=1}^{N} \psi(\frac{\vec{\eta}_j - \vec{b}_i}{\sigma}) \tag{5}$$

where$N$ is the number of data points. $C(\sigma, \vec{b}_i)$ is the wavelet coefficient at the node $\vec{b}_i$ for the scale σ. $C(\sigma, \vec{b}_i)$ is in fact the sum of the weights w_j of the data points.

Considering the mexican hat function we obtain:

$$C(\sigma, \vec{b}_i) = \sum_{j=1}^{N} w_j$$

with:

$$w_j = (n - \frac{r_{ij}^2}{\sigma^2})\exp(-\frac{r_{ij}^2}{2\sigma^2}) \tag{6}$$

r_{ij} being the distance between the ith node $\vec{b}_i$ and the jth point $\vec{\eta}_j$. This distance is expressed on the basis of the metric fitted to the problem. Because of the admissibility condition eq.(3) it is easy to understand that a node centered on a uniformly populated area will have a near zero wavelet coefficient since the positive weights will be compensated by the negative ones, provided that the size of this area is big when compared to the studied scale. On the other hand a studied-scale-sized structure (resp. hole), will give wavelet coefficient strongly positive (resp. negative). In this way, a map of wavelet coefficients is built, in which all the positive coefficients are indicators of structures.

The problem is to distinguish the physical structures from the structures due to chance. In order to do this, the idea is to build a "pseudo-random" distribution in the sense that the local density is conserved. Being given a certain number of boxes in the real distribution, the pseudo-random one is built by putting at random the same number of points in each corresponding box. Then the wavelet coefficient maps are computed for both distributions. The histogram of the wavelet coefficients of the pseudo-random distribution is derived in order to determine the value of the coefficient for which 99.5%, for example, of the other coefficients have a lower value than this threshold. It means that there are less than 5 chance over 1000 to find a higher value than this one in the pseudo-random wavelet coefficient map. By reporting this threshold in the real distribution wavelet coefficient map and by retaining all the coefficients higher than this value, it is easy to understand that the high value of these retained coefficients has less than 5 over 1000 risks to be due to chance. Consequently, the structures associated to these kept coefficients have less than 5 over 1000 risks to be due to chance. This way of proceeding shows how it is possible to give a degree of significance versus chance to the detected structures. From the retained coefficients it is then possible to label all the asteroids whose weights have been taken into account in the computation of this coefficient and a significant structure can be so evidenced. By applying this procedure to a set of different scales with the suited networks, we get a hierarchy of structures with a degree of significance versus chance.

The two next steps are the choice of a metric well fitted to the considered problem and the way to cut the hierarchy in order to pass from structures to families.

The former problem is the same already encountered in the framework of the Hierarchical Clustering method; the solution adopted is the same, so the definition of distance in the space of orbital proper elements is still given by equation (1) introduced in Section 1.3. This metric has been adopted in the definition of the mexican hat function (eq. 4) and the analysis has been performed for the set of scale σ: 60, 80, 110, 160, 230, 320, 450, 640, 900 m/s. The ratio close to $\sqrt{2}$ between two successive scales is a condition imposed when using the mexican hat as analyzing wavelet, in order to loose the least of information between two successive scales (see Daubechi 1990). Significant structures are evidenced at each scale.

On the other hand, it is easy to understand that a criterion of cutting the hierarchy has to be chosen, since different structures can merge into a unique one for a large enough scale. The chosen criterion for passing from the structure to families, lies on the philosophy of the wavelet transform that is the additional information got between two successive scales. The largest scale to consider in the studied area in order to distinguish families is the one for which the number of asteroids added at this scale to the already identified structures, is either 0 or increases drastically. The first case indicates that the structures are well isolated and the second one indicates that the next scale is too large and bridges are made between structures which merge into a big cluster.

1.5. DISCUSSION

The two methods give results in a very good agreement for what concerns both the number of families found in the different zones of the belt and also, for each family, the percentage of common objects, i.e., the asteroids which have been found to belong to the same family on the basis of both WAM and HCM. Both methods identify, from a data set of 4100 asteroids, about twenty families with a high quantified level of significance versus chance fluctuations. The agreement achieved by these two methods gives confidence on the reliability of the resulting asteroid families. Of course, these families have been defined only in a dynamical and statistical sense, thus it is clear that a deeper study on the taxonomy (cosmochemical classification) of the members of these supposed families is obviously necessary. This will allow us to check whether these clusterings have cosmochemical compatible members , i.e., they are not in conflict with what we know about asteroid composition and physics of catastrophic breakups.

The next step of this analysis has consisted in a check of the reliability of both the HCM and the WAM by means of suitable numerical simulations. The idea has been to create some "artificial families" located in different "artificial background". The aim was to analyze the efficiency of the two methods in various situations completely known *a priori* and reproducing some of the most common situations which can be encountered in the real belt (Bendjoya *et al.* 1993). The results seem to indicate that the methods described in the present paper can be considered sufficiently efficient for allowing us to believe that a significant improvement has been now achieved on the subject of asteroid families identification. Moreover, the fine tunings developed from these simulations have led to improvements in the procedures which have been applied in a new classification, based on a new data set of about 6500 asteroids, performed by means of both methods (Zappalà et al. 1993, Bendjoya 1993). The data rely on an improved proper element set (Milani and Knezevic 1992). Once again a good agreement between both methods is pointed out and an optimized list of families is actually to be proposed.

The overall physical properties of the families identified in the asteroid belt can now be analyzed and can give us important information on the collisional history of the belt itself and on the kinds of outcomes that are formed as a consequence of catastrophic breakups of asteroidal bodies.

If the search of order is at the basis of the computation of proper elements and therefore, of the identification of families, the existence of chaos in some regions of the phase space is responsible of different phenomena. For example, chaos allows comets, coming either from the Oort cloud or from the Kuiper belt, to become observable. The same scenario occurs for meteorites which follow chaotic routes. Mappings are, so far, the main tools to study and model such processes. We will present in the following sections some basic recipes to "cook" such mappings.

2. STOCHASTIC MAPPINGS IN ASTRODYNAMIC

2.1. INTRODUCTION

Both the existence of Kirkwood gaps and the transfer of comets into observable orbits have been the main motivations for building mappings since they have many advantages over numerical integrations, mainly for what concerns computing time and accuracy. However, even if the brutal force of super computer may in the near future solve these problems, building mappings will still be a challenge to researchers in Celestial Mechanics. Indeed, building mappings brings a deeper understanding of the general behaviour of non linear dynamical systems. In the particular case of mappings devised to study cometary motion, the origin of chaos is the key point.

2.2. EXOGENOUS STOCHASTIC MAPPINGS

2.2.1. Monte Carlo Mapping of long period comets dynamics. While it has recently been realized that an important part of the dynamics of Oort cloud comets arises from regular motion in the Galactic tidal field (Heisler and Tremaine 1986), a decisive role is nevertheless played by individual stellar encounters. Being given the physical parameters of such an encounter, its effect on the cometary motion is fully determined. However, the parameters of individual stellar encounters are unpredictable, thus stellar perturbations impose a stochastic variation on the cometary orbits.

In Monte Carlo simulations of stellar perturbations (see, e.g., Weissman 1982; Remy and Mignard 1985) the dynamical evolution of a cloud of comets is studied as follows. At a given starting epoch, each comet is initialized by choosing a set of orbital elements. These are perturbed by the gravitational effect of passing stars. The geometrical parameters of the stellar encounters are chosen at random. During the passage of a star, a comet receives a heliocentric impulse through the interaction of the star with the comet and the Sun. This induces a change in the cometary orbital elements, so these are updated and the comet moves along a new Keplerian ellipse until the next encounter with a random star. In Weissman's procedure the impulses are taken in an even more simplified way from a pre-determined distribution. In other words, a stochastic mapping is iterated where the perturbations caused by random stars impose a stochastic process on the cometary orbital elements, which, therefore, undergo a random walk. It is obvious that all orbits are chaotic and correspondingly the largest LCE is strictly positive. The stochasticity is exogenous since the stellar encounters occur at random.

2.3 ENDOGENOUS STOCHASTIC MAPPINGS

2.3.1. Monte Carlo mapping of short-period comet dynamics. As comets are captured into short-period orbits, i.e. with orbital period of the same order as those of the perturbing planets, the situation changes in a fundamental way. We now have to deal with only the intrinsic stochasticity of a dynamical system which can usually be approximated by a three-body problem: Sun-planet-comet (Rickman and Froeschlé 1988). This stochasticity derives mainly from close encounters with the planet, and during the intervals between such encounters the cometary motion is quasi-regular and predictable. In more precise terms, this means that the motion is stochastic only over time scales longer than the typical interval between encounters with the perturbing planet. Thus the phase space domain of short-period comets presents chaotic and ordered regions in an intricate mixture.

Stochastic modelling of this motion is justified by the shadowing principle only if the time step is chosen long enough. Thus, we may consider the following procedure (Rickman and Vaghi 1976; Froeschlé and Rickman 1980): we pick a large number of initial cometary orbits at random into boxes of the (Q, q) plane (Q aphelion and q perihelion distances and integrate these with Jupiter moving with an elliptic orbit as the perturbing planet over a time Δt, which was chosen to be one unperturbed period. From the resulting set of perturbations, we construct stochastic orbital evolutions by picking independently up, a random perturbation, within a corresponding box, for each successive interval Δt. For these evolutions, in order to simulate real ones, we would have a constraint on the choice of Δt so that a minimum time step for the random walk could in principle be defined. However, there are practical problems of such a procedure which are not easy to solve. These are related to the question of the definition of the perturbation samples: how should one choose the "boxes" of phase space where the initial conditions are to be picked up ? As already mentioned, the dimension of the boxes can be reduced by symmetry arguments and in the present case a critical point is that short-period comet orbits are known to be generally of low inclination. One is therefore close to a planar problem with a four-dimensional phase space. Within the limits chosen for the inclination, the choice of inclination and orientation of the nodal line is statistically immaterial. Furthermore, since we are dealing both with Jupiter-crossing or Jupiter-tangent orbits and Jupiter's eccentricity is quite small, there is an approximately circular symmetry such that encounters with the planet can occur with equal probability, independently of the orientation of the apsidal line. The choice of the latter is hence also immaterial, and obviously the time-related parameter expressing Jupiter's position, at the time of the cometary perihelion passage, is the real stochastic variable of the problem which should be taken at random with its true probability distribution. We are left with two orbital parameters which can be taken as, e.g., semi-major axis and eccentricity $(a,\ e)$, or aphelion and perihelion distances $(Q,\ q)$.

As already noticed, boxes in the $(Q,\ q)$-plane were considered in the Monte Carlo simulations mentioned above. Let us note that a further reduction of the dimensions appears feasible. The Tisserand criterion for a co-planar cometary orbit :

$$T = \frac{2aJ}{Q+q} + 2\sqrt{\frac{2q\ q}{aJ(Q+q)}} = constant$$

might be used to restrict the random walk $(\Delta Q,\ \Delta q)$ to one-dimensional curves, and the Monte Carlo simulation would then consist of a set of independent, parallel simulations for the different values of T. In each of these one would consider a suitable orbital parameter varying along the curve, such as the inverse semi-major axis $z = 1/a$, and there would be a random walk with step size distribution $f_i(\Delta z)$ computed for interval $[z_{i-1},\ z_i]$ along the z axis. However, although small (Froeschlé and Rickman 1981), the perturbations ΔT in the elliptic restricted problem are indeed important for the outcome of the low-velocity encounters with Jupiter occurring in low-eccentricity planet-tangent orbits, and those encounters appear essential for the dynamical transfer of comets.

Quite obviously the number of $(Q,\ q)$-boxes is limited in practice by the requirement to compute a sample of perturbations large enough to give a fair representation of the dynamics of the region in question. E.g., pronounced non-Gaussian tails are known to exist in the perturbation distributions and the sample must extent far enough into these tails to cover their significant parts. This means that the number of orbits to be integrated in each box may be very large [for a possible way to reduce this problem, see Rickman and Froeschlé (1983)]. At this time it was difficult to go much beyond the dynamical resolution of the 30 boxes. On the other hand, this already guarantees a certain "dynamical homogeneity" in the sense that the perturbation distribution over a short interval of time (one orbital period) should not vary too much from one side of a box to the opposite one.

However, within all the boxes we can expect to find smaller regions corresponding to "resonant strips" with $a \approx a_{res}$, where the comets are close to a simple mean motion resonance with Jupiter. Within such a region there may be slow circulation of the critical argument, in which case it is extremely difficult to define a mean interval between encounters and to find an integration interval Δt which is everywhere sufficient. But actually the situation is even worse, because there also exist integrable regions of phase space corresponding to stable librations of the critical argument, where encounters with Jupiter never occur.

The main result of a Monte Carlo simulation of cometary dynamics is a picture of the distribution of comets over the various orbits connected by the stochastic transfer process in question. If the short-period comets are viewed in their most general framework, the dynamics includes perturbations by all the planets, as well as non gravitational perturbations, and it would then be interesting to estimate the number of comets trapped in such quasi-stable resonant regions. How this goal would be achieved is not yet clear: the problem is to find an appropriate definition of the perturbation sample boxes and corresponding integration intervals for such a detailed investigation. If, on the other hand, one considers the dynamical transfer in the three-body problem (Sun-Jupiter-comet), the problem is instead to reach what may be called "topological homogeneity" (Froeschlé and Rickman 1988): comets should then follow only chaotic routes and the relevant boxes from which the initial conditions are to be picked are in fact the intersections of the usual $(Q,\ q)$-boxes with the chaotic part of phase space. If this restriction is not taken into account, the rate of the transfer is artificially slowed down by inclusion of irrelevant trappings or an overestimated probability of small perturbations.

2.3.2. Markov chain modelling

a) Short-period comet dynamics. If indeed we consider the orbital distribution of comets as the principal result to be obtained, we can again argue that the random walks by individual sample comets considered in the Monte Carlo simulation contain too much information: the only interesting quantity is the number of comets in each $(Q,\ q)$-box either as a function of time or in a steady state. The same perturbation samples used for the Monte Carlo simulation can then be used to calculate "jump probabilities" between the various $(Q,\ q)$-boxes over a common time interval Δt. Defining the state vector $\mathbf{n}$ of the cometary population to be the set of numbers of comets in the different boxes, and calling P_{ij} the jump probability from box i to box j, n evolves according to a Markov chain:

$$\mathbf{n}(t+\Delta t) = \mathbf{n}(t).P$$

where $P = (P_{ij})$ is the transformation matrix. This method was first used in cometary dynamics by Rickman and Froeschlé (1979) for the same domain of the $(Q,\ q)$-plane as in the above-mentioned Monte Carlo simulation and later on for Oort cloud dynamics by Lago and Casenave (1983).

This Markov method has the advantage of extreme efficiency, in particular for finding steady-state solution where we have just to solve :

$$\mathbf{n}_{ss} = \mathbf{n}_{ss}.P$$

i.e., a system of linear equations. Its main disadvantage is that, just like for the Monte Carlo simulation, the number of $(Q,\ q)$-boxes is limited by the large number of integrated orbits required to obtain accurate estimates of the jump probabilities. If e.g., the 30 boxes of Rickman and Froeschlé are considered, the information obtained on the $(Q,\ q)$ distribution is strictly limited to the 30 sample points represented in the state vector.

b) The long-term dynamical behaviour of small bodies in the Kuiper belt. Again, like Rickman and Froeschlé (1979) but from a different point of view, in order to study the slow diffusion of small bodies in the Kuiper belt, i.e. to determine the time scales of this process, Levinson (1991) has used Markov chain methods. Here we are not seeking any more steady state solutions, but estimates of diffusion times which should be of the order of the age of the solar system.

The $(Q,\ q)$ plane is also divided into small bins for which transition probabilities are estimated through sample of 100 particles in each bin, which are integrated for approximately 100 periods. In addition to the bins within the Kuiper belt, which is a transient region, two special absorbing bins act as border edges, an inner edge $q < 30$ AU where the objects become Neptune crossers and an outer edge left as a free parameter with corresponding probability $P_{ii} = 1$ and $P_{ij} = 0$ for $i \neq j$.

The probabilities matrix takes the form:

$$P = \begin{pmatrix} I & O \\ R & Q \end{pmatrix}$$

and the fundamental matrix

$$M = (M_{ij}) = (I - Q)^{-1}$$

gives the average number of time steps a particle spends in transient bin j before it is absorbed if it started in i. Then $t_i = \sum_j M_{ij}$ gives the average time a particle spends in all transient bins before absorption. Furthermore, the variance of the particle lifetime is $v = (2M - I)t - s$ where $s_i = t_i^2$. Finally $f = MR$ gives the probability that a particle starting in transient bin i enters absorbing bin j. Using these tools Levinson found that the Kuiper belt is a good candidate as a source of short-period comets. However, in addition to the already discussed possible drawbacks of the method (size number of bins, time steps length) the stochasticity underlying the Markov process is indeed very small and may only be due to undetected long-period oscillations in the behaviour of q and Q.

In the previous analysis such a low stochasticity was indeed present as an artifact of the method, but in the study concerning short-period comets dynamics a strong stochasticity due to strong interactions between the comet and Jupiter was mainly responsible for the diffusion, i.e. big-jumps occurred within Δt and therefore such an artifact could not invalidate the results.

2.4. AN OTHER APPROACH THE SYNTHETIC MAPPING

The validity of stochastic mappings is not yet unquestionable since it depends obviously on the intrinsic stochasticity of the region of the phase space under study. An other method also, purely numerical, has been used by Froeschlé and Petit (1990) who built a mapping valid everywhere in the phase space, following an idea already used by Varosi et al. (1987) but in the framework of non-Hamiltonian systems (i.e., systems where attractors do exist). The method consists of coarse-graining the phase space surface of section and then interpolating the value of the image of a point. Linear interpolation requires a rather fine graining of the phase space, hence it is necessary to compute a lot of points on the grid. However, Taylor expansions of order 3 and 5 can provide very good results as long as symmetrical interpolation formulae are applied, for which it is necessary to use an extended grid. Since there are cases where one cannot cross a given limit, asymmetrical interpolation formulae have been tested, but their accuracy was found to be inferior. Therefore Petit and Froeschlé (1991, paper II) have developed another type of interpolation, where the information, including that on the gradients, is stored to the same level of accuracy only for the nearest-neighbouring vertexes. Thus, not only images of vertices are computed, but also tangential mappings at each vertex.

There are in any case two key parameters: the number of bins in each direction N = (total number of cells)$^{1/D}$, where $D(=2$ and 3 in papers I and II) is the dimension of the surface of section, and M the order of the Taylor expansion. In order to explore the validity of the synthetic approach, we have applied our method in two cases:

1) An algebraic area-preserving mapping for which the computation of orbits is very fast. This allows one to follow a large number of orbits and to carry out enough iterations for a meaningful comparison.

2) A special case of the restricted three body problem, already studied by Duncan et al. (1989).

However, contrary to stochastic mappings when the number of degrees of freedom D is greater than 3, the method becomes tedious. We are presently investigating mixed methods.

3. CONCLUSION

If the beautiful mathematical machinery of celestial mechanics is at the basis of the asteroid family story, the chaotic routes have however close connections with the finding of proper elements. Actually, the same machinery, but used in a complementary manner, gives the location and even the size of both mean motion and secular resonances and therefore the regions of the phase space where they overlap and generate chaos. Indeed slow chaotic diffusion, inducing increases of eccentricities result of such resonant interactions and are good candidates for bringing material to the inner solar system. This material must be understood as meteorites resulting of asteroid break up or as Kuiper belt comets. Close approach with planets are at the origin of another source of chaos for which only crude models have been worked out so far. A better understanding of interactions between resonances and strong gravitational perturbations appears to be crucial for future models of chaotic routes through the solar system.

REFERENCE

Arnold, J.A., 1969, in "A. J". 38 : 1235
Bendjoya, Ph.., Slézack, E., Froeschlé, Cl. : 1991, A & A , 251 pp. 312–330
Bendjoya Ph., Cellino A., Froeschlé Cl., Zappalà V. : 1993 A & A 272, 651-670
Bendjoya Ph. : 1993 : A & A, in press
Brouwer, D., 1951, in "A. J." 56 : 9
Carusi, A., Massaro, E., 1978, in A& Suppl." 34 : 81
Daubechie, I., 1990 I.E.E.E. Trans. on Information Theory
Ducan, M., Quinn, T., Tremaine, S. : 1989, Icarus 82, 402-418

Froeschlé, C., Rickman, H : 1981, Icarus 46, 400-414
Froeschlé, C., Rickman, H. : 1988, Celest. Mech. 43, 265-284
Froeschlé. C., Petit, J.M. : 1990, A&A 238, 413-423
Hadjidemetriou, J. : 1991, Celest. Mech 56, 563-599
Heisler, J., Tremaine, S. : 1986, Icarus 65, 13-26
Hirayama, K., 1918 : "A. J." 31 : 185
Kozai, Y., 1979 in "Asteroids", T. Gehrels. Ed., pp. 334-358, University of Arizona Press, Tucson
Lago, B., Cazenave, A. : 1983, Icarus 53, 68-83
Levinson, H.F. : 1991, AJ., Vol 102, N° 2, pp 787-794
Lindblad. B.A., Southworth, R.B., 1971 in "Physical Studies of Minor Planets", T. Gehrels, Ed. pp. 337-352, NASA SP-267
Meyer, Y., 1989, in "Wavelets, time-frequency methods and phase space", J.M. Combes, A. Grossmann, Ph. Tchamitchian, Eds., Springer
Milani, A., Knež ević, Z. : 1990, in "Celest. Mech.", 49 : 247
Milani, A. Farinella, P., Knež ević, Z : 1991 in "Compte rendu de la quinzieme école de printemps d'astrophyique de Goutelas" Eds. Benest, D. and Froeschlé Cl. (Frontières).
Milani, A., Knež ević, Z. : 1992 Icarus, 98 : 211-232
Murtagh, F., Heck, A., 1987 "Mutivariable data analysis", Reidel
Petit, J.M., Froeschlé, C. : A&A, in press
Remy, F., Mignard, F. : 1985, Icarus 63, 1-19
Rickman, H., Froeschlé C. : 1979, AJ. 84, 1910-1917
Rickman, H., Froeschlé, C. : 1983, Moon and Planets 28, 69-86
Rickman, H., Froeschlé, C. : 1988, Celest. Mech. 43, 243-263
Rickman, H., Vaghi, S. : 1976, A&A 51, 327-342
Varosi, F., Gebogi, V., Yorke. J.A. : 1987, Phys. Lett. A 124, 59-64
Weissman, P.R. : 1982, in Comets (ed. L.L. Wilkening), Univ. Arizona Press, Tucson, pp. 637-658
Williams, J.G. : 1979, in "Asteroids", T. Gerhels, Ed., pp.1040-1063 University of Arizona Press, Tucson
Zappalà, V., Cellino, A, Farinella, P., Kneževič, Z. : 1990 in "A. J.", 100, 2030
Zappalà V., Cellino A. Farinella P., Milani A.: 1993, A. J. in press

SUCCESSIVE ELIMINATION OF HARMONICS: A WAY TO EXPLORE THE RESONANT STRUCTURE OF A HAMILTONIAN SYSTEM

A. Morbidelli

Observatory of Nice - CERGA department
BP. 229, 06304 Nice Cedex 4 – France

1. INTRODUCTION

This paper deals with the problem of exploring Hamiltonian systems with semi–numerical methods. I come back to an original idea of Delaunay, consisting in successively eliminating all the perturbation harmonics up to a given order. Here the elimination is performed via the introduction of suitable Arnold action–angle variables, along the lines of the well known theorem of Arnold (1963b), with numerical evaluation of the action integrals. More precisely, given the Fourier expansion of the Hamiltonian, I first ignore all the harmonics in the perturbation except a single one; this is an integrable system and the introduction of Arnold action–angle variables allows the complete elimination of the considered harmonic without remainder terms. Next, I compute the Fourier expansion of the till now ignored terms in the new variables. This algorithm can be implemented on computer and iterated so that one attains a global description of the dynamics. By the way, this allows to point out all the power of the method of Delaunay, which seems to have been almost forgotten.

The algorithm of successive elimination of harmonics is in some sense similar to the renormalization scheme (Escande and Doveil, 1981). Nevertheless the approach is different, since the renormalization scheme points to the description of a torus with a given rotational number, while our algorithm aims to provide a global description of the phase space, although fragmented in different topological regions, according to the location of the main resonances. Moreover, the renormalization scheme takes into account at any step only two harmonics, while in our algorithm all the harmonics are taken into account and they give a contribution in the computation of the Fourier expansion of the perturbation with respect to the new angles.

From a practical viewpoint the method proposed here seems to be particularly indicated for the analysis of systems characterized by a strong coupling among degrees of freedom; indeed, in these cases, the algorithm imposes a fragmentation of the phase space into different "topological regions", according to the location of the main resonances. Moreover, at each step, it compels to leave out of study portions of the phase space (which turn out to be chaotic regions), while elsewhere it allows to define new variables, suitable for the description of the invariant tori in the regular regions.

A quantitative analysis of this method of successive elimination of harmonics allows to give it a complete theoretical support. More precisely, one can show that a suitable adaptation of the method of Delaunay can be used to recover, in a common environment,

Hamiltonian Mechanics, Edited by J. Seimenis,
Plenum Press, New York, 1994

both the KAM theorem and part of the Nekhoroshev theorem. In particular we have proved (Morbidelli and Giorgilli, 1993):

(i) for a sufficiently small perturbation and for an arbitrary time T, there exists a large open region of the phase space which is foliated in n–dimensional tori; these tori are almost invariant up to time T, in the sense that an orbit starting on a torus remains in a small neighbourhood of it for $|t| < T$.

(ii) in the limit of the time T going to infinity, the region above shrinks to a set of invariant tori (which is neither open nor dense but large in measure). This set can be identified with the invariant tori of the KAM theory.

(iii) if one requires that the open region above contains open balls of a given (small) radius ϵ, then the time T must be finite. However, T can be as large as $\exp(1/\epsilon)^a$, with a suitable exponent a. This result looks as a bridge between the KAM theory and the Nekhoroshev theory, at least for what concerns the nonresonant region.

Although from the strictly mathematical viewpoint this result is not completely new, it partially fills the gap between practical computations and theoretical results in perturbation theory.

The most interesting open problems in Celestial Mechanics nowadays are multi-resonance ones, namely problems characterized by the presence of several resonances with strong interaction among them. The method of successive elimination of harmonics turns out to be very suitable to investigate these kinds of problems.

It is well known that the dynamics in the asteroid belt is strongly affected by two kinds of resonances: mean motion resonances (i.e. commensurabilities between the orbital period of the asteroid and of the planets), and secular resonances (i.e. commensurabilities among the precession rates of the orbits of the asteroid and of the planets); both these classes of resonances have been explored in detail, but very little is known about the regions where they interact strongly, cross over, or overlap. Moreover, the picture is complicated by the presence of a third resonance, the so called Kozai resonance, where the argument of perihelion is locked. The method of successive elimination of harmonics has been recently applied to investigate some of these cases. In the paper by Morbidelli, Scholl and Ch. Froeschlé (1993) it is proved that secular resonances are deviated to large inclination approaching the 2/1 mean motion commensurability, so that they do not affect the motion of the existent asteroids. In the paper by Morbidelli and Moons (1993), the role of secular resonances inside mean motion commensurabilities in the Sun-Jupiter-Saturn-Asteroid space problem is investigated. This work points out, for the first time, dynamical differences between the 3/2 mean motion commensurability (occupied by the Hilda group of asteroids) and the 2/1 mean motion commensurability (the puzzling Hecuba gap in the asteroids distribution).

All the results outlined in this paper have been already published in separate papers. The aim of this review is to provide a general panoramic on the method of successive elimination of harmonics, skipping all details. In the next section, I will summarize the main motivations which force to look for a new practical perturbation method such as this one. In section 3 I will outline the algorithm and show how it can be used to find out chaotic and quasi–regular regions. In section 4 I will discuss the theoretical implications and finally in section 5 I will describe how the method of successive elimination of harmonics has been used to study multi–resonance cases in the Sun-Jupiter-Saturn-Asteroid space problem.

2. MOTIVATIONS OF THIS WORK

This section is devoted to revisit the main practical perturbation approaches commonly used in modern Celestial Mechanics; the analysis of their limits will motivate the search for a "new" approach like the one of successive elimination of harmonics.

The most usually applied perturbation method is the Lie algorithm, implemented by expanding the Hamiltonian in Taylor and/or Fourier series and by using powerful algebraic manipulators. In such an approach, one encounters typically two problems:

1) one has to deal with explicit expansions with a *local* character; for instance, one considers the neighbourhood of an equilibrium point or of a torus with a given rotational number. In Celestial Mechanics, where, following Tisserand and Poincaré, the perturbation is usually expanded in power series of the eccentricity and of the

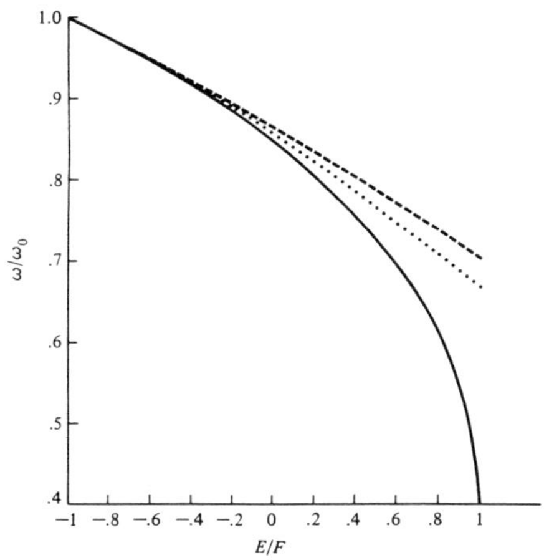

Figure 1. The frequency–energy relation for the pendulum, in the region of libration (-1 is the energy of the stable equilibrium point, 1 the energy of the separatrix). The solid line is the exact result; the dashed line is obtained using first order perturbation theory. The dotted line is the result from second order perturbation theory. The agreement is good only close to the equilibrium point. (From Lichtemberg and Lieberman, 1983).

inclination, the results are limited to the neighbourhood of planar circular orbits; a good approximation for planetary theories, but a poor one to study asteroid dynamics. Local expansions around a non zero value e_*, i_* have been computed in some cases, for example by Ferraz–Mello (1989), but only up to a low order.

2) the Lie algorithm uses in an essential way the approximation given by the unperturbed Hamiltonian which is unchanged at every step of the algorithm. The unperturbed Hamiltonian is often chosen to be the linear one, in order to deal with fixed frequencies (fundamental for algebraic manipulators); this imposes many limits when the dynamics is strongly non linear, such as close to separatrices. A paradoxical example of these limits is provided by trying to recover the integrable dynamics of a pendulum starting from the harmonic oscillator approximation, as shown in figure 1, taken from Lichtemberg an Liberman's book (1983). Therefore, one can easily imagine the difficult task of studying a perturbed pendulum with this approach. Unfortunately this is the typical case in Celestial Mechanics any time one deals with resonance and multi-resonance problems.

This evident limit in dealing with strongly non–linear dynamics, pushed Henrard (1990) to study the possibility of semi-numerical implementation of the Lie algorithm by using Arnold action angle variables. In 1963 Arnold proved the so called Arnold–Liouville theorem. This theorem states that, given an integrable Hamiltonian $H_0(p,q)$, there exists a set of canonical action–angle variables J, ψ such that the Hamiltonian is independent of the new angles, namely $H_0 \equiv H_0(J)$. If the phase space of the integrable Hamiltonian is divided into different topological regions (libration, positive and negative circulation, in the case of a pendulum), different Arnold variables must be introduced in each region. Unfortunately, Arnold action–angle variables are not known explicitly, except in some particular cases (like that of the pendulum, and they involve elliptic functions); however they can be computed, for each orbit of the integrable Hamiltonian, in a numerical way. The implementation of the Lie algorithm, on the contrary, requires explicit algebraic expressions. Nevertheless Henrard (1990) has shown that the first order of the Lie algorithm can still be computed in a semi–numerical way. Although limited to be a first order perturbation method, Henrard's approach often turns out to be a significant improvement. This is due to the fact that it allows to take completely into account the dynamics of the most complex integrable part of the Hamiltonian, despite of its non-linearity; for example, in the case of a perturbed pendulum, it allows to consider as an integrable part the full Hamiltonian of the pendulum, instead of its harmonic oscillator approximation, and therefore is quite accurate even close to the pendulum separatrices. Figure 2 gives an example of the power of this implementation in the case of the Uranus–Miranda–Umbriel problem. This problem is a two degrees of freedom one, with angles σ and ν conjugated to actions S and N, and is characterized

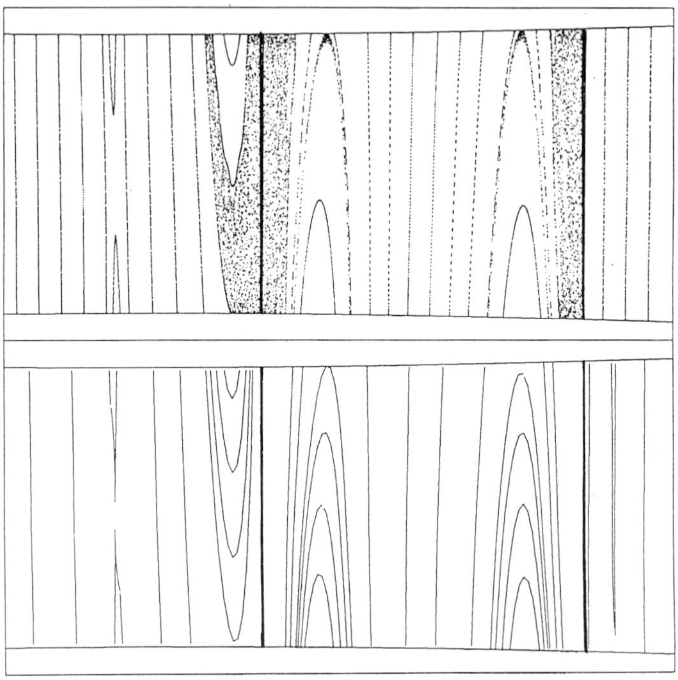

Figure 2. Above the theoretical result after introduction of suitable action–angle variables for the main pendulum like resonance and first order analysis of the perturbation; below the Poincaré section of the numerical integration of the full Hamiltonian. The two thick vertical lines are the section of the separatrix of the pendulum term with argument 2σ. The coordinates are $N - S$ (x–axis) and $\sqrt{S}\cos(\sigma + \nu)$ (y–axis). (Moons and Henrard, 1993.)

by a main resonance (pendulum–like one) with argument 2σ and several secondary resonances. The picture above shows the dynamics of the secondary resonances as computed theoretically in the three dynamical regions of the main resonance (separated by the two thick vertical lines); the picture below shows the numerical Poincaré section of the full Hamiltonian. As one sees the agreement is very good, although nothing allows to predict the correct location and size of the chaotic layer. This is precisely due to the fact that this is only a fist order perturbation approach, and the mutual interaction among the main resonance and the secondary resonances is completely neglected.

This example shows clearly that the use of Arnold action-angle variables is the right way to approach strongly non–linear dynamics; however it is necessary to find a practical way to go beyond the first order results. The Lie's algorithm is certainly unsuitable for this task. I will show in the next section how the old idea of Dealunay of successive elimination of harmonics can be implemented in Arnold variables up to an arbitrary order, and in section 4 how this can be proved to be a genuin perturbation theory also from the quantitative viewpoint.

3. THE ALGORITHM OF SUCCESSIVE ELIMINATION OF HARMONICS

The basic idea of the algorithm of successive elimination of harmonics is very simple and intuitive. Let us consider,for example, a two degrees of freedom Hamiltonian system $H(I_1, \phi_1, I_2, \phi_2)$ with Fourier expansion

$$H(I_1, \phi_1, I_2, \phi_2) = H_{0,0}(I_1, I_2) + \sum_{n,m} h_{n,m}(I_1, I_2) \cos(n\phi_1 + m\phi_2) \ , \tag{1}$$

(here I assume, for simplicity, that only cos components are present in the expansion). The first step is to choose one particular harmonic, $\overline{n}, \overline{m}$ say, and consider the partial

Hamiltonian

$$H_{0,0}(I_1, I_2) + h_{\overline{n},\overline{m}}(I_1, I_2)\cos(\overline{n}\phi_1 + \overline{m}\phi_2) \ . \tag{2}$$

The choice of the harmonic is in principle arbitrary but it is evident that, for a good performance of the algorithm, one should preliminarly analyze the dynamical effect produced *separately* by each perturbation harmonic and select the most relevant one. The Hamiltonian (2) is integrable since it has as many constants of motion as degrees of freedom. Therefore, the Arnold–Liouville theorem can be applied; new action–angle variables J_1, J_2, ψ_1, ψ_2 can be introduced in order to transform (2) into

$$H'_{0,0}(J_1, J_2) \ .$$

As it is well known, different sets of Arnold action–angle variables must be introduced in the different dynamical regions of (2), i.e., the regions of circulation and libration of the critical angle $\overline{n}\phi_1 + \overline{m}\phi_2$, separated by singular surfaces. This imposes a fragmentation of the phase–space: in each part, the algorithm proceeds, from now onwards, separately.

The next step is to transform the remainder

$$\sum_{n\neq\overline{n},m\neq\overline{m}} h_{n,m}(I_1, I_2)\cos(n\phi_1 + m\phi_2) \ ,$$

in the new variables and expand it in Fourier series with respect to the new angles ψ_1 and ψ_2. The new Hamiltonian

$$H'_{0,0}(J_1, J_2) + \sum_{n,m} h'_{n,m}(J_1, J_2)\cos(n\psi_1 + m\psi_2)$$

has exactly the same form as (1), identifying J_1, J_2, ψ_1, ψ_2 with I_1, I_2, ϕ_1, ϕ_2. Therefore, the algorithm is ready for iteration.

At this point some considerations are in order. The first one concerns the semi–numerical implementation of the algorithm. For a general Hamiltonian as (2), the Arnold action–angle variables are not known explicitly. However, the phase–space of (2) is foliated into invariant tori, and each torus is a point in the new action space (J_1, J_2). Moreover, the new angles are linear functions of time, and their frequencies are the proper frequencies of the torus. Therefore, on each invariant torus of (2), it is possible to compute by numerical integration the transformation $I_1 = I_1(J_1, \psi_1, J_2, \psi_2)$, $I_2 = I_2(J_1, \psi_1, J_2, \psi_2)$, $\phi_1 = \phi_1(J_1, \psi_1, J_2, \psi_2)$ and $\phi_2 = \phi_2(J_1, \psi_1, J_2, \psi_2)$. Consequently, it is possible to compute numerically the coefficients $h'_{n,m}$ of the Fourier expansion of the new Hamiltonian. The Hamiltonian is then numerically represented on a grid of points in the action–space, and, by interpolation, can be computed in any desired point. In this way, one can construct a sort of semi–numerical manipulator. Each step of the algorithm of successive elimination of harmonics just requires to define a grid in the new action space and code $H'_{0,0}$ and $h'_{n,m}$ on the new grid. This eliminates all the difficulties in dealing with Arnold action–angle variables, opening the way to study also strongly non–linear perturbed dynamics.

A second remark concerns the efficiency of the algorithm in exploring the dynamical properties of the system in study, with a particular care to the detection of chaotic layers and the description of regular regions. At each step, the algorithm imposes a fragmentation of the phase–space, according to the location of the resonance corresponding to the harmonic being eliminated. Step after step, the geography of resonances is in this way detected, and appropriate action–angle variables are introduced, in each dynamical region of the phase–space, for the description of the invariant tori. Moreover, the computation of the new Fourier expansion of the Hamiltonian at each step, allows to detect the correct location and strength of high order secondary resonances. The Chirikov criterion of resonance overlapping can be confidently applied to these secondary resonances in order to detect the location and size of chaotic layers. An example is provided by figure 3. The figure shows the location and size of secondary resonances detected after the successive elimination of the 4 main harmonics in the Uranus–Miranda–Umbriel problem. The four harmonics eliminated correspond to the main pendulum–like resonance with argument 2σ and to the secondary resonances with argument $\sigma - \nu$, $3\sigma - \nu$ and $4\sigma - 2\nu$. Figure 3 shows the secondary resonances and their overlapping between the resonances $3\sigma - \nu$ and $4\sigma - 2\nu$ (on the left) and between the resonances $4\sigma - 2\nu$ and $\sigma - \nu$ (on the right). The $3\sigma - \nu$ resonance should be considered here as part of the

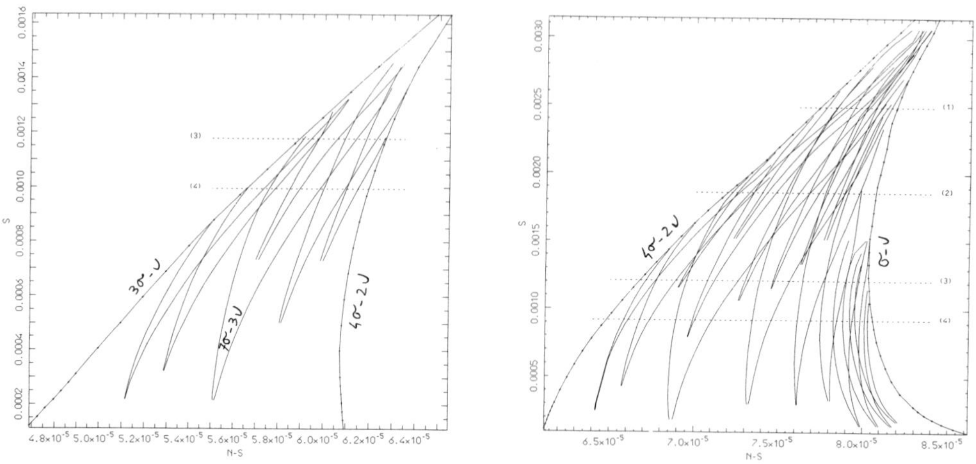

Figure 3 Location and amplitude of secondary resonances in the Uranus–Miranda–Umbriel problem: on the left, the portion of the action–space in between the $3\sigma-\nu$ and $4\sigma-2\nu$ resonances; on the right, the same between the resonances $4\sigma-2\nu$ and $\sigma-\nu$. The 4 dashed lines denote the energy levels corresponding to the numerical Poincaré sections of figure 4.

main chaotic layer. The 4 dashed lines denote the energy levels corresponding to the 4 numerical Poincaré sections of Figure 4. Believing in Chirikov overlapping criterion and looking at Figure 3, one can predict that the chaotic layer should extend to the $\sigma-\nu$ resonance on the first two levels (compare with pictures (a) and (b) of figure 4), while it should be bordered by the $4\sigma-2\nu$ resonance on level 3 (compare with picture (c), fig. 4) and to the $7\sigma-3\nu$ resonance on level 4 (picture (d), fig. 4). The application of Chirikov overlapping criterion to high order secondary resonances, after the elimination of the main harmonics of the problem, seems to be an accurate way to discriminate between chaotic layers and regular regions. The improvement with respect to the first order result of figure 2 (which refers to the same problem) is evident.

For a deeper discussion about the semi–numerical implementation of the algorithm, practical convergency and the detection of chaos, I refer to my extended paper (Morbidelli, 1993).

4. QUANTITATIVE PERTURBATION THEORY BY SUCCESSIVE ELIMINATION OF HARMONICS

Although the tests of application of the algorithm of successive elimination of harmonics are encouraging, it seems not immediately evident, from a theoretical viewpoint, that the method is based on a sufficiently established mathematical background. For instance, the choice of the order of elimination of the harmonics is rather arbitrary, and is justified only in a qualitative fashion. Moreover, there is no evidence *a priori* that the already eliminated harmonics do not reappear with bigger and bigger coefficients: this is indeed the typical problem in classical expansions. In the paper by Morbidelli and Giorgilli (1993), we study this problem from a rigorous point of view, at least for what concerns the non–resonant part of the phase–space. In the following I just outline the approach we have followed, state our final theorems and discuss their implications.

The main technical tool is the analysis of the Arnold action–angle variable transformation for the elimination of one harmonic. As is usually necessary in quantitative perturbation theory, we extend all the functions and variables in a suitable complex domain in order to be able to apply the well known results of the theory of analytic functions. Indeed the transformation to Arnold action–angle variables is analytic and close to the identity, and we provide estimates on the size of the "deformation". Moreover, we estimate the loss of phase space volume due to the introduction of the new variables; this loss of volume is due partly to the deformation of coordinates induced

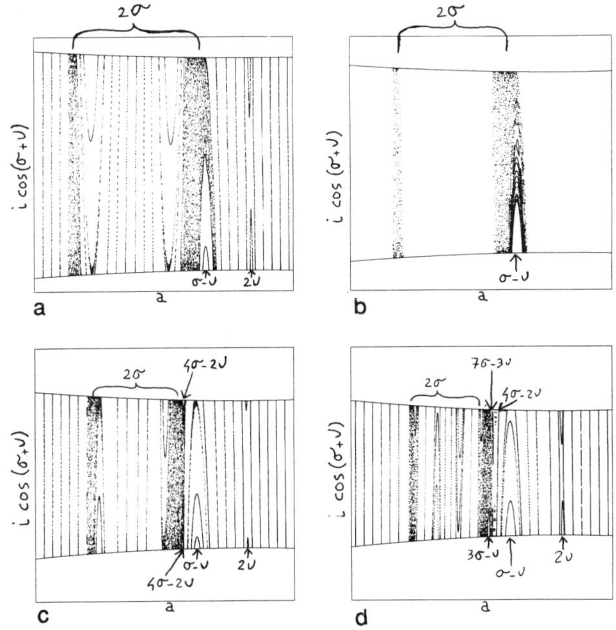

Figure 4 Numerical Poincaré sections of the Uranus–Miranda–Umbriel problem corresponding to the 4 energy levels denoted in figure 3. (courtesy of M. Moons and J. Henrard)

by the transformation, and partly to the elimination of the "resonant strip" around the resonance corresponding to the considered harmonic.

With this basic result, we proceed in analyzing the successive elimination of harmonics. Given the Hamiltonian

$$H(p,q) = H_0(p) + \epsilon H_1(p,q) \ , \qquad \text{with} \qquad H_1(p,q) = \sum_{k \in \mathbf{Z}^n} h_k(p) \cos{(k \cdot q)}$$

the elimination of the harmonic k via the introduction of the new action–angle variables R, r gives the new Hamiltonian

$$H'(R,r) = H_0'(R) + \epsilon H_1'(R,r) + \epsilon^2 \mathcal{R}(R,r) \ ,$$

where H_1' has the same Fourier expansion as H_1 but without the harmonic k which has been eliminated, and $\mathcal{R}$ contains all the terms generated by the introduction of the new variables in each term of the original Fourier expansion of H_1. We provide an estimate of the size of $\mathcal{R}$ after the elimination of the resonant strip $|k \cdot \partial H_0/\partial p| < \gamma$ with a suitable γ.

The analyticity of the Hamiltonian H_1 implies that the coefficients $h_k(p)$ decay exponentially fast, namely that $|h_k| \sim e^{-|k|\sigma}$ for some positive σ. Thus, we proceed with the elimination of harmonics disregarding all the new terms in $\mathcal{R}$ generated by the transformation until $|k|$ reaches a value K_0 roughly of size $|\ln \epsilon|$. At that point we are left with the part of H_1 of size ϵ^2, namely of the same size as $\mathcal{R}$. Then, we expand also $\mathcal{R}$ in Fourier series, reorder all the terms, and start the game again with a perturbation of size ϵ^2. The elimination of all the harmonics satisfying $|k| \leq K_0$ is what we call a quadratic step, because of its evident connection with the quadratic method commonly used in the proof of the KAM theorem.

The final part of our analysis of the successive elimination of harmonics is the iteration of quadratic steps and the estimate of the total loss of volume of the original domain in the phase–space.

With this approach, we are able to prove the two theorems stated below, where we use the following notations:

- in $\mathbf{C}^n$ we introduce the distance $|p - p'| = \max_j |p_j - p'_j|$.
- for an open subset $\mathcal{G}$ of $\mathbf{C}^n$ we denote by $\mathcal{G} - \delta$, with positive δ, the subset of $\mathcal{G}$ of the points which are contained in $\mathcal{G}$ together with a δ–neighbourhood; a δ–neighbourhood is an open ball of radius δ in the distance above.
- considering the immersion of $\mathbf{T}^n$ in $\mathbf{C}^n$, with coordinates q, we denote by $\mathbf{T}^n_\sigma$ the strip $|\mathrm{Im} q| \le \sigma$.
- we denote by $\mathcal{D} = \mathcal{G} \times \mathbf{T}^n_\sigma$ the basic domain of our construction. Moreover, we denote by $\mathcal{D} - (\delta_1, \delta_2) = (\mathcal{G} - \delta_1) \times \mathbf{T}^n_{\sigma - \delta_2}$. By abuse of notation we also write $\mathcal{D} - \delta = (\mathcal{G} - \delta) \times \mathbf{T}^n_\sigma$.
- we are also interested in the volume of the real part of either $\mathcal{D}$ or $\mathcal{G}$. By $\mathrm{Vol}\,\mathcal{D}$ and $\mathrm{Vol}\,\mathcal{G}$ we denote such a volume. We remark that, from $\mathcal{D} = \mathcal{G} \times \mathbf{T}^n_\sigma$, the relation $\mathrm{Vol}\,\mathcal{D} = (2\pi)^n \,\mathrm{Vol}\,\mathcal{G}$ holds.
- if $f(p, q)$ is an analytic function defined on $\mathcal{D}$, we denote with $\|f\|_\mathcal{D}$ the supremum norm, i.e., $\|f\|_\mathcal{D} = \sup_\mathcal{D} |f(p, q)|$. For a vector function $g = (g_1, \ldots, g_n)$, the norm $\|g\|_\mathcal{D}$ is defined as the maximum of the norms of the components. For vectors $v \in \mathbf{C}^n$ the notation $\|v\|$ denotes instead the Euclidean norm.

THEOREM 1 *Let the Hamiltonian*

$$H(p, q) = H_0(p) + \epsilon H_1(p, q)$$

be real analytic in $\mathcal{D} = \mathcal{G} \times \mathbf{T}^n_\sigma$, *the set* $\mathcal{G}$ *being open in* $\mathbf{C}^n$; *assume that, for some positive constants* $\Lambda, \theta, \Theta, \epsilon$, *and* σ *with* $0 < \theta \le \Theta < +\infty$, *the Hamiltonian* H *satisfies:*

a) $\left\| \frac{\partial H_0}{\partial p} \right\|_\mathcal{D} \le \Lambda$

b) $\theta \|v\| \le \|A(p) v\| \le \Theta \|v\|$ *for any* $v \in \mathbf{C}$, *where* $A(p) = \left(\frac{\partial^2 H_0}{\partial p_i \partial p_j} \right)$

c) $\|H_1\|_\mathcal{D} \le \epsilon$

d) $\left\| \frac{\partial H_1}{\partial p} \right\|_\mathcal{D} \le \epsilon^{3/4}$

Assume moreover that, for some positive constants ρ, *and* D_V, *the domain* $\mathcal{G}$ *satisfies:*

e) $\mathrm{Vol}(\mathcal{G} \setminus (\mathcal{G} - d)) \le d D_V \,\mathrm{Vol}\,\mathcal{G}$ *for* $0 < d < \rho$.

Then there exist constants $\bar{\epsilon}, C^*, C^*_\sigma, V$ *and a positive function* $T_*(\epsilon)$ *such that the following statement holds true:*

For any $\epsilon < \bar{\epsilon}$ *and for any* $T > T_*(\epsilon)$, *having defined* $\eta = \epsilon^{1/4} (T_*/T)^{1/3}$, *there exists an open domain* $\mathcal{D}'(\epsilon, T) \subset \mathcal{D}$ *such that:*

1) $\mathcal{D}'(\epsilon, T)$ *is made up of* n*–dimensional tori* $\mathcal{T}$ *parameterised by action variables* P *canonically conjugated to angles* Q, *with* $p = P + f(P, Q), q = Q + g(P, Q)$ *and*

$$\|f\|_{\mathcal{D}'} \le 2C^* \epsilon^{1/2} \ , \quad \|g\|_{\mathcal{D}'} \le 2C^*_\sigma \epsilon^{1/4} \ ;$$

2) $\mathrm{Vol}(\mathcal{D}' - \eta) \ge \left\{ 1 - V \epsilon^{1/16} \left[1 - \left(\frac{T_*}{T} \right)^{1/6} \right] \right\} \mathrm{Vol}\,\mathcal{D}$

3) $\forall (P_0, Q_0) \in \mathcal{D}' - \eta$, *the orbit* $P(t), Q(t)$ *with initial data* P_0, Q_0 *satisfies*

$$|P(t) - P_0| \le \eta \quad \text{for } |t| \le T$$

THEOREM 2 *In the same hypotheses of theorem 1 there exist a constant* $\bar{\epsilon}_1 \le \bar{\epsilon}$, *with* $\bar{\epsilon}$ *as in theorem 1, and a constant* α *not dependent on* ϵ *such that for any* $\epsilon < \bar{\epsilon}_1$ *the domain*

$$\mathcal{D}'(\epsilon, T) \qquad \text{with } T = \frac{2}{\sigma \epsilon^{3/4}} \exp \left[\frac{3}{8} \left(\frac{\alpha}{\epsilon} \right)^{1/(3n-2)} \right]$$

contains balls of radius ϵ.

The picture provided by Theorem 1 is the following: given an arbitrary diffusion time T, there exist an open portion of the phase–space foliated into tori which are quasi–invariant up to time T. This means that it is possible to introduce suitable Arnold action–angle variables such that the actions do not change more than $\eta \sim 1/T^{1/3}$ in a time T. Some regions of the phase–space, however, do not fulfill this property: these are the chaotic zones associated to the resonances up to a certain order (a limit of the theorem is that also regular resonant islands are considered as chaotic zones). Give now a sequence of diffusion times $T_1, \ldots, T_n$, with $T_1 < T_2 < \ldots < T_n$, and look for the sets $\mathcal{D}'_1, \ldots, \mathcal{D}'_n$ of quasi–invariant tori associated to $T_1, \ldots, T_n$; it is evident that $\mathcal{D}'_1 \supset \mathcal{D}'_2 \supset \ldots \supset \mathcal{D}'_n$, the difference being due to the chaotic regions associated to resonances of higher and higher order. In the limit $T \to \infty$, one finds the set $\mathcal{D}'_\infty$ of invariant tori, after the elimination of all the resonances: this is the KAM theorem. Theorem 1 brings into evidence that the diffusion is arbitrary slow, provided one starts close enough to an invariant torus of the KAM theory.

The existence of regions of quasi invariant tori with arbitrary slow diffusion suggests a possible connection with the Nekhoroshev's theorem, which is established by our Theorem 2. The key point here is that Nekhoroshev's result is global in the action space, while our theorem gives results on a quite strange subset of the phase space. A minimal connection between Nekhoroshev and KAM theory requires that the open domain $\mathcal{D}'(T)$ of theorem 1 also contains balls of a suitable radius. From a practical viewpoint this could be considered as a request of compatibility with the errors in the determination of the initial conditions. As a matter of fact, such a request produces an upper bound on the time T in our statement, which turns out to be essentially the exponential time appearing in Nekhoroshev theory.

5. AN APPLICATION TO THE ASTEROID BELT

The algorithm of successive elimination of harmonics, being particularly suitable for studying th strong perturbations induced by resonances, can be used to throw a new light in the most complex problems of dynamics in the asteroid belt, in particular the regions of resonance interaction.

The recent studies on the asteroid belt have pointed out that the classical restricted three body model is not adequate to give a realistic description of the real dynamics, even from the qualitative point of view. Indeed, a realistic model must take into account that the orbits of all the planets, under their mutual interaction, are forced to precede and deform over time scales of some ten thousand years. Therefore, even the most simple model should take into account that the elements of Jupiter (i.e. the eccentricity e', the inclination i', the longitude of perihelion ϖ', and the longitude of node Ω') change according to the linear theory of Laplace, which gives:

$$\begin{aligned} e' \cos \varpi' &= m_{5,5} \cos(g_5 t + \lambda_5^0) + m_{5,6} \cos(g_6 t + \lambda_6^0) \\ e' \sin \varpi' &= m_{5,5} \sin(g_5 t + \lambda_5^0) + m_{5,6} \sin(g_6 t + \lambda_6^0) \\ \sin \frac{i'}{2} \cos \Omega' &= n_{5,6} \cos(s_6 t + \mu_6^0) \\ \sin \frac{i'}{2} \sin \Omega' &= n_{5,6} \sin(s_6 t + \mu_6^0) \ . \end{aligned}$$

In this way, one gets what could be called the "secular restricted three body problem", which is much more complex than the usual one since:

i the inclination of Jupiter is not zero and Ω' rotates, which implies that the reduction to the *planar* problem can just be considered as a first step of a perturbation approach which assumes $n_{5,6}$ as a small parameter;

ii the system is time dependent through three incommensurable frequencies, which is equivalent to increasing by three the number of degrees of freedom;

iii a new class of resonances enters the problem, changing the dynamical picture even from the qualitative point of view. These are the secular resonances, i.e. commensurabilities among the frequencies of precession of the longitudes of perihelion and of the node of the asteroid and the basic frequencies of the planetary system g_5, g_6 and s_6.

Far from all the main resonances, the secular restricted three body problem is still quasi–integrable, and accurate quasi–constant of motion can be computed; these are called *proper elements* by the astronomers and allow the identification of the asteroid families (see Bendjoya, 1993, for a review).

Also secular resonances are quasi–integrable and semi–analytic predictions on the motion of secular resonant bodies can be provided (Morbidelli, 1993b). Secular resonances are very important from the astronomical point of view, since they are an efficient mechanism for the dynamical transport of asteroid fragments to the Earth, i.e. a source of meteorites (Farinella et al, 1993).

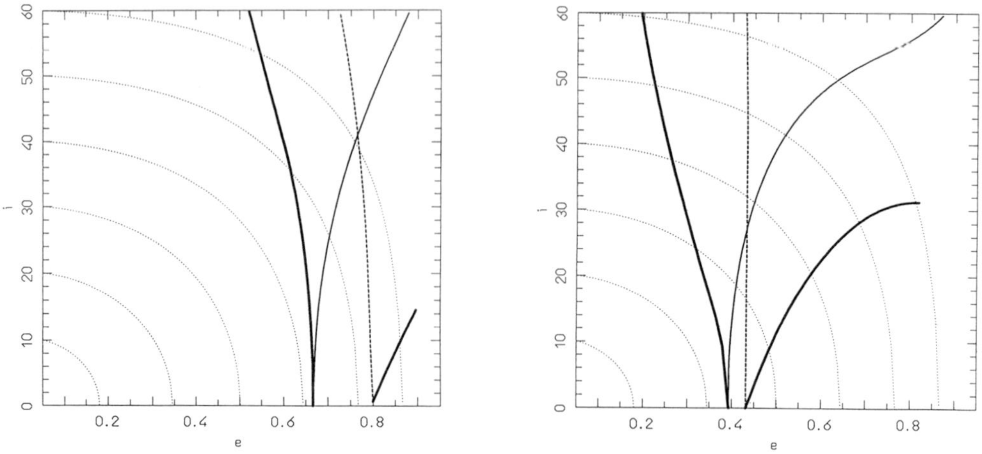

Figure 5 Dynamics outside of the reference plane, for the 2/1 (left) and 3/2 (right) mean motion resonances: the solid curve is the stable family of equilibrium points of the argument of perihelion ($\omega = \pi/2, 3\pi/2$); the dashed line is the unstable one ($\omega = 0, \pi$); the two thick lines denote the separatrices of the region of ω–libration at $\omega = \pi/2$; the dotted curves are levels $N \equiv \sqrt{a}(p+q/p - \sqrt{1-e^2}\cos i$=constant. The dynamics evolve, in first approximation, on a line N=constant.

Mean motion resonances (i.e. commensurabilities between the orbital period of the asteroid and that of Jupiter) turn out to be much more complex in the "secular three body problem" than in the "usual" three body problem. In particular secular resonances can exist inside mean motion resonances, so that these regions turn out to be multi–resonance ones, with possible large chaotic layers. It is therefore necessary to reconsider the mean motion resonances in the framework of the "secular" model; by the way, this could give new light to the problem of the origin of Kirkwood gaps, which is still partially unsolved. The description of the global dynamics of mean–motion resonances in the framework of a realistic model is actually the frontier of our knowledge on the asteroid belt.

We have applied the method of successive elimination of harmonics in order to achieve an at least preliminar global exploration (Morbidelli and Moons, 1993). The application of the algorithm is not straightforward, since the problem, after the averaging over the fast angle, has still 5 degrees of freedom; nevertheless, with some simplifications, we have performed a limited number of steps, which allowed to achieve some interesting results.

First we have eliminated the harmonic related to the properly called mean motion resonance, with argument $\sigma = (p+q/q)\lambda' - p/q\lambda - \varpi$ for a $p+q/p$ resonance, where λ is the mean longitude of the asteroid and λ' the mean longitude of Jupiter. This harmonic exists also in the classical circular three body problem.

Furthermore, we have eliminated the harmonic related to the argument of perihelion of the asteroid ω. This is also a resonant harmonic. Indeed, the argument of perihelion circulates with negative derivative at small eccentricity, librates at moderate eccentricity, and circulates in the opposite direction at large eccentricity. Figure 5 shows the location and the amplitude of the ω–libration resonance on the e, i plane for orbits with small amplitude of σ–libration for the 2/1 and 3/2 mean motion resonances.

Afterwards, we have devoted our attention to the harmonics which enter the problem because of the secular precession of Jupiter's orbit. The ν_5 ($\dot{\varpi} = g_5$) and the ν_6 ($\dot{\varpi} = g_6$) secular resonances occur only at small inclination. On the plane $i = 0$ we

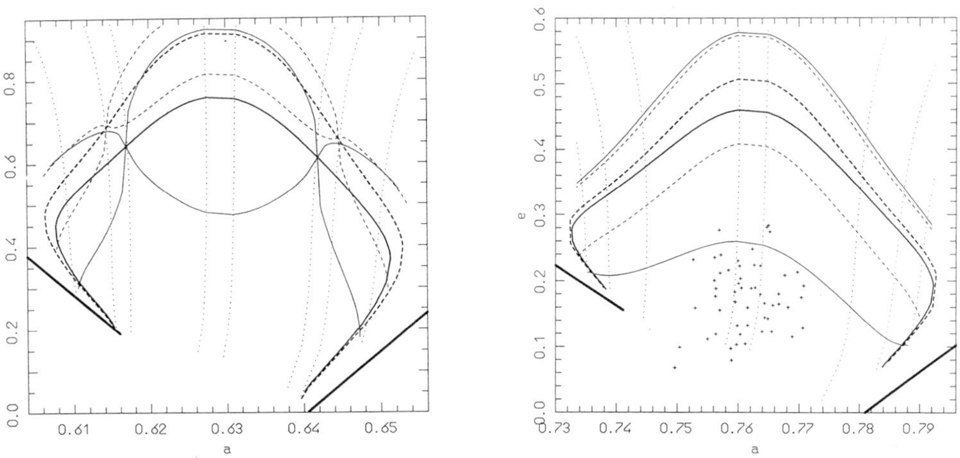

Figure 6 Secular resonances in the 2/1 (left) and 3/2 (right) commensurability on the plane $i = 0$: the solid bold line marks the location of the ν_5 resonance, the dashed bold line the location of the ν_6; the two thin solid lines and the two thin dashed lines denote the separatrices of ν_5 and ν_6 respectively. In the 2/1 mean motion resonances, the two secular resonances seem to "shrink"; this is due to the fact that the amplitudes go to zero while the resonances change topology, exchanging stable and unstable equilibrium points. In the 3/2 mean motion resonance the asteroids of the Hilda group (crosses) avoid the secular resonances. Also the asteroids which seem to appear inside the ν_5 secular resonance are in fact phase–protected .

have computed their location and their amplitude for all σ–librating orbits. The results are shown in figure 6; the two resonances overlap, and we have found the existence of large chaotic layers, by numerical integration of the reduced system where only the two secular resonances are taken into account, after the elimination of the σ–harmonic and of the ω–harmonic.

The ν_{16} secular resonance ($\dot{\Omega} = s_6$) occurs outside of the plane $i = 0$. We have computed the location and amplitude of this resonance on the e, i plane, as shown in figure 7. The most relevant result is that the distribution of the asteroids in the Hilda group (3/2 mean motion commensurability) seems to be bounded by the ν_{16} resonance. Conversely, a group of asteroids in the 2/1 mean motion commensurability, with a distribution similar to the Hildas, would be crossed by the ν_{16} resonance. This fact alone does not explain the existence of the gap in correspondence with the 2/1 mean motion commensurability; the effect of the ν_{16} resonance is to pump up the inclination, and a mechanism to get rid of asteroids with high inclination is not yet clear. However, this is the first time that a dynamical difference between the 2/1 and 3/2 mean motion commensurabilities is pointed out. The picture is not complete yet: due to the large number of degrees of freedom, the existence and the effect of secondary resonances among them have not been analyzed till now. We plan to do this work in the future.

6. CONCLUSIONS

The method of successive elimination of harmonics, which is based on an original idea of Delaunay, can be implemented on computer with semi–numerical techniques, and iterated in order to eliminate all the main perturbation terms. This approach turns

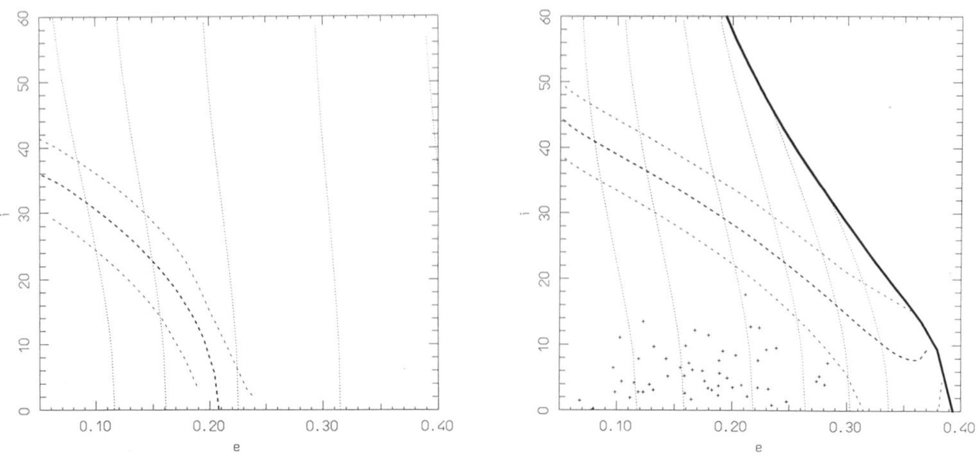

Figure 7 The location of the ν_{16} secular resonance (bold dashed line) and its separatrices (thin dashed lines). The case of the 2/1 mean motion commensurability is on the left, that of the 3/2 on the right. The thick curve denote the separatrix of the ω–resonance. Note how the asteroids of the Hilda group (crosses) are bounded by the secular resonance; conversely the ν_{16} resonance, being located at smaller eccentricity, would cross a group with a similar distribution in the 2/1 commensurability.

out to be particularly suitable for the exploration of the dynamics of strongly non–linear systems, with non negligible coupling among the degrees of freedom. In particular it allows to detect the geography of the main resonances, and to detect the location and size of the macroscopic chaotic layers.

A theoretical analysis of the algorithm proves that it can bee considered as a genuine perturbation theory, with connection to the well known KAM and Nekhoroshev results.

The application of the method of successive elimination of harmonics has already allowed to investigate some complicated problems of asteroid dynamics, characterized by the co–presence of mean motion resonances, secular resonances, and ω–resonance.

REFERENCES

Arnold, V.,I., (1963a): "Small denominators and problems of stability of motion in classical and celestial mechanics.", *Russ. Math. Surveys*, **18**, 85–191.

Arnold, V., I., (1963b): "On a theorem of Liouville concerning integrable problems of dynamics", *Sib. mathem. zh.*, **4**, 2.

Bendjoya, Ph., (1993): "A classification of 6479 asteroids into families by means of wavelets clastering method.", *Astron. Astrophys.*, in press.

Born, M., (1967): "The Mechanics of the Atom", Frederick Ungar Publ. Co.

Chirikov, B., V., (1979), *Phys. Reports.*, **52**, 265.

Delaunay, C., (1867): "Théorie du mouvement de la Lune", *Mem. Ac. Sc.*, Paris, **29**.

Escande, D., F., and Doveil, F., (1981), *J. Stat. Phys.*, **26**, 257.

Escande, D., F., and Doveil, F., (1981), *Phys. Lett.*, **83A**, 307.

Farinella, P., Gonczi, R., Froeschlé, Ch. and Froeschlé C., (1993): "The injection of asteroid fragments into resonances", *Icarus*, **101**, 174–187.

Ferraz–Mello, S., (1989): "A semi–numerical expansion of the averaged disturbing function for some very–high–eccentricity orbits.", *Celest. Mech.*, **45**, 65–68.

Henrard, J., (1990): " A semi–numerical perturbation method for separable Hamiltonian systems.", *Celest. Mech.*, **49**, 43–67.
Moons, M., and Henrard, J., (1993): "Surfaces of sections in the Miranda–Umbriel 1/3 inclination problem", submitted to *Celest. Mech.*.
Lichtenberg, A.J., and Lieberman, M.A., (1983): " Regular and stochastic motion.", *Springer Verlag* ed.
Morbidelli, A., (1993a): "On the successive eliminations of perturbation harmonics.", *Celest. Mech.*, **55**, 101–130.
Morbidelli, A., and Giorgilli, A., (1992): "A quantitative perturbation theory by sucessive elimination of harmonics.", *Celest. Mech.*, **55**, 131–159.
Morbidelli, A., (1993b): "Asteroid resonant proper elements", *Icarus*, in press.
Morbidelli, A., and Moons, M., (1993): "Secular resonances in mean motion commensurabilities: the 2/1 and 3/2 cases, *Icarus*, **102**, 316–332.
Morbidelli, A., Scholl, H., and Froeschlé, Ch., (1993): "The location of secular resonances close to the 2/1 commensurability", *Astron. Astrophys.*, in press.
Nekhoroshev, N., N., (1977): "Exponential estimates of the stability time of near–integrable Hamiltonian systems.", *Russ. Math. Surveys*, **32**, 1-65.
Nekhoroshev, N., N., (1979): "Exponential estimates of the stability time of near–integrable Hamiltonian systems, 2.", *Trudy Sem. Petrovs.*, **5**, 5-50.
Poincaré, H., *Les methodes nouvelles de la mechanique celeste*, Gauthier–Villars, Paris.
Tisserand, M. F., (1882): *Ann. Obs. Paris*, **16**, E1.

PERIODIC SOLUTIONS OF NONLINEAR SCHRÖDINGER EQUATIONS AND THE NASH-MOSER METHOD

Walter Craig
Department of Mathematics
Brown University
Providence, RI 02912

C. Eugene Wayne
Department of Mathematics
Pennsylvania State University
University Park, PA 16802

1. Introduction

In this paper we construct time-periodic solutions of the nonlinear Schrödinger equation

$$\begin{aligned} -i\partial_t u =& (-\partial_{xx} u + v(x))u + g(u,\overline{u},x) \quad , \\ 0 <& x < \pi \quad , \quad t \in \mathbb{R} \quad . \end{aligned} \tag{1.1}$$

satisfying either periodic or Dirichlet boundary conditions. We assume that the nonlinearity g is of the form $g(u,\overline{u},x) = \partial_{\overline{u}}\mathcal{G}(u,\overline{u},x)$, for some real valued function $\mathcal{G}$. This equation can then be viewed as a Hamiltonian system with infinitely many degrees of freedom, and the problem of time periodic solutions is related to perturbation theory in the neighborhood of elliptic stationary points of such systems. In contrast to the case of finite dimensional Hamiltonian systems, even the problem of periodic solutions for (1.1) exhibits the phenomenon of small denominators. Our results imply that the effect of these small denominators may be overcome, resulting in the construction of Cantor-like families of periodic solutions in a neighborhood of the equilibrium $u = 0$. We will present the details here only for the case of periodic boundary conditions, since this case is technically more difficult, allows for interesting resonances between the linear modes, and because the case of Dirichlet boundary conditions has already been treated by Kuksin (1988, 1993) using KAM methods. We note that in the special case in which $\mathcal{G}$ depends on u and $\overline{u}$ only through the combination $|u|^2$, one does not encounter small denominators, and one can construct periodic solutions with an ordinary implicit function theorem.

The methods used here are similar to those we have used earlier to construct solutions of nonlinear wave equations (Craig & Wayne (1991a)). We reduce the problem to an implicit function theorem for functions defined on a lattice, which is then solved using a Nash-Moser type of analysis. The treatment here is considerably simpler than that of Craig & Wayne (1991a), however. In part this reflects the simpler structure of (1.1) as compared with the wave equation. A greater part of the improvement results from two technical innovations. First, we introduce new norms (related to those used in Delatte (1992)) which considerably simplify the relationship between the solution $u(x,t)$ of (1.1) and its lattice representation. Secondly, we use

Hamiltonian Mechanics, Edited by J. Seimenis,
Plenum Press, New York, 1994

ideas of Pöschel to streamline the inversion of the linearized operators which arise in the Nash-Moser method.

We assume that the "potential" $\mathcal{G}(z_1, z_2, x)$ from which the nonlinear term is derived, is periodic in x with period π, analytic in the region $\{(z_1, z_2, x) | \ |z_1| < 1 \ , \ |z_2| < 1 \quad |\mathrm{Im}\, x| < \overline{\sigma}\}$, and that $\mathcal{G}(z_1, z_2, x) \approx \mathcal{O}_4(z_1, z_2)$. (Here, $\mathcal{O}_4(z_1, z_2)$ means that the sum of the exponents of z_1 and z_2 is at least four for all non-zero terms in the Taylor series for $\mathcal{G}$.) Denote the set of all functions $\mathcal{G}$ obeying these hypotheses, and equipped with the topology of uniform convergence as $\mathcal{A}_{\overline{\sigma}}$. We also assume that the potential $v(x)$ is analytic in $\{x \mid |\mathrm{Im} x| < \overline{\sigma}\}$, and periodic with period π. Denote the set of such potentials, again with the topology of uniform convergence, by $\mathcal{V}_{\overline{\sigma}}$.

Remark. The assumption that $\mathcal{G} \approx \mathcal{O}_4(z_1, z_2)$ means that our nonlinear term g is cubic in $u, \overline{u}$. We could also handle the case of quadratic nonlinearity, but in order to simplify the exposition we don't provide the details here.

Note that if $\psi_{j_0}(x)$ is an eigenfunction of the Sturm-Liouville operator $-d^2/dx^2 + v(x)$, with eigenvalue ω_{j_0}, then $u^0_{j_0}(x,t;r) = r\psi_{j_0}(x)\exp(i\omega_{j_0} t)$ is a time periodic solution of the linearized version of (1.1). Our main result is that there are generally Cantor families of nonlinear solutions close to these linear solutions.

Theorem 1.1. *Consider (1.1) with periodic boundary conditions. Fix $j_0 \geq 1$. There exists an open dense set $\mathcal{O}_{j_0} \subset \mathcal{V}_{\overline{\sigma}} \times \mathcal{A}_{\overline{\sigma}}$ such that if $(v, \mathcal{G}) \in \mathcal{O}_{j_0}$, then (1.1) has time periodic solutions close to $u^0_{j_0}$. More precisely, there exists $r_0 > 0$, and a C^3 function $\Omega(r)$, defined on $[0, r_0)$, and a set $\mathcal{C} \subset [0, r_0)$ of positive measure, such that if $r \in \mathcal{C}$, there exists a time-periodic solution $u(x,t;r)$ of (1.1) with frequency $\Omega(r)$, such that*

$$\begin{aligned} |u(x,t;r) - r\psi_{j_0}(x)\exp(i\Omega(r)t)| < Cr^2 \quad , \\ |\Omega(r) - \omega_{j_0}| < Cr^2 \quad . \end{aligned} \tag{1.2}$$

Remark. In fact, with some extra care the function $\Omega(r)$ can be chosen to be C^∞ (see Craig & Wayne (1991a)). Once again, in the interest of simplicity of exposition, we forego stating the strongest possible result.

2. Reduction to a lattice problem

We now reduce the question of the existence of periodic solutions of (1.1) to a question about functions in $\ell^2(\mathbb{N} \times \mathbb{Z})$. Begin by rescaling the time variable in (1.1) as $t \to \xi/\Omega$, where Ω has the dimensions of frequency. Then (1.1) becomes

$$-i\Omega\partial_\xi u = (-\partial_{xx} + v(x))u + g(u, \overline{u}, x) \quad , \tag{2.1}$$

Let $\{\psi_j(x)\}_{j=0}^\infty$ be the eigenfunctions of $\mathcal{L} = (-d^2/dx^2 + v(x))$ with periodic boundary conditions and let $\{\omega_j\}$ be the corresponding eigenvalues. Assuming that (2.1) does have a periodic solution, we can expand it in an eigenfunction-Fourier series as

$$u(x,\xi) = \sum_{j \in \mathbb{N},\ k \in \mathbb{Z}} \hat{u}(j,k)\psi_j(x)e^{-ik\xi} \quad . \tag{2.2}$$

If we substitute this *Ansatz* into (2.1) we find that the coefficients $\hat{u}(j,k)$ must satisfy the nonlinear equation

$$\begin{aligned} F(\hat{u})(j,k) = W(\hat{u})(j,k) + V(\Omega)\hat{u}(j,k) = W(\hat{u})(j,k) + (\omega_j - k\Omega)\hat{u}(j,k) = 0 \ ; \\ j \in \mathbb{N} \ k \in \mathbb{Z} \quad . \end{aligned} \tag{2.3}$$

The term $W(\hat{u})$ arises from expanding the nonlinear term in (2.1) in an eigenfunction-Fourier series. We will comment on some of its properties later.

We will use a Nash-Moser implicit function theorem to show that (2.3) has a solution. We begin by splitting (2.3) into a finite dimensional and an infinite dimensional piece in a way reminiscent of the Lyapunov-Schmidt procedure. We wish to construct perturbations of the linear solutions $u^0_{j_0}$. In particular, if $\Omega = \omega_{j_0}$, then $V(\omega_{j_0})(j_0, 1) = 0$. Define $N = \{(j,k) \in \mathbb{N} \times \mathbb{Z} \mid V(\omega_{j_0})(j,k) = 0\}$. Generically, $\dim(N) = 1$, and it is on that case that we will focus here. However, in the case of periodic boundary conditions, if the potential $v(x)$ is independent of x, (and in some other cases,) the eigenvalues come in pairs, and $\dim(N) \geq 2$. We discuss this **resonant** case briefly in Appendix A. (See also Craig & Wayne (1991) and Craig, Kuksin & Wayne (1993).)

Given N, let Q be the orthogonal projection onto $\ell^2(N) \subset \ell^2(\mathbb{N} \times \mathbb{Z})$, and let $P = Q^\perp$. Then (2.3) is equivalent to the pair of equations

$$PF(\hat{u}) = 0 \ ; \ QF(\hat{u}) = 0 \ . \tag{2.4}$$

The solutions we construct will be perturbations of the solution of the linear equation. In the lattice representation the linear equation is $V(\Omega)\hat{u} = 0$. If $N = \{\mathbf{l}_1, \mathbf{l}_2, \ldots, \mathbf{l}_n\}$, we can solve this equation by taking $\Omega = \omega_{j_0}$ and $\hat{u} = \phi(\mathbf{p}) = \sum_{i-1}^{n} p_i \delta_{\mathbf{l}_i}$, with $p_i \in \mathbb{C}$ and $\delta_\mathbf{l}$ the Kronecker delta function at the lattice site $\mathbf{l}$. Note that if $(l_{i1}, l_{i2}) = \mathbf{l}_i$ then $\phi(\mathbf{p})$ is the lattice representation of $u(x,\xi) = \sum_{i=1}^{n} p_i \psi_{l_{i1}}(x) \exp(i l_{i2} \xi)$, *i.e.* the solution of the linearized differential equation.

We will construct solutions of (2.4) of the form $\hat{u}(\mathbf{p}, \Omega) = \phi(\mathbf{p}) + u(\mathbf{p}, \Omega)$, with $u(\mathbf{p}, \Omega) \in \ell^2((\mathbb{N} \times \mathbb{Z}) \backslash N)$. The function u will be constructed inductively as a sequence of functions $u_n(\mathbf{p}, \Omega) \in \ell^2(B(L_n) \backslash N)$, where $B(L) = \{(j,k) \in \mathbb{N} \times \mathbb{Z} \mid j + |k| \leq L\}$, and $L_n = 2^n L_0$, with L_0 some large constant. (In particular, we require $N \subset B(L_0)$.) In order to start the induction we require a finite number of non-resonance conditions to hold among the eigenvalues of $\mathcal{L}$.

Definition 2.1. *A sequence* $\{\omega_j\}_{j=0}^{\infty}$ *is* (d_0, L_0) *nonresonant with* ω_{j_0}, *if for all* $(j,k) \in B(L_0) \backslash N$ *we have* $|k\omega_{j_0} - \omega_j| > d_0$.

Consider the set of potentials $v \in \mathcal{V}_{\overline{\sigma}}$. Then the sequences of eigenvalues $\{\omega_j\}$ for "most" operators $\mathcal{L} = -d^2/dx^2 + v(x)$ will be nonresonant, in a sense the following proposition makes precise.

Proposition 2.2. *Fix any* $L_* > 0$ *and* $1/2 < \eta < 1$. *There is an open and dense subset* $\mathcal{U}_{j_0} \subset \mathcal{V}_{\overline{\sigma}}$ *such that if* $v \in \mathcal{U}_{j_0}$, *then the eigenvalues of* $-d^2/dx^2 + v(x)$ *with either Dirichlet or periodic boundary conditions are* $(L_0^{-\eta}, L_0)$ *non-resonant with respect to* ω_{j_0} *for some* $L_0 > L_*$.

Proof. The proof is almost identical to the proof of Proposition 2.5 of Craig & Wayne (1991a) and hence we will not repeat it.

We can now solve the infinite dimensional part of the Lyapunov-Schmidt decomposition.

Theorem 2.3. *Consider (2.3) arising from (1.1). Fix* $3/5 < \eta < 1$. *Then there exists* L_*, *(depending on* $g(z_1, z_2, x)$*) such that if the frequency sequence* $\{\omega_j\}$ *is* (d_0, L_0) *non-resonant with respect to* ω_{j_0}, *for some* $d_0 \geq L_0^{-\eta}$ *and* $L_0 \geq L_*$, *then there exists* $r_0^p, r_0^\Omega > 0$, *a function* $u(\mathbf{p}, \Omega) \in \ell^2((\mathbb{N} \times \mathbb{Z}) \backslash N)$, *which is* C^3 *on the set*

$$\mathcal{N}_0 = \{(\mathbf{p}, \Omega) \mid ||\mathbf{p}|| < r_0^p, |\Omega - \omega_{j_0}| < r_0^\Omega\} \ ,$$

and satisfies $u(\|\mathbf{p}\|\exp(i\theta),\Omega)(j,k) = \exp(ik\theta)u(\|\mathbf{p}\|,\Omega)(j,k)$. There also exists a closed set $\mathcal{N} \subset \mathcal{N}_0$ such that for $(\mathbf{p},\Omega) \in \mathcal{N}$,

$$P[W(\phi(\mathbf{p}) + u(\mathbf{p},\Omega)) + V(\Omega)u(\mathbf{p},\Omega)] = 0 \ .$$

The proof of this theorem is given in section 4.

We now turn to the finite dimensional part of the problem. Its solution requires that a non-degeneracy condition hold on the nonlinear term. The form of this nondegeneracy condition depends on the resonant set N. For the remainder of this section we assume that $N = \{(j_0, 1)\}$. Arguments similar to those used to prove Proposition 2.2 show that this is true for an open and dense set of potentials. Since $\dim(N) = 1$, $\mathbf{p} = r\exp(i\theta)$. In Appendix A we discuss the case $\dim(N) = 2$ briefly.

Definition 2.4. *Define* $\mathbf{l}^* = (j_0, 1)$. *If* $t = \langle \delta_{\mathbf{l}^*}, \partial_r^3 W(\phi(r))|_{r=0}\rangle \neq 0$, *we say that (2.3) satisfies the twist condition with twist* t.

Here $\langle\cdot,\cdot\rangle$ is the ℓ^2 lattice inner product. If we expand the "potential" $\mathcal{G}(z_1,z_2,x) = a_{4,0}(x)z_1^4 + a_{3,1}(x)z_1^3 z_2 + a_{2,2}(x)z_1^2 z_2^2 + a_{1,3}(x)z_1 z_2^3 + a_{0,4}(x)z_2^4 + \ldots$, then the nonlinear term in the equation has the form $g(z_1,z_2,x) = a_{3,1}(x)z_1^3 + 2a_{2,2}(x)z_1^2 z_2 + 3a_{1,3}(x)z_1 z_2^2 + 4a_{0,4}(x)z_2^3 + \ldots$. One then computes that $t = 12\pi\int_0^\pi a_{2,2}(x)|\psi_{j_0}(x)|^4\,dx$. Clearly, by changing $a_{2,2}(x)$ by an arbitrarily small amount (*e.g.* by letting $a_{2,2}(x) \to a_{2,2}(x) + \epsilon$) we can make this integral non-zero. Thus we have,

Theorem 2.5. *There exists an open and dense subset of* $\mathcal{A}_{\overline{\sigma}}$ *such that if* $\mathcal{G}$ *is an element of this subset, (2.3) satisfies the twist condition.*

We can now solve the remainder of the Lyapunov-Schmidt problem.

Theorem 2.6. *Assume that the hypotheses of Theorem 2.3 are satisfied, that* $N = \{(j_0,1)\}$, *and that (2.3) satisfies the twist condition. Then there exists a* C^3 *function* $\Omega = \Omega(\|\mathbf{p}\|)$ *such that*

$$Q(W(\phi(\mathbf{p}) + u(\mathbf{p},\Omega(\|\mathbf{p}\|)) + V(\Omega(\|\mathbf{p}\|))\phi(\mathbf{p})) = 0 \ ,$$

and the set $\{\|\mathbf{p}\| \mid (\mathbf{p},\Omega(\|\mathbf{p}\|)) \in \mathcal{N}\}$ *has positive measure.*

If we set $\hat{u}(\mathbf{p},\Omega)(j,k) = (\phi(\mathbf{p}) + u(\mathbf{p},\Omega(\|\mathbf{p}\|))(j,k)$, and define $u(x,\xi;\|\mathbf{p}\|) = \sum_{j,k}\hat{u}(j,k)\psi_j(x)\exp(ik\xi)$, then Theorem 2.3 and Theorem 2.6 imply that $u(x,\xi;\|\mathbf{p}\|)$ is a (weak) solution of (1.1), periodic in time with frequency $\Omega(\|\mathbf{p}\|)$. In fact, all claims in Theorem 1.1 except for (1.2) follow from Theorem 2.3 and Theorem 2.6. We will prove these estimates in sections 4 and 5 where we also prove Theorem 2.3 and Theorem 2.6 and in addition show that $u(x,\xi;\|\mathbf{p}\|)$ is analytic as a function of (x,ξ).

3. Function Spaces

We will solve (2.3) in certain subspaces of $\ell^2(\mathbb{N}\times\mathbb{Z})$. Given m and s greater than or equal to zero, define Hilbert spaces

$$H^{m,s} = \{\hat{u} \in \ell^2(\mathbb{N} \times \mathbb{Z}) \mid \|\hat{u}\|^2_{m,s} = \alpha \sum_{\mathbf{l} \in \mathbb{N}\times\mathbb{Z}} (1 + |\mathbf{l}|^2)^s |\hat{u}(\mathbf{l})|^2 e^{2m|\mathbf{l}|} < \infty\} \ .$$

The constant α (which depends on m and s but not on u) is fixed below. The associated inner products will be denoted $\langle \cdot, \cdot \rangle_{m,s}$. We will denote the ordinary ℓ^2-norm by $\| \cdot \|_{0,0}$, or simply by $\| \cdot \|$. The ℓ^1 norm on the lattice will be denoted by $| \cdot |$.

Note that given any $\hat{u} \in H^{m,s}$, we can define a periodic function of x and ξ by

$$u(x, \xi) = \sum_{j,k} \hat{u}(j,k) \psi_j(x) e^{ik\xi} \tag{3.1}$$

Define $\|u\|_{m,s} = \|\hat{u}\|_{m,s}$. Suppose that $\hat{u}$ and $\hat{v}$ are elements of $H^{m,s}$, let u and v be defined as in (3.1), and let $(uv)(x,\xi)$ be their product. The following proposition expresses a particularly useful property of these norms.

Proposition 3.1. *If $m < \overline{\sigma}$, and $s > 1$, then there exists $\alpha = \alpha(m,s)$, such that $\|(uv)\|_{m,s} \le \|u\|_{m,s} \|v\|_{m,s}$.*

Proof. The proof is reminiscent of the proof that the Sobolev spaces are Banach algebras for appropriate values of the indices. Details are given in Appendix B.

This proposition has the following important corollary. Let $(f(u(x,\xi), \overline{u}(x,\xi), x))\hat{}$ be the function in $\ell^2(\mathbb{N} \times \mathbb{Z})$ obtained by making an eigenfunction-Fourier series expansion of $f(u(x,\xi), \overline{u}(x,\xi), x)$.

Corollary 3.2. *If $f \in \mathcal{A}_{\overline{\sigma}}$, $m < \overline{\sigma}$, $s > 1$, and if $\|\hat{u}\|_{m,s} < 1/2$, then $(f(u(x,\xi), \overline{u}(x,\xi), x))\hat{} \in H^{m,s}$. Indeed, if $|f(z_1, z_2, x)| \le C|z_1|^{n_1}|z_2|^{n_2}$ for all $|z_i| < 1$, then $\|(f(u(x,\xi), \overline{u}(x,\xi), x))\hat{}\|_{m,s} \le C\|\hat{u}\|^{n_1+n_2}_{m,s}$.*

Proof. The proof (see Appendix B) is a straightforward consequence of Proposition 3.1.

Suppose T is a linear operator mapping $H^{m,s}$ to itself. The natural generalization of the operator norms used in Craig & Wayne (1991a) to these new spaces is

$$\|T\|^2_{m,s} = \max\{\sup_{\mathbf{l}_1} \sum_{\mathbf{l}_2} (1 + |\mathbf{l}_1 - \mathbf{l}_2|^2)^s e^{2m|\mathbf{l}_1 - \mathbf{l}_2|} |T(\mathbf{l}_1, \mathbf{l}_2)|^2 \ ,$$
$$\sup_{\mathbf{l}_2} \sum_{\mathbf{l}_1} (1 + |\mathbf{l}_1 - \mathbf{l}_2|^2)^s e^{2m|\mathbf{l}_1 - \mathbf{l}_2|} |T(\mathbf{l}_1, \mathbf{l}_2)|^2\} \ .$$

As in Craig & Wayne (1991a), if $\|T\|^{op}$ denotes the "ordinary" operator norm, we have $\|T\|^{op} \le \|T\|_{m,s}$. More relevant to our needs is the following. Suppose that $\hat{w}, \hat{v} \in H^{m,s}$. Let $w(x,\xi)$ and $v(x,\xi)$ be defined as in (3.1). Define $\widehat{W} : H^{m,s} \to H^{m,s}$ by $\hat{W}\hat{v} = (wv)\hat{}$.

Proposition 3.3. *Under the hypotheses of Proposition 3.1, there exists a constant C_{op}, independent of $\hat{w}$, such that $\|\widehat{W}\|_{m,s} \le C_{op} \|\hat{w}\|_{m,s}$.*

Proof. See Appendix B.

The value of this proposition is that if we wish to consider the norm of the lattice operator $(\widehat{D_u f})\hat{v}$, we can think of this in the (x,ξ) variables where it is just $D_u f(u(x,\xi),\overline{u}(x,\xi),x)v(x,\xi) + D_{\overline{u}} f(u(x,\xi),\overline{u}(x,\xi),x)\overline{v}(x,\xi)$. But we can compute the norm of the product of these two functions using Proposition 3.1 and Corollary 3.2, and we obtain

Corollary 3.4. *Under the hypotheses of Corollary 3.2, there exists a constant C_f such that $\|\widehat{D_u f}\|_{m,s} < C_f$. If $|f(z_1,z_2,x)| \le C|z_1|^{n_1}|z_2|^{n_2}$ for all $|z_i| < 1$, then $\|\widehat{D_u f}\|_{m,s} < C_f\|u\|_{m,s}^{n_1+n_2-1}$.*

We will also need the following very elementary property of these norms.

Lemma 3.5. *Let Π_L be the orthogonal projection onto $\ell^2(B(L)) \subset \ell^2(\mathbb{N}\times\mathbb{Z})$. Then $\|(\mathbf{1}-\Pi_L)u\|_{m,s} \le e^{-\mu L}\|u\|_{m-\mu,s}$.*

Proof. The proof follows immediately from the definition of the norm.

We will construct solutions $\hat{u} = \phi(\mathbf{p}) + u(\mathbf{p},\Omega) \in H^{m,s}$, and will demand that they be analytic on certain domains. Recall that $\mathbf{p} = (p_1,p_2,\ldots,p_n)$. Each $p_i \in \mathbb{C}$, which for the moment we identify with $\mathbb{R}^2$. Let $\mathcal{N} \subset (\mathbb{R}^2)^n \times \mathbb{R}$. If ρ^p, ρ^Ω are non-negative constants, define a complex neighborhood of $\mathcal{N}$ to be, $D(\mathcal{N};\rho^p,\rho^\Omega) = \{(\mathbf{z},\tilde{\Omega}) \in (\mathbb{C}^2)^n \times \mathbb{C} \mid \sqrt{\|\mathbf{z}-\mathbf{p}\|^2} < \rho^p, |\Omega - \tilde{\Omega}| < \rho^\Omega\}$, for some $(\mathbf{p},\Omega) \in \mathcal{N}$.

4. The P-equation

In this section we solve the equation

$$P(W(\hat{u}) + V(\Omega)\hat{u}) = 0 \tag{4.1}$$

The presence of small denominators necessitates the use of a Nash-Moser approach. As in Craig & Wayne (1991a), we solve the problem inductively in a sequence of boxes $B(L_n) = \{(j,k) \in \mathbb{N}\times\mathbb{Z} \mid j + |k| \le L_n\}$, where $L_n = 2^n L_0$, with L_0 some large fixed constant. Define $\Pi_{\overline{L}}$ to be orthogonal projection onto $\ell^2(B(L)\backslash N)$.

We begin the induction by studying

$$\Pi_{\overline{L_0}} P(W(\hat{u}) + V(\Omega)\hat{u}) \equiv W_0(\hat{u}) + V_0(\Omega)\hat{u} = 0 \ . \tag{4.2}$$

We will write the solution of this problem in the form $\hat{u} = \phi(\mathbf{p}) + u_0(\mathbf{p},\Omega)$, with $u_0 \in \ell^2(B(L_0)\backslash N)$.

We can rewrite (4.2) as the fixed point problem

$$u_0(\mathbf{p},\Omega) = \frac{-1}{V_0(\Omega)} W_0(\phi(\mathbf{p}) + u_0(\mathbf{p},\Omega)) \ . \tag{4.3}$$

Suppose $u_0 \in H^{m,s}$, $m < \overline{\sigma}$, $s > 1$, and $\|u_0\|_{m,s} < 1/2$. By Corollary 3.2, $W_0(\phi(\mathbf{p}) + u_0(\mathbf{p},\Omega)) \in H^{m,s}$, and from the hypotheses on the nonlinear term we see that $\|W_0(\phi(\mathbf{p})+u_0(\mathbf{p},\Omega))\|_{m,s} \le C_f(\|\mathbf{p}\| + \|u_0\|_{m,s})^3$. On the other hand since $V_0(\Omega)$ is a diagonal operator with $|V_0(\Omega)(j,k)| = |j\Omega - \omega_k|$, and since the frequency sequence $\{\omega_j\}$ is (d_0,L_0) non-resonant with ω_{j_0}, we have $\|(V_0(\Omega))^{-1}\|_{m,s} \le 2/d_0$, so long as $|\Omega - \omega_{j_0}| < d_0/(2L_0)$. Thus, if $\|u_0\|_{m.s} \le 8C_f\|\mathbf{p}\|^3/d_0$, and if $\|\mathbf{p}\|^2 < d_0/(16C_f)$, we have $\|W_0(\phi(\mathbf{p})+u_0(\mathbf{p},\Omega))\|_{m,s} \leq 8C_f\|\mathbf{p}\|^3/d_0$, and (4.3) maps this set into itself. To see that the transformation is a contraction there we note that

$$\begin{aligned}
&\|W_0(\phi(\mathbf{p}) + u^0(\mathbf{p},\Omega)) - W_0(\phi(\mathbf{p}) + \tilde{u}^0\mathbf{p},\Omega))\|_{m,s} \\
&= \| \int_0^1 D_u W_0(\phi(\mathbf{p}) + \tilde{u}^0 + \tau(u^0 - \tilde{u}^0))(u^0(\mathbf{p},\Omega) - \tilde{u}^0(\mathbf{p},\Omega))d\tau \|_{m,s} \\
&\leq \sup_{0\leq\tau\leq 1} \|D_u W_0(\phi(\mathbf{p}) + \tilde{u}^0 + \tau(u^0 - \tilde{u}^0))\|_{m,s} \|u^0(\mathbf{p},\Omega) - \tilde{u}^0(\mathbf{p},\Omega))\|_{m,s} \ .
\end{aligned}$$

By Corollary 3.4, and the hypothesis that $\max(\|u^0\|_{m,s}, \|\tilde{u}^0\|_{m,s}) \leq 8\|\mathbf{p}\|^3/d_0$, we find that $\|D_u W_0(\phi(\mathbf{p}) + \tilde{u}^0 + \tau(u^0 - \tilde{u}^0)\|_{m,s} \leq 2C_f\|\mathbf{p}\|^2$. But if $\|\mathbf{p}\|^2 < (2C_f)^{-1}$, this is a contraction. Thus, we have

Proposition 4.1. *If the frequency sequence $\{\omega_j\}$ is (d_0, L_0) nonresonance with ω_{j_0}, then there exist constants $c_p > 0$ and $L_0 > 0$, such that if $\|\mathbf{p}\| < \sqrt{c_p d_0}$, and $|\Omega - \omega_{j_0}| < d_0/2L_0$, then (4.2) has a unique solution, $\hat{u} = \phi(\mathbf{p}) + u^0$, which satisfies $\|u^0\|_{m,s} \leq 8C_f\|\mathbf{p}\|^3/d_0$.*

We can conclude immediately from the analyticity of (4.2) (and the Implicit Function Theorem), that u^0 is an analytic function of $(\mathbf{p},\Omega)$. Furthermore, the argument above is insensitive to whether or not $\mathbf{p}$ and Ω are real or complex. Thus we can extend the solution to a complex neighborhood. We will also need some information about derivatives of this solution which we immediately obtain from Cauchy's theorem. In particular, we require the C^1 norm of u^0 to be bounded by 1, and this requires us to shrink slightly the size of the domain for the $\mathbf{p}$-variables. Define $\mathcal{N}_0 = \{(\mathbf{p},\Omega) \in \mathbb{R}^{2n} \times \mathbb{R} \mid |\mathbf{p}| < r_0^p = (c_p d_0)^{4/3} \ , \ |\Omega - \omega_{j_0}| < r_0^\Omega = d_0/L_0\}$. Set $\rho_0^p = r_0^p$ and $\rho_0^\Omega = r_0^\Omega$. Then Proposition 4.1 has the following corollary.

Corollary 4.2. *If the frequency sequence $\{\omega_j\}$ is (d_0, L_0) nonresonance with ω_{j_0}, then there exist constants $c_p > 0$ and $L_0 > 0$, such that there exists a unique solution $\hat{u} = \phi(\mathbf{p}) + u^0(\mathbf{p},\Omega)$ to (4.2), defined and analytic on $D(\mathcal{N}_0; \rho_0^p, \rho_0^\Omega)$. Furthermore, if $8^2C_f(r_0^p)^3L_0 < d_0^{(7/3)}$, the C^1 norm of u^0 is bounded by 1.*

Remark. We could estimate higher derivatives of u^0 in a similar fashion.

We note an additional useful property of the solution u^0. If $u(x,t)$ is a solution of (1.1), so is $u^\theta(x,t) = u(x,t+\theta)$ for any θ. In terms of the Fourier-eigenfunction expansion of the problem, this means that if $\hat{u}(j,k)$ is a solution of (2.3), then so is $\exp(ik\theta)\hat{u}(j,k)$, from which we see that if $T_\theta\hat{u}(j,k) \equiv \exp(ik\theta)\hat{u}(j,k)$, $F(T_\theta\hat{u})(j,k) = \exp(ik\theta)F(\hat{u})(j,k)$. We wish to relate this to the dependence on the parameter $\mathbf{p}$. Suppose that $N = \{(j_1,k_1),\ldots,(j_n,k_n)\}$. Define $T_\theta\mathbf{p} = \{\exp(ik_1\theta)p_1,\ldots$ $\ldots,\exp(ik_n\theta)p_n\}$. Note that $\phi(T_\theta\mathbf{p})(j,k) = ((T_\theta\phi)(\mathbf{p}))(j,k)$, justifying this abuse of notation. But since $V_0(\Omega)$ is independent of $\mathbf{p}$, this implies the following corollary

Corollary 4.3. *Under the hypotheses of Corollary 4.2,*

$$u^0(T_\theta\mathbf{p},\Omega)(j,k) = (T_\theta u^0(\mathbf{p},\Omega))(j,k) \ .$$

We take the solution u^0 as a first approximation to the solution of (4.1) and then refine that approximation iteratively using Newton's method. To explain the iteration suppose we have an approximate solution $\tilde{u} \in \ell^2(B(\tilde{L})\backslash N) \cap H^{\tilde{m},s}$. We assume that this approximation satisfies:

H1. $\tilde{u}(\mathbf{p},\Omega)$ is C^3 as a function of $(\mathbf{p},\Omega) \in \mathcal{N}_0$ and analytic on $D(\tilde{\mathcal{N}};\tilde{\rho}^p,\tilde{\rho}^\Omega)$, with $\|\tilde{u}(\mathbf{p},\Omega)\|_{\tilde{m},s} \leq C\|\mathbf{p}\|$. Suppose further that it satisfies $\tilde{u}(T_\theta\mathbf{p},\Omega)(j,k) = e^{ik\theta}\tilde{u}(\mathbf{p},\Omega)(j,k)$. Finally assume that the C^1 norm of $\tilde{u}$ is bounded by 1 and that

$$\|PF(\phi(\mathbf{p})+\tilde{u})\|_{\tilde{m},s} \leq \tilde{\epsilon}\|\mathbf{p}\|^3 \quad . \tag{4.4}$$

To apply Newton's method we linearize the equation about this approximate solution and then invert the linear operator. In controlling the inverse a special role is played by sites $\mathbf{l}$ at which $|V(\Omega)(\mathbf{l})|$ is small. Define a site $\mathbf{l} \in \mathbb{N}\times\mathbb{Z}$ to be a **singular site** if $|V(\Omega)(\mathbf{l})| < .01$. Our second inductive hypothesis is:

H2. If $A \subset B(3\tilde{L})$ is any lattice region free of singular sites, and if $u' \in \ell^2((\mathbb{N}\times\mathbb{Z})\backslash N)$ satisfying $\|u'-\tilde{u}\|_{\tilde{m},s} \leq \tilde{\rho}/2$, then

$$\|((D_u PF(\phi(\mathbf{p})+u'))|_{\overline{B(\tilde{L})\cup A}})^{-1}\|_{\tilde{m},s} \leq \tilde{X} \quad . \tag{4.5}$$

Our next approximation to the solution will be defined in the larger space $\ell^2(B(2\tilde{L})\backslash N)$. In order to construct it we must study the spectrum of $(D_u PF)|_{\overline{B(2\tilde{L})}}$, and omit from $\tilde{\mathcal{N}}$ those values of $(\mathbf{p},\Omega)$ for which this spectrum is too near to zero. First remark that since $V(\Omega)(j,k) = j\Omega - \omega_k$, and since the asymptotics of the eigenvalues of the Sturm-Liouville problem (Pöschel & Trubowitz (1987), Craig & Wayne (1991a)) imply that $\omega_{2k},\omega_{2k-1} \approx (2k)^2 + \text{constant}$, for $k \geq 1$, we see that if (j,k) and (j',k') are two singular sites with $k < k'-1$, then $|j-j'|+|k-k'| \approx \mathcal{O}(k)$. Thus, singular sites either come in pairs $(j,k),(j,k+1)$, or they are widely separated (*i.e.* by distances of order their distance from the origin.) Define each such site or pair of sites to be a **singular region**. Thus, we have proven the following lemma

Lemma 4.4. *There exists $C > 0$, such that if $\mathbf{l}$ and $\mathbf{l}'$ are singular sites in different singular regions in $B(\tilde{L})^c$, then $|\mathbf{l}-\mathbf{l}'| \geq C\tilde{L}$.*

We note that the fact that any two singular regions are widely separated for any choice of the frequency Ω is one of the reasons that the Schrödinger equation is easier to treat than the wave equation. In Craig & Wayne (1991a), we were forced to restrict our attention to Diophantine frequencies Ω in order to insure that the singular regions remained widely separated.

The phase invariance of the solution set has useful consequences for the spectrum of the linearized operator. Let $T_\theta\mathbf{p}$ be as above. Then for any $E \subset \mathbb{N}\times\mathbb{Z}$, we have

Lemma 4.5. *The spectrum of the operator $D_u PF|_E(\phi(T_\theta\mathbf{p})+\tilde{u}(T_\theta\mathbf{p},\Omega))$ is independent of θ.*

Proof. Note that $T_\theta PF(\phi(\mathbf{p})+\tilde{u}(\mathbf{p},\Omega)) = PF(\phi(T_\theta\mathbf{p})+\tilde{u}(T_\theta\mathbf{p},\Omega))$. Differentiating with respect to u gives

$$T_\theta D_u PF(\phi(\mathbf{p})+\tilde{u}(\mathbf{p},\Omega)) = D_u PF(\phi(T_\theta\mathbf{p})+\tilde{u}(T_\theta\mathbf{p},\Omega))T_\theta \quad . \tag{4.6}$$

Thus, $D_u PF((\phi(\mathbf{p})+\tilde{u}(\mathbf{p},\Omega))$ and $D_u PF(\phi(T_\theta\mathbf{p})+\tilde{u}(T_\theta\mathbf{p},\Omega))$ are unitarily equivalent.

Recall that $D_u PF|_E(\phi(\mathbf{p})+\tilde{u}(\mathbf{p},\Omega))\cdot v = (D_u W|_E(\phi(\mathbf{p})+\tilde{u}(\mathbf{p},\Omega))+V(\Omega)|_E)\cdot v$, and Corollary 3.4 implies $\|D_u W|_E\|_{\tilde{m},s} \leq C_f\|\mathbf{p}\|^2$. In particular, since $\|\mathbf{p}\| < Cr_0$, we have

Lemma 4.6. *There exists $c_{r_0} > 0$ such that if $r_0 < c_{r_0}$, and $E \subset \mathbb{N} \times \mathbb{Z}$ intersects no singular region, then* $\text{spec}(D_u PF|_E(\phi(\mathbf{p}) + \tilde{u}))$ *is bounded away from zero by at least .005. Similarly, if E contains only one singular region, then* $\text{spec}(D_u PF|_E(\phi(\mathbf{p}) + \tilde{u}))$ *has at most two eigenvalues within a distance .005 of zero.*

Proof. This follows from the fact that we can compute $\text{spec}(V(\Omega)|_E)$ and if

$$\|D_u W|_E\|_{\tilde{m},s} \leq C r_0^2 < .005 \quad ,$$

the spectrum of $D_u PF|_E$ can differ from that of $V(\Omega)|_E$ by at most that amount.

Now suppose that $\tilde{\ell} << \tilde{L}$ and consider $\text{spec}(D_u PF|_{C_{\tilde{\ell}}(S)})$ where S is some singular region in $B(2\tilde{L}) \backslash B(\tilde{L})$ and $C_{\tilde{\ell}}(S) = \{(j,k) \mid \text{dist}(S,(j,k)) \leq \tilde{\ell}\}$. By the preceding lemma, and our observation that singular regions are widely separated (*i.e.* by $\mathcal{O}(\tilde{L})$), $(D_u PF|_{C_{\tilde{\ell}}(S)})$ has at most two eigenvalues with absolute value less than .005. We wish to show that by removing a small set of $(\mathbf{p}, \Omega)$, we can insure that neither of these eigenvalues gets too close to zero.

First, for any $v \in \ell^2(C_{\tilde{\ell}}(S))$, the quadratic form $\langle v, D_u PF|_{C_{\tilde{\ell}}(S)} v\rangle_{0,0}$ is a monotonic function of Ω if r_0 is sufficiently small. (This follows just by differentiating with respect to Ω.) Thus, the eigenvalues of $D_u PF|_{C_{\tilde{\ell}}(S)}$ are also monotonic in Ω. Furthermore, this monotonicity property implies that for changes in Ω by an amount $\Delta\Omega$, the corresponding eigenvalue moves by an amount at least $(\Delta\Omega)\tilde{L}/2$, again assuming r_0 is sufficiently small. Thus, if we label the eigenvalues of $D_u PF|_{C_{\tilde{\ell}}(S)}$ as $e_1(\mathbf{p},\Omega) \leq e_2(\mathbf{p},\Omega) \leq \ldots$, and if for some $\tilde{\delta} < 1$ there exists $e_i(\mathbf{p},\Omega)$ with $|e_i(\mathbf{p},\Omega)| < \tilde{\delta}$, then there exists some Ω^* with $|\Omega - \Omega^*| < 2\tilde{\delta}/\tilde{L}$, such that $e_i(\mathbf{p},\Omega^*) = 0$. Since the eigenvalues of a self-adjoint matrix are continuous functions of parameters of the matrix, there will be a continuous surface Z_i^0 such that $Z_i^0 = \{(\mathbf{p}^*,\Omega^*) \mid e_i(\mathbf{p}^*,\Omega^*) = 0\}$. Furthermore, by Lemma 4.5, Z_i^0 is invariant under the rotation T_θ. If we delete from $\tilde{\mathcal{N}}$ any points $(\mathbf{p},\Omega)$ with $\text{dist}((\mathbf{p},\Omega), Z_i^0) < \tilde{\delta}/(2\tilde{L})$, then $|e_i(\mathbf{p},\Omega)| \geq \tilde{\delta}$ for any remaining parameter values $(\mathbf{p},\Omega)$. For $\mathbf{p} = 0$, the nonlinear component $D_u W \equiv 0$, so the surface Z_i^0 passes through the point $\Omega^* - \omega_j/k = 0$ for some $(j,k) \in S$.

We next consider the shape of the surface Z_i^0 for $\mathbf{p} \neq 0$. Note that the monotonicity of the quadratic form considered in the previous paragraph implies that there exists a constant $C > 0$ such that if $(\mathbf{p}_1, \Omega_1)$ and $(\mathbf{p}, \Omega_1)$ are any two points in Z_i^0, then $|\Omega_1 - \Omega_2| < C(|\|\mathbf{p}_1\| + \|\mathbf{p}_2\||)|\|\mathbf{p}_1\| - \|\mathbf{p}_2\||/\tilde{L}$. This follows by noting that if the inequality were reversed, the monotonicity of the quadratic form would force $e_i(\mathbf{p}_1,\Omega_1) \neq e_i(\mathbf{p}_2,\Omega_2)$. From this estimate we see that Z_i^0 is contained in a parabolic cone of the form $(\omega_j/k) \pm C\|\mathbf{p}\|^2/\tilde{L}$. Thus, combining all these observations we find:

Proposition 4.7. *Consider all possible singular regions in $B(2\tilde{L})\backslash B(\tilde{L})$. There exists a constant $C > 0$, such that for any singular region, there are at most two zero surfaces Z_i^0, each of which is contained in a parabolic cone of the form $(\omega_j/k) \pm C\|\mathbf{p}\|^2/\tilde{L}$ for some $(j,k) \in B(2\tilde{L})\backslash B(\tilde{L})$. There exists $c_{r_0} > 0$ (independent of $\tilde{\mathcal{N}}$) such that if $r_0 < c_{r_0}$ and if we delete from $\tilde{\mathcal{N}}$ all points $(\mathbf{p},\Omega)$ with $\text{dist}((\mathbf{p},\Omega), Z_i^0) < \tilde{\delta}/(2\tilde{L})$ for all zero surfaces Z_i^0, then for any remaining $(\mathbf{p},\Omega)$, and for any singular region S, we have $\text{dist}(\text{spec}(D_u PF|_{C_{\tilde{\ell}}(S)}), 0)) \geq \tilde{\delta}$.*

We can repeat this excision process, considering this time not the spectrum of $D_u PF|_{C_{\tilde{\ell}}(S)}$, but $D_u PF|_S$. If we excise as above the values of $(\mathbf{p},\Omega)$ for which the eigenvalues of $D_u PF|_S$ are too close to zero, we obtain a set which we define to be

$\tilde{\tilde{\mathcal{N}}}$, such that for $(\mathbf{p}, \Omega) \in \tilde{\tilde{\mathcal{N}}}$,

$$\mathrm{dist}(\mathrm{spec}(D_u PF|_{C_{\tilde{\imath}}(S)}), 0)) \geq \tilde{\delta} \text{ and } \mathrm{dist}(\mathrm{spec}(D_u PF|_S), 0)) \geq \tilde{\delta} .$$

Let us specialize for a moment to the case $\dim(N) = 1$. Fix a constant $C_t \in \mathbb{R}\backslash 0$. We say that a set $\tilde{\mathcal{N}}$ has the **intersection property** if there exists $0 < \gamma(\overline{s}) < 1$, such that for any surface of the form $\Omega = \omega_{j_0} + C_t\|\mathbf{p}\|^2(1 + \mathcal{S}(\|\mathbf{p}\|))$, with the C^1 norm of $\mathcal{S}$ bounded by $\overline{s} < \infty$ on $\mathcal{N}_0$, one has $\mathrm{meas}(\{\|\mathbf{p}\| \in (0, r_0^p) \mid (\mathbf{p}, \Omega(\|\mathbf{p}\|)) \in \tilde{\mathcal{N}}\} \geq (1 - \gamma(\overline{s}))r_0$. We call $1 - \gamma(\overline{s})$ the **intersection constant** of $\tilde{\mathcal{N}}$.

There are at most $C\tilde{L}^2$ singular regions in the annulus $B(2\tilde{L})\backslash B(\tilde{L})$. Thus, given any surface $(\mathbf{p}, \Omega(\|\mathbf{p}\|))$ of the form above the total measure of the excised set of $\|\mathbf{p}\|$ is bounded by $C\tilde{L}^2\sqrt{\tilde{\delta}}$, for some constant C depending on C_t, provided $\tilde{L} >> 1$. This implies the following result.

Proposition 4.8. *Suppose that* $\dim(N) = 1$, *and that* $\tilde{\mathcal{N}}$ *satisfies the intersection property with intersection constant* $(1-\gamma(\overline{s}))$. *There exists* $C > 0$ *(depending on* C_t*), such that if* $\tilde{L}$ *is sufficiently large (independent of* $\tilde{\mathcal{N}}$*) and* $C\tilde{L}^2\sqrt{\tilde{\delta}} < \tilde{\gamma}(1-\gamma(\overline{s}))r_0$ *for some* $0 < \tilde{\gamma} < 1$, *then* $\tilde{\tilde{\mathcal{N}}}$ *satisfies the intersection property with intersection constant* $(1 - \tilde{\gamma}(\overline{s}))(1 - \tilde{\gamma})$.

The domain $\tilde{\tilde{\mathcal{N}}}$ is the one on which we construct our improvement to the approximate solution $\phi(\mathbf{p}) + \tilde{u}(\mathbf{p}, \Omega)$. We do so using the following inductive lemma.

Lemma 4.9. *Suppose that* $\tilde{u}$ *is an approximate solution of (4.1) satisfying hypotheses H1 and H2. There exist constants* C_s, C_X, c_ρ *and* γ *such that if we define* $\Theta = C_r r_0^2 \tilde{X}\tilde{\delta}^{-1}\tilde{\mu}^{-6}\exp(-\tilde{\mu}\tilde{\ell}/2)$, $\tilde{\tilde{X}} = C_X(1 - \Theta)^{-1}\gamma^s r_0^2 \tilde{X}\tilde{\delta}^{-1}\tilde{\mu}^{-6}\exp(\gamma)$ *and* $\tilde{\tilde{\rho}} = \min(\tilde{\rho}, c_\rho/(\tilde{L}\tilde{\tilde{X}}))$ *and if* $\tilde{\mu}\tilde{L} > 1$, $\Theta < 1$, *and* $2\tilde{\tilde{X}}\tilde{\epsilon}r_0^3 < \tilde{\tilde{\rho}}$, *then there exists a function* $\tilde{v}$, *analytic on* $D(\tilde{\tilde{\mathcal{N}}}; \tilde{\tilde{\rho}}, \tilde{\tilde{\rho}})$, *and* C^3 *on* $\mathcal{N}_0$, *with* $\|\tilde{v}\|_{\tilde{m}-3\tilde{\mu},s} \leq \tilde{\tilde{X}}\tilde{\epsilon}\|\mathbf{p}\|^3$, *and* $\tilde{v}(T_\theta\mathbf{p}, \Omega) = T_\theta\tilde{v}(\mathbf{p}, \Omega)$, *such that if* $\tilde{\tilde{u}} = \tilde{u} + \tilde{v}$ *and* $\tilde{\tilde{\epsilon}} = C\tilde{L}(\tilde{\epsilon}^2 + \tilde{\epsilon}\exp(-\tilde{\mu}\tilde{L}))(1 + \tilde{\mu}^{-1}\tilde{X})^2$, *then* $\|PF(\phi(\mathbf{p}) + \tilde{\tilde{u}})\|_{\tilde{m}-4\tilde{\mu},s} \leq \tilde{\tilde{\epsilon}}\|\mathbf{p}\|^3$ *for* $(\mathbf{p}, \Omega) \in D(\tilde{\tilde{\mathcal{N}}}; \tilde{\tilde{\rho}}, \tilde{\tilde{\rho}})$. *Furthermore, if* $A \subset B(6\tilde{L})$ *contains no singular sites, and* $\|\tilde{\tilde{v}}\|_{\tilde{m}-3\tilde{\mu},s} < \tilde{\tilde{\rho}}/2$, *then* $\|(D_u PF|_E(\phi(\mathbf{p}) + \tilde{\tilde{u}} + \tilde{\tilde{v}}))^{-1}\||_{\tilde{m}-3\tilde{\mu},s} \leq \tilde{\tilde{X}}$, *for all* $E \subset (B(2\tilde{L}) \cup A)\backslash N$ *and* $(\mathbf{p}, \Omega) \in D(\tilde{\tilde{\mathcal{N}}}; \tilde{\tilde{\rho}}, \tilde{\tilde{\rho}})$.

Proof. See Appendix C.

We now use this lemma to complete the construction of a solution of $PF(\phi(\mathbf{p}) + u) = 0$. The approximate solution $u^0(\mathbf{p}, \Omega)$ constructed above satisfies the hypothesis H1 with $\tilde{L} = L_0$, $\tilde{\mathcal{N}} = \mathcal{N}_0$, and $(\tilde{\rho}^p, \tilde{\rho}^\Omega) = (\rho_0^p, \rho_0^\Omega)$, except for (4.4) which we now verify. If we choose $m_* < \overline{\sigma}$, $s > 1$, we can compute

$$\begin{aligned}\|PF(\phi(\mathbf{p}) + u^0)\|_{m_*,s} &= \|(\mathbf{1} - \Pi_{B(L_0)})PF(\phi(\mathbf{p}) + u^0)\|_{m_*,s} \\ &\leq \exp(-\mu^* L_0)\|PF(\phi(\mathbf{p}) + u^0)\|_{m_*-\mu_*,s} \leq Ce^{-\mu_* L_0}\|\mathbf{p}\|^3 \quad ,\end{aligned}$$

for any $\mu_* < m_*$. Take $\mu_* = m_*/2$, $m_0 = m_*/2$, and $\epsilon_0 = \exp(-m_0 L_0)$, and we see that (4.4) holds with $\tilde{\epsilon} = \epsilon_0$.

Next consider (4.5). Suppose $A \subset B(3L_0)$ is any region free of singular sites. Then $D_u PF|_{A\cup\overline{B(L_0)}}(\phi(\mathbf{p}) + u') = D_u PW|_{A\cup\overline{B(L_0)}}(\phi(\mathbf{p}) + u') + V(\Omega)|_{A\cup B(L_0)}$. Then $\|u'\|_{m_0,s} \leq \|u_0\|_{m_0,s} + \|u' - u_0\|_{m_0,s} \leq 8r_0^3/d_0 + \rho_0^p = 8r_0^3/d_0 + (c_p d_0)^{4/3}$. Hence, by Corollary 3.4, $\|D_u PW|_{A\cup\overline{B(L_0)}}(\phi(\mathbf{p}) + u')\|_{m_0,s} \leq C_f(r_0 + 8r_0^3 d_0^{-1} + (c_p d_0)^{4/3})^2$.

If $r_0 < \sqrt{d_0} << 1$, this is bounded by $C(r_0^2 + c_p d_0)$. But $\|(V(\Omega)|_{A\cup B(L_0)})^{-1}\|_{m_0,s} \leq C d_0^{-1}$, by hypothesis. Thus we have

Lemma 4.10. *Let $A \subset B(3L_0)$ be free of singular sites. There exists a constant $C > 0$ such that if c_p is sufficiently small, then $\|D_u PF|_{A\cup\overline{B(L_0)}}(\phi(\mathbf{p})+u'(\mathbf{p},\Omega)))^{-1}\|_{m_0,s} \leq C d_0^{-1}$ for all $(\mathbf{p},\Omega) \in D(\mathcal{N}_0; \rho_0^p, \rho_0^\Omega)$ and all u' satisfying $\|u' - u_0\|_{m_0,s} \leq \rho_0^p = (c_p d_0)^{4/3}$.*

Proof. Just expand $(D_u PW + V(\Omega))^{-1}$ with a Neumann series.

This allows us to start the induction. Define $L_n = 2^n L_0$, $\ell_n = L_n^\alpha$, (with $\alpha << 1$), $\delta_n = L_n^{-\beta}$, $\mu_n = m_0/(32(1+n^2))$, $m_{n+1} = m_n - 4\mu_n$, $X_0 = C d_0^{-1}$, $\Theta_n = C r_0^2 X_n \delta_n^{-1} \mu_n^{-6} \exp(-\mu_n \ell_n/2)$, $X_{n+1} = C(1-\Theta_n)^{-1}\gamma^s \exp(\gamma) r_0^2 X_n \delta_n^{-1} \mu_n^{-6}$, $\rho_{n+1} = \min(\rho_n^p, \rho_n^\Omega, c_p/(L_n X_n))$, and $\epsilon_{n+1} = C L_n(\epsilon_n^2 + \epsilon_n \exp(-\mu_n L_n))(1 + \mu_n^{-1} X_n \delta_n^{-1})^2$. Finally, we define $\mathcal{N}_{n+1}$ by making deletions from $\mathcal{N}_n$ as in Proposition 4.7.

Some simple calculations show that $\delta_n \approx 2^{-n\beta} L_0^{-n\beta}$, while $X_n \approx (n!)^{12} r_0^{2(n-1)} L_0^{n\beta} d_0^{-1}$. On the other hand, there exists $\kappa' > 1$ such that $\exp(-\mu_n L_n) \leq \exp(-\kappa'^n m_0 L_0)$, which if inserted into the inductive definition of ϵ_n implies $\epsilon_{n+1} \leq \exp(-\kappa^n m_0 L_0)$, for some (slightly smaller) $\kappa > 1$, provided L_0 is sufficiently large. We also find that $\Theta_n \approx (n!)^{12} 2^{n\beta} L_0^{n\beta} r_0^{2(n-1)} \exp(-\frac{m_*}{32(n^2+1)} 2^{n\alpha\beta} L_0^{\alpha\beta})$. For L_0 sufficiently large we have $\Theta_n \leq 1/2 < 1$ for all n. Similarly, one has $\epsilon_n r_0^3 < \rho_{n+1} X_{n+1}^{-1}/2$ for all n, again if L_0 is sufficiently large.

Thus, we may apply Lemma 4.9 infinitely often to construct a sequence of approximate solutions $u_{n+1} = u_n + v_n$, defined and analytic on $D(\mathcal{N}_{n+1}; \rho_{n+1}, \rho_{n+1})$. For $(\mathbf{p},\Omega) \in D(\mathcal{N}_{n+1}; \rho_{n+1}, \rho_{n+1})$, v_n satisfies $\|v_n\|_{m_{n+1},s} \leq \epsilon_n X_{n+1} \|\mathbf{p}\|^3 \to 0$, as $n \to \infty$. In the gaps in parameter space that were removed in Proposition 4.7, v_n is smoothly interpolated to zero. From this estimate, and Cauchy's theorem, we conclude that the C^1 norm of u_n is bounded by 1 on $\mathcal{N}_0$, again provided that L_0 is sufficiently large (independent of n.) Morover, $\|PF(\phi(\mathbf{p}) + u_{n+1})\|_{m_{n+1},s} \to 0$ as $n \to \infty$ for $(\mathbf{p},\Omega) \in \mathcal{N}_{n+1}$ Thus, if we set $\mathcal{N} = \cap_{n=0}^{\infty} \mathcal{N}_n$, and let $u = \lim_{n\to\infty} u_n$, we have proven Theorem 2.3. Note that $\lim_{n\to\infty} m_n \geq m_0/2$. Thus, $u \in H^{m_0/2,s}$. Thus if we construct a function of (x,ξ) from these Fourier-eigenfunction coefficients, as in (2.2), it will be an analytic function due to the exponential decay of the coefficients.

5. The Q-equation

Let $u(\mathbf{p},\Omega)$ be the solution of (4.1) constructed in the previous section. It remains to solve

$$QF(\phi(\mathbf{p}) + u(\mathbf{p},\Omega)) = 0 \quad . \tag{5.1}$$

In this section we consider only the case $N = \{(j_0, 1)\}$. Then (5.1) is equivalent to

$$G(\mathbf{p},\Omega) = W(\phi(\mathbf{p}) + u(\mathbf{p},\Omega))(j_0,1) + (\Omega - \omega_{j_0})\mathbf{p} = 0 \quad .$$

Note that the Ω-axis, $\{\mathbf{p} = 0, \Omega\}$ solves this equation. However, we're looking for "non-trivial" (*i.e.* $\mathbf{p} \neq 0$) solutions. Theorem 2.3 implies $G(|\mathbf{p}|e^{i\theta},\Omega) = e^{i\theta}G(|\mathbf{p}|,\Omega)$, so we can consider $G = G(r,\Omega)$ as a function of two real variables, since if $G(r,\Omega) = 0$, then $G(re^{i\theta},\Omega) = 0$. Explicit computation implies $\partial_r G|_{r=0,\Omega=\omega_{j_0}} = \partial_\Omega G|_{r=0,\Omega=\omega_{j_0}} = 0$, $\partial_r \partial_\Omega G|_{r=0,\Omega=\omega_{j_0}} = 1$, and $\partial_r^2 G|_{r=0,\Omega=\omega_{j_0}} = \partial_\Omega^2 G|_{r=0,\Omega=\omega_{j_0}} = 0$.

An application of the Morse lemma now implies that the zero set of G near $(0, \omega_{j_0})$ is the union of the Ω-axis and a (smooth) curve $(r, \Omega(r))$ with $\Omega(0) = \omega_{j_0}$. To better understand the nature of this solution curve consider

$$G(r, \Omega(r)) = (\Omega(r) - \omega_{j_0})r + W(\phi(r) + u(r, \Omega(r))(j_0, 1)) = 0 \ .$$

Differentiating this expression with respect to r gives $\partial_r \Omega|_{r=0} = 0$, $\partial_r^2 \Omega|_{r=0} = -(1/3)\partial_r^3 W(\phi(r))(j_0, 1)|_{r=0} = -t/3$, and $\partial_r^3 \Omega|_{r=0} = -(1/4)\partial_r^4 W(\phi(r)+ u(r, \Omega(r))(j_0, 1))|_{r=0} \equiv \overline{\Sigma} < \infty$. Thus, $\Omega(r) = \omega_{j_0} - tr^2/6 + \mathcal{O}(r^3)$, for r sufficiently small, from which we conclude

Lemma 5.1. *There exists $r^* > 0$, and $\overline{s} > 0$ (dependent on $g(u, \overline{u},, x)$), and a function $\Omega(r) = \omega_{j_0} - \frac{1}{6}tr^2(1 + \mathcal{S}(r))$, defined for $0 \leq r \leq r^*$, with $|\mathcal{S}|_{C^1} \leq \overline{s}$ and $QF(\phi(r) + u(r, \Omega(r)) = 0$.*

Remark. If we let $\mathbf{p} = re^{i\theta}$, then the covariance of the function $u(\mathbf{p}, \Omega)$ with respect to rotations in the $\mathbf{p}$-plane implies

Corollary 5.2. *Under the hypotheses of the previous lemma, one also has $QF(\phi(\mathbf{p}) + u(\mathbf{p}, \Omega(|\mathbf{p}|)) = 0$.*

To complete the proof of Theorem 2.6 we need only verify that the set of $|\mathbf{p}|$ such that $(\mathbf{p}, \Omega(|\mathbf{p}|)) \in \mathcal{N}$ has positive measure. For this we use Proposition 4.8. Define $\gamma_n = 1/(32n^2)$, for $n = 1, 2, \ldots$. Note that if we set $C_t = -t/6$, and recall that $r_0^p \approx d_0^{4/3}$, $r_0^\Omega \approx d_0 L_0^{-1}$, and $d_0 \approx L_0^{-\eta}$, with $3/5 < \eta < 1$, our original domain has intersection constant equal to 1, if L_0 is sufficiently large (depending on t.) By Proposition 4.8, there exists a constant $C(t) > 0$, such that if $C(t)L_n^2\sqrt{\delta_n} < \gamma_n \prod_{j=1}^{n-1}(1 - \gamma_j)r_0^p$, for $n = 1, 2, \ldots, m$, then $\mathcal{N}_m$ will have the intersection property with intersection constant $\prod_{j=1}^{m}(1 - (32j^2)^{-1})$. Recall that $r_0^p = (c_p d_0)^{4/3}$, $d_0 = L_0^{-\eta}$, and $\delta_n = L_n^{-\beta}$. Thus, if $\beta > 5 + 3\eta$ and L_0 is sufficiently large, $C(t)L_n^2\sqrt{\delta_n} < \gamma_n \prod_{j=1}^{n-1}(1 - \gamma_j)CL_0^{-1/2}r_0^p$ for all n and $\mathcal{N} = \cap_{j \geq 0}\mathcal{N}_j$ will have the intersection property with intersection constant $\prod_{j+1}^{m}(1 - (32j^2)^{-1})$. Thus, we have proven

Proposition 5.3. *If the hypotheses of Theorem 2.3 are satisfied, if $N = \{(j_0, 1)\}$, and if (2.3) satisfies the twist condition, then for L_0 sufficiently large, there exists a smooth function $\Omega(r)$ such that $QF(\phi(re^{i\theta}) + u(re^{i\theta}, \Omega(r)) = 0$ and* meas$(\{r \in [0, r_0^p] \mid (r, \Omega(r)) \in \mathcal{N}\}) \geq 1/2$.

This implies Theorem 2.6, completing the present section.

Appendix A: The Resonant Case

Generically the spectrum of the Sturm-Liouville operator $\mathcal{L}$ is simple, and therefore $\dim(N) = 1$. However, there are important cases in which there are linear resonances, and $\dim(N) \geq 2$. We will focus on the case in which $v(x) = m^2$, a constant, in which all except the lowest eigenvalue of $\mathcal{L}$ has multiplicity 2, with eigenvector and eigenvalue pairs: $\psi_0(x) = 1/\sqrt{\pi}$, $\omega_0 = m^2$, $\psi_{2j-1}(x) = \exp(2ijx)/\sqrt{\pi}$, $\omega_{2j-1} = 4j^2 + m^2$, and $\psi_{2j}(x) = \exp(-2ijx)/\sqrt{\pi}$, $\omega_{2j} = 4j^2 + m^2$, if $j \geq 1$. The analysis of the "P-equation"is unchanged in this case, but the solution of the "Q-equation" gives rise to interesting differences. We discuss the results of that analysis here, omitting the details of the calculations in the interest of brevity.

Consider a given ω_{j_0}. If $j_0 > 1$, then as noted above, we have $\omega_{j_0} = \omega_{j_0 \pm 1}$ *i.e.* we have a 1 : 1 resonance. Assume without loss of generality that $\omega_{j_0} = \omega_{j_0+1}$. For all except countably many values of m^2, $\dim(N) = 2$. A fact, related to Proposition 2.2 is that for all $L_* > 0$, there is an open set $\mathcal{M}$ of full measure of parameter values m^2, for which the frequency sequence above is $(L_0^{-\eta}, L_0)$ nonresonant with respect to ω_{j_0}.

We will consider the twist conditions and construction of Cantor families of periodic solutions in the situation in which $m^2 \in \mathcal{M}$, and the nonlinearity g has the form

$$g(u, \overline{u}, x) = \partial_{\overline{u}} \mathcal{G}(u, \overline{u}, x) \quad .$$

We assumed that $\mathcal{G}(u, \overline{u}, x) \approx \mathcal{O}_4(u, \overline{u})$, so we can expand $\mathcal{G}(u, \overline{u}, x) = a_{4,0}(x)u^4 + a_{3,1}(x)u^3\overline{u} + \ldots + a_{0,4}\overline{u}^4 + \mathcal{O}_5(u, \overline{u})$. It turns out that if we perturb about the frequency $\omega_{j_0} = j^2 + m^2$, only the coefficient $a_{2,2}(x)$ enters the twist condition. In fact, we can say even more. If we write $a_{2,2}(x) = \sum_n \alpha_n \exp(2inx)$ (since $a_{2,2}(x + \pi) = a_{2,2}(x)$), only α_0, $\alpha_{\pm 2j_0}$, and $\alpha_{\pm 4j_0}$ enter the twist condition! We assume for simplicity that $\alpha_{\pm 4j_0} = 0$, although it is not a technical requirement. Then, without loss of generality, we can choose α_{2j_0} to be real, with $\alpha_{2j_0} = \alpha_{-2j_0}$. Straightforward calculations, similar to those in Craig & Wayne (1991), show that the following families of periodic solutions are obtained.

If $a(x) = \alpha_0$, a constant, then the spatial translations are an additional symmetry in the problem. Solutions of (1.1) come in Cantor families which are foliated by two dimensional tori, representing one solution along with its space and time translations. Parameterize the resonant plane by $\mathbf{p} = (p_1, p_2) \in \mathbb{C}^2$, $u(x, \xi; \mathbf{p}) = (p_1 \exp(ij_0 x + i\xi) + p_2 \exp(-ij_0 x + i\xi))/\sqrt{\pi}$.

The solution family of the nonlinear problem is close to the solutions of the linear problem in the sense that

$$|u(x, \xi; \mathbf{p}) - (p_1 e^{ij_0 x + i\Omega(\|\mathbf{p}\|)t} + p_2 e^{-ij_0 x + i\Omega(\|\mathbf{p}\|)t})/\sqrt{\pi}| < C\|\mathbf{p}\|^2 \quad ,$$

for $\|\mathbf{p}_1\| = \|\mathbf{p}_2\|$, and $|\Omega(\|\mathbf{p}\|) - (j_0^2 + m^2)| < C\|\mathbf{p}\|^2$. The twist condition is directly related to the nonvanishing of the second term in the dependence of the frequency, Ω, on the amplitude,

$$\Omega(\mathbf{p}) = (j_0^2 + m^2) + \frac{3}{\pi}\alpha_0 \|\mathbf{p}\|^2 + \mathcal{O}(\|\mathbf{p}\|^4) \quad .$$

Therefore it is automatically satisfied as long as α_0 does not vanish.

If $a_{2,2}(x)$ is not constant, and the Fourier coefficients α_0 and α_2 do not vanish, then the symmetry of spatial translation is broken. Two families of periodic solutions remain, again in the form of Cantor sets foliated by circles. Straightforward calculations show that the first family has the form

$$|u_1(x,t;\mathbf{p}) - \frac{2\mathbf{p}}{\sqrt{\pi}}\cos(j_0x)e^{i\Omega(\|\mathbf{p}\|)t}| < C\|\mathbf{p}\|^2 \quad ,$$

$$\Omega(\|\mathbf{p}\|) = (j_0^2 + m^2) + \frac{1}{\pi}(3\alpha_0 + 4\alpha_2)\|\mathbf{p}\|^2 + \mathcal{O}((\|\mathbf{p}\|)^4) \quad ,$$

for $\mathbf{p} \in \mathbb{C}$. The second family has a similar form

$$|u_2(x,t;\mathbf{p}) - \frac{2\mathbf{p}}{\sqrt{\pi}}\sin(j_0x)e^{i\Omega(\|\mathbf{p}\|)t}| < C\|\mathbf{p}\|^2 \quad ,$$

$$\Omega(\|\mathbf{p}\|) = (j_0^2 + m^2) + \frac{1}{\pi}(3\alpha_0 - 4\alpha_2)\|\mathbf{p}\|^2 + \mathcal{O}(\|\mathbf{p}\|^4) \quad .$$

The twist condition is tantamount to the nonvanishing of the quantities $(3a_0 \pm 4a_2)$ respectively.

In general, if $\dim(N) = n$, then at least n many families of periodic solutions will exist, as long as a nondegeneracy condition holds, analogous to the twist condition. Further details of the solution set can also be derived from perturbation theory, but in the interest of brevity we do not pursue these at this time.

Appendix B: Norm Inequalities

We begin by proving Proposition 3.1. Our proof of this proposition is similar to Pöschel's proof (Pöschel (1993)) of Proposition 2.1 of Delatte (1992). We begin by recalling the following property of the eigenfunctions $\{\psi_j(x)\}$ proven in Proposition 7.1 of Craig & Wayne (1991a).

Proposition 7.1. *If we expand $\psi_k(x) = \sum_m \hat{\psi}_k(m)\exp(imx)$, then for $\sigma_* < \overline{\sigma}$, there exists $C > 0$ independent of k such that $|\hat{\psi}_k(m)| \leq C\exp(-\sigma_*|k - |m||)$.*

We first note that an elementary, albeit somewhat messy, calculation shows that

$$\begin{aligned}|\widehat{(uv)}(\mathbf{k})| &= |\sum_{\mathbf{l},\mathbf{m}} \int \hat{u}((l_1,l_2))\psi_{l_1}(x)e^{il_2\xi}\hat{v}((m_1,m_2))\psi_{m_1}(x)e^{im_2\xi}\overline{\psi_{k_1}(x)e^{-ik_2\xi}}dx d\xi \,| \\ &= |\sum_{\substack{l_1,l_2,m_1\\ q_1,r_1}} \hat{u}((l_1,l_2))\hat{v}((m_1,k_2-l_2))\hat{\psi}_{m_1}(q_1-r_1)\hat{\psi}_{l_1}(q_1)\overline{\hat{\psi}_{k_1}(r_1)}|\end{aligned}$$

Applying the estimates of Proposition 7.1, we see that this last expression is bounded by

$$C^3 \sum_{\substack{l_1,l_2,m_1\\ q_1,r_1}} |\hat{u}((l_1,l_2))||\hat{v}((m_1,k_2-l_2))|e^{-\sigma_*(||q_1-r_1|-|m_1||+||l_1|-|q_1||+||k_1|-|r_1||)}$$

Inserting this estimate into the definition of the norm, we have

$$\begin{aligned}\|(uv)\|_{m,s}^2 \leq C^6\alpha\sum_{\mathbf{k}}(1+|\mathbf{k}|^2)^s e^{2m|\mathbf{k}|} &\sum_{\substack{l_1,l_2,m_1,l_1',l_2',m_1'\\ q_1,r_1,q_1',r_1'}} |\hat{u}((l_1,l_2))||\hat{u}((l_1',l_2'))| \\ &|\hat{v}((m_1,k_2-l_2))||\hat{v}((m_1',k_2'-l_2'))|e^{-\sigma_*(||q_1-r_1|-|m_1||+||l_1|-|q_1||+||k_1|-|r_1||)} \\ &e^{-\sigma_*(||q_1'-r_1'|-|m_1'||+||l_1'|-|q_1'||+||k_1'|-|r_1'||)}\end{aligned}$$

If $\mu = \sigma_* - m$, then

$$e^{m|\mathbf{k}|}e^{-m|\mathbf{l}|}e^{-m|(m_1,k_2-l_2)|}e^{-\sigma_*(||q_1-r_1|-|m_1||+||l_1|-|q_1||+||k_1|-|r_1||)}$$
$$\leq e^{-\mu(||q_1-r_1|-|m_1||+||l_1|-|q_1||+||k_1|-|r_1||)} ,$$

and the same inequality holds if $\mathbf{l} \to \mathbf{l}'$, $m_1 \to m_1'$, $q_1 \to q_1'$ and $r_1 \to r_1'$. Inserting this estimate into the previous inequality we find that $\|(uv)\|^2_{m,s}$ can be bounded by

$$C^6\alpha \sum_{\mathbf{k}} \sum_{\substack{l_1,l_2,m_1,l_1',l_2',m_1' \\ q_1,r_1,q_1',r_1'}} (1+|\mathbf{l}|^2)^{s/2}e^{m|\mathbf{l}|}|\hat{u}((l_1,l_2))|(1+|\mathbf{l}'|^2)^{s/2}e^{m|\mathbf{l}'|}$$
$$|\hat{u}((l_1',l_2'))|(1+|(m_1,(k_2-l_2))|^2)^{s/2}e^{m|(m_1,k_2-l_2)|}|\hat{v}((m_1,k_2-l_2))|$$
$$(1+|(m_1',(k_2'-l_2'))|^2)^{s/2}e^{m|(m_1',k_2'-l_2')|}|\hat{v}((m_1',k_2'-l_2'))|(1+|\mathbf{k}|^2)^s(1+|\mathbf{l}|^2)^{-s/2}$$
$$(1+|\mathbf{l}'|^2)^{-s/2}(1+|(m_1,(k_2-l_2))|^2)^{s/2}(1+|(m_1',(k_2'-l_2'))|^2)^{s/2}$$
$$e^{-\mu(||q_1-r_1|-|m_1||+||l_1|-|q_1||+||k_1|-|r_1||)}e^{-\mu(||q_1-r_1|-|m_1||+||l_1|-|q_1||+||k_1|-|r_1||)}$$

We now apply the Cauchy-Schwartz lemma to the sum over l_1, l_2, m_1, q_1, and r_1, and to the sum over l_1', l_2', m_1', q_1' and r_1'. The preceding expression is then bounded by

$$C^6\alpha \sum_{\mathbf{k}} \{ (\sum_{\substack{l_1,l_2,m_1 \\ q_1,r_1}} (1+|\mathbf{l}|^2)^s e^{2m|\mathbf{l}|}|\hat{u}((l_1,l_2))|^2(1+|(m_1,(k_2-l_2)|^2)^s e^{2m|(m_1,k_2-l_2)|}$$
$$|\hat{v}((m_1,k_2-l_2))|^2 e^{-\mu(||q_1-r_1|-|m_1||+||l_1|-|q_1||+||k_1|-|r_1||)})^{1/2}$$
$$\times (\sum_{\substack{l_1,l_2,m_1 \\ q_1,r_1}} (1+|\mathbf{k}|^2)^s(1+|\mathbf{l}|^2)^{-s}(1++(m_1,(k_2-l_2))|^2)^{-s}$$
$$e^{-\mu(||q_1-r_1|-|m_1||+||l_1|-|q_1||+||k_1|-|r_1||)})^{1/2}$$
$$\times (\text{same expression with } l_1,l_2,m_1,q_1,r_1 \text{ replaced by } l_1',l_2',m_1',q_1',r_1')\} \tag{7.1}$$

An elementary calculation proves

Lemma 7.2. *If $s > 1$ and $\mu > 0$, there exists $C(\mu) > 0$ such that for any* $\mathbf{k}$

$$\sum_{\substack{l_1,l_2,m_1 \\ q_1,r_1}} (1+|\mathbf{k}|^2)^s(1+|\mathbf{l}|^2)^{-s}(1+|(m_1,(k_2-l_2))|^2)^{-s}$$
$$e^{-\mu(||q_1-r_1|-|m_1||+||l_1|-|q_1||+||k_1|-|r_1||)} \leq C(\mu) .$$

Thus, (7.1) is bounded by

$$C^6\alpha C(\mu)^2 \sum_{\mathbf{k}} \{ (\sum_{\substack{l_1,l_2,m_1 \\ q_1,r_1}} (1+|\mathbf{l}|^2)^s e^{2m|\mathbf{l}|}|\hat{u}((l_1,l_2))|^2(1+|(m_1,(k_2-l_2)|^2)^s$$
$$e^{2m|(m_1,k_2-l_2)|}|\hat{v}((m_1,k_2-l_2))|^2 e^{-\mu(||q_1-r_1|-|m_1||+||l_1|-|q_1||+||k_1|-|r_1||)})^{1/2}$$
$$\times (\text{same expression with } l_1,l_2,m_1,q_1,r_1 \text{ replaced by } l_1',l_2',m_1',q_1',r_1')\}$$
$$\leq C^6\alpha C(\mu)^2 \sum_{\substack{k_1,k_2,l_1,l_2 \\ m_1,q_1,r_1}} (1+|\mathbf{l}|^2)^s e^{2m|\mathbf{l}|}|\hat{u}((l_1,l_2))|^2(1+|(m_1,(k_2-l_2)|^2)^s$$
$$e^{2m|(m_1,k_2-l_2)|}|\hat{v}((m_1,k_2-l_2))|^2 e^{-\mu(||q_1-r_1|-|m_1||+||l_1|-|q_1||+||k_1|-|r_1||)}) .$$

The last inequality in this expression results from applying the Cauchy-Schwartz lemma to the sum over **k**. Summing over k_1, r_1, q_1, m_1 and k_2, and l_1 and l_2, this expression is bounded by $C'(\mu)\alpha\|u\|_{m,s}\|v\|_{m,,s}$, and Proposition 3.1 follows

We now turn to the proof of Corollary 3.2. First note that since $f(z_1, z_2, x)$ is an analytic function of z_1, z_2, and x,

$$f(u(x,\xi),\overline{u}(x,\xi),x) = \sum_{k_1,k_2}^{\infty} \frac{1}{k_1!k_2!} a_{k_1,k_2}(x)(u(x,\xi))^{k_1}(\overline{u}(x,\xi))^{k_2} \ .$$

Since the Fourier-eigenfunction expansion is a linear operation,

$$\|(f(u(x,\xi),\overline{u}(x,\xi),x))\hat{}\|_{m,s} \le \sum_{k_1,k_2}^{\infty} \frac{1}{k_1!k_2!}\|(a_{k_1,k_2}(x)(u(x,\xi))^{k_1}(\overline{u}(x,\xi))^{k_2})\hat{}\|_{m,s} \ .$$

If $m < \overline{\sigma}$, Cauchy's theorem implies that there exists a constant $C(f, \overline{\sigma} - m)$, such that $\|a_{k_1,k_2}\|_{m,s} \le C(f, \overline{\sigma} - m)k_1!k_2!$. Applying Proposition 3.1, we see that $\|(f(u(x,\xi),\overline{u}(x,\xi),x))\hat{}\|_{m,s} \le \sum_{k_1 k_2}^{\infty} C(f, \overline{\sigma} - m)\|\hat{u}\|_{m,s}^{(k_1+k_2)}$. Since $\|\hat{u}\|_{m,s} \le 1/2$, the first assertion in Corollary 3.2 follows. If $|f(z_1, z_2, x)| \le C|z_1|^{n_1}|z_2|^{n_2}$,, we can pull out a factor of $\|\hat{u}\|_{m,s}^{(n_1+n_2)}$, and the sum still converges. This implies the remaining claim in Corollary 3.2.

Finally, we turn to Proposition 3.3. An explicit calculation shows that

$$\begin{aligned}|\widehat{W}(\mathbf{k},\mathbf{m})| =& |\sum_{l_1,q_1,r_1} \hat{w}(l_1, k_2 - m_2)\hat{\psi}_{m_1}(q_1 - r_1)\hat{\psi}_{l_1}(q_1)\overline{\hat{\psi}_{k_1}(r_1)}| \\ \le& C^3 \sum_{l_1,q_1,r_1} |\hat{w}(l_1, k_2 - m_2)| e^{-\sigma_*(||q_1-r_1|-|m_1||+||l_1|-|q_1||+||k_1|-|r_1||)} \ ,\end{aligned}$$

where the last inequality follows from Proposition 7.1. Inserting this into the first of the two expressions whose maximum gives $\|\widehat{W}\|_{m,s}^2$, we find it is bounded by

$$\begin{aligned}&\sup_{\mathbf{k}} \sum_{\mathbf{m}} (1 + |\mathbf{k} - \mathbf{m}|^2)^s e^{2m|\mathbf{k}-\mathbf{m}|} \sum_{\substack{l_1,q_1,r_1 \\ l_1',q_1',r_1'}} |\hat{w}(l_1, k_2 - m_2)||\hat{w}(l_1', k_2' - m_2)| \\ &\qquad e^{-\sigma_*(||q_1-r_1|-|m_1||+||l_1|-|q_1||+||k_1|-|r_1||)} e^{-\sigma_*(||q_1'-r_1'|-|m_1||+||l_1'|-|q_1'||+||k_1|-|r_1'||)} \ .\end{aligned}$$

This expression can be bounded in a very similar fashion to the sum which arose in the proof of Proposition 3.1. We leave the details to the reader. The second of the two expressions in the definition of $\|\widehat{W}\|_{m,s}^2$ is bounded in an identical fashion, except that the roles of **k** and **m** are reversed. This completes the proof of Proposition 3.3.

Appendix C: The Inductive Step

We suppose throughout this appendix that the hypotheses of Lemma 4.9 hold. We begin by stating an estimate on the inverse of the linearized operator.

Theorem 8.1. *Let $A \subset B(6\tilde{L})$ be a lattice region containing no singular sites. There exist constants $C > 0$ and $\gamma > 0$, such that if $\|\overline{v}\|_{\tilde{m}-3\tilde{\mu},s} < \tilde{\rho}$, then*

$$\|(D_u PF|_{(B(2\tilde{L})\cup A)\setminus N}(\phi(\mathbf{p}) + \tilde{u} + \overline{v}))\|_{\tilde{m}-3\tilde{\mu},s} \leq C\gamma^s \exp(\gamma)(1-\Theta)^{-1} C_f r_0^2 \tilde{X} \tilde{\delta}^{-1} \equiv \tilde{\tilde{X}},$$

for all $(\rho, \Omega) \in D(\tilde{\tilde{\mathcal{N}}}; \tilde{\tilde{\rho}}, \tilde{\tilde{\rho}})$.

Proof. We prove this theorem in Appendix D using Pöschel's coupling lemma (Pöschel (1990)).

We now construct our improvement to the approximate solution $\phi(\mathbf{p}) + \tilde{u}$. We wish to solve $PF(\phi(\mathbf{p}) + \tilde{u} + \tilde{v}) = 0$, for $\tilde{v} \in \ell^2(B(2\tilde{L})\setminus N)$. Define $\Pi_{\overline{B(2\tilde{L})}}$ to be orthogonal projection onto $\ell^2(B(2\tilde{L})\setminus N)$. Applying Newton's method we find

$$\tilde{v} = (\Pi_{\overline{B(2\tilde{L})}} D_u PF(\phi(\mathbf{p}) + \tilde{u})\Pi_{\overline{B(2\tilde{L})}})^{-1}\Pi_{\overline{B(2\tilde{L})}} PF(\phi(\mathbf{p}) + \tilde{u}) \ . \tag{8.1}$$

For $(\mathbf{p}, \Omega) \in D(\tilde{\tilde{\mathcal{N}}}; \tilde{\tilde{\rho}}, \tilde{\tilde{\rho}})$, Theorem 8.1 and (4.4) imply $\|\tilde{v}\|_{\tilde{m}-3\tilde{\mu},s} \leq \tilde{\tilde{X}}\tilde{\epsilon}\|\mathbf{p}\|^3$. Also, (4.6) and hypothesis H1 imply that $\tilde{v}(T_\theta \mathbf{p}, \Omega) = T_\theta \tilde{v}(\mathbf{p}, \Omega)$. Note that $\tilde{v}$ is also analytic on this domain so using Cauchy's theorem we can bound its derivatives on $D(\tilde{\tilde{\mathcal{N}}}; \tilde{\tilde{\rho}}/2, \tilde{\tilde{\rho}}/2)$, and then the Whitney extension theorem allows us to immediately extend $\tilde{v}$ to a C^3 function on $\mathcal{N}_0$ with C^3 norm bounded by the C^3 norm of $\tilde{v}$ on $D(\tilde{\tilde{\mathcal{N}}}; \tilde{\tilde{\rho}}/2, \tilde{\tilde{\rho}}/2)$. With a little more work we could extend $\tilde{v}$ to a C^∞ function on $\mathcal{N}_0$ as in Craig & Wayne (1991a).

We now estimate

$$\begin{aligned} PF(\phi(\mathbf{p}) + \tilde{u} + \tilde{v}) =& \Pi_{\overline{B(2\tilde{L})}}[PF(\phi(\mathbf{p}) + \tilde{u} + \tilde{v}) - (PF(\phi(\mathbf{p}) + \tilde{u}) \\ &+ D_u PF(\phi(\mathbf{p}) + \tilde{u})\tilde{v})] + (\mathbf{1} - \Pi_{\overline{B(2\tilde{L})}})PF(\phi(\mathbf{p}) + \tilde{u} + \tilde{v}) \ . \end{aligned} \tag{8.2}$$

Note that

$$\begin{aligned} \|(\mathbf{1} - \Pi_{\overline{B(2\tilde{L})}})PF(\phi(\mathbf{p}) + \tilde{u} + \tilde{v})\|_{\tilde{m}-4\tilde{\mu},s} \leq& \|(\mathbf{1} - \Pi_{\overline{B(2\tilde{L})}})PF(\phi(\mathbf{p}) + \tilde{u})\|_{\tilde{m}-4\tilde{\mu},s} \\ &+ \|(\mathbf{1} - \Pi_{\overline{B(2\tilde{L})}}) \int_0^1 PF(\phi(\mathbf{p}) + \tilde{u} + \tau\tilde{v})\tilde{v} d\tau\|_{\tilde{m}-4\tilde{\mu},s} \end{aligned}$$

By Lemma 3.5, $\|(\mathbf{1} - \Pi_{\overline{B(2\tilde{L})}})PF(\phi(\mathbf{p}) + \tilde{u} + \tilde{v})\|_{\tilde{m}-4\tilde{\mu},s} \leq \exp(-4\tilde{\mu}\tilde{L})\tilde{\epsilon}\|\mathbf{p}\|^3$. On the other hand, there exists $C > 0$, such that $\|(\mathbf{1} - \Pi_{\overline{B(2\tilde{L})}})D_u PF(\phi(\mathbf{p}) + \tilde{u} + \tau\tilde{v})\tilde{v}\|_{\tilde{m}-4\tilde{\mu},s} \leq C\tilde{\mu}^{-1}\tilde{\tilde{X}}\tilde{\epsilon}\|\mathbf{p}\|^3 \exp(-\tilde{\mu}\tilde{L})$, if $\tilde{L} > \tilde{\mu}^{-1}$.

Turning to the first term on the right hand side of (8.2), Taylor's theorem implies that $PF(\phi(\mathbf{p})+\tilde{u}+\tilde{v}) - (PF(\phi(\mathbf{p})+\tilde{u}) + D_u PF(\phi(\mathbf{p})+\tilde{u})\tilde{v})] = \int_0^1 \int_0^t d\tau D_u^2 PF(\phi(\mathbf{p})+\tilde{u}+\tau\tilde{v})\tilde{v}\tilde{v}$. By Corollary 3.4, and the explicit form of $V(\Omega)$, we can bound $\|\Pi_{\overline{B(2\tilde{L})}} D_u^2 PF(\phi(\mathbf{p})+\tilde{u}+\tau\tilde{v})\tilde{v}\tilde{v}\|_{\tilde{m}-4\tilde{\mu},s} \leq C\tilde{L}(\tilde{\tilde{X}}\tilde{\epsilon}\|\mathbf{p}\|^3)^2$. Combining these estimates we see that $\|PF(\phi(\mathbf{p})+\tilde{u}+\tilde{v})\|_{\tilde{m}-4\tilde{\mu},s} \leq (1+\tilde{\mu}^{-1}\tilde{\tilde{X}})^2(\tilde{\epsilon}\exp(-\tilde{\mu}\tilde{L})+\tilde{\epsilon}^2)\|\mathbf{p}\|^3 C\tilde{L}$. This completes the proof of the first assertion in Lemma 4.9. The second claim follows from Theorem 8.1 if we note that for any $\tilde{\tilde{v}}$, with $\|\tilde{\tilde{v}}\|_{\tilde{m}-3\tilde{\mu},s} < \tilde{\tilde{\rho}}/2$, $\phi(\mathbf{p}) + \tilde{\tilde{u}} + \tilde{\tilde{v}} = \phi(\mathbf{p}) + \tilde{u} - \overline{v}$, with $\overline{v} = \tilde{v} + \tilde{\tilde{v}}$, and $\|\overline{v}\|_{\tilde{m}-3\tilde{\mu},s} \leq \|\tilde{v}\|_{\tilde{m}-3\tilde{\mu},s} + \|\tilde{\tilde{v}}\|_{\tilde{m}-3\tilde{\mu},s} \leq \tilde{\tilde{X}}\tilde{\epsilon}\|\mathbf{p}\|^3 + \tilde{\tilde{\rho}}/2 \leq \tilde{\tilde{\rho}}$. This completes the proof of Lemma 4.9.

Appendix D: The Inverse of the Linearized Operator

In this appendix we prove Theorem 8.1 using ideas of Pöschel (1990). We use the notation and terminology of this reference without comment.

Given a lattice region Λ, with zero metric set Z, and a linear operator S on $\ell^2(\Lambda)$, define

$$|||S|||^2_{m,s,Z} = \max(\sup_{\mathbf{l}} \sum_{\mathbf{l}'} (1 + \text{dist}_Z(\mathbf{l}, \mathbf{l}')^2)^{s/2} |S(\mathbf{l}, \mathbf{l}')|^2 e^{2m \text{dist}_Z(\mathbf{l},\mathbf{l}')},$$
$$\sup_{\mathbf{l}'} \sum_{\mathbf{l}} (1 + \text{dist}_Z(\mathbf{l}, \mathbf{l}')^2)^{s/2} |S(\mathbf{l}, \mathbf{l}')|^2 e^{2m \text{dist}_Z(\mathbf{l},\mathbf{l}')}) \ .$$

Note that if $Z = \emptyset$, this reduces to our previous norm. Consider $D_u PF = D_u PW + PV(\Omega)$. By Corollary 3.4 if $m < \overline{\sigma}$ and $s > 1$, $|||D|_u PW|||_{m,s,Z} \leq C_f r_0^2$. We now appeal to

Lemma 9.1. (Pöschel's Coupling Lemma) *Let $\Lambda = \mathcal{A} \cup \mathcal{B}$, and let $Y \subset \mathcal{A}$, $Z \subset \mathcal{B}$ be their respective zero metric sets. Assume that $|||(D_u PF|_{\mathcal{A}})^{-1}|||_{m,s,Y} \leq X_{\mathcal{A}}$ and that around each point $\mathbf{l} \in \Lambda \backslash \mathcal{A}$ there is a neighborhood $V \subset \mathcal{B}$ such that $|||(D_u PF|_V)^{-1}|||_{m,s,Z} \leq X_{\mathcal{B}}$, with $\text{dist}_Z(\mathbf{l}, \Lambda \backslash V) \geq r$. Without loss of generality we assume that $X_{\mathcal{A}}^{-1}, X_{\mathcal{B}}^{-1} \leq C_f r_0^2$. If there exists $0 \leq \mu \leq m$ such that $\Theta = (C_f r_0^2)^2 X_{\mathcal{A}} X_{\mathcal{B}} \exp(-\mu r) < 1$, then*

$$|||(D_u PF|_\Lambda)^{-1}|||_{m-\mu,s,Z\cup Y} \leq 2(1-\Theta)^{-1}(C_f r_0^2) X_{\mathcal{A}} X_{\mathcal{B}} \ .$$

Proof. The proof is almost identical to that in Pöschel (1990) and we do not repeat it.

We now apply this as follows. Assume that $(\mathbf{p}, \Omega) \in \tilde{\tilde{\mathcal{N}}}$, let $A \subset B(6\tilde{L}) \backslash N$ be any lattice region free of singular sites, and let $E \subset (B(2\tilde{L}) \cup A) \backslash N$. Let $S = \cup_j S_j$, where the union runs over all singular regions in $B(2\tilde{L}) \backslash B(\tilde{L})$.

Now suppose that U is a lattice region containing no singular sites. Then $(D_u PF|_U)^{-1} = (D_u PW|_U + PV(\Omega)|_U)^{-1}$. Since $\|D_u PW|_U\|_{\tilde{m},s} \leq C_f r_0^2$ and $\|(V(\Omega)|_U)^{-1}\|_{\tilde{m},s} \leq 100$, a Neumann series implies that for r_0 sufficiently small, $\|(D_u PF|_U)^{-1}\|_{\tilde{m},s} \leq 200$, from which we can conclude that for any sites $\mathbf{l}, \mathbf{l}' \in U$, $|D_u PW|_U(\mathbf{l}, \mathbf{l}')| \leq C_f r_0^2 (1 + |\mathbf{l} - \mathbf{l}'|^2)^{-s/2} \exp(-\tilde{m}|\mathbf{l} - \mathbf{l}'|)$ and $|(D_u PF|_U)^{-1}(\mathbf{l}, \mathbf{l}')| \leq 200(1 + |\mathbf{l} - \mathbf{l}'|^2)^{-s/2} \exp(-\tilde{m}|\mathbf{l} - \mathbf{l}'|)$. On the other hand, we know that for $(\mathbf{p}, \Omega) \in \tilde{\tilde{\mathcal{N}}}$, $\text{dist}(\text{spec}(D_u PF|_S), 0) \geq \tilde{\delta}$, $\text{dist}(\text{spec}(D_u PF|_{C_{\tilde{\ell}}(S)}), 0) \geq \tilde{\delta}$, and $C_{\tilde{\ell}}(S)$ contains no other singular sites than those in S. This will hold for our problem since $\tilde{\ell} << \tilde{L}$. Given these facts, we can prove

Lemma 9.2. *For r_0 sufficiently small, $\tilde{\ell} << \tilde{L}$ and $F \subset C_{\tilde{\ell}}(S)$, there exists $C > 0$ such that $|(D_u PF|_{C_{\tilde{\ell}}(S) \cap F}(\phi(\mathbf{p}) + \tilde{u}))^{-1}(\mathbf{l}, \mathbf{l}')| \leq C \tilde{\delta} \tilde{\mu}^{-4} (1 + |\mathbf{l} - \mathbf{l}'|^2)^{-s/2} \exp(-(\tilde{m} - \tilde{\mu})|\mathbf{l} - \mathbf{l}'|)|$.*

Proof. The proof is very similar to the proof of Lemma 5.4 of Craig & Wayne (1991a) and we do not repeat it. Inserting this bound on the matrix elements into the definition of $\| \cdot \|_{\tilde{m},s}$ yields the following easy corollary.

Corollary 9.3. *Under the hypotheses of Lemma 9.2* $\|(D_u PF|_{C_{\tilde{\ell}}(S)\cap F})^{-1}\|_{\tilde{m}-2\tilde{\mu},s} \leq C\tilde{\delta}^{-1}\tilde{\mu}^{-6}$.

Now choose $\mathcal{A} = (E\backslash S)\cap B(3\tilde{L})$, with $Y = \emptyset$ and let $\mathcal{B} = (S\cap E)\cup(E\cap B(2\tilde{L})^c)$, with $Z = S\cap E$, so that $\Lambda = E$. By hypothesis H2

$$|||(D_u PF|_{\mathcal{A}})^{-1}|||_{\tilde{m}-2\tilde{\mu},s,Y} = \|(D_u PF|_{\mathcal{A}})^{-1}\|_{\tilde{m}-2\tilde{\mu},s} \leq \tilde{X} = X_{\mathcal{A}} \ .$$

Next, if $\mathbf{l} \in (E\backslash A)$, we must consider two possibilities. If $\ell \in S_j \subset S$, choose $V = C_{\tilde{\ell}}(S_j)\cap E$. From Corollary 9.3, $\|(D_u PF|_V)^{-1}\|_{\tilde{m}-2\tilde{\mu},s} \leq C\tilde{\delta}^{-1}\tilde{\mu}^{-6} = X_{\mathcal{B}}$. Note that since $C_{\tilde{\ell}}(S_j)$ contains no singular sites other than those in S_j, $\mathrm{dist}_Z(\mathbf{l}, \Lambda\backslash V) \geq \tilde{\ell} = r$. If on the other hand, $\ell \in (B(3\tilde{L})^c \cap E)$ we choose $V = C_{\tilde{\ell}}(\mathbf{l}) \cap E$. Since all sites in Z are a distance $\mathcal{O}(\tilde{L}) >> \tilde{\ell}$ from $\mathbf{l}$, we have $\mathrm{dist}_Z(\mathbf{l}, \Lambda\backslash V) \geq \tilde{\ell} = r$. We then note that since A contains no singular sites, a Neumann series like we used above to bound $\|(D_u PF|_U)^{-1}\|_{\tilde{m},s}$ implies that $|||(D_u PF|_V)^{-1}|||_{\tilde{m},s,Z} \leq 200$. Thus we set $X_{\mathcal{B}} = \max(C\tilde{\delta}^{-1}\tilde{\mu}^{-6}, 200) = C\tilde{\delta}^{-1}\tilde{\mu}^{-6}$, possibly by increasing the size of C. Then if we set $\Theta = C_{r_0}^2 \tilde{X}\tilde{\mu}^{-6}\tilde{\delta}^{-1}\exp(-\tilde{\mu}\tilde{\ell})$, and if $\Theta < 1$, we have $|||(D_u PF|_\Lambda)^{-1}|||_{\tilde{m}-3\tilde{\mu},s,Z\cup Y} \leq 2(1-\Theta)^{-1}Cr_0^2\tilde{X}\tilde{\mu}^{-6}\tilde{\delta}^{-1}$.

To complete the proof we must relate the $|||\cdot|||$ norm to the $\|\cdot\|$ norm. The distance between singular regions in $B(2\tilde{L})\backslash B(\tilde{L})$ is $\mathcal{O}(\tilde{L})$. Thus there exists some constant γ, such that the number of singular regions in $B(2\tilde{L})\backslash B(\tilde{L})$ is bounded by γ for any Ω (and for any $\tilde{L}$). Each singular region contains at most two sites, thus we conclude that $\mathrm{dist}_{Z\cup Y}(\mathbf{l}, \mathbf{l}') \geq |\mathbf{l}-\mathbf{l}'| - \gamma$, for any points $\mathbf{l}, \mathbf{l}' \in E$. Inserting this information into the definition of the norms we see that

$$\begin{aligned}\|(D_u PF|_\Lambda)^{-1}\|_{\tilde{m}-3\tilde{\mu},s} &\leq C\gamma^s e^\gamma |||(D_u PF|_\Lambda)^{-1}|||_{\tilde{m}-3\tilde{\mu},s,Z\cup Y} \\ &\leq C\gamma^s(1-\Theta)^{-1}e^\gamma r_0^2\tilde{X}\tilde{\mu}^{-6}\tilde{\delta}^{-1} \equiv \frac{1}{2}\tilde{\tilde{X}} \ .\end{aligned}$$

We now extend this in two ways. First we allow $(\mathbf{p}, \Omega) \in D(\tilde{\tilde{\mathcal{N}}}; \tilde{\tilde{\rho}}, \tilde{\tilde{\rho}})$. Secondly, we evaluate the derivative $D_u PF$ at $\phi(\mathbf{p}) + \tilde{u} + \tilde{v}$, for some $\tilde{v}$ analytic on $D(\tilde{\tilde{\mathcal{N}}}; \tilde{\tilde{\rho}}, \tilde{\tilde{\rho}})$ and satisfying $\|\tilde{v}\|_{\tilde{m}-\tilde{\mu},s} < \tilde{\tilde{\rho}}$. If $(\mathbf{p}, \Omega) \in D(\tilde{\tilde{\mathcal{N}}}; \tilde{\tilde{\rho}}, \tilde{\tilde{\rho}})$, there exists $(\overline{\mathbf{p}}, \overline{\Omega}) \in \tilde{\tilde{\mathcal{N}}}$ such that $\sqrt{\|\mathbf{p}-\overline{\mathbf{p}}\|^2 + |\Omega-\overline{\Omega}|^2} < \tilde{\tilde{\rho}}$. Thus, $\phi(\mathbf{p}) + u(\mathbf{p}, \Omega) = \phi((\overline{\mathbf{p}}) + u(\overline{\mathbf{p}}, \overline{\Omega}) + (\phi(\mathbf{p}) - \phi(\overline{\mathbf{p}})) + (u(\mathbf{p}, \Omega) - u(\overline{\mathbf{p}}, \overline{\Omega})) \equiv \phi(\overline{\mathbf{p}} + u(\overline{\mathbf{p}}, \overline{\Omega}) + \delta\phi + \delta u)$. Furthermore, since the C^1 norm of u is bounded by 1, $\|\delta u\|_{\tilde{m}-\tilde{\mu},s} \leq \tilde{\tilde{\rho}}$, and $\|\delta\phi\|_{\tilde{m}-\tilde{\mu},s} \leq M\tilde{\tilde{\rho}}$, where $M = \dim(N)$.

Taylor's theorem implies that

$$\begin{aligned}(D_u PF|_E(\phi(\overline{\mathbf{p}} + u(\overline{\mathbf{p}}, \overline{\Omega}) + \delta\phi + \delta u + \overline{v}))^{-1} &= (D_u PF|_E((\phi(\overline{\mathbf{p}} + u(\overline{\mathbf{p}}, \overline{\Omega})) \\ &+ \int_0^1 D_u^2 PF|_E((\phi(\overline{\mathbf{p}} + u(\overline{\mathbf{p}}, \overline{\Omega}) + \tau(\delta u + \delta\phi + \overline{v}))(\delta u + \delta\phi + \overline{v})d\tau)^{-1}\end{aligned} \tag{9.1}$$

Corollary 3.4 and the fact that $E \subset B(6\tilde{L})$ imply that $\|D_u^2 PF|_E(\ldots)\|_{\tilde{m}-3\tilde{\mu},s} \leq C\tilde{L}\tilde{\tilde{\rho}}$. Thus, if $\tilde{\tilde{\rho}} < C/(\tilde{L}\tilde{\tilde{X}})$, we can invert the operator in (9.1) via a Neumann series and we obtain $\|(D_u PF|_E(\phi(\overline{\mathbf{p}} + u(\overline{\mathbf{p}}, \overline{\Omega}) + \delta\phi + \delta u + \overline{v}))^{-1}\|_{\tilde{m}-3\tilde{\mu},s} \leq \tilde{\tilde{X}}$. This completes the proof of Theorem 8.1.

Acknowledgements

This work was supported in part by the National Science Foundation (grants # DMS-8920624 and # DMS-9203359). WC wishes to thank the Forschungsinstitut für Mathematik, ETH Zürich, for its hospitality. CEW wishes to acknowledge the hospitality of the Mathematics Department, Brown University and he also thanks J. Pöschel for pointing out that the norms used in Delatte (1992) might be of use in this problem.

References

Craig, W. & Wayne, C. E., 1991: Nonlinear waves and the 1:1:2 resonance. Proceedings: Singular limits of dispersive equations, ENS-Lyon eds. N. Ercolani, D. Levermore and D. Serre (in press)

Craig, W. & Wayne, C. E., 1991a: Newton's method and periodic solutions of nonlinear wave equations. Preprint: to appear in Comm. Pure Appl. Math.

Craig, W., Kuksin, S. B., & Wayne, C. E., 1993: Unstable periodic orbits in nonlinear wave equations. In preparation

Delatte, D., 1992: On normal forms in Hamiltonian dynamics; a new approach to some convergence questions. Preprint of the Forschungsinstitut für Mathematik, ETH Zurich

Kuksin, S.B., 1988: Perturbation of quasi-periodic solutions of infinite - dimensional Hamiltonian systems. Izv. Akad. Nauk SSSR Ser. Math. **52**, 41-63

Kuksin, S.B., 1993: Nearly integrable infinite dimensional Hamiltonian systems. Preprint of the Forschungsinstitut für Mathematik, ETH Zurich

Pöschel, J., 1990: On Fröhlich Spencer estimates of Green's function. Manuscripta Math. **70**, 27-37

Pöschel, J., 1993: Private communication to C. E. Wayne, February, 1993

Pöschel, J. and Trubowitz, E., 1987: "*Inverse Spectral Theory*", Academic Press; Boston, MA

ON THE TENDENCY TOWARD ERGODICITY WITH INCREASING NUMBER OF DEGREES OF FREEDOM IN HAMILTONIAN SYSTEMS

Lyman Hurd

Iterated Systems, Corp.
5550A Peach Tree Parkway, Suite 545
Norcross, GA 30092

and

Celso Grebogi[a] and Edward Ott[b]

Laboratory for Plasma Research
University of Maryland
College Park, Maryland 20742-3511

ABSTRACT

Numerical experiments on a symplectic coupled map system are performed to investigate the tendency for global ergodic behavior of typical Hamiltonian systems as the number of degrees of freedom N is increased. As N increases, we find that the fraction of phase space volume occupied by invariant tori decreases strongly. Nevertheless, due to observed very long time correlated behavior, a conclusion of effective gross ergodicity cannot be confirmed, even though extremely long numerical runs were employed.

The basic assumption in statistical mechanics is that of ergodicity over the phase space hypersurface determined by the global constants of the motion (e.g., total energy, total angular momentum, etc.). On the other hand, studies of Hamiltonian systems with few degrees of freedom (e.g., two) typically reveal the presence of invariant KAM tori in addition to chaotic orbits; and the existence of KAM tori yields motion that is grossly different from that assumed in statistical mechanics. A natural supposition reconciling the above contradictory views might be that, as the number of degrees of freedom is increased, the tendency for global ergodicity increases. By "tendency for global ergodicity" we mean that, for systems with many degrees of freedom (the situation of interest in statistical mechanics), the overwhelming majority of initial conditions would be ergodic over effectively all of the area of the phase space hypersurface determined by the global constants of the motion.

a. and Department of Mathematics, and Institute for Physical Science and Technology.
b. and Department of Physics and Astronomy, and Department of Electrical Engineering.

Hamiltonian Mechanics, Edited by J. Seimenis,
Plenum Press, New York, 1994

The purpose of this paper is to present numerical experiments which attempt to test this supposition in a specific case. In particular, we study a symplectic map system (the symplectic condition insures that the dynamics is Hamiltonian). A closely related work is that of Falcioni et al. (1991). For other previous relevant works on Hamiltonian dynamics in higher number of degree of freedom systems see Kaneko and Bagley (1985) and Gyorgyi et al. (1989), and the discussion and references in the book by Lichtenberg and Lieberman (1983). The main result of the present paper is that, for the system we study, the fraction of orbits on tori decreases very strongly as the number of degrees of freedom is increased, but there is still no conclusive evidence for effectively complete global ergodicity even over the very long times investigated in our numerical experiments. The latter is due to the extremely long time-scales, insensitive to machine precision, observed in the numerical experiments.

The system we studied derives from the standard map,

$$\begin{aligned} x' &= x + y, \\ y' &= y + k \sin x'. \end{aligned} \tag{1}$$

In these coordinates the map can be considered as a map of the two-torus T^2, $0 \leq x < 2\pi$ and $0 \leq y < 2\pi$.

Given a positive integer N, consider the space $(T^2)^N$ thought of as 2n-tuples $(x_0, y_0, x_1, y_1, \ldots, x_{N-1}, y_{N-1})$. We define a coupled standard map allowing symmetric bidirectional nearest neighbor interactions,

$$\begin{aligned} x'_i &= x_i + y_i, \\ y'_i &= y_i + K \sin x'_i + CK \sin(x'_i - x'_{i-1}) + CK \sin(x'_i - x'_{i+1}), \end{aligned} \tag{2}$$

where the indices are taken modulo N and x_i, y_i are taken modulo 2π. Here C is the coupling parameter to nearest neighbors. Letting $K = k/(2C+1)$, Eqs. (2) reduce to Eqs. (1) for $N = 1$. We call k the nonlinearity parameter. This map is symplectic since it can be obtained from the generating function,

$$F(x, x') = \frac{1}{2} \sum_{i=1}^{n} (x'_i - x_i)^2 + K \cos x'_i + CK \cos(x'_i - x'_{i+1}). \tag{3}$$

One checks readily that $y_i = \partial F / \partial x_i$, $y'_i = -\partial F / \partial x'_i$.
The original aim of our numerical experiments was the exploration of the relative measure of KAM tori as a function of the number of coupled maps. To this end, we first note that motion on KAM surfaces is quasiperiodic with all Lyapunov exponents zero, while motion not on KAM surfaces typically is chaotic and has at least one positive Lyapunov exponent. Thus we proceed as follows (see also Falcioni et al.). A cutoff value ϵ for an orbit to be considered quasiperiodic was set and the number of initial conditions with largest Lyapunov exponent (LE) less than ϵ counted. The run consisted of taking m initial conditions uniformly distributed in the $2N$-torus and iterating them approximately 10^6 times along with their tangent vectors to compute their LE's. A cutoff value $\epsilon = 0.005$ for the largest LE was set below which an orbit was considered quasiperiodic, and the ratio of the number of quasiperiodic initial conditions to the total number of initial conditions was returned.

When the coupling coefficient is zero, the volume of the KAM tori decays exponentially with N. In particular, if f denotes the fraction of phase space occupied by KAM tori for a single standard map, Eq. (1), then the fraction of the phase space $(T^2)^N$

for N uncoupled maps for which motion in the $2N$ variables $(x_0, y_0, \ldots, x_{N-1}, y_{N-1})$ is quasiperiodic is f^N. When $C > 0$, the rate of decay was observed to increase dramatically. Results for the parameter values $C = 0.5$, $k = 0.3$ are displayed in Table 1. In this table the estimated measure of quasiperiodic (QP) initial conditions (second column) is the fraction of 8192 randomly chosen initial conditions yielding LE's less than ϵ.

Table 1. Fraction of Initial Conditions Yielding Quasiperiodic (QP) Orbits

Maps	Estimated Measure of QP Initial Conditions	Iterations
1	1.000	10^6
2	0.403	3.25×10^6
3	0.048	3.25×10^6
4	0.002	3.25×10^6
5	0.000	1.25×10^6
6	0.000	10^6
7	0.000	10^6

Figures 1 show histograms of the observed distribution of maximum LE's for the 8192 randomly chosen initial conditions for $N = 2, 3, \ldots, 7$ coupled maps.

The case of three maps is presented twice with different numbers of iterations for the same set of data. The observed peaks get sharper but the effect is very slow. In most cases the following phenomena were noted:

1. The number of initial conditions following within the ϵ bound for quasiperiodicity decreases rapidly as the number of maps increases.

2. The observed peaks grew sharper with repeated iteration—but very slowly.

3. The histograms with more than one peak preserved those peaks and they individually got sharper.

These observations might lead one to conjecture that each peak represents a distinct ergodic component with its own maximum LE.

We now discuss the behavior of six individual orbits for the $N = 3$ case, where the orbits are chosen so that their maximum LE's calculated after 3.25×10^6 lay in distinct regions of interest of the histogram in Fig. 1(c). The calculated LE values for these six orbits are indicated by the arrows labeled with the letters (a)–(f) along the axis of Fig. 1(c). The projection of these orbits onto the first two components (x_0, y_0) are plotted for 10^4 iterations in Figs. 2(a)–(f).

Distinct orbits appeared to stay constrained in a fixed region of phase space, and this was also true when time series of 10^5 iterations were plotted.

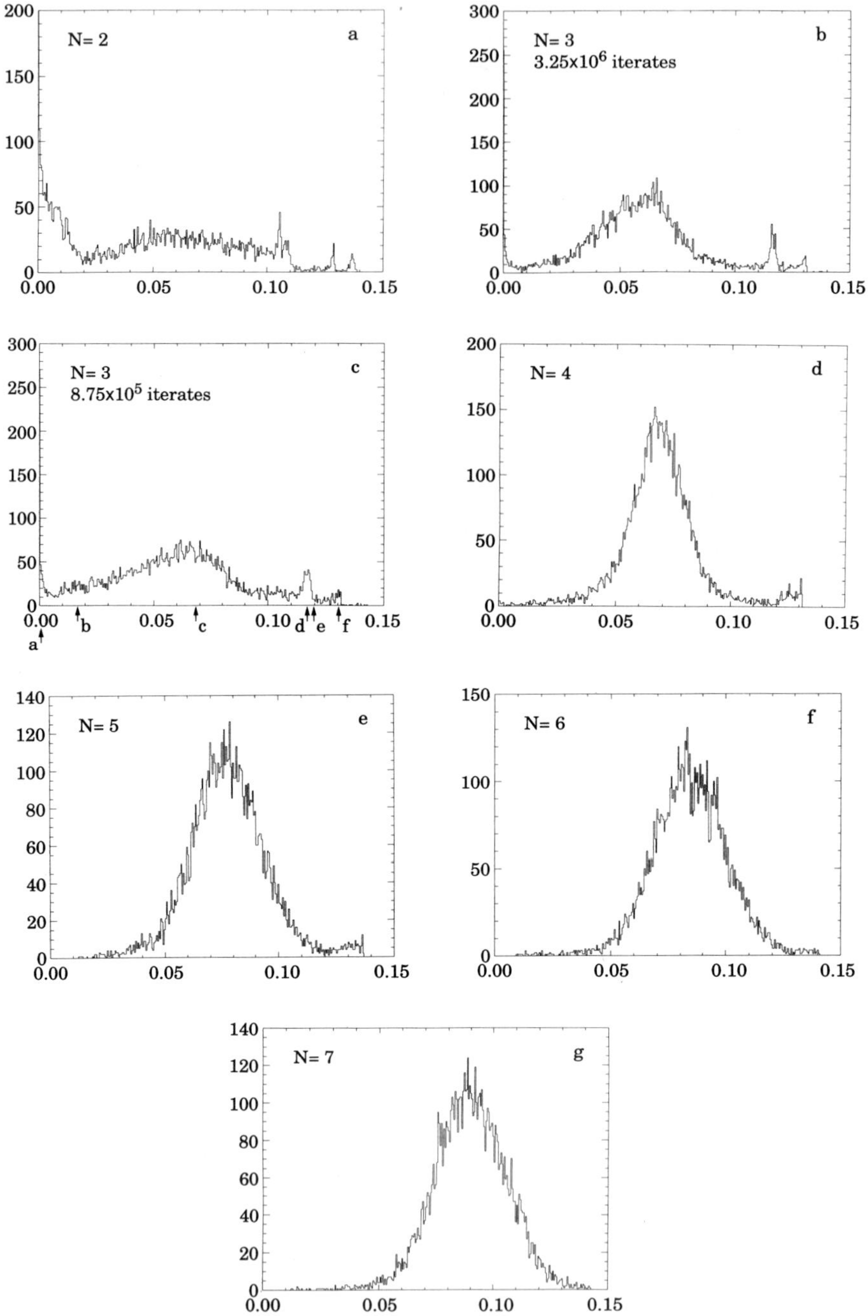

Figure 1. Histograms of maximum Lyapunov exponents for 8192 initial conditions and $N = 2, 3, \ldots, 7$. For (a) the value in the first histogram bin is about 2700, far off the scale shown. In general, the value in the first bin is an estimate of the number of quasiperiodic orbits.

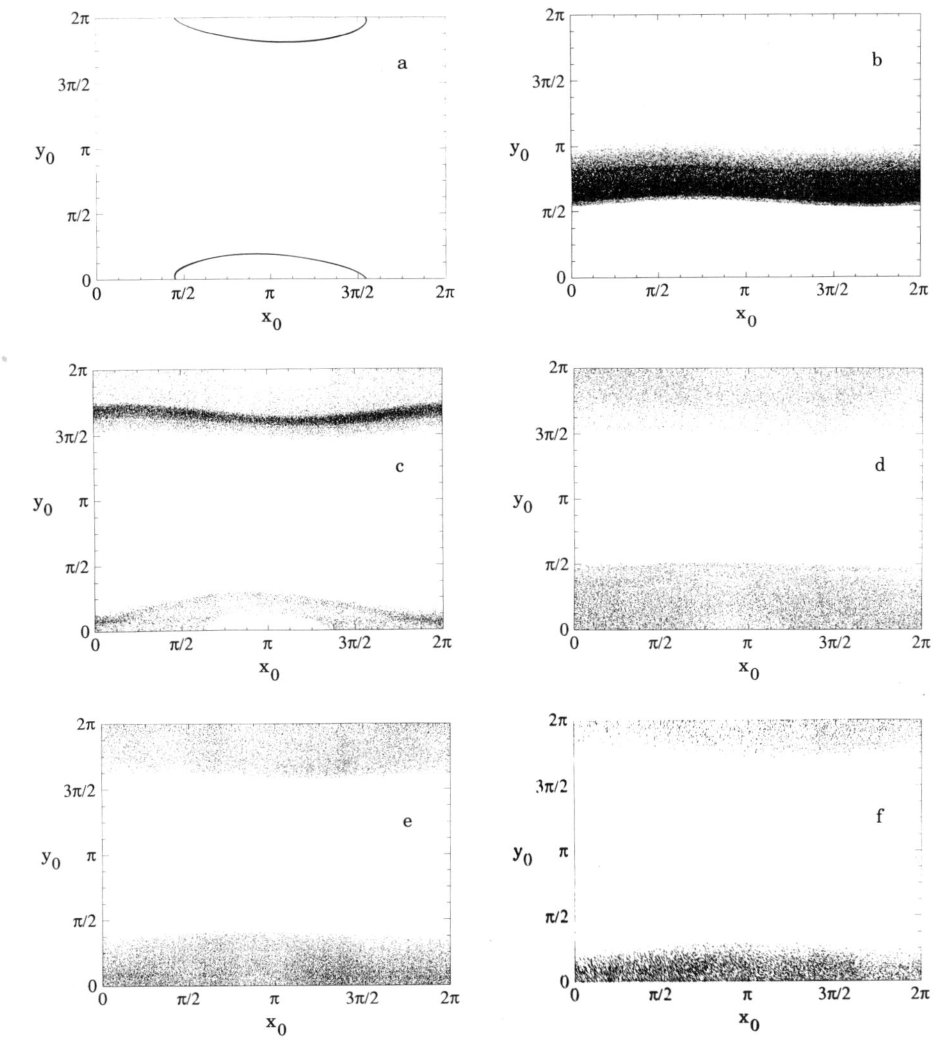

Figure 2. Projection of six individual orbits for $N = 3$ onto the plane corresponding to the first two components. 10^4 iterations are plotted. The calculated maximum LE's for these orbits are (a) 8×10^{-5} (quasiperiodic), (b) 0.0166, (c) 0.0676 [corresponding to the lower LE peak in Fig. 1(c)], (d) 0.1170 [corresponding to the higher LE peak in Fig. 1(c)], (e) 0.1191, and (f) 0.1300. These LE values are indicated along the horizontal axis of Fig. 1(c).

Lyapunov exponents were then computed for each of these orbits for a much greater period of time (3×10^8 iterations). The results are shown in Fig. 3 where the letters (b)–(f) labeling the curves correspond to the orbits shown in Figs. 2(b)–(f) and the arrows shown along the horizontal axis of Fig. 1(c). The first initial condition, which was presumed quasiperiodic, remained stable during the whole process, and remained at a highly stable value. The remaining four, however, appear to have started to converge slowly to a new common value.

Figures 4 break down the curves in Fig. 3 [plus orbit (a)] giving the cumulative LE and a "local" LE which is calculated in 500,000 iterate bursts. Observe the great stability of initial conditions (a) and (f).

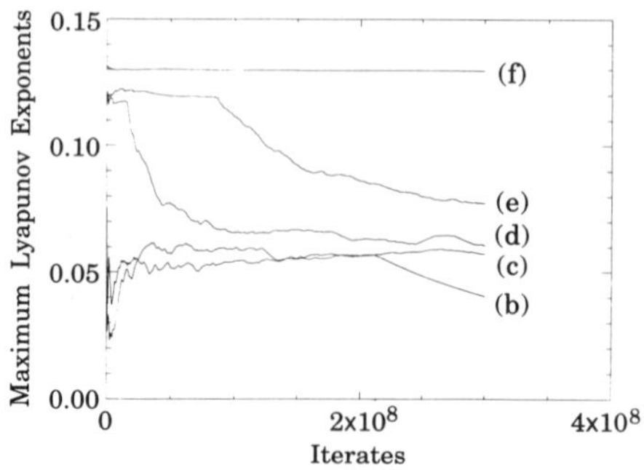

Figure 3. Maximum calculated LE as a function of the number of map iterations for the five orbits corresponding to Figs. 2(b)–2(f).

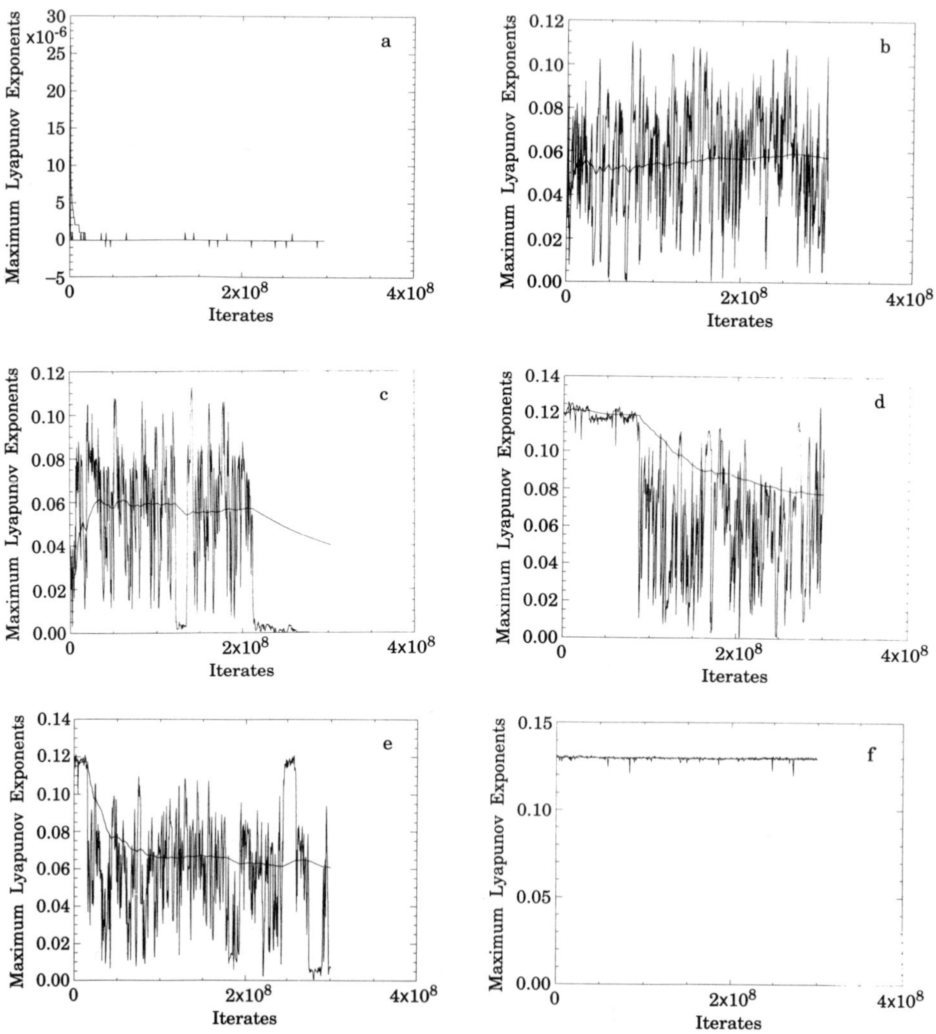

Figure 4. Maximum calculated LE as a function of the number of map iterations for the five orbits corresponding to Figs. 2(b)–2(f).

Further studies were conducted for a variety of initial conditions and various behaviors were observed.

1. Some initial conditions appeared to lead to orbits whose LE's showed a great deal of stability (they remained essentially unchanged over the observed time scale).

2. Some initial conditions showed a high degree of stability at one value of the maximum LE but then "leaked" into a regime with a different LE.

3. Some initial conditions alternated between chaotic behavior and behavior very close to quasiperiodic.

One effect of these observations was to call into question the reliability of the LE calculations in general. Many of these calculations seemed to be stable for greater than 10^6 iterates before changing value. Given the relative rarity of these "leaks," it was impractical to assign any numerical value to this diffusion.

The histogram calculations were performed on a Connection Machine using (of necessity) single-precision arithmetic. The orbit calculations were performed on a DecStation 3100 using double precision. To examine the effect of machine precision several of the long-term LE calculations were done at both single and double precision. The observed behavior was qualitatively the same; the observed leakage between regions of different LE occurred in each case (at slightly different iterates).

This work was supported by the Office of Naval Research (Physics Branch) and by the Department of Energy (Office of Scientific Computing and Office of Energy Research).

REFERENCES

Falcioni, M., Marini Bettolo Marconi, U., and Vulpiani, A., 1991, Phys. Rev. A **44**, 2263.

We had completed our research at the time of the publication of the paper of Falcioni et al. Because of the similarity of that work and ours, in this Letter we shall be somewhat briefer than we otherwise might have been, and will also emphasize that part of our work which is different from that of Falcioni et al.

Gyorgyi, G., Ling, F. A. and Schmidt, G., 1989, Phys. Rev. A **40**, 5311.

Kaneko, K. and Bagley, R. J., 1985, Phys. Lett. A **110**, 435.

Lichtenberg, A. J. and Lieberman, M. A., 1983 *Regular and Stochastic Motion* (Springer-Verlag, Berlin).

GIBBSIAN CHECK OF THE VALIDITY OF GIBBSIAN CALCULATION THROUGH DYNAMICAL OBSERVABLES

Dominique F. Escande,[1] Holger Kantz,[2] Roberto Livi,[3,5] and Stefano Ruffo[4,5]

[1]Equipe Turbulence Plasma, Institut Méditerranéen de Technologie
Technopôle de Château Gombert
F-13451 Marseille Cedex 13, France, and
DRFC
CE – Cadarache
F-13108 Saint Paul Lez Durance, France
[2]Fachbereich Physik
Bergische Universität – Gesamthochschule Wuppertal
Gauss Strasse 20, D-42097 Wuppertal 1, Deutschland
[3]Dipartimento di Fisica, Universitá di Firenze
Largo E. Fermi 2, I-50125 Firenze, Italia
[4]Dipartimento di Energetica, Universitá di Firenze
Via S. Marta 3, I-50100 Firenze, Italia
[5]Istituto Nazionale di Fisica Nucleare Sez. di Firenze
and Istituto Nazionale di Fisica della Materia Unitá di Firenze
Largo E. Fermi 2, I-50125 Firenze, Italia

A key question in statistical physics is the agreement between time and ensemble averages. Moreover, in many applications (e.g. molecular dynamics[1]), one needs a fast relaxation in time with a limited number of degrees of freedom. The theory of dynamical systems tells us that this happens if chaos is strong, i.e. a generic orbit fills most of the available phase space in a short time. The conditions for this to happen are difficult to establish in high dimension. In this contribution we address this question in the context of Hamiltonian dynamics, choosing a specific coupled rotators model on a 1-D lattice[2]. We develop simple *ensemble* calculations of dynamical chaos indicators, which enable us to identify the energy region where the desired fast relaxation takes place.

We consider a chain of N coupled rotators defined by the Hamiltonian

Hamiltonian Mechanics, Edited by J. Seimenis,
Plenum Press, New York, 1994

$$H = \sum_{i=1}^{N}\{\frac{p_i^2}{2} + (1 - \cos(q_{i+1} - q_i))\} = K + U \quad , \tag{1}$$

where periodic boundary conditions are assumed: $q_1 = q_{N+1}, \quad p_1 = p_{N+1}$. K and U stand for the kinetic and potential part of H, respectively. In order to match with the assumption of Gibbs calculation, the total momentum of the chain is taken as zero. The qualitative behaviour of the dynamics is ruled by the energy density $\mathcal{E} = H/N$. When $\mathcal{E} \gg 1$ each rotator is almost freely rotating since the coupling potential U is negligible with respect to K, this regime is hereafter referred to as weak coupling. A strong coupling regime corresponds to $\mathcal{E} \ll 1$. Then the masses are coupled through almost linear forces, since the cosines of Eq.(1) have small arguments, and the motion is quite close to that of a chain of harmonic oscillators. The weak-coupling and strong-coupling regimes are close to be integrable, and chaos is weak in either regime. As a result the statistical description should be appropriate only out of these regimes.

Indeed this is confirmed by comparing the numerical calculation of the time-averaged potential energy density and the specific heat C_V (as given by the fluctuations of K [3]), to the Gibbsian predictions for these observables [4]. They should coincide for large N and large averaging times. Figure 1 displays the results of this comparison for various values of a temperature T, naturally defined as $T(t) \equiv 2 < K >_t /N$, where $< \bullet >_t$ stays for the finite time average $t^{-1}\int_0^t \bullet dt'$. Here $t = 35000$ and $N = 128$. These two finite values already give a nice convergence towards the Gibbsian estimate for intermediate values of T: there exists a statistical regime. The relaxation of the finite time average is fast in this regime, while, when T becomes small or large, the finite time average does not relax. Note that for small T the agreement with the Gibbsian estimate of the average potential energy derives trivially from the virial theorem. Analogously for large T the kinetic energy has small relative fluctuations which imply the Gibbsian value 0.5 for the specific heat. (Observe that in Ref. [4] the specific heat was computed through the fluctuations of the total energy of a part of the chain, simulating a canonical ensemble calculation. This calculation does not agree with the method we use here, which we take from Ref. [3] in the high temperature region; the different results are a consequence of the fact that the system is not ergodic.) In Figure 2 we show the convergence time t_x of the numerical estimates of both $< U >$ and C_V to within x percent of the Gibbsian estimate for several values of x. The convergence time changes suddenly in the vicinity of $T_w \approx 10$ and $T_s \approx 0.09$, that we define as the weak and strong coupling thresholds, respectively. In the following we show that there is a dynamical explanation for these thresholds and that a Gibbsian calculation predicts them.

We first deal with the weak-coupling limit. There it is natural [5,6] to single out a series of pendula (primary resonances) related to the N cosines of Hamiltonian (1). Assume $p_{j+1}(\tau) \simeq p_j(\tau)$ at some time τ for some j; then the phase of the j-th cosine is almost stationary, and the dynamics should be strongly affected by the corresponding nonlinear resonance [6,7]. Define a unitary matrix $\hat{\mu}$ with row-vectors R_k and let the components $R_{1,i}$ of R_1 be all zero except $R_{1,j+1} = -R_{1,j} = 1/\sqrt{2}$. Define $\mathbf{p}^r$ in terms of the momentum vector $\mathbf{p}$ at time τ

$$\mathbf{p}^r = (p_1(\tau), p_2(\tau), \ldots, p_{j-1}(\tau), p_*, p_*, p_{j+2}(\tau), \ldots, p_N(\tau)) \quad , \tag{2}$$

where $p_* = (p_{j+1}(\tau) + p_j(\tau))/2$, and the canonical transformation $(\mathbf{p}, \mathbf{q}) \to (\mathbf{I}, \mathbf{\Psi})$ with generating function

$$F(\mathbf{I}, \mathbf{q}) = \sum_i (p_i^{\tau} + \sum_k I_k \mu_{ki}) q_i = (\mathbf{p}^{\tau} + \mathbf{I}\hat{\mu})\mathbf{q} \quad . \tag{3}$$

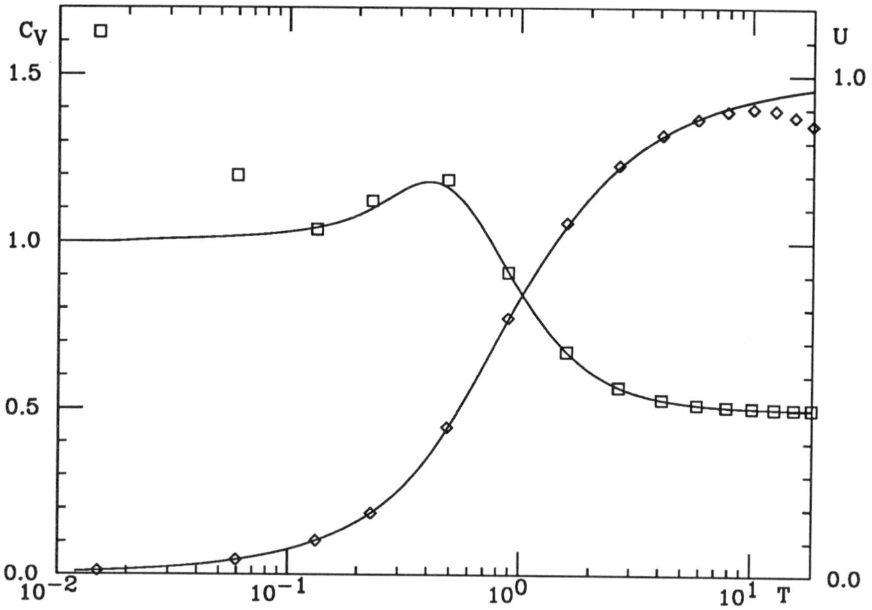

Figure 1. Plot of C_V and $< U >$ vs. T. Numerical estimates: C_V (□) and $< U >$ (◇); Gibbsian predictions (solid lines).

Then $I_1(t) = (p_{j+1}(t) - p_j(t))/\sqrt{2}$ and $\Psi_1(t) = \mathbf{R}_1 \cdot \mathbf{q} = (q_{j+1}(t) - q_j(t))/\sqrt{2}$, and the part of the new Hamiltonian containing only I_1 and Ψ_1 is

$$H_1(I_1, \Psi_1) = \frac{I_1^2}{2} - \cos(\sqrt{2}\Psi_1) \quad , \tag{4}$$

which corresponds to a pendulum whose phase space displays a "cat's eye", the signature of a nonlinear resonance [6,7]. Indeed the corresponding behavior is found numerically on I_1 and Ψ_1 when the dynamics generated by the Hamiltonian (1) is computed. In the sequel, resonance j is said to be *active* when the orbit is inside its cat's eye.

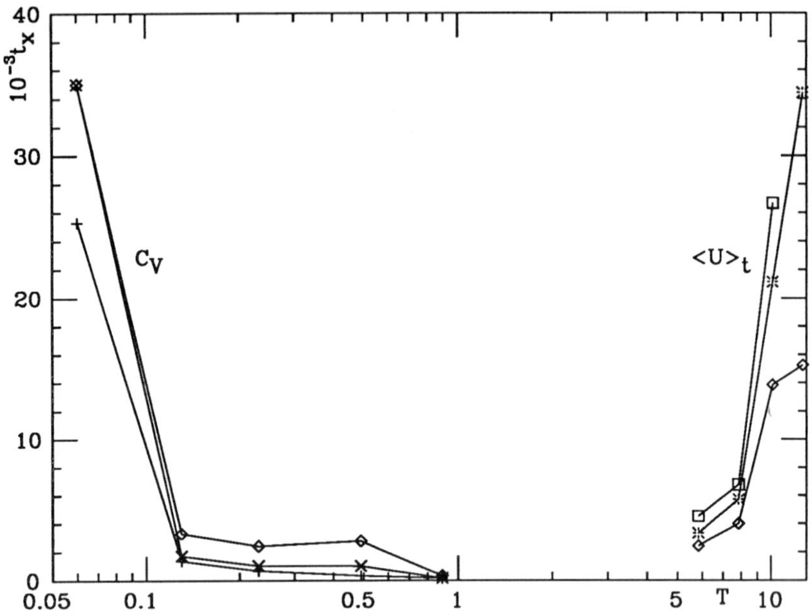

Figure 2. Plot of first time t_x when the numerical estimates of C_V ($T < 1$) and $< U >_t$ ($T > 5$) reach the Gibbsian prediction up to an x% error: $x = 20\%$ (+), $x = 10\%$ (×), $x = 5\%$ (◇), $x = 3\%$ (∗), $x = 2\%$ (□). The upper point to the left is a lower bound.

When $\mathcal{E}$ decreases, the (I_1, Ψ_1) dynamics looks numerically quite similar to that of a pendulum perturbed by a time-dependent potential: chaotic separatrix crossings occur, and are related to the existence of a stochastic layer. The origin of this low-dimensional behavior may be suspected to stem from nearby resonances [6,7], i.e. from resonances produced by the cosines with arguments where q_j or q_{j+1} are present. Hamiltonian (1) has only two such resonances: those with indices $j-1$ and $j+1$. They encompass four degrees of freedom (from $j-1$ to $j+2$) which define a sub-phase-space S_j. Define

$$\mathbf{R}_2 = (0, \ldots, 0, p_{j-1}(\tau), p_*, p_*, p_{j+2}(\tau), 0, \ldots, 0)/\mathcal{N} \quad , \tag{5}$$

where $\mathcal{N} = \sqrt{p_{j-1}^2(\tau) + 2p_*^2 + p_{j+2}^2(\tau)}$. In the case where $p_{j+1}(\tau) = p_j(\tau)$, $\mathbf{R}_2$ is the projection of the normal to the energy surface onto the momentum subspace with indices $i = j-1, \ldots, j+2$.

The part of the new Hamiltonian containing the two degrees of freedom (I_i, Ψ_i), $i = 1, 2$ is

$$\begin{aligned} H_2(I_1, \Psi_1; I_2, \Psi_2) &= H_1 + \frac{1}{2}I_2^2 + \mathcal{N} I_2 - \cos(-\Psi_1/\sqrt{2} + \omega_+ \Psi_2 + \rho_+) \\ &\quad - \cos(-\Psi_1/\sqrt{2} + \omega_- \Psi_2 + \rho_-) \end{aligned} \tag{6}$$

where

$$\Psi_2 = \mathbf{R}_2 \cdot \mathbf{q} \quad , \quad I_2 = (\mathbf{p} - \mathbf{p}^\tau) \cdot \mathbf{R}_2, \tag{7}$$

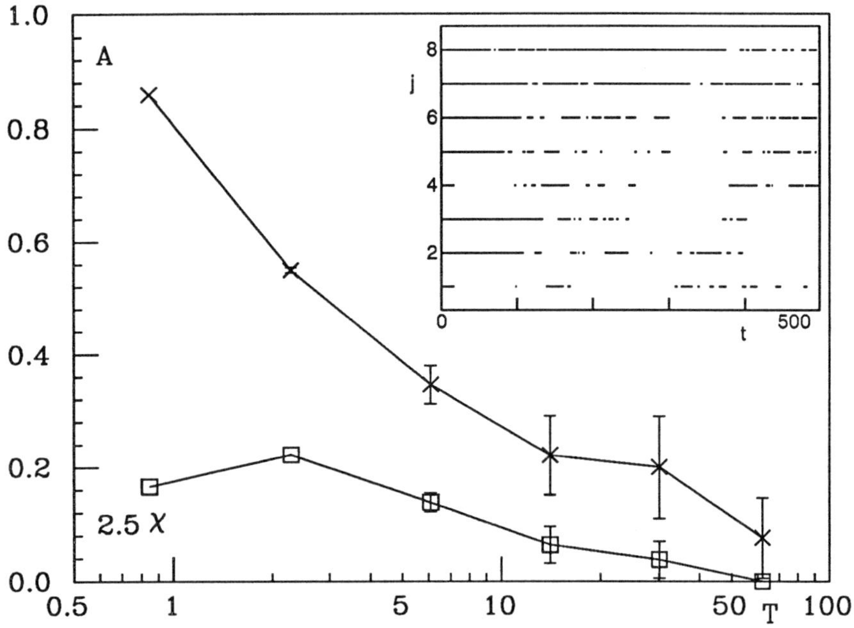

Figure 3. Plot of the activity A and crossing rate χ vs. temperature. The error bars indicate the variance over the 8 initial conditions. The insert shows the labels of active resonances vs. time.

and

$$\rho_+ = q_{j+2} - q_{j+1} + \Psi_1/\sqrt{2} - \omega_+\Psi_2 \quad , \qquad \rho_- = q_j - q_{j-1} + \Psi_1/\sqrt{2} - \omega_-\Psi_2 \quad ,$$
$$\omega_+ = R_{2,j+2} - R_{2,j+1} \quad , \qquad \omega_- = R_{2,j} - R_{2,j-1}. \tag{8}$$

Ψ_2 evolves like $\mathcal{N}t$ if $I_2 \ll \mathcal{N}$. Moreover, if $\rho_\pm$ are constant the dynamics defined by H_2, hereafter called H_2-dynamics, is that of a 1.5 degrees of freedom system (time dependent one degree of freedom). Indeed, we checked on the dynamics defined by (1) that Ψ_2 increases linearly and $\rho_\pm$ vary slowly with respect to Ψ_2 over long time intervals (of order 10^2 also for T as small as ~ 5) as long as p_{j-1} and p_{j+2} stay at their value at time τ. This is consistent with a perturbative calculation of $\rho_\pm$ at first order in the coupling potential U. Note that resonances with indices 2 units away from j may be active without directly affecting the dynamics of (I_1, Ψ_1).

Numerically I_2 fluctuates with a slow drift. If, at time t, I_2 becomes as large as $\pm\mathcal{N}/2$, we define a new $\tau = t$, a new $\mathbf{p}^\tau$ and consequently a new approximating Hamiltonian (6). Then, if $p_{j+1}(\tau) \simeq p_j(\tau)$ we observe the same behaviour, otherwise I_2 becomes quickly large and we go on redefining τ until a resonant behaviour sets in again. As a result, H_2-dynamics may be found in dynamics generated by (1) for a sizeable fraction of time even in the statistical regime. Trapping in either of the nonlinear resonances related to the two cosines of Eq. (6) was also observed.

Since Hamiltonian H_2 gives a good description of the projection of dynamics generated by (1) onto the subspace S_j, in the weak coupling limit one can use the expression for the separatrix given by Hamiltonian (4) in S_j to detect when a resonance is active. We define the mean activity parameter A as

$$A = \frac{\sum_j (time \quad resonance \quad j \quad is \quad active)}{(total \quad time) \times N}, \tag{9}$$

and the crossing rate χ as the mean number of separatrix crossings per unit time. In Fig. 3 we show both A and χ as functions of T, averaged over 8 random initial conditions for $N = 8$. When T is large the system is unlikely to be in a resonant state and furthermore the energy transfer among the rotators is negligible. Thus A and χ are small. Decreasing the temperature below $T \approx 10$ one observes an increase of both A and χ indicating the onset of large scale chaos. In the insert of Fig. 3 we plot the labels of active resonances *vs.* time during a run for $T = 3.5$: the frequent switching of the trajectory from one resonance to another, or its simultaneous presence in two nearby resonances makes explicit the dynamical mechanism responsible for the existence of large scale chaos.

As observed above, H_2-dynamics corresponds to that of a 1.5 degrees of freedom Hamiltonian system. In such a system, the breakup of the last KAM torus between two primary resonances corresponds to a qualitative change: chaos connects the two resonances. Following Chirikov[6], it is natural to obtain an estimate of breakup threshold by computing the one of H_2-dynamics when either of the two cosines of Eq.(6) is suppressed. Then, only two nonlinear resonances of H_2 survive and one obtains the so-called paradigm Hamiltonian [7] whose threshold corresponds to $s_\pm \simeq 0.7$, where $s_\pm$ is the overlap parameter between the primary resonance of H_1 and each of those with index $\pm$

$$s_\pm = \frac{4}{\omega_\pm \mathcal{N}} \quad . \tag{10}$$

Note that $s_\pm$ is a dynamical observable of the full dynamics. Its probability distribution P may be computed in the canonical ensemble. It is

$$P(s) = \frac{s_{max}}{\sqrt{\pi} s^2} \exp\left(-\frac{s^2_{max}}{s^2}\right) \quad . \tag{11}$$

The maximum of the distribution is at $s_{max} = \sqrt{16/(3T)}$. P is shown in Fig. 4 for three different values of T. In the large T limit the most probable overlap is small, and it increases as T decreases.

When $s_\pm$ becomes typically larger than 0.7, we say the system is in the fast chaotic regime. Then chaotic transport in the full dynamics directly connects one primary nonlinear resonance to the next one. Otherwise this connection only occurs through the slower Arnold diffusion, which needs the crossing of the stochastic layers of higher order resonances. Thus, it is natural to expect the weak-coupling threshold to correspond to that of the fast chaotic regime. Indeed a rough estimate of this bound is obtained by demanding that the maximum of P occurs at $s_{max} = 0.7$. This yields $T = 10.9$, which confirms the numerical results for the threshold of large scale chaos shown in Fig. 3 and is close to the weak coupling threshold T_w. As a result we have evidenced that this threshold corresponds to that to large-scale chaos in the low dimensional dynamics described by H_2.

The strong-coupling threshold has a simple dynamical justification, since chaos is known to exist for T large enough. Indeed, it is expected to occur when the nonlinearity of the force between rotators comes into play and induces chaos among the vibrational

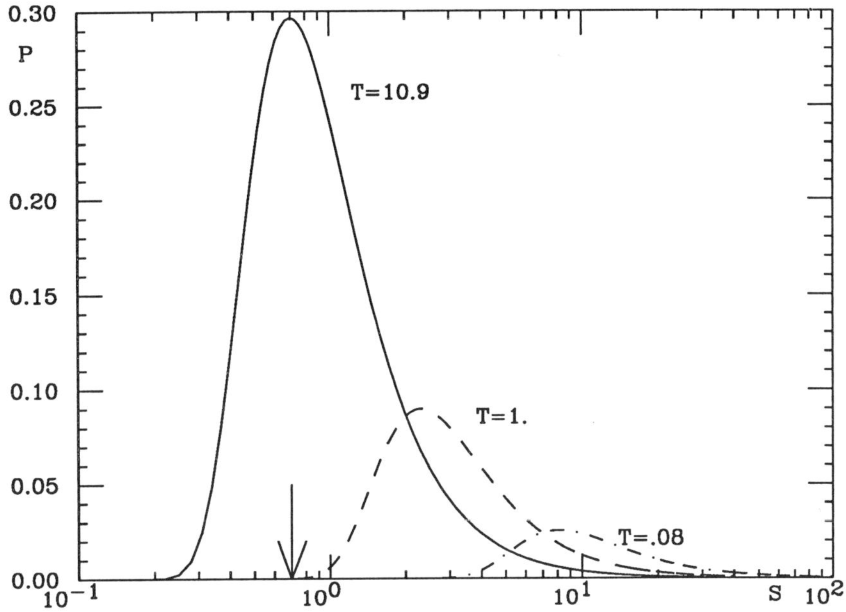

Figure 4. Plot of $P(s)$ vs. s for $T = 10.9$, $T = 1.$ and $T = 0.08$; $s = 0.7$ is indicated by an arrow.

eigenmodes, whereas the integrable Hamiltonian at low temperatures is linear in the actions. This chaos happens as a consequence of the same mechanism of resonance overlap described for the weak-coupling side, but applied to the different and much more complex case of the system of coupled anharmonic oscillators (so far no convincing result has been obtained in this case). The nonlinearity of the force also affects the Gibbsian estimate of the specific heat $C_V(T)$ and makes it depart from the harmonic equipartition law $C_V(T) = 1$. The non-linear threshold as indicated by the deviation of the Gibbsian $C_V(T)$ from its low temperature limit can be obtained as the intersection of the tangent of the left inflection point of $C_V(T)$ *vs.* T in linear scale with the line $C_V(T) = 1$. This yields $T \approx 0.08$, which corresponds to the estimate of T_s obtained from the time relaxation properties of the specific heat $C_V(T)$ (see Fig. 2). The transition to the strong-coupling limit may also be understood from the point of view of H_2-dynamics. Indeed, this limit corresponds to a strong resonance-overlap induced by a very slow time-dependence of H_2. In this regime chaos becomes quite slow and inefficient.

To summarize, we considered a chain of 128 coupled rotators, and found a fast relaxation towards their Gibbsian expectation of the mean energy $U(T)$ and of the specific heat $C_V(T)$, as computed by time-average, for a certain domain of the scaled temperature T. This domain was shown to agree with that where the anharmonicity of the chain is strong enough, as measured through the behavior of the Gibbsian expectation of $C_V(T)$, and where large-scale chaos is dominant in 1.5 degree-of-freedom Hamiltonian systems defined from the full dynamics through geometrical arguments relying upon the concept of nonlinear resonance. The dominance of chaos is obtained by computing the Gibbsian probability distribution function of a corresponding stochasticity parameter, and by comparing its support to the chaotic domain. A more detailed and extensive account of the results presented in this contribution is in preparation [8].

ACKNOWLEDGEMENTS

We thank Y. Elskens for fruitful discussions and for his careful reading of the manuscript. We thank the Institute for Scientific Interchange in Torino for hospitality during the workshops *Complexity and Evolution* 1991 and 1992 and their participants for useful discussions. (D.E.) acknowledges financial support from the Universitá di Firenze, (S.R.) and (R.L.) from the Université de Provence and from the Istituto Nazionale di Fisica Nucleare. (H.K.) thanks the European Community for partial financial support.

REFRENCES

1. G. Ciccotti and W.G. Hoover, eds., "Molecular Dynamics Simulation of Statistical-Mechanical Systems", North Holland, Amsterdam (1986);
   G. Ciccotti, D. Frenkel and I.R. McDonald, eds., "Simulation of Liquids and Solids", North Holland, Amsterdam (1987).
2. G. Benettin, L. Galgani and A. Giorgilli, Classical Perturbation Theory for Systems of Weakly Coupled Rotators, *Nuovo Cimento*, **B89**: 89 (1985) and: Numerical Investigations on a Chain of Weakly Coupled Rotators in the Light of Classical Perturbation Theory, *ibid.* **B89**: 103 (1985).
3. J. L. Lebowitz, J. K. Percus, L. Verlet, Ensemble Dependence of Fluctuations with Application to Machine Computations, *Phys. Rev.* **153**: 250 (1967).
4. R. Livi, M. Pettini, S. Ruffo and A. Vulpiani, Chaotic Behaviour in Nonlinear Hamiltonian Systems and Equilibrium Statistical Mechanics, *J. Stat. Phys.* **48**: 539 (1987).
5. H. Poincaré, "Les Méthodes Nouvelles de la Mécanique Céleste", Gauthier-Villars, Paris (1892, 1893, 1899).
6. B. V. Chirikov, A universal instability of many-dimensional oscillator systems, *Phys. Rep.* **52**: 263 (1979).
7. D. F. Escande, Stochasticity in Classical Hamiltonian Systems: Universal Aspects, *Phys. Rep.* **121**: 166 (1985).
8. D.F. Escande, H. Kantz, R. Livi, and S. Ruffo, Gibbs Calculation of Dynamical Observables, to appear in *J. Stat. Phys.*

ADIABATIC INVARIANTS AND TIME SCALES FOR ENERGY SHARING IN MODELS OF CLASSICAL GASES

Giancarlo Benettin

Università di Padova
Dipartimento di Matematica Pura e Applicata
Via G. Belzoni, 7
35131 Padova, Italy

1. MOTIVATION

The purpose of this paper is twofold. On one hand I would like to illustrate an application of the theory of adiabatic invariants — essentially, Nekhoroshev–like exponential laws — to classical statistical mechanics. On the other hand, I would like to revisit some ideas of distinguished physicists, like Jeans and Landau, who long time ago, much before modern perturbation theory, introduced exponential laws in connection with problems of adiabatic invariance (Jeans,[1,2] 1903 and 1905; Landau and Teller,[3] 1936). These authors worked only heuristically, as physicists often do, but their ideas are definitely deep, and as we shall see, at least in some cases their heuristic arguments can be turned into effective proofs.

The model problems we shall here consider come from gas dynamics. In the simplest case (Fig. 1a,b) a fast vibrating or rotating molecule collides with a fixed smooth potential barrier. More complicated scattering processes, with two or more fast internal degrees of freedom (Fig. 1c,d), can also be considered: such collisions are clearly important, nevertheless we shall refer to them only occasionally, and focus instead the attention to the simpler (but nontrivial) case of a single fast degree of freedom. The easiest example is probably the scattering of a rotator in one dimension; the Hamiltonian has the form

$$H(p,q,I,\varphi) = \frac{I^2}{2J} + \frac{p^2}{2m} + V(q,\varphi) \ , \qquad I \in \mathbb{R} \ , \varphi \in S^1 \quad (p,q) \in \mathbb{R}^2 \ , \tag{1.1}$$

Hamiltonian Mechanics, Edited by J. Seimenis,
Plenum Press, New York, 1994

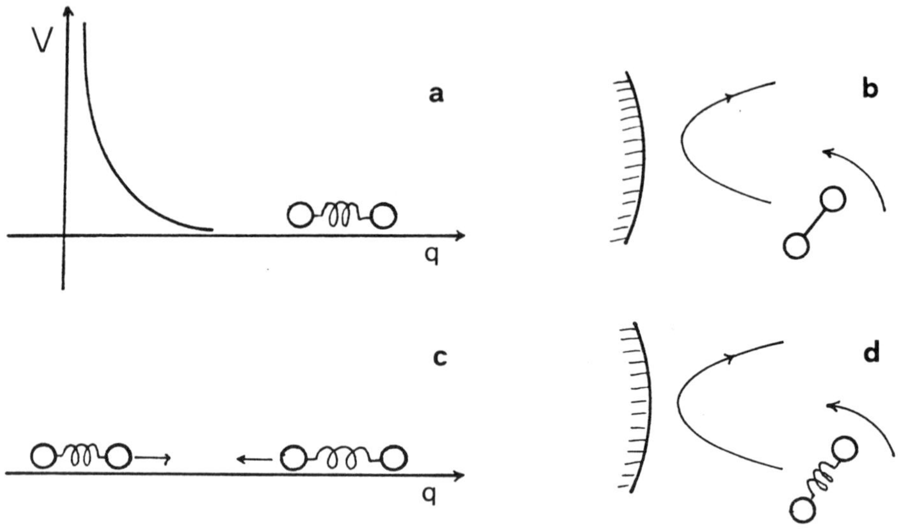

Figure 1 – Elementary examples of collision processes.

where m and J denote respectively the mass and the inertia moment of the rotator, while V is the interaction potential; V is assumed to be regular, in fact analytic, and to decay faster than $1/q$ at infinity.

The basic physical quantity one wants to know, in all of these scattering processes, is the amount of the energy exchange ΔE between the fast and the slow degrees of freedom; in the above example,

$$\Delta E = \frac{1}{2J}[I(+\infty)^2 - I(-\infty)^2] \,. \tag{1.2}$$

Due to the presence of fast phases, and relatively slow motion of the center of mass, this is clearly a problem of adiabatic invariance. Besides its mathematical interest, this question turns out to be rather crucial in the framework of the classical ergodic problem, that is (in the original sense by Boltzmann) the problem of the dynamical foundation of classical statistical mechanics.

As is well known, the essence of statistical equilibrium in classical systems is the equipartition principle: all degrees of freedom, independently of their nature and peculiarities (like their own frequency), take part to the same extent to the energy sharing, and in fact share all of their own energy with the remaining ones. The consequences, as is well known, are striking: in particular, the *thermodynamical* (or internal) energy of the system can be identified with the overall *mechanical* energy of the components; the temperature is simply interpreted as the average kinetic energy of the molecules; and a relevant quantity like the specific heat C_v is immediately computed, by simply counting the number of degrees of freedom in the system. Consider, to be definite, the case of a diatomic gas. In principle, one has seven contributions per molecule to the energy of the system (three for translations, two for rotations, one kinetic and one potential for the internal vibration), so one expects the classical value $C_v = \frac{7}{2}R$ (R is here the so-called constant of gases). On the contrary, experience shows that, at ordinary temperatures, the vibrational energy (although large, for fast vibrating molecules) is practically constant, and not shared with other degrees of freedom; one commonly says it is *frozen*. At lower temperatures, the rotational energy too is frozen, and molecules behave thermodynamically (but certainly not mechanically) as unstructured point particles. One could also recall other freezing phenomena, taking place in solids or in the blackbody, and make for them similar considerations: in all examples it appears that high frequency degrees of freedom do not take part significantly to the energy exchanges, and thus to thermodynamics. Let us stress that such failure of the equipartition principle is usually interpreted as an essential failure of classical mechanics; in fact, freezing phenomena are often referred to as typical quantum phenomena, with no classical analog at all.

This is perhaps a too quick conclusion, which does not take into account a basic question, namely the *time scale* one needs, in order significant energy exchanges, and thus energy equipartition, possibly occur. The possible existence of *very* large time scales for the energy exchanges between fast and slow degrees of freedom was stressed by Boltzmann[4] (1895) and Jeans[1,2] (1903, 1905). In their own words, such time scale could be of "days or years",[4] or "hundreds of centuries",[2] that is so large with respect to any reasonable experimental time, that to any practical purpose the energy stored in fast degrees of freedom should be considered as constant, or frozen.

Jeans, in particular, considers the case of a purely vibrating molecule in one dimension, with Hamiltonian of the form

$$H(p,q,\eta,\xi) = \frac{1}{2}(\eta^2 + \omega^2\xi^2) + \frac{p^2}{2m} + V(q,\xi) \ , \tag{1.3}$$

where ξ represents the internal oscillating degree of freedom (say the excess length of the spring), and η is the corresponding momentum. After very interesting heuristic considerations, which unfortunately we cannot enter here, Jeans proposes for the energy exchange an exponential law of the form

$$\Delta E \simeq \mathcal{E}\, e^{-\tau\omega} \ , \tag{1.4}$$

where $\mathcal{E}$ is some natural (microscopic) unit of energy, and τ is some constant of the order of the collision time.

Let us stress some possible consequences of the above exponential law. Clearly, the number N of collisions one needs in order an appreciable amount of energy is exchanged, and the corresponding time scale T, are given by

$$N \simeq e^{+\tau\omega} \ , \qquad T \simeq T_0\, e^{+\tau\omega} \ , \tag{1.5}$$

where T_0 is a typical time between collisions. These estimates, if correct, lead to very sharp phenomena: assume, as is reasonable, $T = 10^{-12}$ sec, $\tau = 10^{-13}$ sec, and denote by ω_0 the value of ω such that $T = 1\,\mathrm{sec}$, i.e. $\omega_0 = 12 \log 10\, \tau^{-1} \simeq 2.8 \times 10^{14}\,\mathrm{sec}^{-1}$; then one has

$$\begin{array}{lll} \text{for} \quad \omega = \frac{1}{2}\omega_0 & T = 10^{-6}\,\mathrm{sec} & \\ \text{for} \quad \omega = \omega_0 & T = 1\,\mathrm{sec} & \\ \text{for} \quad \omega = 2\omega_0 & T = 10^{12}\,\mathrm{sec} & (\simeq 3 \times 10^4\,\mathrm{years}) \ . \end{array} \tag{1.6}$$

A similar behavior is clearly a promising candidate as a classical analog of quantum freezing, and it is certainly worthwhile to further investigate it.

2. RESULTS FROM PERTURBATION THEORY

Exponential laws are now rather well understood to be a natural outcome of perturbation theory. The essential ideas are due to Nekhoroshev and Neishtadt; special results, adapted in particular to problems of molecular collisions, can be found in Ref.[5-7]. For example, a rough statement adapted to Hamiltonian (1.1), taken from Ref.[7], is the following:

Proposition. *Consider the Hamiltonian (1.1), and assume:*

i. *V is analytic in a suitable complex extension of the real domain.*

ii. *$\omega := I(-\infty)/J$ is large, namely*

$$\omega > \tau^{-1} := C \max(|p(-\infty)|/m, \sqrt{2\|V\|/m}), \tag{2.1}$$

where C is some constant, and $\| \,.\, \|$ denotes a convenient norm.

iii. *V decays sufficiently rapidly for $|q| \to \infty$, say faster than $|q|^{-1}$.*

Then there exist a constant $\mathcal{E}$, such that ΔE defined as in (1.2) satisfies the exponential estimate

$$|\Delta E| < \mathcal{E}\, e^{-\tau\omega} \ . \tag{2.2}$$

For a more precise statement, see Ref.[7].

Similar results have been obtained[6] for Jeans problem, more generally for a system of the form

$$H = \frac{1}{2} \sum_{j=1}^{n} (\eta_j^2 + \omega_j^2 \xi_j^2) + \mathcal{H}(p,q) + V(p,q,\xi) \ , \tag{2.3}$$

representing, in particular, the collinear collision of n oscillating molecules. In suitable assumptions, and for identical molecules (equal frequencies), one still gets an exponential law of the form (2.2); E is in this case the overall vibrational energy.*

Thus, one can say that perturbation theory supports the Jeans conjecture, or even proves it, at least for the above oversimplified models. Nevertheless, perturbative results are open to criticism, for three reasons:

i. The constants one gets are bad (τ is too small, $\mathcal{E}$ is too large). This is not surprising, and is due to the well known "pessimism" of perturbation theory, but certainly prevents the use of perturbative results in realistic situations. In particular, for the above problem of the collision of n vibrating molecules, one gets $\tau \sim n^{-2}$; should this be the case, one would conclude that many–body collisions, although statistically much less favourite, nevertheless are practically the only responsible of energy exchanges in the gas.

ii. The result is, in a sense, unnatural: as is typical of perturbation theory, the exponential law appears in a rather indirect way. In particular, the very essential constant τ is not clearly connected to the parameters of the system (it comes from a rough estimate of the rate of divergence of an asymptotic series).

iii. Finally, the estimates are strictly one sided: one can use the exponential law to say that energy exchanges, if any, are small; certainly not to assure the presence of suitable energy exchanges, eventually leading the system to statistical equilibrium.

A different approach to the problem will be introduced in the next Section.

---

* Due to the resonance, exchanges of vibrational energy between the different oscillators are easy; only the overall vibrational energy is almost conserved.

3. THE LANDAU–TELLER APPROXIMATION

In 1936 Landau and Teller[3] proposed a very crude approximation for adiabatic invariants, which later on was applied by Rapp[8] to a problem of molecular collisions. The Hamiltonian considered by Rapp is

$$H = \frac{1}{2}(\eta^2 + \omega^2\xi^2) + \frac{p^2}{2} + (1+\xi)e^{-q} \ ; \tag{3.1}$$

the corresponding equations of motion, in second order form, are

$$\ddot{q} - e^{-q} = \xi e^{-q} \ , \qquad \ddot{\xi} + \omega^2\xi = e^{-q} \ . \tag{3.2}$$

The approximation consists in neglecting the r.h.s. of the former equation, and using the solution $\widehat{q}(t)$ of the resulting approximate equation in the r.h.s. of the latter; this becomes forced linear, and is easily solved. As a result, one finally gets

$$\Delta E = c_0|J|^2 + c_1|J|\cos\varphi^0 \ , \tag{3.3}$$

where φ^0 is a convenient asymptotic phase of the oscillator, and

$$J = \int_{-\infty}^{\infty} e^{-\widehat{q}(t)} e^{-i\omega t} \mathrm{d}t \ . \tag{3.4}$$

By the usual trick of the shift of the integration path, one immediately gets

$$J \sim e^{-\tau\omega} \ , \tag{3.5}$$

τ being the width of the analyticity strip of $\widehat{q}(t)$, thought of as a function of the complex variable t. Jeans exponential law appears then in the form

$$\Delta E = \mathcal{E}_0 e^{-2\tau\omega} + \mathcal{E}_1 e^{-\tau\omega}\cos\varphi^0 \ . \tag{3.6}$$

As is remarkable, the average of the energy exchange on the phase of the incoming molecule is much smaller, for large ω, than the oscillating part. In a statistical problem, the average is certainly the most relevant quantity: however, the oscillating part is also important, as it produces very relevant fluctuations (for an analysis of this delicate point, see Ref.[9]).

A more general (but essentially equivalent) approximation scheme, working for Hamiltonians of the form

$$H(p,q,I,\varphi) = h(I) + \frac{|p|^2}{2m} + V(q,I,\varphi) \ , \tag{3.7}$$

with V decaying to zero for $|q| \to \infty$, is the following:[10]

i. Let

$$V(q,I,\varphi) = U_I(q) + W(q,I,\varphi) \ , \qquad U_I(q) := \int_0^{2\pi} V(q,I,\varphi)\mathrm{d}\varphi \ ; \tag{3.8}$$

forget W, and consider the reduced system

$$\mathcal{H}_I(p,q) = \frac{|p|^2}{2} + U_I(q) \ , \tag{3.9}$$

where I is thought of as a parameter. Denote by $\widehat{p}(t)$, $\widehat{q}(t)$ the solution for given initial data.

ii. Use this approximated solution, together with the free motion $\widehat{I}(t) = I^0$, $\widehat{\varphi}(t) = \varphi^0 + \omega(I^0)t$, in the equation for I, namely

$$\dot{I} = -\frac{\partial W}{\partial \varphi}(\widehat{q}(t), \widehat{I}(t), \widehat{\varphi}(t)) \; ; \tag{3.10}$$

solve and compute ΔE.

As a result, one gets ΔE in the form of a Fourier series:

$$\Delta E = E_0 + \sum_{k>0} E_k \cos(k\varphi^0 - \alpha_k) \; , \tag{3.11}$$

with exponentially small coefficients E_k, precisely

$$E_k = \mathcal{E}_k e^{-\lambda_k \omega} \; , \qquad \lambda_k = \begin{cases} k\tau & \text{for } k > 0 \\ 2 & \text{for } k = 0 \end{cases} \; , \tag{3.12}$$

τ being, as before, the width of the analyticity strip of the approximated solution. Apart from the highly inessential (because extremely small) high Fourier components, the result, as expected, is identical to (3.6).

The Landau–Teller approximation looks rather interesting, since in a very short elementary way it leads to the exponential laws. Moreover, the most relevant constants, namely the coefficients λ_k of ω, are well identified, and directly connected to the Hamiltonian, via the analyticity properties of the unperturbed motion $\widehat{p}(t)$, $\widehat{q}(t)$.

Two obvious questions naturally pose, namely whether the approximation can be considered as reliable, and whether some at least of the results can be turned (within suitable assumptions) into effective proofs. Let us stress that this is not obvious, and in principle, it is not even clear that the approximation has any meaning at all, since for example the term ξe^{-q}, which was disregarded at the very beginning in (3.2), is only of order ω^{-1}, and thus much greater, so to speak, than the final result.

As a matter of fact, a *very* accurate numerical test[10] suggest that the result (3.12) is indeed excellent. The test is performed as follows: on the one hand, by the usual numerical simulation one measures ΔE; more precisely, for given values of the asymptotic velocity v and angular velocity ω, one repeats several times the experiment with different phases φ_0, in order to obtain* the Fourier components E_k of ΔE. On the other hand, one independently computes the distance to the real axis of the closest singularity of $\widehat{p}(t)$, $\widehat{q}(t)$, in the complex t plane. This is easy, since one has[3]

$$\tau = \left| \, \mathrm{Im} \int_{q'}^{q''} \frac{dq}{\sqrt{2m^{-1}(\widehat{E} - U_I(q))}} \, \right| \; , \tag{3.13}$$

where $\widehat{E}$ is the energy of the reduced system, q' is the inversion point, that is the (real) solution of $\widehat{E} - U_I(q) = 0$, and q'' is the dominant singularity of U_I, i.e. the one which gives the smallest value of τ ($q'' = \infty$ is also possible). The integral is easily evaluated numerically (for special potentials, an analytic estimate is also possible).

---

* This point, however, is quite delicate, because of the very rapid decay of Fourier components with k, for large ω. See Ref. [10] for details.

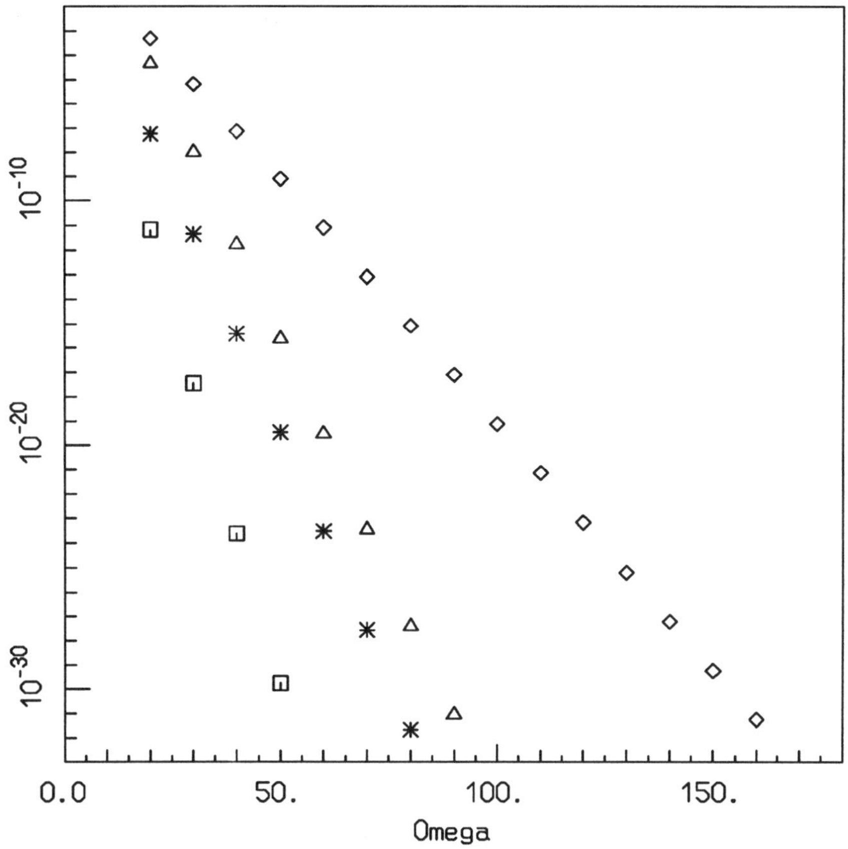

Figure 2 – The amplitude of the Fourier components $|E_0|$ (triangles), $|E_1|$ (diamonds), $|E_2|$ (stars) and $|E_3|$ (squares), as function of frequency ω, in a typical case.

A typical result of the test is reported in Fig. 2, where the Fourier amplitudes E_k are plotted vs. ω, in semilog scale, for $0 \leq k \leq 3$. The data refer to Hamiltonian (1.1), with interaction potential of the form

$$V(q, \varphi) = V_0 \frac{e^{-r/R}}{r/R} , \qquad r = q + \tfrac{R}{10} \cos \varphi ; \tag{3.14}$$

the initial asymptotic velocity was $v = \sqrt{2}$ (i.e., translational energy one) in natural units, namely such that $m = 1$, $R = 1$ and $V_0 = 1$. In these conditions the integral (3.13) gives $\tau \simeq 0.471$, while an easy fit of the straight lines in the figure gives the coefficients λ_k, to be confronted with the theoretical value $k\tau$ (2τ, for $k = 0$). As a result one finds a very good agreement, namely

$$\lambda_0/\tau = 1.92 , \qquad \lambda_1/\tau = 0.988 , \qquad \lambda_2/\tau = 1.98 , \qquad \lambda_3/\tau = 2.99 .$$

One can observe that extremely small energy exchanges are numerically measured (*up to one part over* 10^{30}), and the exponential laws are practically followed for something like 27 orders of magnitude. The test has been repeated with several different choices of the interaction potential and of the initial velocity, with essentially the same result (for details, see Ref. [10]).

An immediate question, in front of these results, is whether the above test itself is *a priori* reliable: indeed, an immediate objection is that the error in energy conservation, at each integration step, is much larger than the smallest energy exchanges one there pretends to measure, so one could even wonder whether the numerical computation has any meaning at all. By the way, the above results have been obtained by means of an apparently poor algorithm, namely the usual leap–frog scheme:

$$p' = p - \varepsilon \frac{\partial V}{\partial q} , \qquad q' = q + \varepsilon p' , \tag{3.15}$$

and similarly for the I, φ variables. The question is delicate, and would need a rather wide discussion. We shall limit ourselves to some essential comments, and skip them to Section 4; a more complete treatment can be found in Ref. [11] (see also Ref. [9], where some preliminary results appeared).

A first natural idea to study rigorously the Landau–Teller approximation is to compare the analytic structure of the approximated solution $\widehat{q}(t)$, entering the approximated equation, with that of the (ω–depending) true solution $q_\omega(t)$, entering the exact equation. More precisely, one would like to study the analyticity strip τ_ω of q_ω, and look whether for large ω one has $\tau_\omega \to \tau$.

Such a study has been done in Ref. [12] for Rapp Hamiltonian (3.1), but only under the very restrictive assumption that asymptotically, at $t = -\infty$, the oscillator is at rest (more precisely: $\eta^2 + \omega^2\xi^2 < e^{-4\tau\omega}$). In this case one proves a proposition, according to which

$$\tau_\omega \geq \tau\big(1 - \mathcal{O}(\omega^{-1/4})\big) , \tag{3.16}$$

but the result does not hold if asymptotically the oscillator is not at rest, nor it holds for a rotator. It is not a question of method of proof: in these other cases, what one would like to prove is apparently wrong, and one has instead $\tau_\omega \to 0$ for large ω.

This is clearly seen numerically in the following way: one makes an ordinary numerical simulation of a collision process, with however integration path shifted at $\mathrm{Im}\, t = \sigma > 0$; any analytic function of coordinates and momenta is then expected to be independent of σ as far as $\sigma < \tau_\omega$, and then have a discontinuity. A sensitive test function is the energy exchange ΔE itself, which is actually found to be almost zero (and constant) for small σ, then suddenly jump to values of order one for higher values of σ. Looking for such discontinuities, one locates τ_ω with rather good precision.* The behavior of τ_ω as function of ω is reported in Fig. 3, which refers to the same condition as Fig. 2. In these conditions one has $\tau \simeq 0.47$.

The fact that τ (the approximation) is better than τ_ω (the real with of the analyticity strip) is, so to speak, embarrassing, and claims for a deeper understanding. Work is in progress, with partial success: quite recently, in particular, Carati [13] developed a new perturbative scheme to determine the energy exchange, and (rigorously) proved that for large ω the approximated value τ, and not τ_ω, gives the coefficient of ω in the exponential law. As is remarkable, this new approach closely follows some ideas by Jeans, and essentially succeeds in turning them into a rigorous proof.

Work is in progress on this point also by other techniques [14], which exploit an essential similarity of the problem at hand with the splitting of separatrices and

---

* As a check of consistency, one can instead look at the behavior of the total energy E, which (although not constant in phase space) is independent of time. As expected, on finds that E is insensitive of σ, and always well conserved, even beyond the analyticity strip.

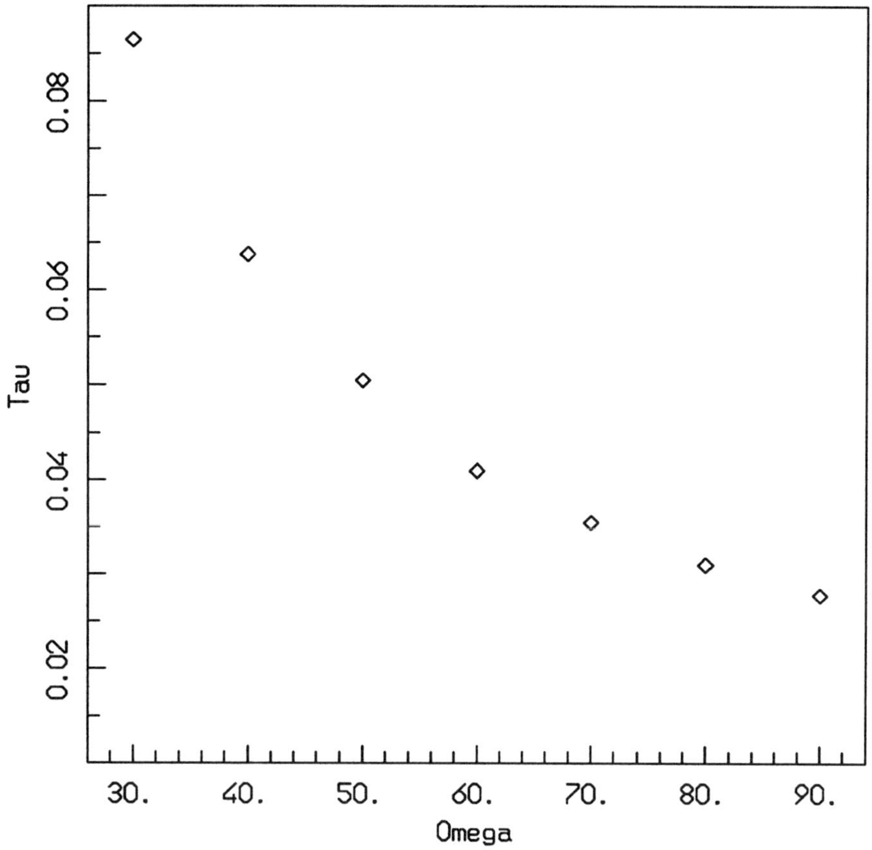

Figure 3 – The analyticity strip τ_ω of the true motion $\widehat{p}(t)$, $\widehat{q}(t)$, as a function of ω, for the scattering of a rotator.

Arnold diffusion. As is quite important, the method leads, in principle, to a two–sided estimate, that is to a prove that the exponential bounds, with τ defined exactly as the with of the analyticity strip of the reduced problem, are in fact optimal.

4. ON THE RELIABILITY OF NUMERICAL RESULTS

As remarked above, the possibility of measuring energy exchanges as small as one part over 10^{-30}, in spite of the apparently much larger errors in energy conservation produced by the numerical algorithm, requires some comments. Let us look more closely to the question of energy conservation. What happens during a numerical simulation of a scattering process, like the one of Fig. 3, is quite unusual: the error

$$\delta E(t) = [E(t) - E(0)]/E(0) \tag{4.1}$$

is (obviously) zero at the very beginning of the computation, when the incoming molecule is very far from the scatterer; then, during the interaction, δE increases up to values, say, of order 10^{-5} (depending on the integration step and on the potential), and performs regular oscillations; *later on, however, the error decreases again,* and typically reduces to less than 10^{-31} at the end of the scattering process, when the interaction potential again vanishes. This behavior is illustrated in Fig. 4, which refers to the potential (3.14), initial velocity $v = \sqrt{2}$ as for Fig. 2, and initial angular velocity $\omega = 30$.

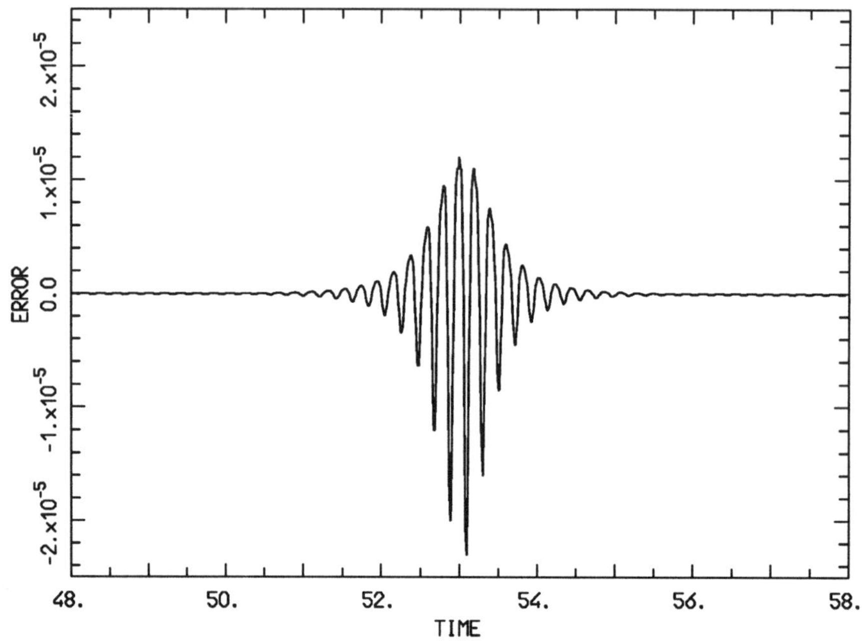

Figure 4 – The relative error in energy conservation as function of time, in a typical case (central part of the collision process). The final value is smaller than 10^{-31}.

Let us discuss this question making reference to Hamiltonians of the form*

$$H(p,q) = \frac{1}{2}\sum_{j=1}^{n} p_j^2 + V(q) \ , \qquad p = (p_1, \dots, p_n) \, , \ q = (q_1, \dots, q_n) \tag{4.2}$$

(some of the coordinates may be angles as well). In the following we shall not pay attention to the value of constants, and indicate (possibly different) constants generically by C.

Assume V is analytic. The leap–frog algorithm is then an analytic map $(p,q) \mapsto \Psi_\varepsilon(p,q)$ close to the identity, of the form

$$\Psi_\varepsilon(p,q) = (p+\varepsilon u(q), q+\varepsilon(p+\varepsilon u(q))) \ , \qquad u = (u_1, \dots, u_n) = (\frac{\partial V}{\partial q_1}, \dots, \frac{\partial V}{\partial q_n}) \ . \tag{4.3}$$

In general, one says that a map Ψ_ε is an algorithm of order s for the Hamiltonian H, if

$$\| \Psi_\varepsilon - \Phi_H^{t=\varepsilon} \| < C \, \varepsilon^{s+1} \ , \tag{4.4}$$

* More general Hamiltonians, including in particular harmonic oscillators written in action-angle variables, could easily be considered. The case of Hamiltonians which are quadratic in momenta is however technically simpler (for example, the leap–frog algorithm is explicit), so we here restrict ourselves to this case. For the general case, see Ref. [11].

where Φ_H^t denotes the flow of H, and $\|\,.\,\|$ is, for example, the sup norm; for the leap–frog algorithm one has $s = 1$. In the following, we shall take $s = 1$. Now, one can prove[15,16,11] that for any analytic mapping Ψ_ε of the form (4.2) there exists an *autonomous* Hamiltonian $K_\varepsilon(p,q)$ which almost interpolates it: precisely, for the above Hamiltonian, one gets

$$|\Phi_{K_\varepsilon}^{t=\varepsilon}(p,q) - \Psi_\varepsilon(p,q)| < C\, e^{-\varepsilon^*/\varepsilon}\, \mathcal{G}(q) \ , \qquad \varepsilon^* > 0 \ , \tag{4.5}$$

where $\mathcal{G}(q)$ is (essentially) $|u(q)|$, i.e. $|\frac{\partial V}{\partial q}(q)|$ (for details, see Ref.[11]). Moreover,[11] in the global case, say when Ψ_ε admits a globally defined generating function, K_ε too turns out to be globally defined, and one gets

$$|K_\varepsilon(p,q) - H(p,q)| < \varepsilon\, \mathcal{F}(q) \ , \qquad \mathcal{F}(q) \leq C \int_{l(q^*,q)} \mathcal{G}(q')\,|\mathrm{d}q'| \ , \tag{4.6}$$

where q^* is any chosen point where the two Hamiltonians are taken to be equal, and $l(q^*,q)$ is any path from q^* to q. Let us notice that in any compact case $\mathcal{F}$ is bounded; in the case of scattering, if the derivatives of V decrease more rapidly than q_j^{-1} when a coordinate q_j goes to infinity, one immediately sees that $\mathcal{F}$ is bounded, and moreover, it vanishes if both $|q^*|$ and $|q|$ go to infinity (for this one simply takes the integration path at infinity). This means that the difference between K_ε and H is everywhere small with ε, and moreover it vanishes everywhere at infinity.

The consequences of inequalities (4.5), (4.6) are quite relevant (see Ref.[11] for details):

i. The mapping Ψ_ε admits an almost–constant of motion, namely K_ε itself: indeed one easily gets

$$|K_\varepsilon(\Psi_\varepsilon^k(p,q)) - K_\varepsilon(p,q)| \ \leq\ C\,\varepsilon\, e^{-\varepsilon^*/\varepsilon} \sum_{i=0}^{k} \mathcal{G}(q^{(k)}) \ , \tag{4.7}$$

where $q^{(k)}$ denotes the q–component of $\Psi_\varepsilon^k(p,q)$, and the oscillation of K_ε turns out to be of order ε up to an exponentially large number of iterations. In the case of scattering the result is better: indeed, the sum at the r.h.s. is bounded, and *the oscillation of K_ε turns out to be exponentially small for any k.*

ii. Up to $t = k\varepsilon = \mathcal{O}(1)$, the orbits generated by the algorithm Ψ_ε and the orbits of the interpolating Hamiltonian K_ε are almost identical:

$$|\Psi_\varepsilon^k(p,q) - \Phi_{K_\varepsilon}^{t=k\varepsilon}(p,q)| \ <\ C\, e^{-\varepsilon^*/\varepsilon} \ , \tag{4.8}$$

even in the case of exponential divergence of nearby trajectories. Notice that, for very reasonable ε, one easily obtains that the r.h.s. is (much) smaller than the computer round–off error (the latter is about 10^{-33}, in quadruple precision).

The latter property, together with (4.6), helps one to understand what really happens when one numerically integrates the equations of motion of a Hamiltonian system, by means of a symplectic algorithm: indeed, one simply replaces the "true" Hamiltonian H by a different Hamiltonian K_ε, ε–close to it, and then iterates "exactly" (namely, within the round–off error) the time–one mapping of K_ε. Now, if the problem at hand is essentially structurally unstable (a small change of the Hamiltonian deeply influences the results), then replacing H by K_ε makes indeed a big difference: in this case, however, we sincerely doubt that a numerical computation, with any integration

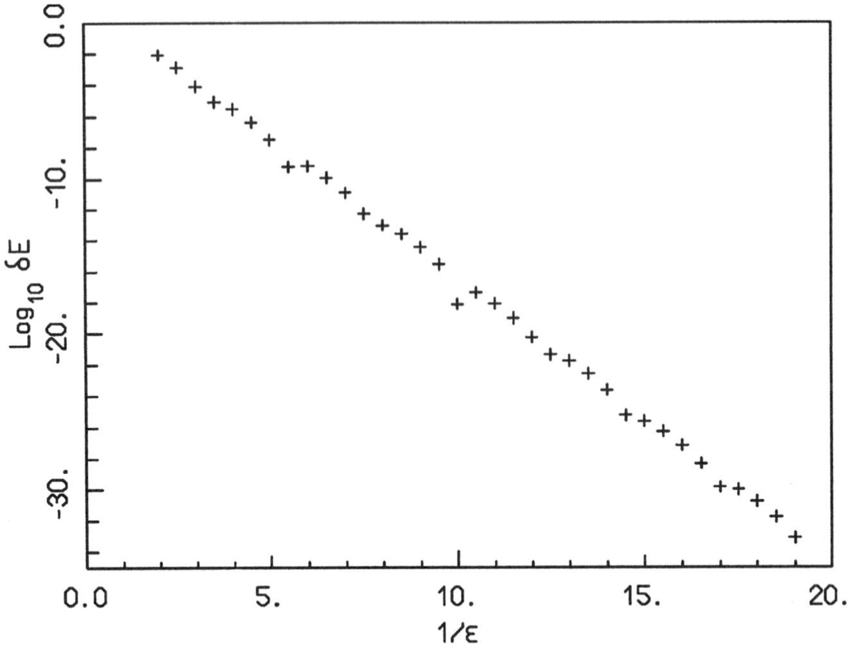

Figure 5. The asymptotic error in energy conservation, δE, as function of the inverse time step $1/\varepsilon$, in semilog scale, for an elementary scattering problem (quadruple precision, approximately 33 significant digits).

method, can be of any interest. On the contrary, if small changes in the Hamiltonian are expected to be not really relevant, at least for the quantities one is looking at, then replacing H by K_ε makes no essential difference, and numerical computation should be considered as essentially reliable. By the way, this is the case of nearly–integrable Hamiltonians treated in perturbation theory: replacing H by K_ε slightly modifies the perturbation, and nothing else.

Inequality (4.7) is instead responsible of the exceptionally good conservation of energy. Indeed, from the trivial inequality

$$\begin{aligned}|H(p(t),q(t)) - H(p(t_0),q(t_0))| < {} & |H(p(t),q(t)) - K_\varepsilon(p(t),q(t))| \\ & + |K_\varepsilon(p(t),q(t)) - K_\varepsilon(p(t_0),q(t_0))| \\ & + |K_\varepsilon(p(t_0),q(t_0)) - H(p(t_0),q(t_0))| \;,\end{aligned} \tag{4.9}$$

for $t = t_0 + k\varepsilon$ one gets

$$\begin{aligned}|H(p(t),q(t)) - H(p(t_0),q(t_0))| < {} & C_1 \varepsilon e^{-\varepsilon^*/\varepsilon} \sum_{i=0}^{k} \mathcal{G}(q^{(k)}) \\ & + \varepsilon[\mathcal{F}(q(t_0)) + \mathcal{F}(q(t))] \;.\end{aligned} \tag{4.10}$$

In the general case one sees that the error in energy conservation remains small with ε, up to an exponentially large number of iterations: *symplectic integration algorithms do not produce any systematic drift in energy, up to exponentially large times.* Consider then the case of scattering: the sum is now uniformly bounded, and for $t_0 \to -\infty$ one gets

$$|E(t) - E(-\infty)| < C\, e^{-\varepsilon^*/\varepsilon} + \varepsilon\, \mathcal{F}(q(t)) \;. \tag{4.11}$$

This inequality clearly explains Fig. 4: the error is practically given by $\mathcal{F}(q(t))$, and is relatively large near the scatterer, while it almost vanishes for $t \to \infty$. Notice that, according to (4.11), the final error in energy conservation is expected to be exponentially small in $1/\varepsilon$. This exponential law is easily observed numerically, as is shown from Fig. 5, where final error in energy conservation is reported in semilog scale versus the reciprocal of the integration step $1/\varepsilon$.

REFERENCES

1. J.H. Jeans, *Phil. Mag.* **6**, 279 (1903).
2. J.H. Jeans, *Phil. Mag.* **10**, 91 (1905).
3. L. Landau and E. Teller, *Physik. Z. Sowjetunion* **10**, 34 (1936), in D.Ter Haar (ed.) *Collected Papers of L. D. Landau*, (Pergamon Press, Oxford, 1965), page 147.
4. L. Boltzmann, *Nature* **51**, 413 (1895).
5. G. Benettin, L. Galgani and A. Giorgilli, *Comm. Math. Phys.* **113,** 87 (1987)
6. G. Benettin, L. Galgani and A. Giorgilli, *Comm. Math. Phys.* **121,** 557 (1989)
7. G. Benettin and F. Fassò, *J. Stat. Phys.* **63**, 737 (1991)
8. D. Rapp, *Journ. Chem. Phys.* **32,** 735 (1960).
9. O. Baldan and G. Benettin, *J. Stat. Phys.* **62**, 201 (1991)
10. G. Benettin, A. Carati and P. Sempio, *On the Landau–Teller approximation for the energy exchanges with fast degrees of freedom*, to appear in *Journ. Stat. Phys.* (1993).
11. G. Benettin and A. Giorgilli, *On the Hamiltonian interpolation of near to the identity symplectic mappings, with application to symplectic integration algorithms*, preprint 1993.
12. A. Carati, G. Benettin and L. Galgani, *Comm. Math. Phys.* **150,** 321-336 (1992).
13. A. Carati, Ph.D. Thesis and paper in preparation.
14. G. Benettin, A. Carati and G. Gallavotti, paper in preparation.
15. J. Moser, *Lectures on Hamiltonian Systems,* in *Mem. American Math. Soc.* No 81, 1 (1968).
16. A.I. Neishtadt, *Prikl. Matem. Mekan.* **45,** 80 (1982) [*PMM U.S.S.R.* **45,** 58 (1982)].

NUMERICAL INTEGRATION OF HAMILTONIAN SYSTEMS IN THE PRESENCE OF ADDITIONAL INTEGRALS: APPLICATION OF THE OBSERVER METHOD

Andrzej J. Maciejewski[1] and Jean-Marie Strelcyn[2,3]

[1]Institute of Astronomy, Nicolaus Copernicus University
87-100 Toruń, Chopina 12/18, Poland

[2]Département de Mathématiques, Université de Rouen,
76821 Mont Saint Aignan Cedex, France, URA CNRS 1378

[3]Laboratoire Analyse, Géometrie et Applications,
URA CNRS 742, Institut Galilée, Département de Mathématiques,
Avenue J.-B.Clement, 93430 Villetaneuse, France

To Janusz Szpotański
with deepest admiration.

INTRODUCTION

The observer method introduced by Busvelle *et al.* (1993), is a simple idea that can be used to improve reliable numerical integration of a system of differential equations in the presence of first integrals. In the cited paper the observer method was already tested on some important examples like the Kepler problem and the Gavrilov-Shilnikov system.

The aim of this paper is to apply and to test the observer method to two more systems which are Hamiltonian and where the method can be applied in many different ways. For this purpose we chose systems with nonlinear phase space and with additional integrals that are polynomials of the order greater than two with respect to momenta.

THE OBSERVER METHOD

Here we sketch the observer method. Details and proofs can be find in the paper of Busvelle *et al.* (1993).

Let us consider the initial value problem for a system of ordinary differential equations written in the vector form

$$\frac{dx}{dt} = F(x), \qquad (1)$$
$$x(0) = x_0$$

where $F \in C^1(U, \mathbb{R}^p)$ and U is an open subset of $\mathbb{R}^p$ and $x_0 \in U$. Let us suppose that $H_1, \ldots, H_q \in C^1(U)$, $1 \leq q < p$, are the first integrals of the system (1) and let us denote

$$H = (H_1, \ldots, H_q)^T, \qquad H \in C^1(U, \mathbb{R}^q),$$

$$D_H(x) = \begin{pmatrix} \frac{\partial H_1}{\partial x_1}, & \ldots, & \frac{\partial H_1}{\partial x_p} \\ \ldots, & \ldots, & \ldots \\ \frac{\partial H_q}{\partial x_1}, & \ldots, & \frac{\partial H_1}{\partial x_p} \end{pmatrix},$$

where ^{T} denotes transposition. Let us consider $x_0 \in U$ and the level set of integrals

$$\Gamma = \{x \in U : H(x) = H_0\},$$

where $H_0 = H(x_0)$. We make two assumptions:

1. The set Γ is compact.
2. Integrals H are functionally independent on Γ.

From the implicit function theorem one easily deduces that for sufficiently small ϵ, $0 < \epsilon \leq \epsilon_0$, the following set

$$\Gamma_\epsilon = \{x \in U; \| H_0 - H(x) \| \leq \epsilon\}$$

is a compact subset of U and integrals $H_1, \ldots, H_q$ are functionally independent on it ($\| \cdot \|$ denotes the Euclidean norm).

The basic idea of the observer method consists of replacing the original initial value problem by the new one

$$\frac{dy}{dt} = F(y) + Ob(y), \qquad (2)$$
$$y(0) = y_0, \qquad (3)$$

where

$$Ob(y) = D_H^T(y) \left[D_H(y) D_H^T(y)\right]^{-1} K(y) \left(H_0 - H(y)\right),$$

and $K(y)$ is a diagonal matrix given by

$$K(y) = \begin{pmatrix} \theta_1 & 0 & \cdots & 0 \\ 0 & \theta_2 & & \vdots \\ \vdots & & \ddots & \\ 0 & \cdots & 0 & \theta_q \end{pmatrix},$$

where $\theta_1, \ldots, \theta_q$ are continuous functions on Γ_ϵ such that inequality

$$\eta = \min_{1 \leq i \leq q} \min_{y \in \Gamma_\epsilon} (\theta_i(y)) > 0$$

holds. The perturbation term $Ob(y)$ is called the observer. Its specific form and assumptions which we made, allow to prove following facts.

1. $\Gamma \subset U$ is an invariant submanifold with respect to the flow of the system (2).

2. Γ is the global attractor of the system (2) restricted to the Γ_ϵ with an exponential rate of convergence of orbits towards Γ, which is exactly controlled.

Moreover, when solving the initial value problem (2)–(3) by any standard one step explicit numerical method of first order, it is possible to prove, choosing appropriate values of the step-size, η and ϵ, that the error of integrals is bounded over infinite time interval. We again refer to the paper of Busvelle *et al.* (1993) for proofs and detailed discussion.

Finally, let us make two remarks concerning numerical integration of the initial value problem (2)–(3).

In the matrix $K(y)$ diagonal elements can be taken constant. Having the system of ODE, the initial condition and chosen integrator it is necessary to find the 'optimal' values of θ_i. Bad choice will not improve the precision of integration.

At first glance it seems that during numerical integration of the system (2) it is necessary to inverse matrix $A = D_H^T(y)D_H(y)$ once per integration step. However the observer can be rewritten as

$$Ob(y) = D_H^T z,$$

where $z = z(y)$ is the solution of the linear system

$$Az = K(y)(H_0 - H(y))$$

which can be solved numericall without an explicit matrix inversion.

THE HEAVY TOP

Equations of Motion

Let us consider a rigid body located in a homogenous gravity field and moving around a fixed point. The fixed inertial frame is chosen in such a way that the gravitational acceleration vector is equal $(0, 0, -g)$. The classical Euler-Poisson equations have the following form

$$\left.\begin{aligned} A\frac{dp}{dt} + (C-B)qr - mg(z_0\gamma_2 - y_0\gamma_3), \\ B\frac{dq}{dt} + (A-C)rp = mg(x_0\gamma_3 - z_0\gamma_1), \\ C\frac{dr}{dt} + (B-A)pq = mg(y_0\gamma_1 - x_0\gamma_2), \end{aligned}\right\} \tag{4}$$

$$\left.\begin{aligned} \frac{d\gamma_1}{dt} &= r\gamma_2 - q\gamma_3, \\ \frac{d\gamma_2}{dt} &= p\gamma_3 - r\gamma_1, \\ \frac{d\gamma_3}{dt} &= q\gamma_1 - p\gamma_2, \end{aligned}\right\} \tag{5}$$

where A, B, C are the principal moments of inertia, m is the mass of the body and g is the gravitational acceleration. (p, q, r), (x_0, y_0, z_0) and $(\gamma_1, \gamma_2, \gamma_3)$ are respectively components of the angular velocity of the body, the radius vector of the mass center of the body and the unit vector along the z-axis of the fixed frame (see e.g. Golubiev 1953). Components of all these vectors are taken with respect to the principal axes reference frame.

These equations possess three classical integrals

$$\frac{1}{2}\left(Ap^2 + Bq^2 + Cr^2\right) + mg(x_0\gamma_1 + y_0\gamma_2 + z_0\gamma_3) = C_1, \tag{6}$$

$$Ap\gamma_1 + Bq\gamma_2 + Cr\gamma_3 = C_2, \tag{7}$$

$$\frac{1}{2}(\gamma_1^2 + \gamma_2{}^2 + \gamma_3{}^2) = C_3, \tag{8}$$

and only in some exceptional cases the fourth first integral, which guarantees the integrability of the system, is known (see Arkhangelski 1977, Arnold *et al.* 1988, Fomenko 1988, Golubiev 1953, Kozlov 1983).

Let us introduce the vector notation

$$\omega = (p, q, r)^T, \qquad u = mg(x_0, y_0, z_0)^T, \qquad e = (\gamma_1, \gamma_2, \gamma_3)^T, \qquad M = \mathbb{I}\omega,$$

$$\mathbb{I} = \begin{pmatrix} A & 0 & 0 \\ 0 & B & 0 \\ 0 & 0 & C \end{pmatrix},$$

and let $[\cdot,\cdot]$, $<\cdot,\cdot>$ denote standard vector and scalar product in $\mathbb{R}^3$, respectively. Now, we can write equations (4), (5) in the compact form

$$\frac{dM}{dt} = [M, \omega] + [e, u], \qquad \frac{de}{dt} = [e, \omega]. \tag{9}$$

We also rewrite integrals (6)–(8) in the following form

$$H = H(M, e) = \frac{1}{2} < M, \mathbb{I}^{-1}M > + < u, e >,$$

$$f_2(M, e) = < M, e >, \qquad f_3(M, e) = \frac{1}{2} < e, e >$$

Equations (9) defined a vector field on $\mathbb{R}^6 = \mathbb{R}^3\{M\} \times \mathbb{R}^3\{e\}$. The common level of integrals

$$M_{23} = \{(M, e) \in \mathbb{R}^6 : f_2(M, e) = C_2, f_3(M, e) = C_3 > 0\},$$

is diffeomorphic to the (co)tangent boundle of a two-dimensional sphere. On it, the above vector field is Hamiltonian with respect to a certain natural Poisson bracket (for details see e.g. Fomenko 1988, Kozlov 1983). The Hamiltonian function for this field is $H(M, e)$ (restricted to M_{23}).

The configuration space of a rigid body with a fixed point is the group $SO(3)$. In the phase space $M = T^*SO(3)$ we can use a local coordinates and write the equations of motion in the form of the classical canonical Hamitonian equations. In analytical studies, mainly two kinds of canonical coordinates are used: the Euler angles and Andoyer-Deprit coordinate.

In terms of the Euler angles (ψ, θ, φ) and momenta conjugated to them $(p_\psi, p_\theta, p_\varphi)$ the Hamiltonian function can be written in the following form

$$H(\varphi, \theta, \psi, p_\psi, p_\theta, p_\varphi) = \frac{1}{2}\left(\frac{M_1}{A} + \frac{M_2}{B} + \frac{M_3}{C}\right) + mg(x_0\gamma_1 + y_0\gamma_2 + z_0\gamma_3), \tag{10}$$

where

$$M_1 = (p_\psi - p_\varphi \cos\theta)\frac{\sin\varphi}{\sin\theta} + p_\theta \cos\varphi, \qquad \gamma_1 = \sin\theta \sin\varphi,$$
$$M_2 = (p_\psi - p_\varphi \cos\theta)\frac{\cos\varphi}{\sin\theta} - p_\theta \sin\varphi, \qquad \gamma_2 = \sin\theta \cos\varphi,$$
$$M_3 = p_\varphi, \qquad \gamma_3 = \cos\theta.$$

Let us note that coordinate ψ is cyclic. Thus, taking p_ψ as the parameter, we can reduce the dimension of the phase space by one degree of freedom.

In terms of the Andoyer-Deprit special coordinates (l, g, h, L, G, H) the Hamiltonian has the form

$$K(l,g,h,L,G,H) = \frac{G^2 - L^2}{2}\left(\frac{\sin^2 l}{A} + \frac{\cos^2 l}{B}\right) + \frac{L^2}{2C} + mg(x_0\gamma_1 + y_0\gamma_2 + z_0\gamma_3), \tag{11}$$

where now

$$\gamma_1 = \sin l(\cos\delta_1 \sin\delta_2 + \sin\delta_1 \cos\delta_2 \cos g) + \cos l \sin\delta_1 \sin g,$$
$$\gamma_2 = \cos l(\cos\delta_1 \sin\delta_2 + \sin\delta_1 \cos\delta_2 \cos g) - \sin l \sin\delta_1 \sin g,$$
$$\gamma_3 = \cos\delta_1 \cos\delta_2 - \sin\delta_1 \sin\delta_2 \cos g,$$

and where

$$\cos\delta_1 = \frac{H}{G}, \qquad \sin\delta_1 = \sqrt{1 - \left(\frac{H}{G}\right)^2}, \qquad \cos\delta_2 = \frac{L}{G}, \qquad \sin\delta_2 = \sqrt{1 - \left(\frac{L}{G}\right)^2}.$$

In these coordinates variables h is cyclic and we can reduce the system by one degree of freedom as in the previous case.

Numerical Tests

For numerical integration we used following three procedure.

1. The fourth order fixed step-size Merson Runge-Kutta method (Hairer *et al.* 1985). We refer to it as the RK integrator.

2. The fourth order two stages implicit Butcher Runge-Kutta method (Butcher 1964) that is known to be symplectic (Sanz-Serna 1991). We will refer to it as the SI integrator.

3. The Burlish-Stoer extrapolation method with adaptive step-size control (Hairer *et al.* 1985). This method we will call the BS integrator.

First, we integrated the Euler-Poisson equations (4)–(5) using the BS procedure with and without the observer corrections, respectively. The observer was applied for three integrals (6)–(8) simultaneously. In all tests of this section integration time was 10000. The following values of parameters where adopted: $mg = 1$, $A = B = x_0 = 1$, $C = 0.51$, $y_0 = z_0 = 0$. That means that we consider a small nonintegrable perturbation of S. Kovalevski case ($A = B = 2C$, $y_0 = z_0 = 0$). The initial conditions were: $M_1 = M_2 = 0.1$, $M_3 = 2$, $\gamma_1 = \gamma_2 = 0.7071\ldots$, $\gamma_3 = 0$. The local precision of integration was 10^{-8}. Both integrations were compared with the results obtained from integration of the original equations (4)–(5) with the local precision 10^{-17}. This last solution will be taken as the 'exact' one. Figure 1 shows the maximal relative errors of the computed orbits for both integrations. In all figures the labels '1' and '2' refer to the integration without and with the observer, respectively. The relative error is defined as $\|x - x_e\|/\|x_e\|$, where x and x_e denote numerical and 'exact' solution, respectively. Figure 1 shows that the application of the observer method improves result of integration quantitatively and qualitatively. It seems that in this example errors grow linearly with time when the observer method is applied. Figures 2–4 show how the application of the observer method improves the preservation of observed integrals.

In our second test we kept the same initial condition and we changed only the value of C to be equal 0.5. For these values of the parameters we have the well known S. Kovalevski integrable case with additional integral of the form

$$C_4 = (p^2 - q^2 - c\gamma_1)^2 + (2pq - c\gamma_2)^2,$$

where $c = mg/C$. We used exactly the same integrations procedures as above. In Figure 5 the maximal relative errors of the computed orbit during integrations are shown. As in the previous example we observe a substantial improvement of the results. In Figure 6 the errors of the integral C_4 are reported. Note, that the observer method improves considerably preservation of this not observed integral. This support an empirically observed fact (see Busvelle *et al.* 1993) that when the observer method is correctly applied, the reliability of the computation of orbits is also improved.

In our next test we integrated Hamiltonian equations of motion corresponding to the Hamiltonian (10). The values of the parameters where the same like in the previous test (i.e. the S. Kovalevski case) and initial condition were: $\theta = \varphi = 1.1$, $p_\theta = 0.1$, $p_\varphi = 1$; the constant value of p_ψ was equal zero. Figure 7 and 8 show the errors in the energy and in the S. Kovalevski integral, respectively. The results were obtained respectively by the RK integrator without, and with the observer method applied to the energy integral. The step size was 0.1 for both integration. Figures 9 and 10 show the same quantities obtained by the SI integrator (continuous lines) and the RK integrator (scattered points) with the observer applied for both integrals. It is easy to to see that in this integration, the observer method and the symplectic integrator give comparable results.

In our last test we integrated the Hamilton equations corresponding to the Hamiltonian (11). The values of the parameters were chosen as follows: $A = 1.5$, $B = 1.6$, $C = 1.7$, $x_0 = z_0 = 0.1$, $y_0 = 0.05$; constant value of H was equal 0.9 and initial conditions were following: $l = g = 0.1$, $L = 0.9$, $G = 1$. Figure 11 shows the error in the energy obtained when we integrated the equations of motion with the RK method, RK method with the observer applied to the integral H and the SI method, respectively. Curves with labels '3' correspond to the symplectic integration.

It should be stressed that the Andoyer-Deprit coordinates are very inconvenient when a fixed step-size method is used for integration. This coordinates have geometrical singularity at $\delta_1 = 0$ and $\delta_2 = 0$. Passages near these singularities and numerical errors causes that in the right hand sides square roots with negative arguments appears. Moreover, the passages near the singularites are sources of rapid error accumulation. As a conseguence of this, the quality of results depends strongly on the choice of initial conditions. The geometrical singularities can make an implicit integration method (like the SI) completely useless. The same coments can be made for any other local coordinate system.

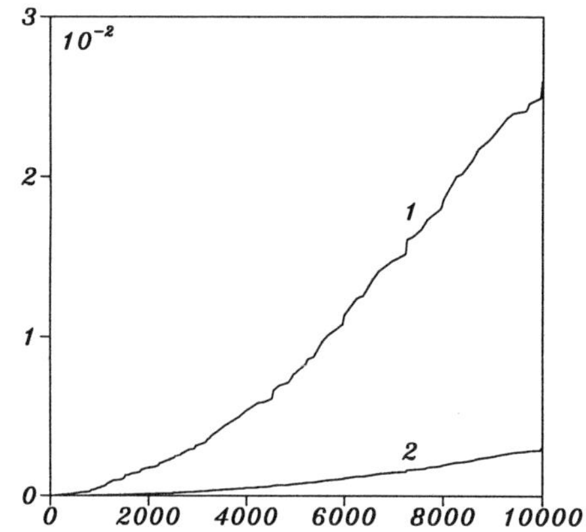

Figure 1. The maximal relative errors of a computed orbit for a small nointegrable perturbation of S. Kovalevski integrable case of the Euler-Poisson equations.

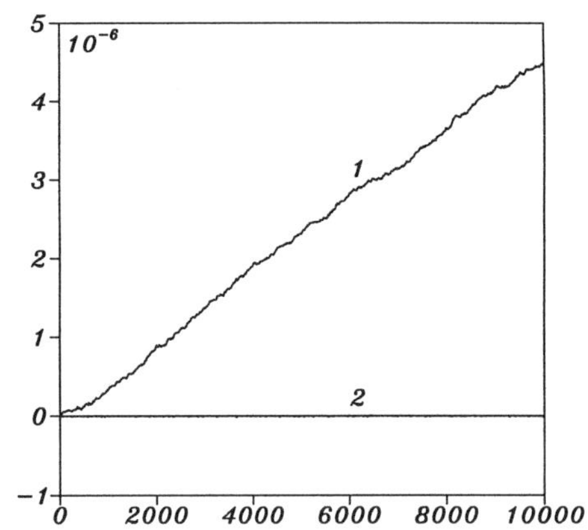

Figure 2. Errors in the first integral H computed along the same orbit as in Fig. 1.

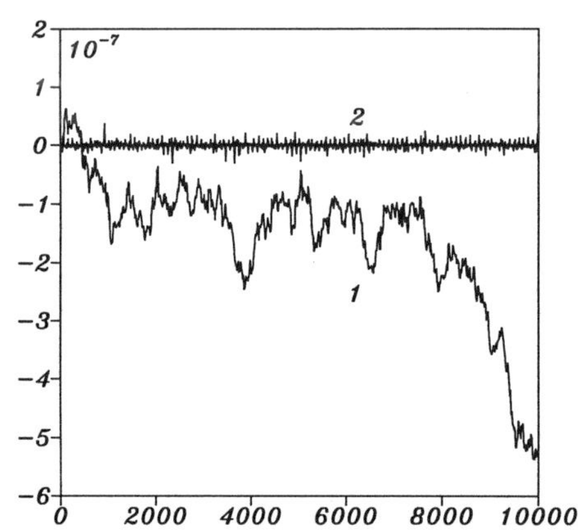

Figure 3. The same as in Fig. 2 but for the first integral f_2.

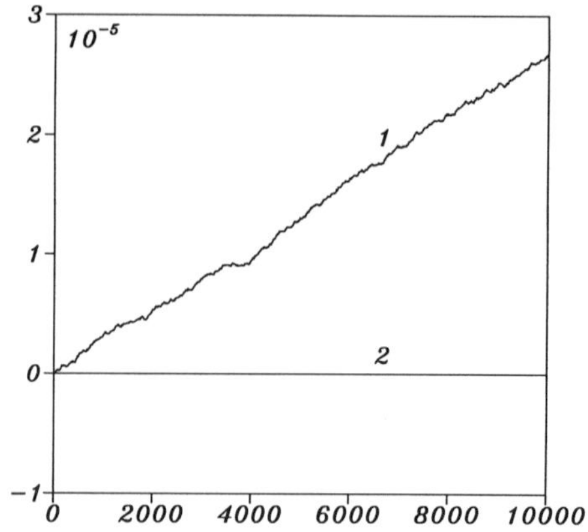

Figure 4. The same as in Fig. 2 but for the first integral f_3.

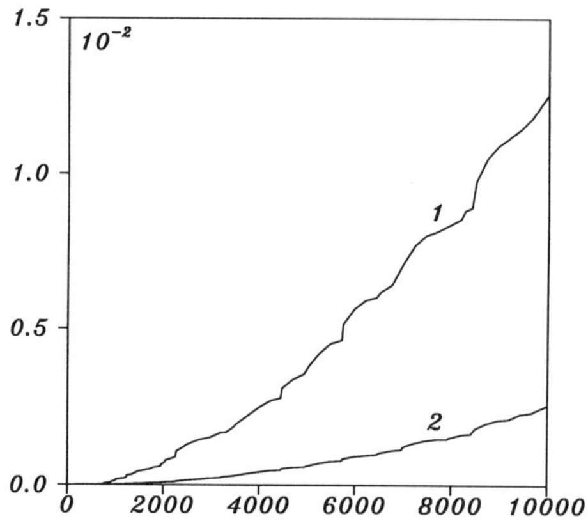

Figure 5. The maximal relative errors of a computed orbit for S. Kovalevski integrable case of the Euler-Poisson equations.

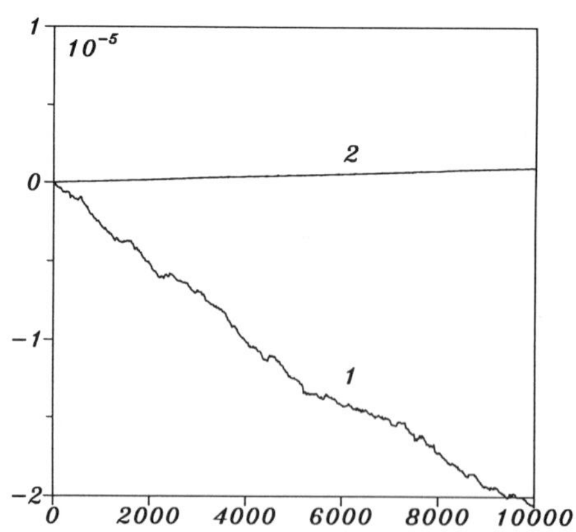

Figure 6. Errors in the S. Kovalevski first integral computed along the same orbit as in Fig. 5.

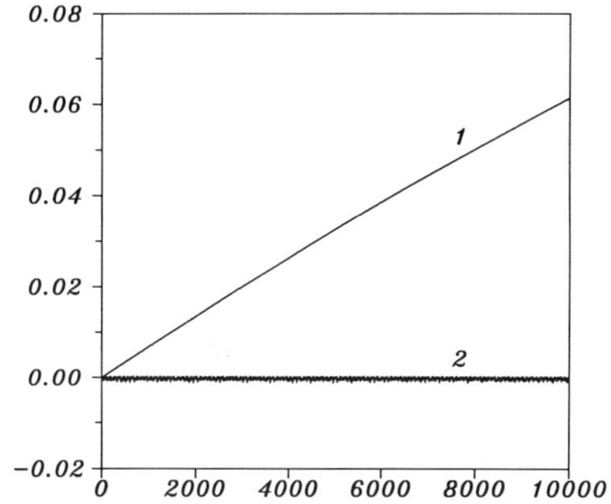

Figure 7. Errors in energy along an orbit of the S. Kovalevski case, computed using the Hamiltonian (10) and the RK integrator, without and with the observer applied to the energy integral.

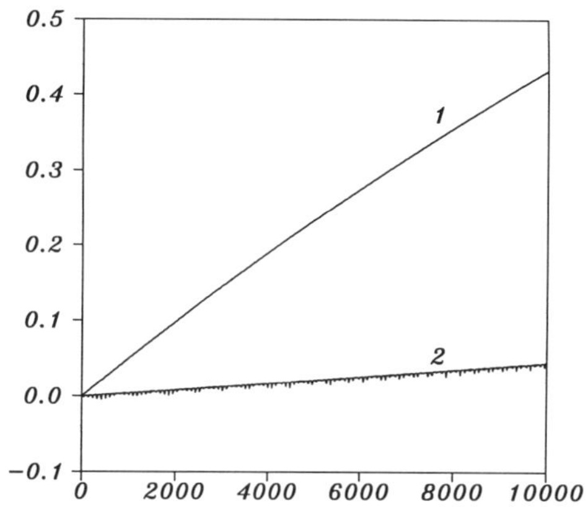

Figure 8. The same as in Fig. 7 but for the S. Kovalevski integral.

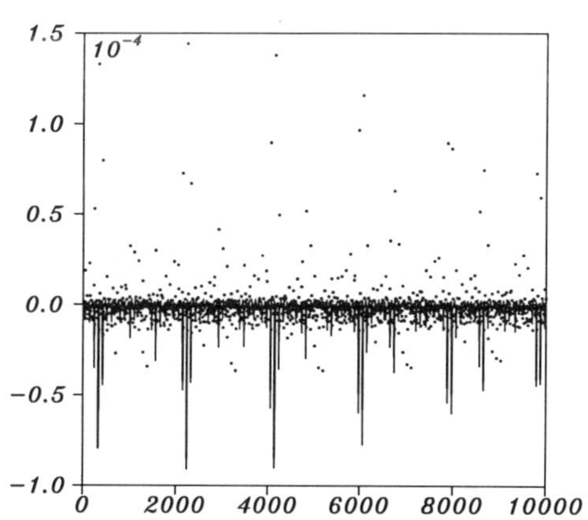

Figure 9. The same as in Fig. 7 but when the observer method is applied together to the energy and the S. Kovalevski integral.

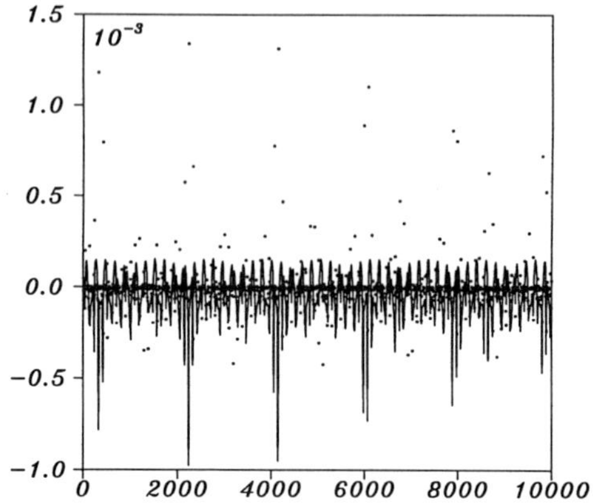

Figure 10. The same as in Fig. 8 but when the observer method is applied together to the energy and the S. Kovalevski integral.

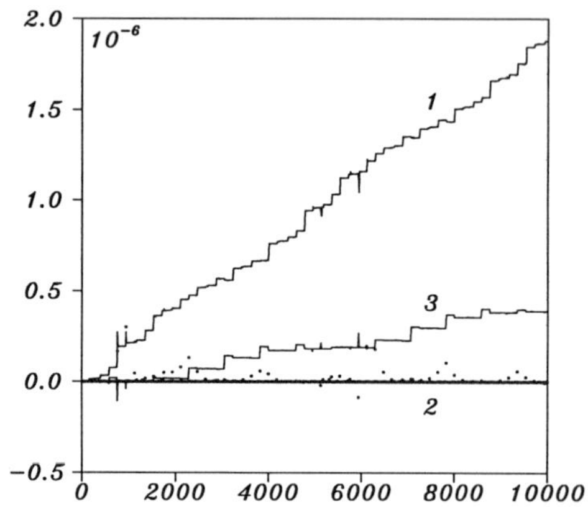

Figure 11. Errors in energy computed for an orbit of the Hamiltonian (11).

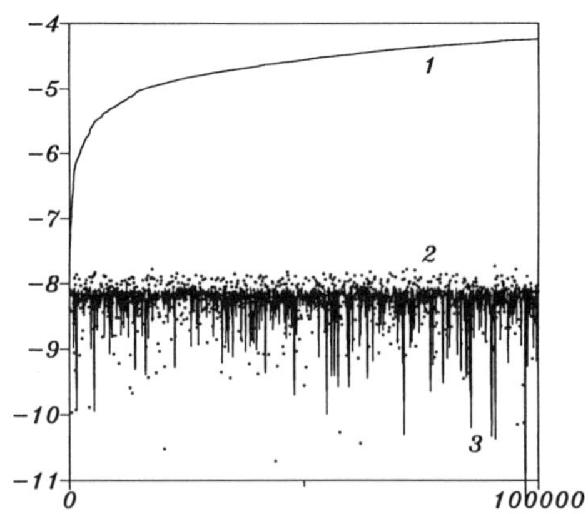

Figure 12. Errors in energy computed for an orbit of the system of four equal vortices.

THE POINT VORTICES

Equations of Motion

Let us consider the motion of N point vortices in a plane stationary flow of an ideal incompressible fluid. Such system can be described in Hamiltonian formalism as follows.

Let us denote

$$(q_1, \dots, q_N, p_1, \dots, p_N) \in \mathbb{R}^{2N}, \qquad z_k = p_k + iq_k, \qquad 1 \le k \le N.$$

The motion of N point vortices is governed by the following equtions

$$\frac{d\bar{z}_k}{dt} = \frac{1}{2\pi i} \sum_{\substack{l \neq k \\ 1 \le l \le N}} \frac{\Gamma_l}{z_l - z_k}, \qquad 1 \le k \le N, \tag{12}$$

where the bar denotes the complex conjugation. The real coefficient Γ_k is called the intensity of the k-th vortex. Equations (12) can be written in the Hamiltonian form

$$\frac{dq_k}{dt} = \{H, q_k\}, \qquad \frac{dp_k}{dt} = \{H, p_k\}, \qquad 1 \le k \le N, \tag{13}$$

with the Hamiltonian

$$H = H(p, q) = \sum_{\substack{k \neq l \\ 1 \le k,l \le N}} \Gamma_k \Gamma_l \ln |z_k - z_l|,$$

and the Poisson brackets $\{\cdot, \cdot\}$ are given by

$$\{f, g\} = \sum_{l=1}^{N} \frac{1}{\Gamma_l} \left(\frac{\partial f}{\partial p_l} \frac{\partial g}{\partial q_l} - \frac{\partial f}{\partial q_l} \frac{\partial g_l}{\partial p_l} \right).$$

The equations (13) are called the Kirchoff equations. Except H they possesses also three other first integrals

$$Q = \sum_{k=1}^{N} q_k, \qquad P = \sum_{k=1}^{N} p_k, \qquad L^2 = \sum_{k=1}^{N} \Gamma_k (q_k^2 + p_k^2).$$

It is well known that for $N \le 3$ the system of N vortices is integrable and that for $N \ge 4$ it is not integrable (see Aref and Pomphrey 1982 for more details and references).

Numerical Results

For numerical investigations we chose the system of four vortices with the same intensities 1. We took the same initial conditions as the C2 initial conditions of the paper of Pullin and Safman (1991), namely: $q_1 = -1.85$, $q_2 = -0.07$, $q_3 = 0.78$, $q_4 = 0.76$, $p_1 = 0.05$, $p_2 = -1.332$, $p_3 = 0.167$, $p_4 = -1.2902$. We integrated equation of motion three times using RK method, RK method with the observer and the SI method, respectively. In all three cases the step of integration was 0.2 and the time of integration was 10^5. The observer method was applied to three integrals: H, L^2 and $Q^2 + P^2$ simultaneously.

Results of computations are shown in Figure 12 and 13 where absolute errors of integrals H, and L^2 respectively are presented in the logarithmic scale. We do not report results for the third integral because for all integrations errors in it is less than 10^{-15}.

Let us note that the application of the observer method improves preservation of H and L^2 by the RK method by about three order. It makes the RK method with the observer almost as good as the symplectic integrator. Note that error in L^2 for the SI integrator is much lower than for the observer method but it systematicaly grows.

Let us note also that the orbit we tested lies in the chaotic region of the phase space. We made also test for an almost periodic orbit an we obtain similar result. We also tried to apply the observer method with the SI Runge-Kutta method however in the case of four equal vortices it do not improve results.

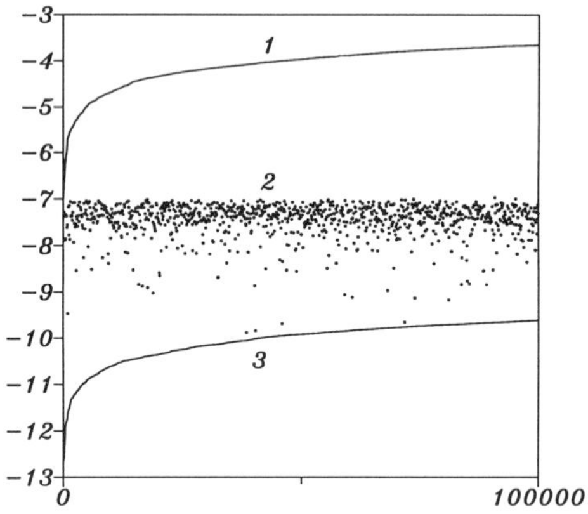

Figure 13. The same as in Fig. 12 but for the integral L^2.

ACKNOWLEDGMENTS

For the first author this work was supported by the grant KBN 22149203. During the final stage of elaboration of this paper the he was supported by the fellowship ERB3510PL9210479 from the European Community's Action for Cooperation in Science and Technology with Central and Eastern European Countries. He acknowledges the Departement of Mathematics of the Rouen University as well as URA CNRS 1378 for their hospitality and exellent working condition. In particular he thanks sincerely Gerard Grancher for his help and attention.

REFERENCES

Aref H. and Pomprey N., 1982, Integrable and chaotic motions of four vortices I. The case of identical vortices, *Proc. R. Soc. London A*, **380**, 359–387.

Arkhangelski Yu.A., 1977, *Analytical Dynamics of the Rigid Body (in Russian)*, (Moscow: Nauka).

Arnold V.I., Kozlov V.V. and Neishtadt A.I., 1988, Mathematical aspects of classical and celestial mechanics, *in Arnold V.I. (Edit.) Dynamical Systems III, Encyclopaedia of Mathematical Sciences*, **3**, (Berlin: Springer Verlag).

Busvelle E., Kharab R., Maciejewski A.J. and Strelcyn J.-M., 1993, Numerical Integration of Differential Equations in the Presence of First Integrals: Observer Method, preprint, *Rouen University, March 1993*, , .

Butcher J.C., 1964, Implicit Runge-Kutta processes, *Math. Comput.*, **18**, 50–64.

Fomenko A.T., 1988, *Integrability and nonintegrability in geometry and mechanics*, (Dodrecht: Kluwer Academic Publishers).

Golubev V.V., 1953, *Lectures on Integration of the Equations of Motion of a Rigid Body about a Fixed Point (in Russian)*, (Moscow: Gostekhizdat). English translation by the Israel Programme od Scient. Transl. (Jerusalem, 1960).

Hairer E., Norsett S.P. and Wanner G., 1987, *Ordinary differential equations I, Nonstiff problem*, (Berlin: Springer Verlag).

Kozlov V.V., 1983, Integrability and nonintegrability in hamiltonian mechanics, *Russian Math. Survey*, **38 (1)**, 1–76.

Pullin D.I. and Saffman P.G., 1991, Long-time symplectic integration: the example of four vortex motion, *Proc. R. Soc. London A*, **432**, 481-494.

Sanz-Serna J.M., 1991, *Symplectic integrators for Hamiltonian systems: an overview, Acta Numerica, (annual volume published by Cambridge University Press)*, **1**, 243–286.

SYMMETRIES AND TOPOLOGY OF DYNAMICAL SYSTEMS WITH TWO DEGREES OF FREEDOM

Valery Kozlov

Department of Mathematics and Mechanics
Moscow State University
Moscow–119899, Russia

INTRODUCTION

Let M be a closed two-dimensional surface viewed as *a configuration space* of a dynamical system with two degrees of freedom. We denote the local coordinates on M by q_1, q_2. In mechanics they are usually called *the generalized coordinates.* Let p_1, p_2 be canonical momenta conjugated with q_1, q_2. Hence,

$$x = (q_1, q_2, p_1, p_2)$$

are local coordinates in *the phase space* T^*M (the cotangent bundle of the surface M). In geometry of the phase space *the symplectic structure*—the closed non-degenerate 2-form

$$\omega = \sum dp_s \wedge dq_s$$

plays an important role.

Let H be a real function on T^*M. To it there uniquely corresponds the vector field v_H defined by the relation

$$\omega(v, \cdot) = -dH. \tag{1.1}$$

The field v is called *Hamiltonian.* It generates a Hamiltonian dynamical system on T^*M

$$\dot{x} = v(x). \tag{1.2}$$

Written in terms of local coordinates q, p, these equations take the customary form of the canonical *Hamilton equations*

$$\dot{q}_s = \frac{\partial H}{\partial p_s}, \quad \dot{p}_s = -\frac{\partial H}{\partial q_s}, \qquad s = 1, 2. \tag{1.3}$$

For a Hamiltonian we can choose some *Riemann metric* on M, i.e., positive definite quadratic form in p_1, p_2 whose coefficients are single-valued functions on M.

Hamiltonian Mechanics, Edited by J. Seimenis,
Plenum Press, New York, 1994

In mechanics, equations (1.3) describe a free motion, while in differential geometry the geodesic equation for the chosen metric. Apart from the solution $q(t), p(t)$, equations (1.3) admit also the solution $q(-t), p(-t)$. Following Birkhoff, such systems are called reversible.

Then the Hamiltonian takes the following form

$$H = \frac{\Lambda}{2}(p_1^2 + p_2^2),$$

where Λ is a positive function of q_1, q_2.

Consider a problem on *symmetries* of the system (1.2), which are generated by the vector fields u on T^*M commuting with the original Hamiltonian field v: $[u, v] = 0$. Such fields will be called *symmetry fields*. It turns out that the existence of symmetry fields depends essentially on topology of the configuration space M. In the sequel all objects are supposed to be infinitely differentiable.

The derivative along the field v has the form

$$L_v = \Lambda p_1 \frac{\partial}{\partial q_1} + \Lambda p_2 \frac{\partial}{\partial q_2} - \frac{1}{2}\frac{\partial \Lambda}{\partial q_1}(p_1^2 + p_2^2)\frac{\partial}{\partial p_1} - \frac{1}{2}\frac{\partial \Lambda}{\partial q_2}(p_1^2 + p_2^2)\frac{\partial}{\partial p_2}.$$

Let u be a vector field with the derivative operator

$$L_u = Q_1 \frac{\partial}{\partial q_1} + Q_2 \frac{\partial}{\partial q_2} + P_1 \frac{\partial}{\partial p_1} + P_2 \frac{\partial}{\partial p_1}.$$

If u is a symmetry field, the operators L_v and L_u commute.

Definition. If Q_1, Q_2 are polynomials in momenta of degree $n-1$ and P_1, P_2 polynomials of degree n, then the field u is a *homogeneous symmetry field* of degree n.

We denote the degree of the homogeneous field u by deg u. In particular, deg $v = 2$. The symmetry field u can be expanded into a formal series in degrees of homogeneous fields:

$$u = u_0 + u_1 + u_2 + \cdots; \qquad \deg u_k = k, \quad k \geq 1.$$

The derivative along the field u_0 has the form

$$a_1 \frac{\partial}{\partial p_1} + a_2 \frac{\partial}{\partial p_2},$$

where a_1, a_2 are functions on M.

Proposition. *Each homogeneous form of the field u is itself a symmetry field.*

The field u_0 is obviously equal to zero. Otherwise, we have $\Lambda = 0$. Therefore, we may restrict ourselves by considering only homogeneous symmetry fields of degree ≥ 1.

The symmetry field u is called *Hamiltonian* if

$$\omega(u, \cdot) = -dF,$$

where F is a single-valued function in the phase space. If F is a homogeneous polynomial in momenta of degree m, then $\deg u = m$. Put $u = v_F$. It is well-known that

$$[v_H, v_F] = v_{\{H,F\}},$$

where $\{\,,\,\}$ is a Poisson bracket. If the fields u and v commute, then $\{H, F\} =$const. Since the Hamiltonian is quadratic, this constant equals zero. Therefore, the Hamiltonian F is an integral of the system (1.2). Integrals that are linear in momenta generate Hamiltonian symmetry fields studied by E.Nother.

The symmetry field u will be called *locally Hamiltonian* if the 1-form $\omega(u, \cdot)$ is closed, but not exact. In this case equations (1.2) admit, as an invariant, a closed 1-form, which can be called *multi-valued integral.*

One should not think that the symmetry fields of the system (1.2) are always Hamiltonian (or locally Hamiltonian). Here is a simple counter example: if $\Lambda =$const, then the quadratic vector field defined by

$$p_2 \frac{\partial}{\partial q_1} - p_1 \frac{\partial}{\partial q_2}$$

is a symmetry field. However, it is not even locally Hamiltonian with respect to the standard symplectic structure ω.

MAIN RESULTS

First we discuss the structure of Hamiltonian symmetry fields. It is obvious that if equations (1.2) admit a smooth integral $F : T^*M \to \mathbf{R}$, then any homogeneous form in its Maclaurin expansion in degrees of momenta is an integral as well.

In [1] it was proved that if the Euler characteristic κ of the surface M is negative, then equations (1.2) do not admit an integral independent of the energy integral H. According to the Gauss–Bonne formula, the condition $\kappa(M) < 0$ is equivalent to the condition for the average Gauss curvature of the Riemannian metric on M to be negative. If the Gauss curvature is negative everywhere, the geodesic flow on M is an Anosov's system, which is known to be ergodic.

Later V. Kolokoltzov [2] presented another proof of the statement given in [1], which is based on introducing the conformal structure on M. This approach goes back to Birkhoff, who investigated the problem on the existence of local integrals of first and second degree in momenta. Birkhoff considered *conditional* integrals whose derivative vanishes for a some value of the total energy $h = h$. He showed that the existence of a conditional linear integral is related to the existence of a *hidden cyclic* coordinate which is not involved in the Hamilton function, while the existence of hidden *separable* variables is responsible for the existence of a quadratic integral. In [4] these Birkhoff's results were generalized, namely, conditions for the global existence of a conditional linear and quadratic integrals of a system on a two-dimensional torus were obtained.

Recently S.V. Bolotin proved that for the case $\kappa(M) < 0$ on isoenergetic surfaces $H =$const> 0 there are unstable closed trajectories with transversally intersecting separatrices. This implies the existence of domains of stochastic behavior and, as a consequence, the absence of an additional integral.

G. Paternian related the nonintegrability of the geodesic flow on a surface with genus > 1 to positive topological entropy of this dynamical system (see [5]). Note that the entropy is positive for the case of smooth surfaces. Nevertheless, one can give an example of a smooth, but not analytic surface of any genus, on which the geodesic flow admits a smooth, but not analytic nonconstant integral on a three-dimensional isoenergetic surface $\{H =$const$> 0\}$. In paper [1] the analytical case was considered.

For $\kappa \geq 0$ the surface M is homeomorphic to sphere of torus. The description of metrics on two-dimensional sphere and torus admitting nontrivial linear and quadratic integrals was given by Kolokoltzov [2], who followed Birkhoff's approach.

Now we discuss in details the case $\kappa = 0$ when $M = \mathbf{T}^2 = \{q_1, q_2 \bmod 2\pi\}$. A Riemannian metric on a torus can be always reduced to the form (1.4) globally. Then the question arises: are there exist polynomial integrals of degree ≥ 3 that are independent from the function H and are not reduced to polynomials of degree ≥ 2 ? S. Bolotin conjectured that there are no such integrals for systems on a torus. This is not proved completely. However, we have the following theorem

Theorem 1. *Suppose that equations* (1.3) *with the Hamiltonian* (1.4) *admit an integral F homogeneous in momenta and independent from the energy integral. Let one of the following conditions are satisfied:*

a) F is an even function in p_1, p_2,

b) F is even in p_1 (p_2) and odd in p_2 (p_2).

Then equations (1.3) *admit an additional integral of degree* ≥ 2.

If M is homeomorphic to the two-dimensional sphere, the situation is different. There are examples of metrics on the sphere, for which the geodesic equations admit integrals of 3-rd and 4-th degrees in momenta that are not reduced to polynomials of lower degrees.

There are no examples of Riemannian metrics on two-dimensional sphere which admit nontrivial integrals of degree ≥ 5. It seems that such metrics do not exist at all.

Now we proceed to symmetry fields of general form. Concerning the case $\kappa(M) < 0$, the author [6] proved that any symmetry field u has the form $f(H)v$. Thus, there are no nontrivial symmetries, and any symmetry field is Hamiltonian with the Hamilton function

$$\int f(H)\, dH.$$

For the case $\kappa(M) = 0$ we have

Theorem 2. *Any symmetry field of first degree is Hamiltonian.*

It generated by a linear integral, and, therefore, is a Noether field.

Theorem 3. *If the Gauss curvature of the metric on a torus does not equal zero identically, any symmetry field of the second degree is Hamiltonian.*

The analytic criterion for the metric (1.4) on the torus to be Euclidean has the form $\Lambda =$const. Taking the field (1.5) as an example, we see that the condition $\Lambda \neq$const in Theorem 3 cannot be omitted.

For symmetry fields of degree ≥ 3 Theorem 2 does not holds. Indeed, let Λ be a nonconstant function periodic only in the angle variable q_2. Then the vector field u of first degree specified by equations

$$q_1' = 1, \quad q_2' = 0, \quad p_1' = 0, \quad p_2' = 0,$$

is a symmetry field. It is obviously Hamiltonian. However, the field hu of third degree also being a symmetry field is not Hamiltonian.

It seems true that if the system with the Hamiltonian (1.4) admits a polynomial symmetry field of degree n that is not collinear to v, then there exists an additional polynomial integral of degree $\leq n$ in momenta.

Studying symmetries of reversible systems on a sphere is a more complicated problem, and we know very little about it.

IRREVERSIBLE SYSTEMS

Now consider more complicated problem with the standard symplectic structure $\sum dp_k \wedge dq_k$ on T^*M replaced by a closed nondegenerate 2-form $\omega + \phi$, where ϕ is a 2-form on M. In local coordinates q_1, q_2 the latter is represented by

$$\lambda(q_1, q_2)\, dq_1 \wedge dq_2.$$

Replacing ω in (1.1) by $\omega + \phi$, we obtain the following Hamilton equations

$$\dot{q}_1 = \frac{\partial H}{\partial p_1}, \quad \dot{q}_2 = \frac{\partial H}{\partial p_2}, \quad \dot{p}_1 = -\frac{\partial H}{\partial q_1} + \lambda \frac{\partial H}{\partial p_2}, \quad \dot{p}_2 = -\frac{\partial H}{\partial q_2} - \lambda \frac{\partial H}{\partial p_1}. \tag{3.1}$$

Following Birkhoff [3], such systems are called *irreversible*. The terms in (3.1) containing λ are *hyroscopic forces.* They arise in various mechanical problems, under transformation to rotating frame of reference, and in systems describing the motion of charged particles in magnetic fields.

Equations (3.1) obviously admit the energy integral. The problem on the existence of other integrals polynomial in momenta p_1, p_2 on closed manifolds was considered by S. Bolotin in [7]. In particular, it was proved that if the system (3.1) on the two-dimensional torus with the Hamiltonian (1.4) has a polynomial integral independent from H, then

$$\int_{T^2} \lambda\, dq_1\, dq_2 = 0. \tag{3.2}$$

Consider a more general problem on the existence of vector symmetry fields of the system (3.1) with polynomial components. In contrast to the reversible case, such symmetry fields are not homogeneous. They can be represented as a sum of homogeneous fields:

$$u = u_m + u_{m-1} + \cdots, \qquad \deg u_k = k,$$

We call the maximal degree m the degree of the field u.

Clearly, u_m is a symmetry field of a reversible system (for $\lambda = 0$). This simple observation enables us to apply the results of Section 2.

Theorem 4. *A nonzero symmetry field of first degree is always locally Hamiltonian. It is Hamiltonian globally provided only the condition* (3.2) *is satisfied.*

For $m \geq 2$ the situation is different. We have the following

Theorem 5. *Let the Hamiltonian field* (3.1) *admit a symmetry field of degree* $m \geq 2$ *with major homogeneous terms of the fields u and v being linearly independent for* $p_1 = 1, p_2 = i$ $(i^2 = -1)$. *Then the condition* (3.2) *is satisfied.*

Apparently, Theorem 5 holds if the fields u and v are independent. Let the field u be Hamiltonian with the Hamilton function $F = F_m + F_{m-1} + \cdots$. Then the condition for the vectors v and u_m to be independent at the point $p_1 = 1, p_2 = i$ is equivalent to the following condition: the function H is not a divisor of the polynomial F_m.

References

1. Kozlov V.V. Topological obstructions to the integrability of natural mechanical systems. Dokl. Akad. Nauk SSSR **249**, 1299-1302 (1979) (in Russian), [English transl.: Soviet Math. Dokl. **20** No 6 (1979)].

2. Kolokoltzov V.N. Geodesic flows on two-dimensional manifolds with an additional first integral that is polynomial in the velocities. Izv. Acad. Nauk SSSR, Ser. Mat. **46**, 994–1010 (1982) (in Russian) [English transl.: Math. USSR Izv. **21** No 2 (1983)].

3. Birkhoff G.D. Dynamical Systems. Amer. Math. Soc. Colloquium Publications **9**, New York 1927.

4. Kozlov V.V. Integrability and nonintegrability in Hamiltonian mechanics. Uspekhi Mat. Nauk **38**, No 1 3–67 (1983) (in Russian) [English transl.: Russian Math. Surveys **38**, No 1 1–76 (1983)].

5. Paternian G. On the topology of manifolds with completely integrable geodesic flows. Ergod. Th. & Dynam. Sys. 1992. V.12. 109–121.

6. Kozlov V.V. Symmetry groups of geodesic flows on closed surfaces. Mat. Zametki **48**, 62–67 (1990) (Russian) [English transl.: Math. Notes **48** (1990), 1119–1122 (1991)].

7. Bolotin S.V. On first integrals of systems with gyroscopic forces. Vestn. Moskov. Univ., Ser. Mat. Mekh. **6**, 75–82 (1984) (in Russian).

VARIATIONAL CRITERIA FOR NONINTEGRABILITY AND CHAOS IN HAMILTONIAN SYSTEMS

Sergey Bolotin

Department of Mathematics and Mechanics
Moscow State University
Moscow–119899, Russia

INTRODUCTION

The goal of this article is to show how existence results obtained by variational methods can be applied to prove chaotic behavior for a class of Hamiltonian systems. The standard tool is the perturbation theory, for example, the Poincaré-Melnikov-Arnold integral. This approach is restricted to the case when the system is in some sense close to integrable. If this is not so, non-perturbative methods are needed. Our method works best for natural analytic systems with two degrees of freedom. We obtain the existence of homoclinic orbits by variational methods and then prove, using analyticity, that they are isolated from purely topological reasons. Of course, it is impossible to prove transversality by variational methods. However, if the homoclinic orbit is obtained by minimizing the action functional, or by the mountain-pass theorem, then it is usually possible to prove a weaker property, that they are transversal in the topological sense. Then we can apply a modification of the well-known results on dynamics near transversal homoclinics.

Recently interesting results on application of variational methods to chaos in multi-dimensional time-periodic Hamiltonian systems were obtained by Sere (1992). He used the approach suggested in (Coti- Zelati, Ekeland, and Sere, 1990) and (Coti-Zelati and Rabinowitz, 1991), where the existence of an infinite number of homoclinics is established (see also the references in the cited papers). However, these results were proved under the assumption that the homoclinic orbit, obtained by variational methods, is isolated as a point in a suitable function space. This assumption cannot be verified for a concrete system, if the perturbation methods fail. Moreover, it never holds for autonomous systems.

In this paper we give some purely geometrical criteria for the existence of an isolated homoclinic. The main results admit partial extension to multi-dimensional and non-autonomous Hamiltonian systems. However, we concentrate on the simplest case of natural Hamiltonian systems with two degrees of freedom and state only results that have concrete applications, without trying to reach maximal generality. This article is partly based on (Bolotin, 1984, 1992).

Hamiltonian Mechanics, Edited by J. Seimenis,
Plenum Press, New York, 1994

TOPOLOGICAL OBSTRUCTIONS TO INTEGRABILITY

Consider a natural Hamiltonian system with configuration space M^m and the Hamiltonian function

$$H = \|p\|^2/2 + V(x)$$

on the phase space T^*M. The kinetic energy is defined by a Riemannian metric on M. We assume that H is analytic. The following theorem provides topological obstacles to integrability.

Theorem 1. *If M is compact and $\dim H_1(M) > m$, then the Hamiltonian system cannot be completely integrable.*

This means that there do not exist m commuting analytic integrals of the system that are not everywhere dependent. This theorem was proved by Taymanov (1985) using variational methods. Later he established that for a completely integrable system $\dim H_k(M) \leq \binom{m}{k}$. The case of two degrees of freedom was studied by Kozlov (1979, 1983). In (Bolotin, 1984) non-compact manifolds and manifolds with boundary were considered. In contrast with the results of these papers, Theorem 1 does not establish the complicated behavior of phase trajectories: nonintegrability is not sufficient for this. The proof is based on obtaining a contradiction to the existence of a foliation by invariant tori. Paternain (1992) proved that under assumptions similar to that of Theorem 1, the topological entropy is positive. For $m = 2$ this was established by Katok (1983). If $m > 2$, then it is not proved that positive topological entropy implies nonintegrability. Moreover, in this case almost nothing is known about the location of the chaotic set.

A more detailed information on the mechanism of nonintegrability and chaotic behavior can be established for $m = 2$. We fix the energy level $H = h > \sup V$. If M is not compact, we require that some conditions at infinity are satisfied. For example, it is sufficient that V is bounded, the Riemannian metric is complete and the length of curves, vending about infinity points and tending to infinity, tend to infinity.

Theorem 2. *If the Euler characteristic $\chi(M)$ is negative, then there are an infinite number of action-minimizing unstable periodic orbits on the energy level $H = h$, and there exist an infinite number of transversal homoclinic orbits to each of them. Therefore, the behavior is chaotic on some invariant set in a neighborhood of these orbits.*

Sketch of the proof. The existence of an infinite number of minimal (action - minimizing) periodic and heteroclinic orbits is proved by variational methods. It turns out that there is a heteroclinic orbit corresponding to any homotopy class of maps of the figure eight into M. If the system is analytic, then the stable W^s and unstable W^u manifolds of minimal periodic orbits are analytic. Thus some of the heteroclinic orbits are isolated and of odd multiplicity (the intersection degree of W^s and W^u is odd) from purely topological reasons. It follows from simple dynamical systems considerations that there exist transversal homoclinic orbits to periodic orbits. It is well known that in this case the motion is chaotic in a certain invariant set in a neighborhood of the homoclinic orbit. This is obtained in (Bolotin 1984, 1992).

Example 1. The n-center problem. Consider a point $x \in \mathbf{R}^2$ moving in the plane in the gravitational field of n fixed centers

$$V(x) = -\sum_{i=1}^{n} \frac{\mu_i}{|x - a_i|}, \qquad \mu_i > 0.$$

The configuration space is $\mathbf{R}^2 \setminus \{a_1, \dots, a_n\}$, and $H = |p|^2/2 + V(x)$. For $n \leq 2$ this system is integrable. It is the Kepler problem for $n = 1$ and the Euler problem of two fixed centers for $n = 2$.

Theorem 3. *For $n > 2$ the system is non-integrable and its behavior is chaotic on some subset of every energy level $H = h \geq 0$ (Bolotin,* 1984).

Proof. Let $\pi : M \to \mathbf{R}^2$ be a two-sheeted covering branched over the points a_i. It follows from the Riemann-Hurvicz formula that the Euler characteristic is $\chi(M) = 2 - n < 0$. The orbits of the n-center problem with fixed energy are covered by orbits of some natural Hamiltonian system on M without singularities.

Locally the covering π is just the Levi-Civita regularization. We can apply Theorem 2 to obtain a lot of transversal homoclinic orbits for the regularized system. It can be proved that most of the constructed orbits do not pass through the points $\pi^{-1}(a_i)$, and thus their images are orbits of the n-center problem without any collisions. This finishes the proof.

The same result holds for any configuration space M, if the potential has $n > 2\chi(M)$ singular points of the Newtonian type.

Example 2. Plane restricted circular problem of $n+1$ bodies. The configuration space and the potential are the same, but the Hamiltonian

$$H = |p|^2/2 - \omega(x_1 p_2 - x_2 p_1) + V(x), \qquad x \in \mathbf{R}^2 \setminus \{a_1, \dots, a_n\},$$

is of the non-reversible type. This leads to some technical complications connected with Finsler geometry of the Jacobi metric. However, for $n > 2$ the assertion of Theorem 3 holds (Bolotin, 1984). It seems that for $n = 2$ there is no chaos on the levels $H = h \geq 0$.

Example 3. Let a particle move in the plane, and the potential $V : \mathbf{R}^2 \to \mathbf{R}$ be bounded from above. Suppose that there exist two periodic orbits γ_1 and γ_2 with energy $h > \sup V$ that are contractible to a point without intersecting each other. Then the motion is chaotic on some subset of the energy level $H = h$. The proof is based on a generalization of Theorem 2 to the case of manifolds with geodesically convex boundary (Bolotin 1984).

Example 4. Consider inertial motion on a sphere (geodesic flow). By the theorem of Lusternik-Shnirelman, there exist three closed geodesics without self-intersections. If they can be contracted to a point without intersecting each other, then the flow is chaotic on some invariant set.

In these examples chaos is induced by topology. However, most significant systems are integrable for certain parameter values, and thus have simple configuration space: $\chi(M) \geq 0$. Then the conditions for chaos are more complicated.

THE CASE OF ZERO EULER CHARACTERISTIC

We start with $\chi(M) = 0$ (greater the Euler characteristic, harder the task). If M is compact and oriented, then M is a torus. Sometimes the case $\chi(M) = 0$

can be reduced to the case $\chi(M) < 0$. For example, it is so if the system has a homotopically trivial periodic orbit on some energy level $H = h > \max V$ (Example 3). Then the behavior of the system on this level is chaotic.

Let P be the point of strict nondegenerate maximum of the potential energy V and let $h = V(P)$. Suppose that the eigenvalues of the Hessian of V at P are distinct. For any curve γ in M denote by

$$S(\gamma) = \int \sqrt{2(h - V(\gamma(t)))}\|\dot{\gamma}(t)\|\, dt$$

the Maupertuis-Jacobi action of γ. Let Γ be a simple homotopy class of closed curves in M, and $\Gamma_P \subset \Gamma$ the set of the curves in Γ passing through P.

Theorem 4. *Suppose that*

$$\inf_{\Gamma} S < \inf_{\Gamma_P} S.$$

Then for small $\varepsilon > 0$ the behavior of the system is chaotic on some invariant subset of the energy level $H = h + \varepsilon$.

Sketch of the proof. Let $\gamma \in \Gamma$ be a minimizer of the action S on Γ. Then γ is weakly hyperbolic and $W^u(P) \cap W^s(\gamma) \neq \emptyset$. Moreover, the intersection consists of at least two geometrically distinct heteroclinic orbits. They are obtained by minimizing the action functional. By topology, the intersection is isolated and of odd multiplicity. Now the theorem follows from a slight generalization of the Turayev-Shilnikov result (1989) on symbolic representation of Hamiltonian systems with several transversal heteroclinic orbits connecting equilibria and periodic trajectories. The chaotic set is contained in a neighborhood of the closure of the heteroclinic orbits.

Example 5. Let a point move on the plane in a 2π-periodic potential field. Suppose that $V(x_1, x_2) = V(-x_1, x_2) = V(x_1, -x_2)$. If $h = V(0) = \max V$ and

$$\int_0^{\pi} \sqrt{h - V(x_1, 0)}\, dx_1 > \int_0^{\pi} \sqrt{h - V(x_1, \pi)}\, dx_1, \qquad \frac{\partial^2 V}{\partial x_1^2}(0) \neq \frac{\partial^2 V}{\partial x_2^2}(0),$$

then the motion is chaotic on some invariant set.

Such systems are often encountered in the perturbation theory. Consider a perturbation of an integrable multi-dimensional system near a resonant torus satisfying two independent resonance relations. By the Delaunay method, the problem is reduced to a small perturbation of a natural Hamiltonian system with the configuration space $\mathbf{T}^2$ and a flat indefinite metric. Under the convexity assumption, this metric is Riemannian.

THE CASE OF POSITIVE EULER CHARACTERISTIC

The next two theorems deal with the most "integrable" case $\chi(M) > 0$. In this case the criteria for chaos must be more complicated, because there are a lot of integrable systems. In general no homoclinics and heteroclinics to periodic orbits can be obtained by variational methods. Therefore, we are left with hyperbolic equilibria.

Let P is the point of strict nondegenerate maximum of V. One can prove by variational methods that there always exist a homoclinic orbit γ to P (Bolotin, 1978). Suppose that we established that γ is transversal (thou usually it is impossible to do by variational methods). Even now the system can be integrable, as shown by the following example.

Example 6. The Neumann problem. Consider a point moving on the standard sphere $S^2 \subset \mathbf{R}^3$ and let the potential energy be

$$V(x) = \frac{1}{2}\sum_{i=1}^{3} a_i x_i^2, \qquad a_1 < a_2 < a_3.$$

It is well known that this system is integrable. Let $P_\pm = \{x_1 = x_2 = 0,\ x_3 = \pm 1\}$ be the points of maximum of V. There exist four obvious orbits γ_i heteroclinic from P_- to P_+. They are half-circles lying in the coordinate planes $x_1 = 0$ and $x_2 = 0$. Devaney (1978) showed that they are transversal in the energy level $H = a_3$.

If we identify the opposite points of the sphere, then the configuration space turns into the projective plane $\mathbf{R}P^2$, and γ_i turn into four transversal homoclinic orbits to the point of maximum P of the potential. However, the system is integrable! The reason for this phenomenon is that the characteristic exponents of the equilibrium P are real (it is always so for the point of maximum of the potential energy) and γ_i are the main asymptotic orbits to $P_\pm$ as $t \to \pm\infty$. If the characteristic exponents are complex, then the existence of one transversal homoclinic implies chaos. However, this is impossible for natural Hamiltonian systems.

In this example $M = \mathbf{R}P^2$, but there exist such systems with $M = S^2$ (for example, the Kovalevskaya top). It follows that to prove chaotic behavior, we need more than just existence of transversal homoclinics. This is not so in the case of time-dependant systems, or homoclinic orbits to periodic solutions. For proving nonintegrability, these cases are much simpler.

First we consider the case $M = \mathbf{R}P^2$. This means that the Hamiltonian H on T^*S^2 is invariant under the involution $x \to -x$. As it is in the Neumann problem, suppose that H is invariant under all involutions $x_i \to -x_i$. Let $P_\pm = \{x_1 = x_2 = 0,\ x_3 = \pm 1\} \subset S^2$ be the points of maximum for V. There are four obvious heteroclinic orbits γ_i joining P_- and P_+ on S^2 (on $\mathbf{R}P^2$ they are homoclinic orbits). Usually two of them are action minimizing among the curves joining P_- and P_+. For the Neumann problem two other have Morse index one.

Theorem 5. *Suppose that the heteroclinic orbits γ_i are strictly locally minimal and the eigenvalues of the Hessian of V at P are different. Then the system possess chaotic behavior on some invariant subset of the level $\{H = h + \varepsilon\}$, $\varepsilon > 0$.*

The proof is based on the existence of additional heteroclinic or homoclinic orbit of odd multiplicity in each of the sectors in S^2 bounded by the heteroclinic orbits γ_i.

Now consider the case $M = S^2$. Let P be the point of strict nondegenerate maximum of V.

Theorem 6. *Suppose that the eigenvalues of the Hessian of V at P are different. There always exist a homoclinic orbit γ to P. Suppose that γ is the point of strict local minimum of the action functional on the space of loops with endpoints in P, and that γ is tangent, as $t \to \pm\infty$, to the eigenvector of the Hessian corresponding to the largest eigenvalue. Then for small $\varepsilon > 0$ the behavior of the system on a subset of the energy level $H = h + \varepsilon$ is chaotic.*

A similar statement holds when γ is tangent to the other eigenvector. The given result is needed for an application in the next section.

Sketch of the proof. Using the local minimality of the homoclinic orbit γ, it can be proved by variational methods that in both components of $S^2 \setminus \gamma$ there exist homoclinic orbits. They are obtained by applying global variational methods to a regularized variational problem, and then passing to a limit. The tangency assumption is used to construct certain geodesically convex (in the Jacobi metric) subsets of these components. The hard part is to prove that the obtained critical points of the action functional are of the mountain-pass type. Since the system is analytic, this implies that the corresponding homoclinic orbits are of odd multiplicity. For establishing chaotic behavior, this is nearly the same as transversality. The constructed homoclinics are tangent to the eigenvector of the Hessian with smaller eigenvalue. Now Theorem 6 follows from a slight modification of the Turayev-Shilnikov theorem (1989). The details of the proof can be found in (Bolotin, 1992).

Of course, the condition of the theorem is rather restrictive, but there exist some non-trivial applications.

APPLICATION TO THE RIGID BODY DYNAMICS

We consider a heavy top. The motion is governed by the Euler-Poisson equations of a heavy rigid body with a fixed point:

$$\dot{M} + [\omega, M] = [x, e], \quad \dot{e} + [\omega, e] = 0, \qquad M = J\omega,$$

where M is the angular momentum, and x the Poisson vector. The phase space is $e(3)^* = \mathbf{R}^6\{M, x\}$ with the Poisson bracket given by the table

$$\{M_i, M_j\} = \varepsilon_{ijk} M_k, \quad \{M_i, x_j\} = \varepsilon_{ijk} x_k, \quad \{x_i, x_j\} = 0,$$

and the Hamiltonian is

$$H = \langle J^{-1}M, M\rangle/2 + V, \qquad V = \langle e, x\rangle,$$

where J is the inertia tensor, and e a constant vector. We study the case $\langle M, x\rangle = 0$, since for $\langle M, x\rangle \neq 0$ the proof of nonintegrability is easier (Kozlov, 1983). The Hamiltonian equations on the level of first integrals

$$N = \{(M, x) : \langle M, x\rangle = 0, \ |x| = 1\} = T^*S^2$$

are reduced to a natural Hamiltonian system with configuration space S^2, potential energy V, and kinetic energy defined by the Riemannian metric

$$\|v\|^2 = \langle Jv, v\rangle/\langle J'x, x\rangle,$$

where J' is the adjoint matrix $J' = J^{-1}\det(J)$. For simplicity, we consider the case when two of the eigenvalues of J are equal, and e belongs to the corresponding two-dimensional eigenspace. Let $a \geq 1/2$ be the ratio of the eigenvalues. We can assume that $J = \mathrm{diag}(a, a, 1)$ and $e = (1, 0, 0)$. There are three well known integrable cases: $a = 1$, $a = 2$ (Kovalevskaya case), and $a = 4$ (Goryachev–Chaplygin case).

Theorem 7. *For $a > 4$ the behavior of the system is chaotic on some subset of $N \cap \{H = h + \varepsilon\}$, where $h = \max V$ and $\varepsilon > 0$.*

It seems impossible to prove this without variational methods. The idea of the proof is as follows. The point of maximum of the potential energy is $P = \{x = e\}$. There are four obvious pendulum type homoclinic orbits to P: Γ and Γ^{-1} are lying in the plane $x_3 = 0$, and another two in the plane $x_2 = 0$. The orbit Γ is tangent to the eigenvector of the Hessian with largest eigenvalue. To calculate the Morse index of the action functional S, we apply the following comparison method.

Suppose that for a parameter value $a = a_0$ the stable and unstable manifolds W^s and W^u are tangent along a trajectory Γ, and the projection $\pi : W^{s,u} \to M$ is non-degenerate on Γ. If $S'_a \geq S'_a\big|_\Gamma$, then $\mathrm{ind}_S(\Gamma) = 0$ for $a \geq a_0$.

Using the integrals of the Goryachev-Chaplygin case, it is possible to show that for $a = 4$ the homoclinic orbit Γ satisfies this condition. Since

$$S(\gamma) = \int \sqrt{2(1-\gamma_1)(\dot{\gamma}_1^2 + \dot{\gamma}_2^2 + a\dot{\gamma}_3^2)/(\gamma_1^2 + \gamma_2^2 + a^{-1}\gamma_3^2)}\, dt,$$

we have $S'_a > 0$, and $S(\Gamma)$ is independent of a. Thus, Γ is a local nondegenerate minimum of the action functional for $a > 4$. It remains to apply Theorem 6.

There exists a similar application to Kirchhoff's equations of a rigid body in an ideal fluid.

References

Bolotin, S.V., 1978, Libration orbits of natural dynamical systems, *Vestnik Moskov. Univ. Ser. I Matem. Mekh.* 6:72.

Bolotin, S.V., 1984, Nonintegrability of the n-center problem for $n > 2$, *Vestnik Moskov. Univ. Ser. I Matem. Mekh.* 3:65.

Bolotin, S.V., 1984, Influence of singularities of the potential energy on the integrability of dynamical systems, *Prikl. Matem. i Mekhan.* 48:356.

Bolotin, S.V., 1992, Variational methods for constructing chaotic motions in the rigid body dynamics, *Prikl. Matem. i Mekhan.* 56:230.

Bolotin, S.V., 1992, Doubly asymptotic orbits of minimal geodesics, *Vestnik Moskov. Univ. Ser. I Matem. Mekh.* 1:92.

Coti-Zelati, V., Ekeland, I., and Séré, E., 1990, A variational approach to homoclinic orbits in Hamiltonian systems, *Math. Ann.* 288:133.

Coti-Zelati, V., and Rabinowitz, P., 1991, Homoclinic orbits for second order Hamiltonian systems possessing superquadratic potentials, *Jour. Amer. Math. Soc.* 4:693.

Devaney, R.L., 1978, Transversal homoclinic orbits in an integrable system, *Amer. J. Math.* 100:631.

Katok, A., 1982, Entropy and closed geodesics, *Ergod. Th. and Dynam. Syst.* 2:339.

Kozlov, V.V., 1979, Topological obstructions to the integrability of natural mechanical systems, *Dokl. Akad. Nauk. SSSR.* 249:1299.

Kozlov, V.V., 1983, Integrability and nonintegrability in classical mechanics, *Uspekhi Mat. Nauk.* 38:3.

Paternain, G.P., 1992, On the topology of manifolds with completely integrable geodesic flows, *Ergod. Th. and Dynam. Sys.* 12:109.

Séré, E., 1992, Looking for the Bernoulli shift, *Preprint,*Universite Paris-Dauphine.

Taymanov, I.A., 1988, On topological properties of integrable geodesic flows, *Mat. Zametki.* 44:283.

Turayev, D.V., and Shilnikov, L.P., 1989, On Hamiltonian systems with homoclinic curves of a saddle, *Dokl. AN SSSR* 304:811.

EXPONENTIALLY SMALL SPLITTING IN HAMILTONIAN SYSTEMS*

Amadeu Delshams† Tere M. Seara†

Abstract

We consider the exponentially small splitting of separatrices of analytic one-degree of freedom classical systems

$$H_0(x,y) = y^2/2 + V(x) \tag{1}$$

under general rapidly periodic Hamiltonian perturbations.

In particular, we focus our attention on analytic perturbations of the following type:

$$H(x,y) = H_0(x,y) + \mu\varepsilon^p G(x, t/\varepsilon),$$

where μ , ε are small parameters.

Let $x_0(t)$ be the separatrix of the Hamiltonian system associated to (1), and assume that $V'(x_0(t))$ and $G'(x_0(t),\tau)$ have poles at the singularities ai of $x_0(t)$:

$$V'(x_0(t)) = O((t-ai)^{r+1}), \quad G'(x_0(t),\tau) = O((t-ai)^{\ell}), \qquad (t \to ai).$$

Then, for p big enough ($p \geq \ell + r - 1$), it is shown that the Poincaré-Melnikov formula gives the correct asymtotic expression for the exponentially small splitting of the separatrices. Several concrete examples are also considered.

1 Introduction and application

The detection of transversal homoclinic intersections associated to weak hyperbolic objects has been lately achieved by several authors ([HMS88], [Ge90],[LST90], [Fo91], [DS92]). So far, the main results concerning o.d.e. have been focused on the forced pendulum equation

$$\ddot{x} + \sin x = \mu\varepsilon^p \sin\frac{t}{\varepsilon}$$

as a model where these phenomena take place.

Even though these results are of great importance, they cannot be applied for more general equations and they are not useful, for instance, when we want to study other equations, like

$$\ddot{q} = -\sin q(1 - \mu\varepsilon^2 \sin\frac{t}{\varepsilon}) \tag{2}$$

*This work has been partially supported by the spanish grant DGCYT PB90–0580.

†Dep. de Matemàtica Aplicada I, Univ. Politècnica de Catalunya, Diagonal 647, 08028 Barcelona.

Hamiltonian Mechanics, Edited by J. Seimenis,
Plenum Press, New York, 1994

which arises when we consider a hamiltonian of the type

$$H = \frac{I^2}{2} + \delta H_1(\varphi, t)$$

near a resonance (see [Nei84]).

In order to give more general results about transversality of the homoclinic intersection we focus our attention on the equation

$$\ddot{x} + f(x) = \mu\varepsilon^p g(x, \frac{t}{\varepsilon}) \tag{3}$$

where ε, μ are independent small parameters and $p \geq 0$. We will assume that this equation verifies:

H1 The unperturbed equation

$$\ddot{x} + f(x) = 0 \ ,$$

considered like a hamiltonian system

$$\begin{aligned} \dot{x} &= y \, , \\ \dot{y} &= -f(x) \, , \end{aligned} \tag{4}$$

has a saddle point at the origin, and there exists a homoclinic solution $\Gamma = \{(x_0(t), y_0(t))\}$ to this point.

H2 f is an entire function, and $y_0(t) = \dot{x}_0(t)$ is analytic on a strip $|\Im t| < a$, with one pole of order r on each line $\Im t = \pm a$.

H3 $g(x, \tau)$ is an analytic, 2π-periodic in τ function, with zero mean: $\int_0^{2\pi} g(x, s) ds = 0$, and $g(x_0(t), \tau)$ has a pole of order ℓ on each line $\Im t = \pm a$.

H4 The $2\pi\varepsilon$-periodic Poincaré-Arnold-Melnikov function

$$M(s, \varepsilon) = \int_{-\infty}^{+\infty} y_0(\sigma + s) g(x_0(\sigma + s), \frac{\sigma}{\varepsilon}) \, d\sigma = \sum_{k \neq 0} M_k(\varepsilon) e^{i\frac{k}{\varepsilon}s}. \tag{5}$$

has non-zero first Fourier coefficients: $M_{\pm 1} \neq 0$.

2 Main results

In order to study the dynamics of the equation (3) it is very convenient to write it like a system

$$\begin{aligned} \dot{x} &= y \, , \\ \dot{y} &= -f(x) + \mu\varepsilon^p g(x, \frac{t}{\varepsilon}) \, , \end{aligned} \tag{6}$$

and introduce the associated Poincaré map defined by:

$$P(x_0, y_0) = (x(2\pi\varepsilon), y(2\pi\varepsilon)) \, ,$$

where $(x(t), y(t) = \dot{x}(t))$ is the orbit of system (6) that begins at (x_0, y_0) when $t = 0$.

For $\mu = 0$ the equation (3) becomes autonomous and therefore the phase portraits of the Poincaré map and of system (4) are identical. In fact this phase portrait is filled up whith the level curves of the hamiltonian

$$h^0(x, y) := \frac{y^2}{2} + V(x) \ , \qquad V'(x) = f(x) \, .$$

If we assume $V(0) = 0$, the homoclinic orbit Γ is contained in the level curve $h^0 = 0$.

For $\mu \neq 0$, the system is also hamiltonian with hamiltonian

$$h(x,y) := h^0(x,y) + \mu\varepsilon^p G(x, \frac{t}{\varepsilon}),$$

with $\partial G/\partial x = -g$, and, in general, the phase portrait of P changes. There exists a hyperbolic fixed point near $(0,0)$ whose stable and unstable curves have to intersect because P is area preserving, but generically they do not coincide. We call $z_h := (x_h, y_h)$ the homoclinic point that is closest to the unperturbed one $(x_0(0), y_0(0))$. The aim of this talk is to give an asymptotic formula for the area S of the loop that remains between the two invariant curves from z_h to their next intersection. There are other interesting quantities like the angle between the curves at z_h, or the maximum distance between them at the loop, that can be also computed, but we focus our attention on S, wich is the only one that is invariant. We can now summarize our results in the next theorem.

Theorem 1 *Assume $p \geq \ell + r - 1$. Then for $\varepsilon \to 0^+$, $\mu \to 0$, the following formula holds:*

$$S = \mu\varepsilon^p \left[\int_{s_0}^{\bar{s}_0} M(\sigma, \varepsilon)\, d\sigma + O(\mu\varepsilon^{\gamma+1}, \varepsilon^2) e^{-a/\varepsilon} \right],$$

where $M(s,\varepsilon)$ is the Poincaré-Arnold-Melnikov function (5), the constant a is defined in **H2** *as the width of the analyticity of $y_0(\tau)$, and $\gamma := p + 1 - r - \ell \geq 0$. The values of s_0, $\bar{s}_0$ are taken as the two zeros of the function $M(s,\varepsilon)$ nearest to zero.*

It has to be noticed that in this kind of singular problems, the function $M(s,\varepsilon)$ depends on ε. In fact, if its first Fourier coefficients $M_{\pm 1}$ are not zero, one can see, using the residues theory, that $M(s,\varepsilon)$ is of order $O(\varepsilon^{1-r}e^{-\frac{a}{\varepsilon}})$, and therefore Theorem 1 assures that it gives the correct result for S like in the regular case. Applying Theorem 1 to the example (2) we get the following corollary.

Corollary 1 *For $\varepsilon \to 0^+$, $\mu \to 0$, the following estimate holds for the equation (2).*

$$S = \frac{4\pi\mu\varepsilon}{\sinh\frac{\pi}{2\varepsilon}}[1 + O(\mu,\varepsilon)],$$

In the next sections we will develop some ideas about the proof of Theorem 1.

3 The averaged system and the Birkhoff normal form

It is necessary a good knowledge of the orbits of equation (3) in order to have good parameterizations of the invariant manifolds and a well defined function that measures the area between them. As a first step, we deal with the averaged system of (6).

Lemma 1 *There exists a canonical, $2\pi\varepsilon$-periodic in t, analytic change Π, given by*

$$\begin{aligned} \bar{x} &= \Pi_1(x,y,\frac{t}{\varepsilon}) = x + O(\mu\varepsilon^{p+2}), \\ \bar{y} &= \Pi_2(x,y,\frac{t}{\varepsilon}) = y - \mu\varepsilon^{p+1}g^1(x,\frac{t}{\varepsilon}) + O(\mu\varepsilon^{p+2}), \end{aligned} \tag{7}$$

wich transforms system (6) to a hamiltonian system of hamiltonian $\mathcal{H}$, where

$$\mathcal{H} = h^0(\bar{x},\bar{y}) + O(\mu\varepsilon^{p+2}).$$

Notice that in the averaged system, $\mathcal{H}$ has the same integrable part h, but the perturbation term is of order $\mu\varepsilon^{p+2}$. This order of perturbation is preserved when one considers the Birkhoff normal form of $\mathcal{H}$.

Theorem 2 (The Birkhoff normal form) *For $\mu\varepsilon^{p+2}$ small enough, the hamiltonian system associated to $\mathcal{H}$ has a $2\pi\varepsilon$-periodic orbit near $(0,0)$:*

$$\bar{\gamma} = (\bar{x}_p(t), \bar{y}_p(t)) = O(\mu\varepsilon^{p+2}) \ , \tag{8}$$

and there exists an analytic canonical change, $2\pi\varepsilon$-periodic in t:

$$\begin{aligned} X &= \Phi(\bar{x}, \bar{y}, \frac{t}{\varepsilon}, \mu, \varepsilon) \, , \\ Y &= \Psi(\bar{x}, \bar{y}, \frac{t}{\varepsilon}, \mu, \varepsilon) \, , \end{aligned} \tag{9}$$

defined in a neighborhood of size $2r$ of $\bar{\gamma}$, with r independent of μ, ε, such that transforms the system associated to $\mathcal{H}$ into its normal form near $\bar{\gamma}$:

$$\begin{aligned} \dot{X} &= F'(XY, \mu, \varepsilon) X \ , \\ \dot{Y} &= -F'(XY, \mu, \varepsilon) Y \ , \end{aligned} \tag{10}$$

that is also a hamiltonian system with hamiltonian

$$\begin{aligned} H(X, Y, \mu, \varepsilon) &= F(XY, \mu, \varepsilon) = \omega XY + O_2(XY) \, , \\ \omega &= F'(0) = \omega_0 + O(\mu\varepsilon^{p+2}) \, , \\ \omega_0 &= \sqrt{-f'(0)} \, . \end{aligned}$$

Composing the two changes (7) and (9), we can go from system (6) to its normal form. Thus we obtain a periodic orbit of system (6) $\gamma = \Pi^{-1}(\bar{\gamma})$, whose size is $O(\mu\varepsilon^{p+1})$. Now we study this normal form in order to obtain the main tools needed for computing the separatrix invariant S.

4 Parameterization of the orbits near γ

It is very easy to obtain parameterizations of the orbits of system (10), because it can be integrated by quadratures. If we perform the canonical change

$$\begin{aligned} s &= S(X, Y) = -\frac{lnY}{F'(XY)} \, , \\ e &= E(X, Y) = F(XY) \, , \end{aligned} \tag{11}$$

in $U = \{(X, Y)/X^2 + Y^2 \leq R^2, Y \neq 0\}$, the new variables verify

$$\begin{aligned} s' &= 1 \, , \\ e' &= 0 \, . \end{aligned}$$

Then $E(X, Y)$, $S(X, Y) - t$ are first integrals of (10), and this gives us a parameterization of the orbits in U:

$$\begin{aligned} X(t, s, e) &= e^{F'(F^{-1}(e))(t+s)} F^{-1}(e) \, , \\ Y(t, s, e) &= e^{-F'(F^{-1}(e))(t+s)} \, . \end{aligned} \tag{12}$$

The local stable manifold associated to the origin can be obtained by putting $e = 0$. The local unstable manifold is not contained in U, but it can still be parameterized by

$$(X^u(t,s), Y^u(t,s)) = (e^{\omega(t+s)}, 0) \ ,$$

Performing the change (9) we obtain parameterizations for the orbits of the system associated to $\mathcal{H}$ near the periodic orbit $\bar{\gamma}$, and adding the change (7), we can parameterize all the orbits of system (6) while they are near the periodic orbit γ. This give us two first integrals

$$\begin{aligned} e(x,y,t,\mu,\varepsilon) = E(\Phi(\Pi_1,\Pi_2,\frac{t}{\varepsilon}),\Psi(\Pi_1,\Pi_2,\frac{t}{\varepsilon})) &= e_0(x,y) + O(\mu\varepsilon^{p+1}) \\ &= h^0(x,y) + O(\mu\varepsilon^{p+1}), \\ s(x,y,t,\mu,\varepsilon) = S(\Phi(\Pi_1,\Pi_2,\frac{t}{\varepsilon}),\Psi(\Pi_1,\Pi_2,\frac{t}{\varepsilon})) &= S_0(x,y) + O(\mu\varepsilon^{p+1}), \end{aligned} \tag{13}$$

which are defined in a neighborhood of the periodic orbit. It is easy to see that the first integral e vanishes in the local invariant manifolds $W^u_{loc}(\gamma)$, $W^s_{loc}(\gamma)$ of the periodic orbit γ.

Let us call $(x^\beta(t,s), y^\beta(t,s))$, $\beta = u, s$, the parameterization of the unstable and stable manifolds obtained through these changes. Of course this two parameterizations are defined while these manifolds are near the periodic orbit, and this happens for different sets of parameters. More especifically, we have

$$\begin{aligned} &(x^u(t,s), y^u(t,s)) \quad \text{defined for} \quad t \in \mathbb{R}\ , \quad s \in \mathbb{C}\ , \quad \Re(t+s) \le -\frac{T}{2}\,; \\ &(x^s(t,s), y^s(t,s)) \quad \text{defined for} \quad t \in \mathbb{R}\ , \quad s \in \mathbb{C}\ , \quad \Re(t+s) \ge +\frac{T}{2}\,; \end{aligned} \tag{14}$$

where the constant T does not depend on ε, μ. These parameterizations verify

$$\begin{aligned} x^\beta(t+2\pi\varepsilon, s) &= x^\beta(t, s+2\pi\varepsilon)\ , \\ x^\beta(t,s) - x_0(t+s) &= O(\mu\varepsilon^{p+2})\ , \quad \beta = u, s\,. \end{aligned} \tag{15}$$

At this moment we have parameterizations for the invariant manifolds but defined for different sets of the parameters t and s. In order to compare these two manifolds one of them is going to be extended near $x_0(t+s)$. However, it has to be noticed that $x_0(t+s)$ has a singularity in the complex field for $t+s = \pm ai$. That means that we will not be able to control the growth of the parameterization for $\Im(t+s) = \pm a$. In the next theorem the unstable manifold is extended for complex values of s but with the restriction $\Im(t+s) \le a - \varepsilon$, $\Re(t+s) \le T$. As we will see later on, this range of values is enough in order to prove Theorem 1.

Theorem 3 (Extension Theorem) *Assume $p \ge \ell + r - 1$. Let $x_0(t+s)$ be the unperturbed separatrix, defined for $t \in \mathbb{R}$, $\Im s < a$. Let $s \in \mathbb{C}$, $|\Im s| \le a - \varepsilon$, and $x^u(t,s)$ the unstable manifold defined in (14).*

Then $x^u(t,s)$ can be extended to the strip $-T \le \Re(t+s) \le T$, and satisfies:

$$\begin{aligned} x^u(t,s) - x_0(t+s) &= O(\mu\varepsilon^{p+1-r-\ell}), \\ \dot{x}^u(t,s) - y_0(t+s) &= O(\mu\varepsilon^{p+1-r-\ell}), \end{aligned}$$

The proof of this general theorem follows the same lines as in the case of the pendulum equation, which is given in [DS92].

5 Distance between the invariant manifolds

Due to the Extension Theorem, the unstable manifold comes back to the neighborhood where the first integrals $e(x, y, \frac{t}{\varepsilon})$, $s(x, y, \frac{t}{\varepsilon}) - t$ exist, and in this way we can evaluate for $\frac{T}{2} \leq t + \Re s \leq T$

$$\begin{aligned} \mathcal{S}^s(r) &:= s(x^s(t,r), y^s(t,r), \frac{t}{\varepsilon}) - t\,, \\ \mathcal{S}^u(r) &:= s(x^u(t,r), y^u(t,r), \frac{t}{\varepsilon}) - t\,. \end{aligned}$$

Since $s - t$ is a first integral of (6) the functions $\mathcal{S}^\beta$, $\beta = s, u$ do not depend on t. We can invert them, using (12) and the Extension Theorem, obtaining

$$\begin{aligned} \mathcal{R}^s(s) &= s\,, \\ \mathcal{R}^u(s) &= s + O(\varepsilon^{p+1-r-\ell})\,. \end{aligned}$$

We can now define the function ψ, that will give us all the information about the splitting:

$$\begin{aligned} \psi(s) &:= e(x^u(t,\mathcal{R}^u(s)), y^u(t,\mathcal{R}^u(s)), \frac{t}{\varepsilon}, \mu, \varepsilon) - e(x^s(t,s), y^s(t,s), \frac{t}{\varepsilon}, \mu, \varepsilon) \\ &= e(x^u(t,\mathcal{R}^u(s)), y^u(t,\mathcal{R}^u(s)), \frac{t}{\varepsilon}, \mu, \varepsilon)\,, \end{aligned}$$

for

$$t \in \mathrm{IR}\,, \quad \frac{T}{2} < \Re(t+s) < T\,, \quad |\Im s| \leq a - \varepsilon\,.$$

Since e is a first integral of system (6), and the manifolds verify (15), ψ does not depend on t and is $2\pi\varepsilon$-periodic in its variable s.

Next proposition relates the function ψ with the well known Poincaré-Arnold-Melnikov integral.

Proposition 1 *For $|\Im s| \leq r_\varepsilon = a - \varepsilon$, we have*

$$\psi(s) = \mu\varepsilon^p[M(s,\varepsilon) + O(\mu\varepsilon^{p+1-r}) + O(\varepsilon)]\,, \tag{16}$$

where $M(s,\varepsilon)$ is the Poincaré-Arnold-Melnikov integral, given at (5).

During the rest of this talk, s will denote a real number. Considering now the Fourier expansion of ψ,

$$\psi(s) = \sum_k \psi_k e^{i\frac{k}{\varepsilon}s}\,.$$

the Fourier coefficients can be computed shifting along complex lines $\Im\sigma = \pm r_\varepsilon$, and using (16) one gets

$$\begin{aligned} \psi_k &= \frac{e^{-|k|\frac{r_\varepsilon}{\varepsilon}}}{2\pi\varepsilon} \int_0^{2\pi\varepsilon} \psi(s \pm i r_\varepsilon) e^{i\frac{k}{\varepsilon}s} ds \\ &= \mu\varepsilon^p M_k(\varepsilon) + O\left(e^{-\frac{|k|}{\varepsilon}a}(\mu^2\varepsilon^{2p+1-r} + \mu\varepsilon^{p+1})\right)\,. \end{aligned}$$

By **H4**, $M_{\pm 1}$ are non zero, and thus

$$\psi(s) = \psi_0 + \mu\varepsilon^p M(s,\varepsilon) + O\left(e^{-\frac{a}{\varepsilon}}(\mu^2\varepsilon^{2p+1-r} + \mu^2\varepsilon^{p+1})\right)\,. \tag{17}$$

Now, we only need to compute ψ_0, and to relate function ψ with the desired quantity S. This is done in the next proposition.

Proposition 2 *The function ψ satisfies the following properties.*

a) There exist s_1 and s_2 such that $\vec{x}^u(t,s_1) = \vec{x}^s(t,s_2)$ (giving an homoclinic connection), and then $\psi(s_1) = 0$,

b) $\psi_0 = \int_{s_1}^{s_1+2\pi\varepsilon} \psi(s)ds = 0$.

c) $s_1 = s_0 + O(\mu\varepsilon^{p+1-r})$, where s_0 is the smallest zero of $M(s,\varepsilon)$ given at (5).

d) $S = \int_{s_1}^{\bar{s}_1} \psi(s)ds$, being s_1 and $\bar{s}_1$ the two zeros of $\psi(s)$ in $[0, 2\pi\varepsilon)$.

Similar properties are proved in [DS92], and they are a consequence of the following facts:

- system (6) is hamiltonian,
- the Poincaré map P is area preserving,
- the change of variables (11) is symplectic.
- equation (17) which relates $\psi(s)$ with $M(s,\varepsilon)$.

Thanks to the previous proposition and formula (17) we can easily derive the proof of Theorem 1. □

References

[DS92] A. Delshams, T. M. Seara. An asymptotic expression for the splitting of separatrices of the rapidly forced pendulum. *Commun. Math. Phys.*, 150:433–463, 1992.

[Fo91] E. Fontich. Exponentially small upper bounds for the splitting of separatrices for high frequency periodic perturbation. *Preprint*, 1991.

[Ge90] V.G. Gelfreich. Splitting of separatrices for the rapidly forced pendulum. *Preprint*, 1990.

[HMS88] P. Holmes, J. Marsden, J. Scheurle. Exponentially small splittings of separatrices with applications to KAM theory and degenerate bifurcations. *Contemporary Mathematics*, 81: 213-244, 1988.

[LST90] V.F. Lazutkin, I.G. Schachmannski, M.B. Tabanov. Splitting of separatrices for standard and semistandard mappings. *Physica D*, 40: 235-248, 1990.

[Nei84] A.I. Neishtadt. The separation of motions in systems with rapidly rotating phase. *J. Appl. Math. Mech.*, 48 (2): 133-139, 1984.

INTEGRABLE AND CHAOTIC BEHAVIOUR IN THE PAUL TRAP AND THE HYDROGEN ATOM IN A GENERALIZED VAN DER WAALS POTENTIAL

David Farrelly,* James E. Howard,* and T. Uzer†

*Department of Chemistry and Biochemistry
Utah State University, Logan, UT 84322-0300, U.S.A.
†School of Physics, Georgia Institute of Technology
Atlanta, GA 30332-0430, U.S.A.

INTRODUCTION

For almost two decades the problem of Rydberg atoms interacting with external fields has proved to be a unique atomic laboratory for the study of chaos.[1-5] While some studies have specialized to interactions with explicitly time dependent microwave fields,[1] others have dealt instead with static fields.[2-4] Many of the most outstanding cases of interest fall into the second category, and, further, their Hamiltonians turn out to be particular limits of the so-called generalized van der Waals Hamiltonian (GVDW)[6-7] (e.g., the notoriously chaotic quadratic Zeeman effect[2-5]). Studies of atomic Rydberg states have clearly led to substantial progress being made in furthering an understanding of the implications of classical chaos in quantum systems. Quite recently, however, investigations of a different problem, the Paul trap,[8-9] suggest that this system might prove of comparable value in the study of chaotic dynamics in atomic systems. In this paper we show that the generalized van der Waals and trap Hamiltonians are special cases of a more general Hamiltonian and, remarkably, they share *identical* integrable limits. Despite their similitude, important differences also exist; the most significant of them being a double well structure in the effective potential energy surface for the Paul trap. This structure gives rise to a new type of two-ion crystal corresponding to a dynamical rather than a static equilibrium, leading to new resonances and modes of regular and chaotic behavior stemming from the dynamics of the associated separatrix layer.

Integrability is clearly a central issue in understanding the origins and implications of chaotic behavior in dynamical systems. Physically interesting integrable systems are rare, and consequently, it stirs up considerable excitement when one is discovered. Accordingly, the recent discovery of three integrable cases for the Paul trap[9] has spurred several theoretical investigations. At the same time, motivated by the discovery of an adiabatic invariant,[6] integrable cases have been identified for a restricted subclass of the generalized van der Waals Hamiltonian using several different methods.[7] Exploiting the connection between the two problems, we are able to obtain new global invariants for the unrestricted GVDW Hamiltonian

Hamiltonian Mechanics, Edited by J. Seimenis,
Plenum Press, New York, 1994

and identify previously unrecognized separable limits of both problems. Of special interest is the observation that one integrable limit of the general Hamiltonian is unrelated to separability of the system in any known coordinate system, and consequently, the corresponding invariants represent an even rarer class of systems; that of non-separable, integrable Hamiltonians. These finding are presented as is a short discussion of the dynamics of two ions in the Paul trap.

THE PAUL TRAP AND GVDW HAMILTONIANS

In this section we obtain the Hamiltonian for the Paul trap in the pseudopotential approximation, and demonstrate that it is a special case of a more general Hamiltonian that includes the GVDW problem. The relative motion of two like-charged ions in a Paul trap is given by[9]

$$\frac{d^2}{dt^2}\begin{bmatrix} x \\ y \\ z \end{bmatrix} + [a + 2q\cos(2t)]\begin{bmatrix} x \\ y \\ -2z \end{bmatrix} = \frac{\mu_x^2}{|\mathbf{r}|^3}\begin{bmatrix} x \\ y \\ z \end{bmatrix} \tag{1}$$

where $\mu_x = \sqrt{a + \frac{1}{2}q^2}$ is the radial secular frequency and a and q are proportional to the static and radiofrequency (rf) field strengths, respectively. Conversion of scaled frequencies to physical units may be accomplished using, e.g., $\mu_x \rightarrow \mu_x\Omega/(4\pi)$ where Ω is the rf-driving frequency. Equation (1) is studied in the "pseudopotential" approximation in which the driving terms that induce a high frequency quiver motion (the *micromotion*) are averaged over using Kapitza's method.[9,10] This yields a time independent Hamiltonian (in cylindrical coordinates and after a further scaling)

$$H^{trap} = \frac{1}{2}(p_\rho^2 + p_z^2) + U^{trap}(\rho, z) \tag{2a}$$

where

$$U^{trap}(\rho, z) = \frac{1}{2}(\rho^2 + \lambda^2 z^2) + \frac{\nu^2}{2\rho^2} + \frac{1}{r} \tag{2b}$$

and $r = \sqrt{\rho^2 + z^2}$, $\lambda = \mu_z/\mu_x$; $\mu_z = \sqrt{2(q^2 - a)}$ is the axial secular frequency, and ν is the scaled z-component of angular momentum (L_z) and is conserved. Baumann and Nonnemacher[10] have shown that the dynamics of two ions in a Penning trap is also governed by eq. (2) except that slightly different meanings are attached to λ and ν. Thus, our results apply equally to the Paul and Penning traps and also to the Combined trap in which both rf and magnetic fields are used.[11]

The generalized van der Waals potential is[6,7]

$$U^{gvdw}(\rho, z) = \frac{\gamma}{2}(x^2 + y^2 + \beta^2 z^2) - \frac{1}{r} \tag{3}$$

where β and γ are dimensionless physical parameters. For $\beta = \sqrt{2}$ we recover the van der Waals potential, for $\beta = 0$ the quadratic Zeeman effect (QZE), and for $\beta = 1$ the spherical QZE. Scaling and transforming (3) to cylindrical coordinates yields the GVDW Hamiltonian

$$H^{gvdw} = \frac{1}{2}(p_\rho^2 + p_z^2) + \frac{1}{2}(\rho^2 + \lambda^2 z^2) + \frac{\nu^2}{2\rho^2} - \frac{1}{r} \tag{4}$$

which differs from H^{trap} only in the sign of the Coulomb term.

The integrability of eq. (4) has been studied both by Lie group and Painlevé analysis.[7] In particular, it has been shown that the equations of motion derived from

(4) in parabolic coordinates with $\nu = 0$ have the Painlevé property for $\lambda = 1, 2$, and $1/2$. Global invariants were obtained for these cases using Noether's theorem, but only for $\nu = 0$.

Equations (2) and (4) can be combined in the form,

$$\mathcal{H} = \frac{1}{2}(p_\rho^2 + p_z^2) + \frac{1}{2}(\rho^2 + \lambda^2 z^2) + \frac{\nu^2}{2\rho^2} + \frac{\sigma}{r} = E \tag{5}$$

where $\sigma = 1$ for the Paul trap and $\sigma = -1$ for the GVDW problem.

The most interesting difference between the two systems is the possibility of a double well in the trap potential whereas the GVDW problem has a single local minimum on the ρ axis for all λ and ν. For the trap it is useful to treat the two cases $\nu = 0$ and $\nu \neq 0$ separately, although each involves a bifurcation that produces a double well in the effective potential $U(\rho, z)$ as λ is varied.

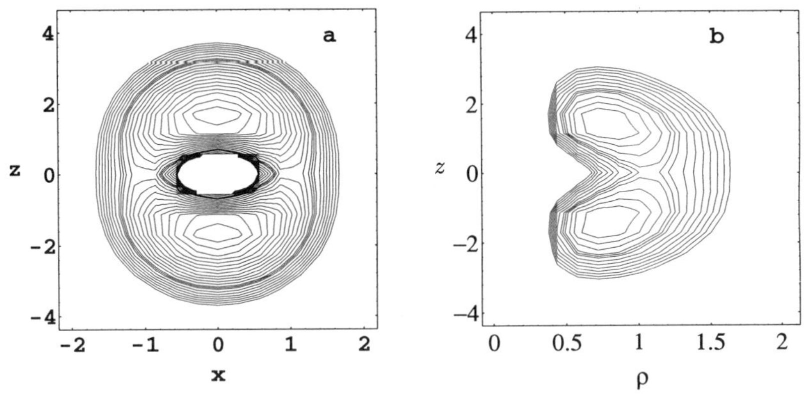

Figure 1. Equipotentials for the trap with $\lambda = 1/2$ and (a) $\nu = 0$, (b) $\nu = 1/2$

When $\nu = 0$ the motion is restricted to a plane and it is preferable to work in Cartesian coordinates (x, z). Thus, for $\lambda > \lambda_c, U(x, z)$ possesses two local minima on the x-axis at $x_0 = \pm 1$ and two saddles on the z-axis at $z_0 = \pm\lambda^{-2/3}$. When λ passes through the critical value $\lambda_c = 1$, the elliptic and hyperbolic points are interchanged through a circle of singularities at unit radius, a process termed a "Birkhoff bifurcation," whch gives rise to the double well lying along the $z-$ axis shown in Fig. 1 (a). For fixed $\nu \neq 0$ and λ greater than a critical value λ_c, given by $1 - \lambda_c^2 = \nu^2 \lambda_c^{8/3}$, $U(\rho, z)$ has a single local minimum on the ρ-axis at ρ_0, given by $\rho_0(\rho_0^3 - 1) = \nu^2$. When λ passes through λ_c this minimum undergoes a pitchfork bifurcation, destabilizing and spawning a pair of off-axis stable minima at $\rho_0 = \sqrt{|\nu|}/(1 - \lambda^2)^{1/4}$, $z_0 = \pm\sqrt{\lambda^{-4/3} - |\nu|(1 - \lambda^2)^{-1/2}}$, as depicted in Fig. 1 (b). Physically, the fixed point at the minimum of either of the two wells corresponds, *(i)* for $|\nu| > 0$, to the ion pair rotating like a dumbell at some fixed angle to the $z-$axis and, *(ii)* for $\nu = 0$ to the ion pair aligning itself either along the $\rho-$axis ($\lambda > \lambda_c$) or along the $z-$axis ($\lambda < \lambda_c$).

INTEGRABLE AND SEPARABLE LIMITS

The first issue is to establish if eq. (5) possesses any separable limits. This can be achieved by considering the potential in eq. (5) to be a superposition of a generalized space oscillator and the Coulomb potential. The Coulomb problem is known to separate in both polar and parabolic coordinates. The oscillator contribution to

$\mathcal{H}$ is clearly separable as written in cylindrical coordinates but may also separate in parabolic or spherical polar coordinates depending on λ. We now consider the separable cases individually.[12,13]

(a) $\lambda = \pm 1$

The Hamiltonian (5) is separable in spherical polar coordinates in this limit. To see this, eq. (5), with $\lambda = \pm 1$, is converted to Cartesian coordinates giving,

$$\mathcal{H} = \frac{1}{2}(p_x^2 + p_y^2 + p_z^2) + \frac{1}{2}(x^2 + y^2 + z^2) + \frac{\sigma}{r} \tag{6}$$

where ν does not now appear explicitly. It is apparent that eq. (6) is separable in spherical polar coordinates (for arbitrary ν) and reduces to a radial equation. The separation constant $\alpha = p_\theta^2 + p_\phi^2 \cot^2\theta$, or equivalently

$$I_1 = (\rho\dot{z} - z\dot{\rho})^2 + \frac{\nu^2 z^2}{\rho^2} \tag{7}$$

is conserved for arbitrary ν. It is worth noting that for $\nu = 0$ integrability coincides with the bifurcation at $\lambda = 1$ that occurs in the effective potential.

(b) $\lambda = \pm 2$

For a 1:2 frequency ratio the transformation to parabolic coordinates,

$$z = (u^2 - v^2)/2, \qquad \rho = uv \tag{8}$$

leads to a complete separation for arbitrary ν. In these coordinates eq. (5) becomes,

$$H = \frac{1}{2(u^2 + v^2)}[p_u^2 + p_v^2 + u^6 + v^6 + 4\sigma + \frac{\nu^2(u^2 + v^2)}{u^2 v^2}] \tag{9}$$

which is clearly separable upon multiplication by $(u^2 + v^2)$. The integral of the motion in this case, I_2, can be obtained as the separation constant of the Hamiltonian in parabolic coordinates. Alternatively, we may use the method of Blümel *et al.*[9] as applied originally to the Paul trap. Using as ansatz the generalized Runge-Lenz vector,

$$\mathbf{I} = \mathbf{p}\mathbf{x}\mathbf{L} - \hat{\mathbf{r}} + \mathbf{C}(\rho, z) \tag{10}$$

where the function $\mathbf{C}(\rho, z)$ is to be determined we find the following global invariant for $\lambda = 2$:

$$I_2 = -\dot{\rho}(\rho\dot{z} - z\dot{\rho}) + \frac{\sigma z}{r} - \rho^2 z + \frac{\nu^2 z}{\rho^2} \tag{11}$$

(c) $\lambda = \pm 1/2$

In the case of a 2:1 ratio of frequencies the system might again be expected to separate using a slightly different set of parabolic coordinates,

$$\rho = (u^2 - v^2)/2, \qquad z = uv, \tag{12}$$

which yields

$$\mathcal{H} = \frac{1}{2(u^2 + v^2)}(p_u^2 + p_v^2 + \frac{u^6}{4} + \frac{v^6}{4} + 4\sigma) + \frac{2\nu^2}{(u^2 - v^2)^2}. \tag{13}$$

However, the term in ν^2 in the Hamiltonian obviously wrecks any chance of separability and the problem thus separates only when $\nu = 0$. Nevertheless, the problem remains integrable for all ν. The issue of integrability of the Paul trap in this limit has recently been the subject of considerable controversy.[10,12–15] An explicit integral of motion, valid for all ν, may be found using the generalized Runge-Lenz approach of Blümel *et al.*[9] and is

$$I_{1/2}^2 = I_\rho^2 + I_\phi^2 + \nu^2 r^2 \tag{14}$$

where

$$I_\rho = \dot{z}(\rho\dot{z} - z\dot{\rho}) + \frac{\sigma\rho}{r} - \frac{1}{4}\rho z^2 + \frac{\nu}{\rho}; \qquad I_\phi = -\frac{\nu}{\rho}(\rho\dot{\rho} + z\dot{z}). \tag{15}$$

This integral was first found for the Paul trap by Blümel *et al.*[9] and analytical and numerical investigations[12–15] have subsequently verified the correctness of their original claim that the limit $\lambda = \pm 1/2$ is integrable for arbitrary ν.

COMPARATIVE DYNAMICS

Figures 2 and 3 show Poincaré sections for the trap and GVDW problems respectively and several values of λ, including the integrable limit $\lambda = 1/2$. Contrast the existence of the double well in the trap when $\lambda < \lambda_c$ with the GVDW problem.

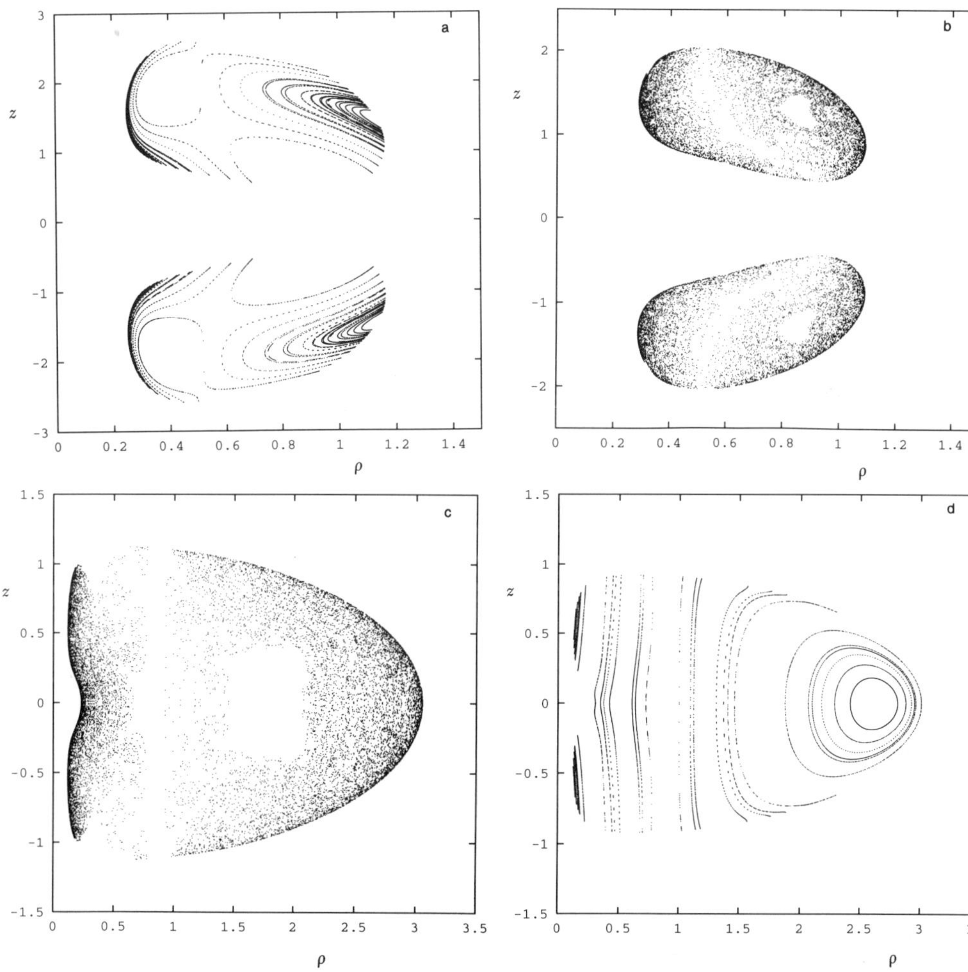

Figure 2. Poincaré surfaces of section ($p_\rho = 0, p_z > 0$) for the trap with $\nu = 0.25$, and, (a) $E = 1.475, \lambda = 0.5$, (b) $E = 1.475, \lambda = 0.6$, (c) $E = 5.0, \lambda = 2.5$, (d) $E = 5.0, \lambda = 3.0$

Note, however, that a dynamical double well seems to exist in the GVDW problem in Fig. 3 (a), corresponding to localization in the positive and negative z directions. This is similar to the situation in the QZE. Depending on energy and angular momentum, the actual dynamics in the trap, below the double well separatrix, may be regular or chaotic. The situation in Fig. 2 (b) falls between the strictly crystalline and cloud (melted crystal) phases in that the individual ions execute chaotic motion while the ion pair itself remains localized in one of the wells. Despite the strongly chaotic nature of the sections at $\lambda = 2.5$, for $\lambda = 3$ in the trap, the motion is startlingly regular although this does not correspond to an integrable limit. This behaviour is apparently due to the existence of an adiabatic invariant. For larger values of λ the GVDW system becomes progressively more chaotic while the trap becomes steadily less chaotic. Thus, while the trap and the GVDW share the same integrable limits, they do not necessarily share similar adiabatic invariants.

DOUBLE WELL DYNAMICS OF THE TRAP

Consideration of the Mathieu stability diagram of the Paul trap reveals that the

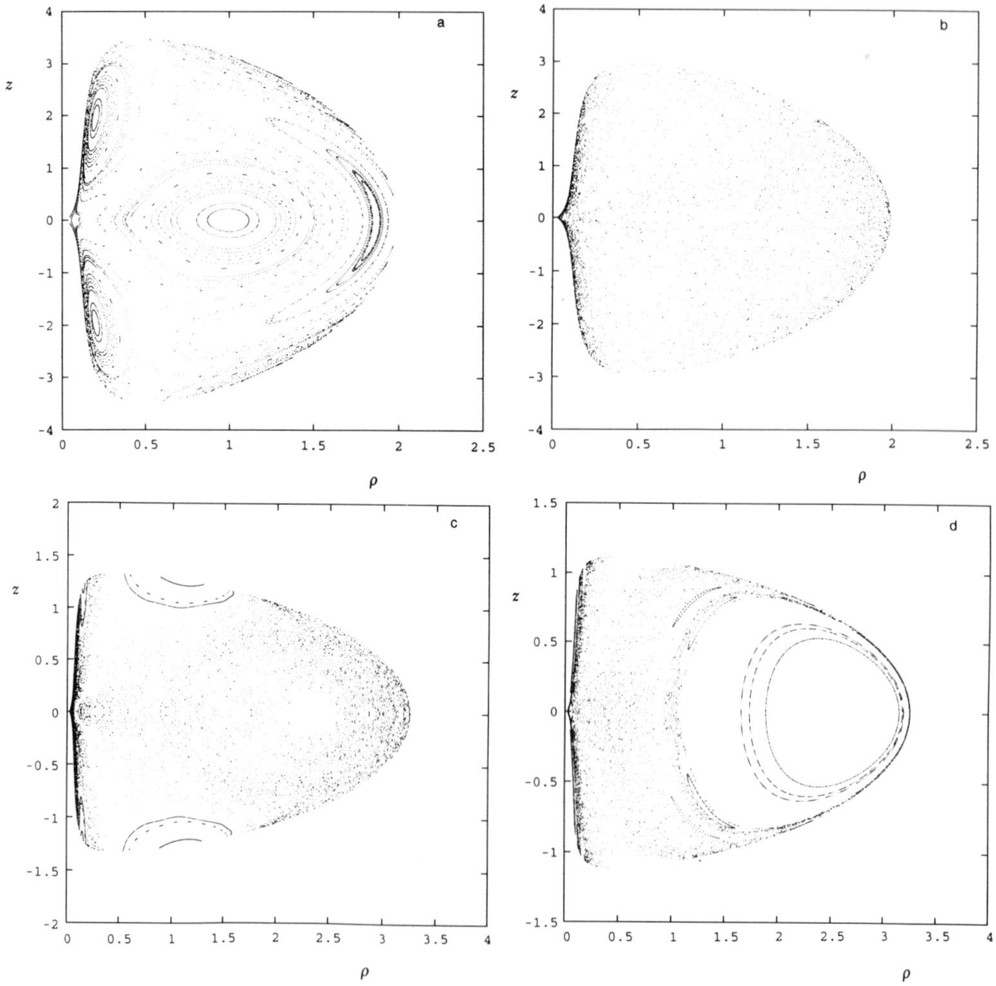

Figure 3. Poincaré surfaces of section ($p_\rho = 0, p_z > 0$) for the GVDW potential with $\nu = 0.25$, and, (a) $E = 1.475, \lambda = 0.5$, (b) $E = 1.475, \lambda = 0.6$, (c) $E = 5.0, \lambda = 2.5$, (d) $E = 5.0, \lambda = 3.0$

double well exists over a significant range of a and q values for which the trap can actually confine ions. For a typical ion pair the effective $\hbar$ is rather small and thousands of vibrational states may be supported by each well, depending on the particulars of the experiment. Thus it should be possible to trap an ion pair below the separatrix of the double well in the small $|\nu|$ regime [see Fig. 1 (b)]. Experimental detection of the double well might be achieved by observing Doppler shifted sidebands in fluorescence spectra or by exciting the vibrational modes of the ion pair.[9,11,13] We focus on the latter and have calculated the two libration frequencies $\Omega_{\pm}(\lambda, \nu)$ associated with the minima typified in Fig. 1 (b). These modes, for $|\nu| > 0$, correspond to librations about circular orbits (steady motion) rather than about static equilibria. In general, for small $|\nu|$ the frequencies undergo an avoided crossing at λ_x, and for $\lambda < \lambda_x$ the normal modes $\Omega_{+(-)}$ lie roughly along the axial (radial) directions; for $\lambda > \lambda_x$ this labelling is reversed. All of this ignores the micromotion which can be substantial in practice. However, integration of the full dynamics[14] generated by eq. (1) indicates that the double well may survive inclusion of the micromotion: four principal types of behaviour occur, assuming the system is started in one of the two wells; *(i)* remaining in the initial well, executing regular or chaotic motion, *(ii)* quasiperiodic or chaotic tumbling between the wells, *(iii)* escape from the initial well, but remaining in the trap while executing chaotic motion, and *(iv)* escape from the well and subsequently the trap due to strong rf heating. Localization and chaotic tumbling are shown in Fig. 4.

Two critical issues remain; the first concerns experimental preparation of the double well, particularly for $|\nu| > 0$. Wineland, *et al.*[16] prepared a "pseudomolecule" consisting of a pair of Hg^+ ions aligned along the z-axis corresponding to the situation (although this was not explicitly pointed out) in Fig. 1 (a). Thus, by radiation pressure it is possible to cool an ion pair into the double well with $\nu = 0$. In order to excite the system into the double well of Fig. 1 (b) we propose use of a Combined trap and the following strategy: initially cool the ion pair into one of the $\nu = 0$ wells, then excite the radial mode using a small ac field while gradually (adiabatically) turning on an axial magnetic field in order to impart angular momentum about the $z-$axis to the ion pair. Subsequently turn off the ac and magnetic fields adiabatically, leaving the ion pair in the well corresponding to $|\nu| > 0$. Trajectory simulations of this process suggest that it might be feasible experimentally.

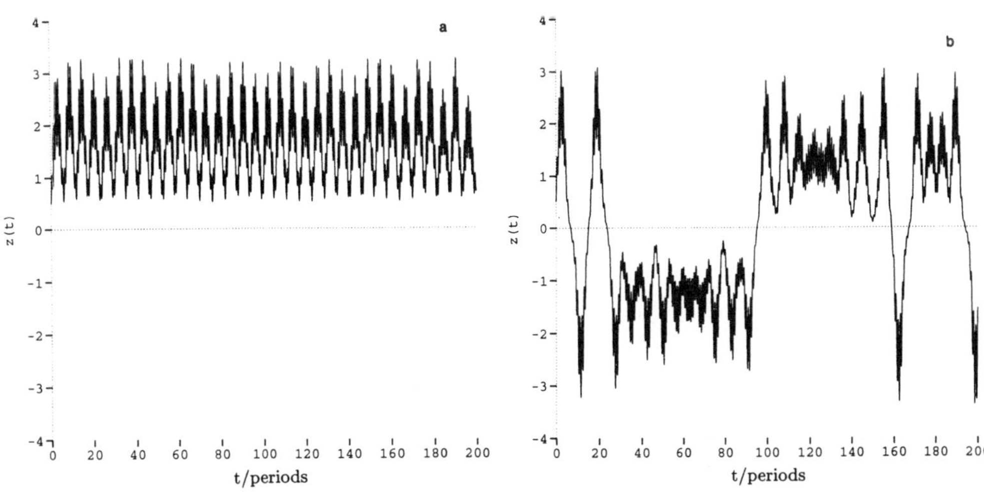

Figure 4. Time histories of z showing, (a) localized motion and, (b) chaotic tumbling between the wells. Generated by integration of eq. (1); $a = 0.07, q = 0.3$

A last issue concerns the relation of these results to the quantum behaviour of the ion pair. Previous quantum simulations of trap dynamics[17] have focussed mainly on single ion motion, for which the time evolution is essentially harmonic. An intriguing possibility is that rf driving in the Paul trap might significantly enhance or modify quantum transport in the chaotic separatrix layer, as has been shown in simulations of a model driven double well.[18] Preparation of the double well in the Paul trap might therefore provide an accessible laboratory in which to engage numerous pressing issues in the classical and quantum behaviour of non-linear dynamical systems.

ACKNOWLEDGEMENTS

Support of this research by NSF and the American Chemical Society and useful conversations with H. Walther, E. Peik and D. J. Wineland are gratefully acknowledged. One of us (TU) is grateful to the NATO Special Program Panel on Chaos, Order, and Patterns for a Research, Study, and Lecture Visit Award.

REFERENCES

1. J.E. Bayfield and P.M. Koch, *Phys. Rev. Lett.*, 33:258 (1974).
2. M.L. Zimerman, M.L. Kash and D. D. Kleppner, *Phys. Rev. Lett.*, 45: 1092 (1980).
3. A. Holle, J. Main, G. Wiebusch, H. Rottke and K.H. Welge, *Phys. Rev. Lett.*, 61:161 (1987).
4. J.C. Gay, in: "*Atoms in Unusual Situations*," J.P. Briand, ed., Proceedings of the NATO Advanced Study Institute, Series B, Physics, Plenum, New York, (1986); 143: 1073.
5. T. Uzer, D. Farrelly, J.A. Milligan, P.E. Raines, and J.P. Skelton, *Science*, 242:41 (1991).
6. Y. Alhassid, E. A. Hinds and D. Meschede, *Phys. Rev. Lett.*, 59:1545 (1987).
7. K. Ganesan and M. Lakshmanan, *Phys. Rev. Lett.*, 62 (1989) 232; *Phys. Rev. A*, 42:3940 (1990); *Phys. Rev. A*, 45:1548 (1992).
8. W. Paul, *Rev. Mod. Phys.*, 62:531 (1990).
9. R. Blümel, C. Kappler, W. Quint, and H. Walther, *Phys. Rev. A*, 40:808 (1989). Erratum: *ibid.*, 46:8034 (1993).
10. G. Baumann and T. F. Nonnenmacher, *Phys. Rev. A*, 46: 2682 (1992).
11. R. Blatt, P. Gill, and R.C. Thompson, *J. Mod. Opt.*, 39:193 (1992): D.J. Bate, K. Dholakia, R.C. Thompson and D.C. Wilson, *J. Mod. Opt.*, 39:305 (1992).
12. D. Farrelly and J.E. Howard, *Phys. Rev. A*, (in press);
13. J.E. Howard and D. Farrelly, *Phys. Lett. A*, (in press);
14. D. Farrelly and J.E. Howard, *Phys. Rev. Lett.*, (submitted):
15. R. Blümel, *Phys. Rev. A*, (in press); M. Moore and R. Blümel, *Phys. Rev. A*, (preprint): R. Blümel, *Phys. Lett. A*, 174:174 (1993)
16. D.J. Wineland, J.C. Bergquist, W.M. Itano, J.J. Bollinger, and C.H. Manney, *Phys. Rev. Lett.*, 59:2935 (1987).
17. L.S. Brown, *Phys. Rev. Lett.*, 66:527 (1991): S. Stenholm, *J. Mod. Opt.*, 39:279 (1992).
18. W.A. Lin and L.E. Ballentine, *Phys. Rev. Lett.*, 65:2927 (1990): D. Farrelly and J. A. Milligan, *Phys. Rev. E* (in press).

RECENT APPLICATIONS OF HAMILTONIAN DYNAMICS TO ACCELERATOR PHYSICS

Ezio Todesco

Department of Physics, University of Bologna
INFN, Sezione di Bologna
Via Irnerio 46, 40126, Bologna, Italy

INTRODUCTION

The use of superconducting magnets in large hadron accelerators has raised new interest in the nonlinear effects in accelerator physics (Scandale and Turchetti, 1990): indeed, both analytical tools and reliable numerical methods are required to design future accelerators. The standard numerical approach is based on the concept of symplectic transfer maps (Iselin and Niederer, 1988; Schmidt, 1990); the corresponding analytical tool is the theory of normal forms (Turchetti, 1988; Bazzani et al., 1993a), which is the natural generalization of canonical perturbation theory for flows to transfer maps. Normal forms for symplectic maps have the big advantage of being easily implementable in computer codes (Servizi and Turchetti, 1984), allowing the automatic computation of high perturbative orders; moreover, even if the series are generically divergent such as in the hamiltonian case, a detailed analysis of the mechanism of divergence was carried out, allowing to use the approximation provided by truncated normal forms in judiciously chosen domains (Bazzani et al., 1993a and 1993b).

In this paper we outline the normal form theory for magnetic lattices, pointing out applications such as the computation of the quantities which characterize the nonlinear motion and the analytical optimization of the lattice parameters. We also review a method which allows to estimate the stability domain of a 2D symplectic map, based on the computation of the invariant manifolds emanating from the unstable fixed point of the map (Giovannozzi, 1992).

The plan of the paper is the following: we first introduce the map formalism for magnetic lattices, pointing out the problems connected with the symplectic truncations and tracking. We then give the definition of normal forms and show how to

Hamiltonian Mechanics, Edited by J. Seimenis,
Plenum Press, New York, 1994

construct perturbative expansions of the most relevant dynamical quantities. In the third section the technique of analytical optimization of the lattice is outlined; finally, in the last section we review a method for the computation of the stability boundary of a 2D map.

MAP FORMALISM FOR MAGNETIC LATTICES

We consider the motion of a single particle in a circular magnetic lattice of L magnetic elements $\mathcal{M}^{(1)},\ldots,\mathcal{M}^{(L)}$. We denote with x, y the horizontal and vertical axes perpendicular to the orbit, and with s the curvilinear coordinate. Neglecting the coupling with the longitudinal motion we analyse the dynamics in the transverse plane (x, y). The conjugate momenta are the dimensionless quantities $p_x \equiv dx/ds, p_y \equiv dy/ds$ and the motion takes place in a four dimensional phase space, whose points are denoted by $\mathbf{x} = (x, p_x, y, p_y)$.

Under various approximations (Steffen, 1985; Ruth, 1985), the motion equations for the betatronic oscillations of a single particle read

$$\begin{cases} \dfrac{d^2x}{ds^2} + \left(\dfrac{1}{\rho(s)^2} - k_1(s)\right) x = \mathrm{Re}\left[\displaystyle\sum_{n=2}^{M} \frac{k_n(s) + ij_n(s)}{n!}(x+iy)^n\right] \\ \\ \dfrac{d^2y}{ds^2} + k_1(s)y = -\mathrm{Im}\left[\displaystyle\sum_{n=2}^{M} \frac{k_n(s) + ij_n(s)}{n!}(x+iy)^n\right], \end{cases} \tag{1}$$

where $k_n(s)$ and $j_n(s)$ are respectively the normal and skew gradients, proportional to the n-th derivatives of the magnetic field $\mathbf{B}$ of the lattice elements, and $\rho(s)$ is the curvature radius in the dipoles.

Since the gradients are well approximated by functions which are constant on a single magnet, it is convenient to introduce the transfer map $\mathbf{M}^{(l)}$ of the magnetic element $\mathcal{M}^{(l)}$. Being s_{l-1} and s_l the extremes of $\mathcal{M}^{(l)}$, we denote by $\mathbf{M}^{(l)}$ the function which transforms the phase space coordinates $\mathbf{x}(s_{l-1})$ to $\mathbf{x}(s_l)$ according to eq. (1):

$$\mathbf{x}(s_l) = \mathbf{M}^{(l)}(\mathbf{x}(s_{l-1})) \qquad \mathbf{M}^{(l)} : \mathbf{R}^4 \to \mathbf{R}^4. \tag{2}$$

The transverse motion equations (1) can be derived by a time–dependent (i.e. s–dependent) two dimensional hamiltonian; due to the conservative character of the flow, $\mathbf{M}^{(l)}$ is a symplectic map, i.e. satisfies the conditions on the Poisson brackets of the four components $M_i^{(l)}$, $i = 1, 2, 3, 4$, of the map

$$\{M_i^{(l)}, M_j^{(l)}\} = 0 \quad \text{except} \quad \{M_{2k}^{(l)}, M_{2k+1}^{(l)}\} = -\{M_{2k+1}^{(l)}, M_{2k}^{(l)}\} = 1 \quad k = 1, 2. \tag{3}$$

In the two dimensional case one can check that the symplectic conditions are equivalent to the area–preserving constraint.

The one turn map $\mathbf{M}$, called the Poincaré map at section $s = s_0$, is given by the composition of the single element maps: $\mathbf{M}$ transforms the phase space coordinates $\mathbf{x}(s_0)$ of a single particle into its coordinates $\mathbf{x}(s_L)$ after one full turn according to the motion equations (1). Using the standard notation of mapping theory, $\mathbf{x}' \equiv \mathbf{x}(s_L)$ and $\mathbf{x} \equiv \mathbf{x}(s_0)$, one can write

$$\mathbf{x}' = \mathbf{M}(\mathbf{x}) \equiv \mathbf{M}^{(L)}(\mathbf{M}^{(L-1)}(\ldots\mathbf{M}^{(2)}(\mathbf{M}^{(1)}(\mathbf{x}))\ldots)). \tag{4}$$

The evaluation of the iterates of the map $\mathbf{M}$ which give the orbit of the single particle in the transverse phase space can be done in different ways. The standard numerical approach is based on tracking (Iselin and Niederer, 1988; Schmidt, 1990): one computes $\mathbf{M}^{(l)}$ using methods which provide an exactly symplectic mapping (such as the thin lens approximation, see Bazzani et al., 1988 and 1993a), and successively applies all the transfer maps to evaluate the orbit. The final data relative to the trajectory in phase space must be post–processed in order to extract the quantities which are relevant to the nonlinear motion; this method, even if has the advantage of conceptually being very simple and easy to implement, can be very time consuming and does not provide any theoretical insight.

The direct computation of the one turn map can be done using polynomial composers: one numerically evaluates the coefficients of the element map $\mathbf{M}^{(l)}$, and then composes the L polynomial symplectic maps $\mathbf{M}^{(1)}, ..., \mathbf{M}^{(L)}$. In general the exact representation of $\mathbf{M}$ is not possible since it is a polynomial whose degree is the product of the degrees of the transfer maps of the individual elements of the lattice. The only exact (up to computer round off errors) computation of $\mathbf{M}$ is achieved by tracking. Indeed, one can easily compute the truncated map by composing and truncating at order N in the $\mathbf{x}$ coordinates the single element maps $\mathbf{M}^{(l)}$. The final result is the numerical value of the coefficients of the polynomial expansion of $\mathbf{M}$ up to the order N. Using the available computer codes (Servizi and Turchetti, 1984; Berz, 1990; Bazzani et al., 1991) one can reach sufficiently high orders (i.e. $N \approx 15 - 20$) for the lattice of a large machine; the constraints are both in the computation time and in the memory required to store all the map coefficients, which for a four dimensional map truncated at order N are $(N+3)(N+2)(N+1)N/6$.

Let us observe that the truncated map is symplectic only up to order N, i.e. it satisfies Poisson brackets only up to terms of order $|\mathbf{x}|^N$. As a consequence, the truncated map cannot be used for tracking, since deviations from symplectic conditions provoke a change in the topology of the orbits, which are spiraling inward or outward for a sufficiently large number of iterations.

Linearizing eq. (1) one obtains the well–known Hill equations (Steffen, 1985), whose solutions are oscillators with both frequencies and amplitudes depending on the coordinate s. The corresponding linear transfer map $\mathbf{L}$ (i.e. the linearization of $\mathbf{M}$ around the origin which is a fixed point) is called the Twiss matrix, and can be diagonalized using a linear transformation $\mathbf{V}$ (Courant–Snyder transformation, see Courant and Snyder, 1958) to the matrix $\Lambda = \mathrm{Diag}(e^{i\omega_1}, e^{-i\omega_1}, e^{i\omega_2}, e^{-i\omega_2})$:

$$\Lambda = \mathbf{V}^{-1}\mathbf{L}\mathbf{V}. \tag{5}$$

The new coordinates $\mathbf{z} \equiv \mathbf{V}^{-1}\mathbf{x}$ are called the complex Courant–Snyder coordinates, and explicitly read $\mathbf{z} = (z_1, z_1^*, z_2, z_2^*)$, where the star denotes the complex conjugate. Since in the $\mathbf{z}$ plane the motion is a direct product of rotations, the linear invariants are $z_1 z_1^*$ and $z_2 z_2^*$.

The truncated nonlinear map $\mathbf{M}$ is transformed through $\mathbf{V}$ to the map $\mathbf{F}$

$$\mathbf{F}(\mathbf{z}) = \mathbf{V}^{-1}\mathbf{M}(\mathbf{V}\mathbf{z}). \tag{6}$$

The linear part of $\mathbf{F} \equiv (F_1, F_1^*, F_2, F_2^*)$ is diagonal and the map truncated at order N explicitly reads

$$\begin{cases} z_1' = e^{i\omega_1}\, z_1 + \sum_{n=2}^{N} [F_1]_n(\mathbf{z}) & z_1^{*\prime} = e^{-i\omega_1}\, z_1^* + \sum_{n=2}^{N} [F_1^*]_n(\mathbf{z}) \\ \\ z_2' = e^{i\omega_2}\, z_2 + \sum_{n=2}^{N} [F_2]_n(\mathbf{z}) & z_2^{*\prime} = e^{-i\omega_2}\, z_2^* + \sum_{n=2}^{N} [F_2^*]_n(\mathbf{z}), \end{cases} \tag{7}$$

where $[F_i]_n$ denote homogeneous polynomials of order n in the variables z_1, z_1^*, z_2, z_2^*; since the map $\mathbf{M}$ is real, the second couple of equations for the components F_1^*, F_2^* are complex conjugated of the first couple and therefore can be neglected.

NORMAL FORM APPROACH

The normal form approach (Turchetti, 1988; Scandale et al. 1992; Bazzani et al., 1988 and 1993b) is the natural generalization of the Courant–Snyder transformation to the nonlinear case: given a 4D map $\mathbf{F}$, with a fixed point in the origin, one looks for a nonlinear change of coordinates $\boldsymbol{\Phi}$ such that the new map, we denote by $\mathbf{U}$, assumes a particularly simple form, i.e. has explicit invariants, iteration, interpolating flows, and has more symmetries than the original map $\mathbf{F}$. The map $\mathbf{U}$ is called normal form; the conjugation equation of a given map with its normal form is formally equal to the similarity transformation (5) which diagonalizes the linear motion

$$\mathbf{U}(\boldsymbol{\zeta}) = \boldsymbol{\Phi}^{-1} \circ \mathbf{F} \circ \boldsymbol{\Phi}(\boldsymbol{\zeta}) \equiv \boldsymbol{\Phi}^{-1}(\mathbf{F}(\boldsymbol{\Phi}(\boldsymbol{\zeta}))), \tag{8}$$

the only difference being that now $\mathbf{F}$, $\mathbf{U}$ and $\boldsymbol{\Phi}$ are not matrices, but nonlinear maps and therefore the multiplication of matrices is replaced by the composition of nonlinear maps denoted by $\circ$. The new coordinates are $\boldsymbol{\zeta} \equiv (\zeta_1, \zeta_1^*, \zeta_2, \zeta_2^*)$; $\mathbf{U}$ is invariant under a group of transformations: this symmetry condition implies that its polynomial expansion only contains some monomials. According to the symmetry groups one can classify the different types of normal forms; it is possible to prove that one can always formally build a normal form $\mathbf{U}$ (i.e. compute the coefficients without caring of the convergence of the series) with respect to the symmetry group generated by the linear part of the map $\boldsymbol{\Lambda}$, or with respect to subgroups of it.

Analytic solutions of the functional equation do not in general exist in open neighbourhoods of the origin: therefore the series $\boldsymbol{\Phi}$ and $\mathbf{U}$ are divergent in the generic case (Benettin et al., 1983; Bazzani et al., 1993b). Nevertheless truncation of the series provide an accurate approximation of the original map $\mathbf{F}$ on a given phase space domain; when the series are truncated at a finite order N then the conjugation of $\mathbf{F}$ with the normal form is no longer exact, being affected by an error term: one has to choose N such that the error term is minimized over the phase space domain of interest.

The symplectic condition on a truncated map is satisfied only up to order N: therefore also the truncated normal form $\mathbf{U}$ is not exactly symplectic: indeed, the only way to investigate the asymptotic dynamics is the construction of the invariants of the map (Turchetti, 1988; Bazzani et al., 1993b), namely functions $I(\boldsymbol{\zeta})$ which are invariant under the action of $\mathbf{U}$; I foliates the space into surfaces (curves for the 2D map) to which the orbits belong, and provides the relevant geometric information. The best strategy consists in choosing among all the invariants I an hamiltonian function H whose orbits interpolate the orbits of $\mathbf{U}$ (namely agree with the orbit of the map at integer times). Truncated invariants are still exactly symplectic and provide approximated interpolation of the orbits.

For sake of simplicity we only consider maps whose eigenvalues are not resonant, i.e. $\omega_1/2\pi$ and $\omega_2/2\pi$ irrational without common divisors. In this case one can build at all perturbative orders the nonresonant normal form $\mathbf{U}$, which is invariant under the group of continuous rotations and therefore reads

$$\begin{cases} U_1(\boldsymbol{\zeta}) = e^{i\Omega_1(\zeta_1\zeta_1^*, \zeta_2\zeta_2^*)}\zeta_1 \\ U_2(\boldsymbol{\zeta}) = e^{i\Omega_2(\zeta_1\zeta_1^*, \zeta_2\zeta_2^*)}\zeta_2, \end{cases} \tag{9}$$

i.e. is an amplitude dependent rotation (Ω_1 and Ω_2 are real functions of $|\zeta_1|^2$ and $|\zeta_2|^2$). This map has explicit invariants: $\rho_1 = \zeta_1\zeta_1^*$ and $\rho_2 = \zeta_2\zeta_2^*$; given ρ_1 and ρ_2 the motion takes place on a two dimensional torus with frequencies $\Omega_1(\rho_1, \rho_2)$ and $\Omega_2(\rho_1, \rho_2)$. Due to the symplectic character of $\mathbf{U}$, the nonlinear frequencies can be expressed in terms of the interpolating hamiltonian $H(\rho_1, \rho_2)$:

$$\Omega_i(\rho_1, \rho_2) = \frac{\partial H}{\partial \rho_i}(\rho_1, \rho_2) \qquad H(\rho_1, \rho_2) = \omega_1\rho_1 + \omega_2\rho_2 + \sum_{k_1+k_2 \geq 2} h_{k_1,k_2}\rho_1^{k_1}\rho_2^{k_2}; \quad (10)$$

the shift in the frequency due to the nonlinear terms is called tune shift, and is a relevant quantity of the nonlinear motion: in fact the tune shift can drive the original nonresonant linear frequencies ω_1, ω_2 on resonant values which endanger the stability of the motion.

Using the inverse conjugating function $\boldsymbol{\Phi}^{-1}$ one can compute both the invariants and the nonlinear tunes of the map $\mathbf{F}$. The function $\boldsymbol{\Phi}$ also allows to compute the 2D surfaces in the space $\mathbf{z}$ where the motion takes place for fixed values of the invariants, and their deformation with respect to the 2D tori.

When the linear frequencies ω_1, ω_2 are close to a resonant condition it is well known that the topology of the orbits changes: close invariant curves are broken into islands (Poincarè–Birkhoff theorem, see Arnold and Avez, 1968), and chaos may appear close to the hyperbolic fixed points. Indeed one can build a normal form which includes chains of islands of a single resonance and all its harmonics (Bazzani et al., 1993b): in this case $\mathbf{U}$ is invariant under a subgroup of discrete rotations. Let us consider for simplicity the 2D case; let ω be a quasiresonant frequency, i.e. $\omega = 2\pi p/q + \epsilon$ with $p, q \in \mathbf{Z}$ and $\epsilon \ll 1$. Then one can conjugate F to a resonant normal form U which has the polynomial structure

$$U(\zeta, \zeta^*) = \sum_{n=0}^{\infty}(\zeta\zeta^*)^n \left[\sum_{l=0}^{\infty} u_{n,l}^{(+)}\zeta^{ql+1} + \sum_{l=0}^{\infty} u_{n,l}^{(-)}\zeta^{*ql-1}\right], \quad (11)$$

where $u_{n,l}^{(+)}$ and $u_{n,l}^{(-)}$ are complex coefficients; U commutes with the finite subgroup generated by rotations by an angle $2\pi/q$ and its orbits are symmetric with respect to it, namely they superimpose identically after a rotation of $2\pi/q$. Interpolation of U is not straightforward such as in the nonresonant case, but one can build the hamiltonian H

$$H(\zeta, \zeta^*) = \sum_{n=0}^{\infty}\sum_{l=0}^{\infty} \left[h_{n,l}^{(+)}(\zeta\zeta^*)^n\zeta^{ql} + h_{n,l}^{(-)}(\zeta\zeta^*)^n\zeta^{*ql}\right] \quad (12)$$

(where $h_{n,l}^{(+)}$ and $h_{n,l}^{(-)}$ are complex coefficients), which provides the relevant information on the nonlinear motion, such as the position of the fixed points, the island width, tune shifts and secondary frequencies (Bazzani et al., 1993a and 1993b).

OPTIMIZATION OF THE NONLINEAR PARAMETERS

In the previous section we have outlined how to construct perturbative series of the most relevant quantities which characterize the nonlinear motion such as the invariants and the tune shifts. The analytic construction of normal forms allows to compute the dependence of the coefficients of those perturbative expansions on the lattice parameters such as the multipolar gradients k_n, j_n (see (1)). This feature allows to carry out effective optimizations of the lattice by analytic tools.

First applications were carried out for a model of the planned Large Hadron Collider in CERN (Scandale, 1989); here one has to correct the nonlinearities of the superconducting dipoles by inserting some magnetic elements (correctors) in order to stabilize the motion. The proposed strategy was based on the minimization of the first two perturbative orders of the tune shift (Scandale et al., 1992). The advantage of this strategy compared to the numerical approach, based on the minimization of the tune shift evaluated through tracking in a few points of the phase space, is relevant: in the first case one can in general analytically solve the systems which give the values of the correctors, and therefore, contrary to the numerical approach, one is sure to compute the global minimum. Moreover the method attacks the tune shift all over the phase space where the normal form approximation is valid, and not only in a few points; finally, the analytic method is much faster, and therefore allows a systematic analysis of the correction schemes.

A generalization of the above outlined procedure which includes also the effect of off-momentum particles (i.e. particles which do not have the nominal energy of the machine) was carried out by Bazzani et al. (1993a) and applied to the LHC lattice by Giovannozzi et al. (1993); computer codes which automatically calculate the best set of correctors are available. Such methods can be generalized to other cases, i.e. the minimization of other quantities which characterize the nonlinear motion such as the deformation of the orbits or the resonance parameters which can be analytically evaluated through the normal form approach.

Since the final aim of the optimization of the lattice is to increase the stability domain in the phase space where one can safely operate with the beam, one has always to check the effectiveness of the correction by evaluating the border of stability; since no analytical tools are available, this estimate is usually carried out through long term tracking. Simulations on the LHC lattice showed that the minimization of the tune shift allows to considerably increase the stability domain of the machine, especially when one considers more realistic models which include other sources of nonlinearities such as random errors in the magnets, off momentum effects and radiofrequency.

DYNAMIC APERTURE IN THE 2D CASE

Since the normal form method is divergent, no conclusion can be drawn for the estimate of the dynamic aperture, i.e. the minimum distance of the stability boundary to the origin. Indeed this quantity is well defined only in the two dimensional case, where the invariant curves are a topological barrier to the motion, and therefore the phase space of the map inside the last invariant curve is stable. On the contrary, in the 4D case a particle whose initial condition is arbitrarily close to the origin can escape to infinity through a net of resonances (Arnold diffusion). Therefore a rigorous definition of the stability domain can only be given for large but finite times, and relative to a given percentage of the set of initial particles.

We consider the simplest nontrivial nonlinear 2D mapping first introduced by Hénon (1969):

$$z' = e^{i\omega}\left[z - \frac{i}{4}(z + z^*)^2\right] \qquad z \in \mathbf{C}, \tag{13}$$

which corresponds to the map of a linear lattice only having a quadratic nonlinearity (i.e. $k_2 \neq 0$) in the one kick approximation; the analytic evaluation of the dynamic aperture of this simple model is a hard problem. In the limit $\omega \to 0$ one can prove

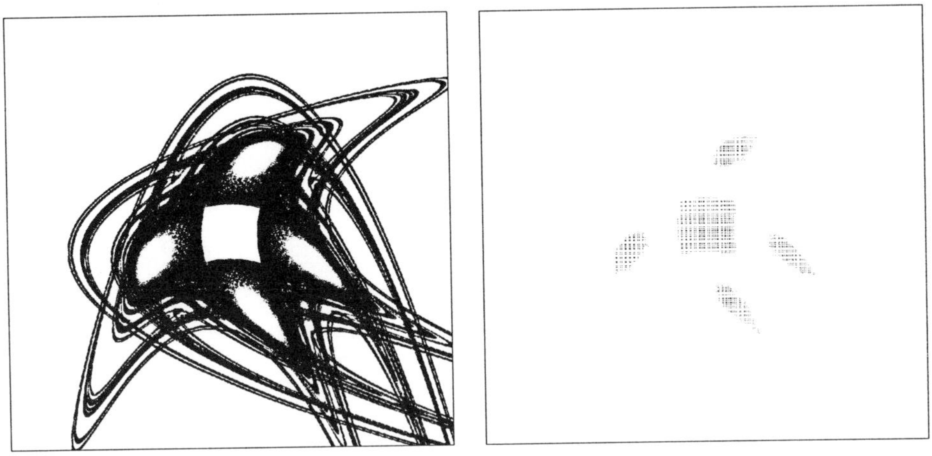

Figure 1. Invariant manifolds of the unstable fixed point of the Hénon map for $\omega/2\pi$=0.255 (left) and basin of stability (right).

that the stability boundary is given by the invariant manifolds emanating from the unstable fixed point of the map, which can be computed as the nontrivial solution of the system $z_0' = z_0$ and explicitly reads

$$z_0 = 2\tan\frac{\omega}{2} + 2i\tan^2\frac{\omega}{2}; \tag{14}$$

indeed for a generic ω there is a strong numerical evidence (Giovannozzi, 1992) that also in this case the stability boundary is determined by the stable and unstable manifolds of the unstable fixed point of the map. Such manifolds can be computed using a numerical method, iterating a given number of initial conditions which lie on the linear manifolds in the neighbourhood of the hyperbolic fixed point. In fig. 1 we draw the stable and unstable manifolds with the related homoclinic tangle for ω close to resonance four; the comparison with the basin of stability shows an excellent agreement, since the invariant manifolds of F enter the chaotic sea and reach the border of stability. A numerical experiment carried out for a wide set of frequencies show that this mechanism allows to reproduce the dynamic aperture of the Hénon map in its dependence on ω. It has also been shown (Giovannozzi, 1992) that the same mechanism works for a generic two dimensional map; a generalization to the higher dimensional case is not trivial.

ACKNOWLEDGEMENTS

The outlined results were obtained by the research group of the University of Bologna directed by Prof. G. Turchetti; major contributions were given by A. Bazzani, M. Giovannozzi and G. Servizi. Special thanks to W. Scandale of the CERN SL division for providing motivations for part of this work and many clarifying discussions.

REFERENCES

Arnold, V. I., and Avez, A., 1968, "Ergodic problems of classical mechanics," V. A. Benjamin, New York.

Bazzani, A., Mazzanti, P., Servizi, G., and Turchetti, G., 1988, Normal forms for hamiltonian maps and non linear effects in a particle accelerator, *Nuovo Cim.* 102B:51–80.

Bazzani, A., and Turchetti, G., 1991, Polyn: a program for algorithmic manipulations of polynomials, *CERN SL report* 91-13 AP.

Bazzani, A., Servizi, G., Todesco, E., and Turchetti, G., 1993a, A normal form approach to the theory of nonlinear betatronic motion, submitted as CERN yellow report.

Bazzani, A., Giovannozzi, M., Servizi, G., Todesco, E., and Turchetti, G., 1993b, Resonant normal forms, interpolating hamiltonians and stability of area preserving maps, *Physica D* 64:66–93.

Benettin, G., Giorgilli, A., Servizi, G., and Turchetti, G., 1983, Resonances and asymptotic behaviour of Birkhoff series, *Phys. Lett.* 95A:11–14.

Berz, M., 1990, Computational aspects of design and simulations: COSY INFINITY, *Nucl. Instrum. and Methods Phys. Res., A* 298:473–9.

Courant, E. and Snyder, H., 1958, Theory of the alternating-gradient synchrotron, *Ann. Phys. (N.Y.)* 3:1–48.

Giovannozzi, M., 1992, Analysis of the Stability Domain for the Hénon Map, *CERN SL report* 92-23 AP.

Giovannozzi, M., and Schmidt, F., 1993, General normal form procedure to correct tune-shift and non-linear chromaticity for large accelerators like the LHC, *in:* "1993 Particle Accelerator Conference", to be published.

Hénon, M., 1969, Numerical study of quadratic area preserving mappings, *Q. Appl. Math.* 27:291–312.

Iselin, F.C., and Niederer, J., 1988, The MAD program 7.2 User reference manual, *CERN LEP report* 88-38 TH.

Ruth, R.D., 1985, Single particles dynamics and nonlinear resonances in circular accelerators, *Lectures notes in physics* 247:37–63.

Scandale, W., 1989, Status report on the design of the LHC lattice, *CERN LHC Note* 68.

Scandale, W., and Turchetti, G., 1990, "Nonlinear problems in future particle accelerators," World Scientific, Singapore.

Scandale, W., Schmidt, F., and Todesco, E., 1992, Compensation of the tuneshift in the LHC, using normal form techniques, *Part. Accel.* 35:53–88.

Schmidt, F., 1990, SIXTRACK: Single particle tracking code treating transverse motion with synchrotron oscillations in a symplectic manner, *CERN SL report* 90-11 AP.

Servizi, G., and Turchetti, G., 1984, A computer program for the Birkhoff series for area preserving maps, *Comput. Phys. Commun.* 32:201–7.

Steffen, K., 1985, Basic course in accelerator optics, *CERN* 85-19:25–63.

Turchetti, G., 1988, Perturbative methods for hamiltonian maps, *in*: "Methods and applications of nonlinear dynamics," A. W. Saenz, ed., World Scientific, Singapore.

SINGULARITY ANALYSIS OF 2D COMPLEXIFIED HAMILTONIAN SYSTEMS

S. Abenda, A. Bazzani

Department of Mathematics, University of Bologna
INFN, Sezione di Bologna
P.zza Porta San Donato n.5, 40126, Bologna, Italy

INTRODUCTION

The connection between the integrability of a differential equation and the nature of the singularities in the complex time of the solutions, has been studied by various authors since Painlevè (Bountis et. al. 1991). Recently it has been considered the connection of the integrability criterion of Painlevè and the algebraic integrability of Arnol'd-Liouville (Arnol'd 1978) for the Hamiltonian systems (M.Adler and P. van Moerbeke 1988, P. van Moerbeke 1987); unfortunately the family of algebraic integrable systems is very poor with respect to integrable systems according to Arnol'd-Liouville. On the other hand some non-integrability criteria have been established for the Hamiltonian systems (V.V. Kozlov 1983).
The study of analyticity properties of normalizing transformation for an area preserving map in a neighborhood of an elliptic fixed point has been recently considered (A.Bazzani and G.Turchetti 1992,1993) in order to connect the singularities in the Action variable and the change of topology due to the non-linear resonances in the phase space. In this paper we consider the construction of global Action-Angle variables for a complexified Hamiltonian system with a cubic potential (the case of a quartic potential can also be studied in the same way): the monodromy of the Action is described by means of a subclass of unimodular transformations, which depends on the algebraic multiplicity of the singular energy value (separatrix); on the other side the Angle variable is simply related to an elliptic integral (C.L. Siegel 1971). The analysis of the Action variable can be generalized to the case of higher degree potentials, while the angle variable becomes an hyperelliptic integral (C.L. Siegel 1971).

We consider also the homological equation (V.I. Arnol'd 1983) of the first order perturbation theory and we study a possible mechanism for the rising of a natural boundary in the analyticity domain of the Angle variable.
The purpose of the analysis would be to classify the global analytical structure of Action-Angle variables for 2-D integrable Hamiltonian systems and to understand the relation between the nonintegrability character and analyticity properties, such as the presence of natural boundaries in the Angle, of the transformation in Action-Angle variables.

CONSTRUCTION OF THE ACTION-ANGLE VARIABLES

Let us consider the following Hamiltonian function

$$H(q,p) = \frac{p^2}{2} + \omega\frac{q^2}{2} - \frac{q^3}{3} \qquad (q,p) \in \mathbb{C}^2 \tag{1}$$

with $\omega \neq 0$. The singular values of the energy are $E = 0, \omega^3/6$ and correspond to the separatrices. From the results of the complex function theory, it is known that the surfaces of constant energy $H(q,p) = E$ where E is not a singular value is an algebraic curve of genus $g = 1$ in the compactified phase space $\tilde{\mathbb{C}}^2$ ($\tilde{\mathbb{C}} = \mathbb{C} \cup \{\infty\}$); as a consequence if we perform the change of variables

$$\begin{aligned} q &= 6Q + \frac{\omega}{2} \\ p &= 6P \end{aligned} \tag{2}$$

we reduce the equation $H(q,p) = E$ to the special form

$$P^2 = 4Q^3 - g_2 Q - g_3, \tag{3}$$

where $g_2 = \omega^2/12$ and $g_3 = (\omega^3/6 - 2E)/36$. The requirement that E is different from a singular value implies that the roots z_1, z_2 and z_3 of equation (3) are distinct; in the sequel we will suppose E not singular. If we consider two copies of the Q-sphere and we introduce a cut along a curve from z_1 to z_2 and a curve from z_3 to the ∞ point on each copy, we can construct the Riemann surface associated to the algebraic function $P = P(Q)$, defined by eq. (3), by gluing the two copies along the cuts: the Riemann surface is a torus, whose fundamental cycles are given by a closed path $\gamma_{1,2}$ around the roots z_1 and z_2 and a closed path $\gamma_{2,3}$ around the roots z_2 and z_3. When the energy E is singular, two roots of eq. (3) collapse and the corresponding fundamental cycle disappears so that the topology of the Riemann surface is changed.
Since the Riemann surface (3) for a fixed value of the energy has two fundamental cycles we define two Actions according to

$$\begin{aligned} I_{1,2} &= \frac{1}{2\pi}\oint_{\gamma_{1,2}} p\,dq = \frac{18}{\pi}\oint_{\gamma_{1,2}} P\,dQ = \frac{36}{\pi}\int_{z_1}^{z_2} \sqrt{4Q^3 - g_2 Q - g_3}\,dQ \\ I_{2,3} &= \frac{1}{2\pi}\oint_{\gamma_{2,3}} p\,dq = \frac{18}{\pi}\oint_{\gamma_{2,3}} P\,dQ = \frac{36}{\pi}\int_{z_2}^{z_3} \sqrt{4Q^3 - g_2 Q - g_3}\,dQ. \end{aligned} \tag{4}$$

When we construct the Action variable in the real phase space, we are considering one of the two fundamental cycles of the Riemann surface.

The Action I is a function of the energy E that corresponds to the Riemann surface and we require $dI/dE \neq 0$ in order to define correctly the transformation in Action-Angle variables. Taking into account the definition of the coefficients g_2 and g_3, from eqs. (4) we obtain

$$\begin{aligned} \frac{dI_{1,2}}{dE} &= \frac{1}{2\pi} \oint_{\gamma_{1,2}} \frac{dQ}{P} = \frac{T_1}{2\pi} \\ \frac{dI_{2,3}}{dE} &= \frac{1}{2\pi} \oint_{\gamma_{2,3}} \frac{dQ}{P} = \frac{T_2}{2\pi} \end{aligned} \tag{5}$$

and the derivative dI/dE is always different from zero if E is not singular: the singular values of the energy are indeed isolated singular points of the function $I = I(E)$ and it is possible to prove the existence of the limit values for the Actions (4) when the energy approach a singular value. The integrals in the r.h.s. of eqs. (5) define the two fundamental periods T_1 and T_2 associated to the Riemann surface. For real values of the energy, one fundamental period is real and the other is purely imaginary if eq. (3) has three real solutions, whereas the fundamental periods are complex conjugated, if eq. (3) has only one real solution. As a consequence in the first case we have one real Action and one imaginary Action, whereas in the second case the Actions are complex conjugated.

In order to study the behaviour of $I(E)$ around the singular points, we consider what happens to the Actions $I_{1,2}$ and $I_{2,3}$ when we make a closed path in the energy complex plane around one of the singular value. Let us suppose that the for $E = 0$ the roots z_1 and z_2 of eq. (3) collapse, then it is easy to realize that the effect of a closed path of E around 0 is to exchange the roots z_1 and z_2, while the root z_3 returns to its initial value; therefore the fundamental cycle $\gamma_{2,3}$ is substituted by the cycle $\gamma_{1,3}$, which is equivalent to the direct sum of the cycles $\gamma_{1,2}$ and $\gamma_{2,3}$. As a consequence the Actions (4) are transformed according to

$$\begin{aligned} I_{1,2} &\to I_{1,2} \\ I_{2,3} &\to I_{1,2} + I_{2,3} : \end{aligned} \tag{6}$$

i.e. when the energy goes around the singular value $E = 0$, the Actions (4) are transformed by means of an unimodular transformation. By using the same argument for the singular value $E = \omega^3/6$, when the roots z_2 and z_3 collapse, we obtain that when the energy makes a closed path around $E = \omega^3/6$ then the Actions (4) are transformed by means of an unimodular transformation

$$\begin{aligned} I_{1,2} &\to I_{1,2} + I_{2,3} \\ I_{2,3} &\to I_{1,2}. \end{aligned} \tag{7}$$

Therefore for a generic closed path in the energy plane the Actions (4) are subjected to an unimodular transformation, which is given by a suitable sequence of transformations (6) and (7). From a geometrical point of view, a closed path in the energy plane transforms a given Riemann surface into itself; since the Riemann surfaces associated to $H(q,p) = E$, are topologically equivalent to tori, this means that the fundamental periods of the transformed surface have to be connected to the initial period by an unimodular transformation; indeed this condition is necessary for the analytical equivalence of two tori. The transformation of the fundamental periods is the same transformation which occurs between the Actions.

This analysis shows how it is possible to define a global analytic not-singlevalue Action variable $I = I(E)$ for the complexified Hamiltonian system (1), since in correspondence to the energy values of the separatrices we have two branch points, which give

rise to an infinite number of sheets connected through the cuts. Then the global Action is a multivalued function of the energy.
To each Action variables (5), we can associate an Angle variable according to

$$\begin{aligned}\phi_1 &= \frac{dE}{dI_{1,2}}\int^q \frac{dq'}{p(q',I_{1,2})} = \frac{2\pi}{T_1}\int^Q \frac{dQ'}{\sqrt{4Q'^3 - g_2Q' - g_3}}\\ \phi_2 &= \frac{dE}{dI_{2,3}}\int^q \frac{dq'}{p(q',I_{2,3})} = \frac{2\pi}{T_2}\int^Q \frac{dQ'}{\sqrt{4Q'^3 - g_2Q' - g_3}}\end{aligned} \tag{8}$$

where the integral is taken on a curve on the Riemann surface $H(q,p) = E$ and T_1,T_2 are the fundamental periods associated to the cycles $\gamma_{1,2}$ and $\gamma_{2,3}$. The factor $2\pi/T_1$ (resp. $2\pi/T_2$) in front of the integral has the effect of normalizing to 2π the increment of ϕ_1 (resp. ϕ_2), when we make a complete turn around $\gamma_{1,2}$ (resp. $\gamma_{2,3}$). This is not obviously true if we integrate on the other cycle; from (8) we obtain the value $2\pi T_2/T_1$ (resp. $2\pi T_1/T_2$) for the increment of ϕ_1 (resp. ϕ_2) when we integrate along the cycle $\gamma_{2,3}$ (resp. $\gamma_{2,3}$). The equations of motion for the Angles (8) reads

$$\begin{aligned}\frac{d\phi_1}{dt} &= \frac{dE}{dI_{1,2}} = \frac{2\pi}{T_1} = \Omega_1(I_{1,2})\\ \frac{d\phi_2}{dt} &= \frac{dE}{dI_{2,3}} = \frac{2\pi}{T_2} = \Omega_2(I_{2,3})\end{aligned} \tag{9}$$

According to the theory of elliptic functions, which can be applied to the integral (8), the function $\phi_1(Q)$ defines a mapping from the Riemann surface of energy E to a parallelogram in the complex ϕ_1-plane whose sides are given by a segment of length 2π parallel to the real axis and a segment of length $2\pi|T_2/T_1|$ parallel to a line directed as $2\pi T_2/T_1$; moreover we have to identify the opposite sides of the parallelogram. The parallelogram previously defined is the fundamental region of the discontinuous group in the ϕ_1-plane

$$\phi_1' = \phi_1 + 2\pi n_1 + 2\pi \frac{T_2}{T_1} n_2 \qquad (n_1,n_2) \in \mathbb{Z}^2 \tag{10}$$

The inverse function $Q = Q(\phi_1)$ is an elliptic function of fundamental periods 2π and $2\pi T_2/T_1$ and the image of the ϕ_1-plane is the universal covering of the Riemann surface $H(Q,P) = E$: according to the definition (8) the function $Q(\phi_1)$ can be explicitly written in term of the Weierstrass elliptic function $\mathcal{P}(w)$ of periods $T_1(I_{1,2})$ and $T_2(I_{1,2})$, according to

$$Q = \mathcal{P}(\frac{T_1}{2\pi}\phi_1). \tag{11}$$

In order to complete the transformation in the Action-Angle variables we observe that

$$P = \frac{dQ}{dt} = \frac{d\phi_1}{dt}\frac{dQ}{d\phi_1} = \frac{2\pi}{T_1}\frac{d}{d\phi_1}\mathcal{P}(\frac{T_1}{2\pi}\phi_1) = \mathcal{P}'(\frac{T_1}{2\pi}\phi_1) \tag{12}$$

where $\mathcal{P}'(w)$ is the derivative of the Weierstrass function. Finally we recover the initial variables (q,p) by means of the transformation (2). We remark that the dependence on the Action variable in the equations (11) and (12) enters through the definition of the fundamental periods of the Weierstrass function. A similar analysis can be carried on using the angle ϕ_2.
Recalling the properties of the function $\mathcal{P}(w)$, it follows that for a fixed value of the energy, the functions $Q(\phi_1)$ and $Q(\Phi_1)$ have a lattice of polar singularities in the

ϕ_1-plane of periods 2π and $2\pi T_2/T_1$, whose positions depends on an arbitrary choice for the initial condition of ϕ_1 in the definition (8). We remark that the singularities in the Angle variable corresponds to the singularities in the complex time, when we perform the analysis a la Painlevé on the solutions of the Hamiltonian system (1).
The final picture for the analyticity properties of the global Action-Angle variables for the complexified Hamiltonian system (1) is the following: we can introduce an Action variable, which gives the analytic continuation of the usual real Action variable, for all the complex surfaces $H(q,p) = E$ except the separatrices; on each surface we can introduce a complex Angle variable, which parametrizes the initial variable (q,p) by means of the elliptic functions (11) and (12) so that we have a lattice of polar singularities in the Angle whose periodicities depends on the energy and consequently on the Action. When we move the energy in the complex plane, the fundamental periods of the Angle variable change with continuity: this implies that Riemann surfaces corresponding to different values of the energy are in general not analytically equivalent (but only topologically equivalent). At the singular values of the energy there are branch points for the function $I = I(E)$: if the energy makes a cycle around the singular values, the Action is changed by means of an unimodular transformation according to eqs. (6) and (7) and the same transformation acts on the fundamental periods, which corresponds to a different choice of the basis vectors associated with the lattice of periods of the elliptic function. As a consequence we change also the Angle variables (8) according to the same unimodular transformations (6) and (7).

FIRST ORDER OF THE PERTURBATION THEORY

Let us consider the perturbed Hamiltonian system

$$H(q,p) = \frac{p^2}{2} + \omega\frac{q^2}{2} - \frac{q^3}{3} + \epsilon f(q)\mathrm{e}^{it} \qquad (q,p) \in \mathbb{C}^2 \tag{13}$$

where $f(q)$ is for instance a polynomial in q. By using the Action-Angle variables for the complexified system (1) introduced in the previous section, we reduce the Hamiltonian (13) to the form

$$H(I,\phi) = H_0(I) + \epsilon f(I,\phi)\mathrm{e}^{it} \tag{14}$$

where the function $f(I,\phi)$ is an elliptic function of the Angle ϕ whose periodicity depends on the Action I (we choose without loss of generality $I = I_{1,2}$ and $\phi = \phi_1$). As a consequence $f(I,\phi)$ has a lattice of polar singularities in the ϕ-plane. We have restricted the dependence on t to a single Fourier components, since the generalization to a trigonometric polynomial is straightforward.
In this section we will consider the first order of the canonical perturbation theory for the complexified system (14) in order to see how the non-integrability character of the dynamics reflects on the singularities of the generating function which solves the Hamilton-Jacobi equation. Let

$$\hat{G}(J,\phi,t) = J\phi + \epsilon G(J,\phi)\mathrm{e}^{it} \tag{15}$$

be the generating function of a canonical transformation from the old variables (ϕ, I) to the new ones (ψ, J), which reduces the Hamiltonian system to an integrable form up to term of order ϵ^2. According to the Hamilton-Jacobi equation at the first order in ϵ, we obtain the following equation for the generating function (15)

$$\Omega(J)\frac{\partial G}{\partial \phi}(J,\phi)\mathrm{e}^{it} + iG(J,\phi)\mathrm{e}^{it} = -f(J,\phi)\mathrm{e}^{it} \tag{16}$$

where $\Omega = dH_0/dI$; the equation (16) is called homological equation. Let us introduce a Fourier expansion for the function $f(J,\phi)$, which is convergent inside a strip in the ϕ-plane, defined by the lattice of singularities of f: for instance, since f is always periodic with period 2π, it can be expanded in a Fourier series in a strip parallel to the real axis according to

$$f(J,\phi) = \sum_k f_k(J)e^{ik\phi}. \tag{17}$$

We can solve the homological equation in each strip where the Fourier expansion is convergent provided that Ω does not satisfy any resonant relation

$$k\Omega(J) + 1 = 0 \qquad k \in \mathbb{Z} \tag{18}$$

and the formal solution reads

$$G(J,\phi) = \sum_k \frac{-f_k(J)}{i(k\Omega(J)+1)}e^{ik\phi}. \tag{19}$$

Moreover if the frequency Ω satisfies a Bruno condition (A. Bruno 1990), it is possible to prove the convergence of the expansion (19) in the same strip in ϕ where the expansion (17) is defined. Our aim is to make an analytic continuation on the ϕ-plane of the solutions (19) defined in different parallel strips. We recall that $\Omega(I)$ is real if the Action I is real: i.e. we are considering the real dynamics; otherwise it is in general a complex number. As a consequence if we consider the unperturbed dynamics (see eq. (9)) $\phi(t) = \phi(0) + \Omega(I)t \quad t \in \mathbb{C}$, the analyticity strip is invariant for the unperturbed dynamics only for a specific direction of the complex time t: for instance if the solution (19) is defined on a strip parallel to the real axis, we have to require that Ωt is real. On the other hand it is easy to define the analytic continuation of the solution $G(J,\phi)$ along the unperturbed orbits, since from the homological equation (16) we obtain

$$G(J,\phi + \Omega(J)t)e^{it} = G(J,\phi) - \int_0^t f(J,\phi + \Omega(J)\tau)e^{i\tau}d\tau. \tag{20}$$

Let us first consider the case Ω complex and $\mathrm{Im}\Omega < 0$, then it is possible to write the solution (19) in the form

$$G(J,\phi) = \frac{1}{\Omega}\int_0^\infty f(J,\phi+\tau)e^{\frac{i\tau}{\Omega}}d\tau \tag{21}$$

where the convergence of the integral in the each strip parallel to the real axis is assured by the exponential since Re $i/\Omega < 0$. Then it is not difficult to prove that the analytic continuation of the function (21) across the line which passes through the poles of $f(I,\phi)$ in the boundary of the initial analyticity strip is impossible: i.e. if we consider two solutions of the homological equation of the form (19) defined in different nearby strips then it is impossible to make an analytical continuation of one solution in the other. Indeed if we consider a closed path around a singular point of $f(I,\phi)$, by using eq. (20) it is easy to proof that to each singular point there corresponds a branch point of logarithmic type for the solution (21); this result is consistent with the singularity analysis in the complex time, of an non-integrable Hamiltonian system (Bountis et al. 1987). If all the cuts are directed along the line which connects the singularities at the border of the analyticity strip, then this line appears as a natural boundary for the analyticity domain of the solution (19). Nevertheless it is possible

to make an analytic continuation of the solution (19) along the trajectory of the unperturbed motion according to eq. (20), but this does not provide in general a periodic solution both in ϕ and t. In any case when we continue the solution (19) along the trajectories of the unperturbed motion for a fixed direction in the complex time t, we have a branch point of logarithmic type at each singularity of $f(I,\phi)$ and the corresponding cuts are parallel to the chosen direction of t; as a consequence if we consider a direction which belongs to the lattice of the singularities then we have only a finite number of such cuts.
In the case $\Omega \in \mathbb{R}$, which corresponds to the real dynamics, we cannot use the equation (21) to represent the solution (19); however since the perturbation is periodic in t, we can introduce a discrete version for the homological equation which corresponds to the homological equation for the Poincarè map of the system. By integrating (20) for an entire period, we obtain

$$G(J,\phi+2\pi\Omega(J)) = G(J,\phi) - \int_0^{2\pi} f(J,\phi+\Omega(J)\tau)e^{i\tau}d\tau. \tag{22}$$

In this case, a direct proof for the existence of a natural boundary for the analytic continuation of the perturbative solution (19) in different strips, preserving the periodicity in ϕ and t, is not available; but this is suggested by the limit $\mathrm{Im}\Omega \to 0$ in the solution (21).
Finally we observe that when we approach to the separatrices, one of the fundamental periods (5), for instance the real one, tends to ∞ so that if we normalize this period to 2π, the other period of the lattice tends to 0: this means that the width of the analyticity strip for the solution (19) tends to 0 when we approach the separatrix.
A numerical analysis on the analyticity domain for KAM tori in the standard map by using Padè approximants, suggests that the picture given by the homological equation (16) may be observed also at higher perturbative orders (L. Billi et al. 1993).

ACKNOWLEDGEMENTS

We would like to thank Prof. G. Turchetti who has suggested this work, for his fundamental contributions during several discussions. We are in debt also with N.N. Nekoroshev for his suggestions when he visited Bologna.

REFERENCES

M. Adler and P. van Moerbeke, 1988, "Algebraic completely integrable systems: a systematic approach," *Perspectives in Mathematics Academic Press, Boston* .

V.I. Arnol'd , 1978, "Mathematical Methods of Classical Mechanics," Springer Verlag, .

V.I. Arnol'd , 1983, "Geometrical Methods in the theory of ordinary differential equations," Springer Verlag .

A. Bazzani and G. Turchetti, 1992, Singularities of normal forms and topology of the orbits in area-preserving maps, *J. Phys. A* 25:L427.

A. Bazzani and G. Turchetti, 1993, Analyticity of normalizing transformations for area-preserving maps, *in*: "Chaotic Dynamics:Theory and Practise," T. Bountis, ed., Springer Verlag .

L. Billi, M. Malvasi and G. Turchetti, 1993, Natural boundaries of normalizing transformations, *Contribution to this volume* .

T. Bountis, V. Papageorgiou and M. Bier, 1987, On the singularity analysis of intersecting separatrices in near-integrable dynamical systems, *Physica D* 24:292-304.

T. Bountis, L. Drossos and I.C. Percival, 1991, Non-integrable systems with algebraic singularities in complex time, *J.Phys. A Gen.* 24:3217-3236.

A.D. Bruno , 1990, A comparison of conditions on small divisors, *IHES* M: 36.

V.V. Kozlov, 1983, Integrability and non-integrability in Hamiltonian Mechanics, *Russian Math. Surveys* 38:1-76.

P. van Moerbeke , 1987, Introduction to Algebraic Integrable Systems and their Painlevé analysis, *in*: "Bowdoin AMS Summer Symposium," , ed., preprint.

C.L. Siegel , 1971, "Topics in Complex Function Theory, vol. 1,2,3," Wiley Interscience.

PERTURBATION THEORY FOR SYSTEMS WITHOUT GLOBAL ACTION–ANGLE COORDINATES

Francesco Fassò

Dipartimento di Matematica dell' Università
Via Sommarive 14 (Povo)
38050 Trento, Italy

INTRODUCTION

Hamiltonian perturbation theory is usually formulated with reference to systems defined in a product space $B \times \mathbb{T}^m$ endowed with a system of action–angle coordinates $I \in B$, $\varphi \in \mathbb{T}^m$, where B is an open set in $\mathbb{R}^m$. This is essentially a 'local' formulation since the phase space of an integrable Hamiltonian system can easily fail to have such a product structure in the large, and correspondingly there exists no single 'global' system of action–angle coordinates.

It turns out that this local formulation is still sufficient to study perturbations of systems without global action–angle coordinates if they are non–degenerate, but that it is ultimately inadequate in case of *degenerate* systems. Degenerate means that motions are quasi–periodic with $n < m$ frequencies, if $2m$ is the dimension of the phase space.

This problem is of interest since there are important degenerate systems which do not possess global action–angle coordinates, noticeably Kepler ($m = 3$ but $n = 1$: all bounded motions are periodic) and 'Euler–Poinsot' (the rigid body with a fixed point and no external torques: $m = 3$, $n = 2$). In the latter case, as explained below, the action–angle coordinates are singular when the angular momentum vector is parallel to the z–axis of a reference frame fixed in space, so that an atlas for the (relevant part of the) phase space consists of at least two charts.

This difficulty is overcome by a 'global' formulation of perturbation theory which refers to the flow on the phase space, rather than to a local coordinate description alone. This has been done by Moser (1970) (see also Cushman, 1984) for the special case of systems with all unperturbed motions periodic, and by the author (Fassò, 1991, 1993) for the general case of quasi–periodic motions. The aim of the present talk is to give a survey of this global construction.

Hamiltonian Mechanics, Edited by J. Seimenis,
Plenum Press, New York, 1994

THE GEOMETRY OF THE PHASE SPACE

A major role in our study is played by the geometry of the fibration by invariant tori of a completely integrable degenerate Hamiltonian system. It is described by generalizations of the Liouville–Arnold theorem due to Nekhoroshev (1972), Duistermaat (1980), and Dazord and Delzant (1987).

Consider a Hamiltonian system with m degrees of freedom whose motions are quasi–periodic with at most $n \leq m$ non–zero frequencies. Let M be the subset of its phase space which is foliated by the invariant tori of dimension n.† According to the just quoted generalizations of the Liouville–Arnold theorem, each such torus has a neighbourhood in M where there exists a system of *generalized action–angle coordinates*: that is, coordinates $(I^i, p^j, \varphi^i, q^j)$, $i = 1, \ldots, n$, $j = 1, \ldots, m-n$, which are such that the φ's are angular coordinates on the tori and the symplectic two–form of M is given by $\sum_i dI^i \wedge d\varphi^i + \sum_j dp^j \wedge dq^j$.

Depending on the geometry of the foliation by invariant tori, it may happen that no single system of these local coordinates can be extended to cover all of M. Nevertheless, one can always construct an atlas for M by matching a suitable number of them. Let M_λ and $(I_\lambda, p_\lambda, \varphi_\lambda, q_\lambda)$ be, respectively, the chart domains and the local coordinate systems of such an atlas, the index λ running in some set Λ. As observed by Nekhoroshev (1972), and as is in fact easily proven, in any non–empty intersection of domains $M_\lambda \cap M_\mu$ the transition functions between such coordinates have the form

$$\begin{aligned} I_\mu &= Z_{\mu\lambda} I_\lambda + z_{\mu\lambda} \\ (p_\mu, q_\mu) &= \mathcal{C}_{\mu\lambda}(I_\lambda, p_\lambda, q_\lambda) \\ \varphi_\mu &= \left(Z_{\mu\lambda}^{-1}\right)^T \varphi_\lambda + \mathcal{F}_{\mu\lambda}(I_\lambda, p_\lambda, q_\lambda) \end{aligned} \tag{1}$$

with some matrix $Z_{\mu\lambda}$ with integer entries and determinant plus or minus one, some vector $z_{\mu\lambda} \in \mathbb{R}^n$, and some maps $\mathcal{C}_{\mu\lambda}$ and $\mathcal{F}_{\mu\lambda}$.

Equations (1) show that M has the structure of a fiber bundle with fibers the invariant tori, and base a manifold B of dimension $2m - n$ with local coordinates $(I_\lambda, p_\lambda, q_\lambda)$, $\lambda \in \Lambda$. We shall call this bundle a *toral bundle of order n*. Equations (1) also show that the base B is foliated by submanifolds of dimension $2m - 2n$, which are locally described by $I_\lambda =$ const and carry local coordinates (p_λ, q_λ); they will be called the *symplectic leaves* of B (in fact, B is a Poisson manifold). The space of the leaves of this foliation is an n–dimensional manifold A, with the 'action' I_λ as local coordinates. We shall call A the *action space* of the toral bundle. We shall denote by $\pi_1 : M \to B$ the bundle projection of the toral bundle, and by $\pi_2 : B \to A$ the map which associates to each point of B the symplectic leaf through it. In local coordinates they are given, respectively, by $(I_\lambda, p_\lambda, q_\lambda, \varphi_\lambda) \mapsto (I_\lambda, p_\lambda, q_\lambda)$ and $(I_\lambda, p_\lambda, q_\lambda) \mapsto I_\lambda$. We shall also use the projection $\pi = \pi_2 \circ \pi_1 : M \to A$.

The Hamilton function of the system is, locally, a function of the actions alone, so Hamilton's equations read

$$\dot{I}_\lambda = 0, \qquad \dot{p}_\lambda = 0, \qquad \dot{q}_\lambda = 0, \qquad \dot{\varphi}_\lambda = \omega_\lambda(I_\lambda), \tag{2}$$

† M is in general a proper subset of the phase space. In particular, all the singularities of the foliation by n–dimensional tori – like isolated equilibria or isolated periodic orbits – are excluded from M. The fact that in a neighbourhood of any such singular leaf action–angle coordinates do not exist has nothing to do with the global problem we are here concerned with, which pertains to how the n–dimensional tori are assembled and twisted.

showing that, locally, there exist systems of $2m - n$ integrals of motion.

In non–degenerate cases ($m = n$) the symplectic leaves are points, and one has standard systems of local action–angle coordinates $(I_\lambda, \varphi_\lambda)$. In degenerate cases, as shown by (2), a symplectic leaf is the set of tori supporting motions with the same frequencies. In principle, the coordinates q_λ need not be angles.

We now give an overview of the question of the existence of global coordinates in a toral bundle, which has a central importance for our purposes; for greater details, see the papers by Nekhoroshev (1977), Duistermaat (1980), and Dazord and Delzant (1987). We say that a toral bundle has *global* generalized action–angle coordinates if it has an atlas with such coordinates made of just one chart. There are many, and important, integrable systems whose fibration by invariant tori does not possess such global coordinates. Examples are the spherical pendulum (Duistermaat, 1980; Cushman, 1983), the Lagrange top (Cushman and Knörrer, 1985) and, among the degenerate systems, Kepler (Moser, 1970; Coffey et al., 1986) and Euler–Poinsot (Fassò, 1993).

In degenerate cases, an obstruction to the existence of global generalized action–angle coordinates comes from the topology of the symplectic leaves: if they do not possess a global system of coordinates (p, q), then no global system of generalized action–angle coordinates exists in M. Noticeably, this is the case when the symplectic leaves are closed manifolds. As discussed later, this fact is the most relevant one for what concerns perturbation theory.

Other obstructions are related to the presence of matrices $Z_{\mu\lambda}$, vectors $z_{\mu\lambda}$, and maps $\mathcal{F}_{\mu\lambda}$ in (1). It can be shown that the generalized action–angle coordinates of a toral bundle can be chosen so that all vectors $z_{\mu\lambda}$ are zero whenever the symplectic two–form of M is exact – the typical case of mechanics. However, such coordinates cannot always be chosen so that all matrices $Z_{\mu\lambda}$ are the identity: when this is possible, the toral bundle is said to have *trivial monodromy.* An important fact is that the monodromy is always trivial, if the action space is simply connected.

If all $Z_{\mu\lambda} = \mathbb{1}$ and all $z_{\mu\lambda} = 0$, then the local action coordinates have identical transition functions: $I_\lambda = I_\mu$. One can then define a map I on the action space whose restriction to the domain of I_λ coincides with I_λ, for any $\lambda \in \Lambda$. However, such a map can fail to be injective. If it is injective, then it provides the action space with a global coordinate system, and we shall say that the toral bundle possesses *global actions.*

Finally, we shall say that the toral bundle has *global angles* if the generalized action–angle coordinates can be chosen so that $\varphi_\mu = \varphi_\lambda$ in every nonempty set $M_\lambda \cap M_\mu$. This depends on monodromy, and also on the presence of functions $\mathcal{F}_{\mu\lambda}$ in (1), which is related to the existence of a global section of the bundle, with certain properties. This implies, in particular, that a toral bundle which is non–trivial as a topological bundle does not possess global angles.

EULER–POINSOT

It is worthwhile to illustrate these ideas on the case of Euler–Poinsot (see Fassò, 1993, for details). We restrict for simplicity to the case of a symmetric body. It is well known that the phase space is $SO(3) \times \mathbb{R}^3$, the state of the body being represented by $(\mathcal{R}, m^b)$, where $\mathcal{R} \in SO(3)$ is a matrix determining the orientation of the body,

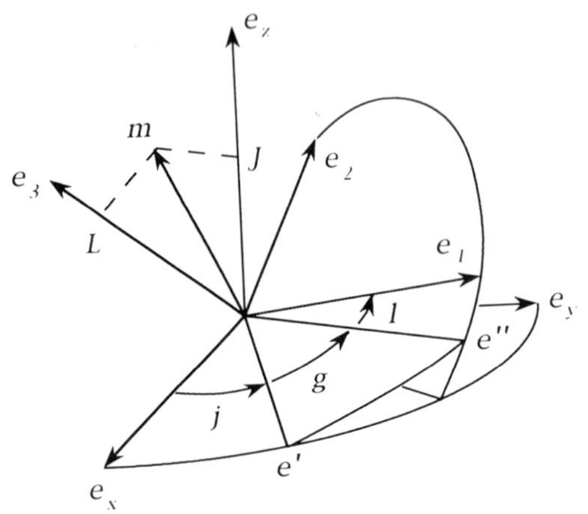

Figure showing the definition of the local generalized action–angle coordinates of the Euler–Poinsot system; e' and e'' are unit vectors in the directions of $e_z \times m$ and of $m \times e_3$, respectively.

and $m^b = (m_1, m_2, m_3)$ is the representative of the angular momentum vector m in an orthogonal reference frame attached to the body, with axes e_1, e_2, e_3 parallel to the inertia axes; let the corresponding inertia moments be a_1, $a_2 = a_1$, a_3.

The angular momentum vector in space $m^s = \mathcal{R}m^b$ and the energy h are integrals of motion. In the symmetric case, the projection $L = m \cdot e_3$ of the angular momentum along the inertia symmetry axis e_3 is an integral of motion, too, although it is not independent of h and m^s: indeed, from $h = (m_1^2 + m_2^2)/2a_1 + m_3^2/2a_3$ one gets

$$h(G, L) = \frac{1}{2a_1} G^2 + \frac{a_1 - a_3}{2a_1 a_3} L^2 \tag{3}$$

where $G = \|m\|$. Thus, the trajectories lie on the level sets of the energy–momentum map $h \times m^s$. Equivalently, but in fact conveniently, one can refer to the modified map

$$\pi_1 = m^s \times L \,.$$

Let M be the subset of the phase space obtained by removing the equilibria and the steady rotations about the inertia symmetry axis of the body, that is the states with either $m = 0$ or m parallel to e_3. A well known argument (Arnold, 1976) shows that, in M, the level sets of π_1 are two–dimensional tori (which collapse into circles where $m \times e_3 = 0$, that is in the complement of M). Thus, π_1 defines a fibration of M with fiber $\mathbb{T}^2$ and base

$$B = \{(m^s, L) \in \mathbb{R}^3 \times \mathbb{R} : \; m^s \neq 0, \; |L| < \|m^s\|\} \,.$$

Notice that such a fibration is topologically non–trivial: this follows from the fact that, for each $m^b \in \mathbb{R}^3$, π_1 acts on the factor $SO(3)$ of M as the Hopf fibration $\mathcal{H} : SO(3) \to S^2$, $\mathcal{R} \mapsto \mathcal{R}m^b$, which is well known to be non–trivial.

It is possible to prove that the bundle $\pi_1 : M \to B$ is in fact a toral bundle of order two. Its action space is

$$A = \{(G, L) \in \mathbb{R}^2 : \; G > 0, \; |L| < G\} \,,$$

and the projection $\pi_2 : B \to A$ is given by $\pi_2(m^s, L) = (\|m^s\|, L)$. Thus, its symplectic leaves are the two–dimensional spheres $\pi_2^{-1}(G, L) = \{(m^s, L) : \|m^s\| = G\}$ which, for each given L, can be identified with the spheres of constant modulus of the angular momentum vector in space.

Both the compactness of the symplectic leaves and the non-triviality of the fibration imply that M does not possess global generalized action–angle coordinates. The *local* generalized action–angle coordinates of M are the well known coordinates by Andoyer (1927). Their construction depends on the (arbitrary) choice of a reference frame fixed in space, with orthogonal axes e_x, e_y, e_z. Denote $J = m \cdot e_z$, and define three angles g, l, j as shown in the Figure. Then, it is a known result that G, L, J, g, l, j are symplectic coordinates in a proper subset of M, precisely in the set M_z obtained removing from M all the states with m parallel to e_z (Deprit, 1967). The actions are G and L, the angles g and l, and J and j are local coordinates on the symplectic leaves: in fact, (J, j) are a sort of cylindric coordinates on the spheres $\|m^s\| = \text{const}$, with polar singularities on the z–axis. Thus, such coordinates cannot be extended to cover all of M. However, since $M = M_z \cup M_{z'}$ for any two non–parallel axes e_z and $e_{z'}$, one constructs an atlas for M by matching two such local coordinate systems. Exploiting their transition functions one sees that M has no monodromy and that it has global actions (but not global angles).

Remark: Usually, Euler–Poinsot is considered on the basis of the Liouville–Arnold theorem. This approach leads one to construct a system of 'action–angle' coordinates which reflects the existence of a fibration by invariant tori of dimension three; these coordinates are precisely the ones due to Andoyer. Such an approach has the serious shortcoming that there is no fibration by three-dimensional invariant tori which is either intrinsically or globally defined in the phase space – only the fibration by two–tori has these properties. Three-tori are constructed by grouping together the two–dimensional ones, but there is no natural way to do that, and singularities are developed: one is indeed approximating spheres by cylinders. This is also the reason why the Andoyer coordinates look like three usual pairs of action–angle coordinates: one of the three angles is a local coordinate on a sphere.

PERTURBATION THEORY ON A TORAL BUNDLE

We say that a Hamiltonian system on a toral bundle M of dimension $2m$ and order n is integrable if its Hamilton function h has the form $h = \tilde{h} \circ \pi$, where $\tilde{h}$ is a function on the action space A of M; if $n < m$, we say that the system is degenerate. The local representatives h_λ of h are then functions of the local actions I_λ alone. Thus, locally, Hamilton's equations take the form (2).

A 'nearly integrable' Hamiltonian system on a toral bundle M is defined by a Hamilton function of the form

$$h + \varepsilon f \tag{4}$$

where $h = \tilde{h} \circ \pi$ is integrable, ε is a small parameter, and $f : M \to \mathbb{R}$ is any function. Our aim is to study such a system by perturbation techniques, that is through the construction of normal forms.

We are interested in the case in which M has no global generalized action–angle coordinates. The question is of course whether it is sufficient to construct a normal form within a single coordinate domain. Since all predictions one obtains this way are valid as long as the system remains within such a domain, this local approach is effective if one can assure the confinement of the system within a chart domain for all the (extremely long) times of interest.

Remarkably, this happens with *non–degenerate* systems. The point is that, on the one hand, the angles φ_λ cannot in any case escape a chart domain, since it contains entirely each torus. On the other hand, the outcome of perturbation theory is precisely to show that the actions I_λ remain almost constant, up to variations small with ε, for a certain time scale – thus, they certainly do not escape the chart domain in such a time, either.

In *degenerate* cases, however, one has in general no control on the variation of the coordinates (p, q) on the symplectic leaves: according to (2) they are slow variables, so their variation is essentially determined by the perturbation. As a consequence, if the symplectic leaves do not have global coordinates (p, q), one is likely not to have any control at all on the escape time. Thus, a 'local' perturbation theory is useless.

To overcome this difficulty, we shall construct the normal forms globally on the manifold. We shall proceed in a rather constructive way: we shall show that one can always construct the normal form Hamiltonians in each chart domain so that they are the local representatives of a Hamilton function which is globally defined on the bundle. As a matter of fact, the usual methods of perturbation theory lead to normal forms which quite naturally fulfill this requirement.

Let us preliminarily make the following basic observation. Since, as just discussed, the actions will be eventually shown to remain almost constant, one can consistently limit himself to work in a subbundle of the toral bundle of the form $\pi^{-1}(U)$, where U is some *small* subset of the action space. In particular, U can be taken to be simply connected, so that the toral bundle $\pi^{-1}(U)$ has no monodromy. This procedure is not restrictive, since one can cover the action space of M with such sets U.

To simplify the exposition, we shall also assume that $\pi^{-1}(U)$ has global actions; this hypothesis is not really restrictive if the symplectic two–form of M is exact, it being sufficient to take U sufficiently small.

However, we do not assume that $\pi^{-1}(U)$ has global generalized action–angle coordinates: as in the Euler–Poinsot case, the coordinates on the symplectic leaves and the angles need not be globally defined.

Resonances and Averaging

A central role in Hamiltonian perturbation theory is played by averaging operations, which in turn depend on the local resonance properties. They can be defined in a straightforward way on the toral bundle $\pi^{-1}(U)$ we are considering here, if this has trivial monodromy. For simplicity, let us also assume that $\pi^{-1}(U)$ has global actions, and let us refer to an atlas of it with all matrices $Z_{\mu\lambda} = \mathbb{1}$, all vectors $z_{\mu\lambda} = 0$ and, moreover, global actions I.

Under these hypotheses, all local representatives of the unperturbed Hamilton function $h = \tilde{h} \circ \pi$ coincide, and the frequency vector $\omega(I) = D\tilde{h}(I)$ is globally defined in the action space U. Thus, exactly as in the local case, we say that a vector $k \in \mathbb{Z}^n$ is a resonance at the point I of the action space U if $\omega(I) \cdot k = 0$.

In the local case, averaging operations are performed by projections onto the subspace of functions having Fourier components only on a subset of $\mathbb{Z}^n$: if $\mathcal{L} \subset \mathbb{Z}^n$, and if f is any function with Fourier components $\langle f \rangle_k$, one defines

$$\Pi_{\mathcal{L}} f\,(I, p, q, \varphi) \;=\; \sum_{k \in \mathcal{L}} \langle f \rangle_k (I, p, q)\; e^{ik\cdot\varphi}\,.$$

In general, one cannot perform Fourier expansions of functions on a toral bundle, but such projections are still meaningful if the monodromy is trivial. Indeed, let f_λ and f_μ be the local representatives of a function f in two charts of the considered atlas of $\pi^{-1}(U)$. Then, it is easy to verify on the basis of (1) that, in the intersection of the domains, their Fourier components are related by

$$\langle f_\mu\rangle_k(I,p_\mu,q_\mu) \;=\; \langle f_\lambda\rangle_k(I,p_\lambda,q_\lambda)\, e^{ik\cdot\mathcal{F}_{\mu\lambda}(I,p_\lambda,q_\lambda)}\,. \tag{5}$$

Conversely, one verifies that (5) is a sufficient condition for a family of functions $\langle f_\lambda\rangle_k$, $\lambda\in\Lambda$, $k\in\mathbb{Z}^n$, to be the Fourier components of the local representatives of a function on the bundle. It follows from here that the projections $\Pi_{\mathcal{L}}f_\lambda$, $\lambda\in\Lambda$, of the local representatives of a function f are the local representatives of a function on the toral bundle, which we denote by $\Pi_{\mathcal{L}}f$.

Normal Forms

We outline now the construction of a first order normal form for system (4). We will not dwell on questions concerning estimates, domains, and so on, since they can be dealt with essentially as in the local case.

We plan to construct a first order normal form 'adapted' to a subset $\mathcal{L}$ of $\mathbb{Z}^n$, that is to eliminate, at first order in ε, all the harmonics of the perturbation belonging to $\mathcal{L}$. Typically, $\mathcal{L}$ consists of a sublattice $\mathcal{L}_*$ of $\mathbb{Z}^n$ and – if a cutoff N is used – of all integer vectors of norm larger than N, that is $\mathcal{L}=\mathcal{L}_*\cup\{k\in\mathbb{Z}^n:\|k\|\geq N\}$. Such a normal form is naturally constructed in a subbundle of the form $\pi^{-1}(R)$, where $R\subset U$ is a sufficiently small set whose points resonate only with vectors in $\mathcal{L}$, that is $\inf_{I\in R}\inf_{k\notin\mathcal{L}}|\langle\omega(I),k\rangle|>0$.

To construct the normal form, one has to construct a symplectic diffeomorphism Φ such that

$$(h\,+\,\varepsilon f)\circ\Phi \;=\; h\,+\,\varepsilon\Pi_{\mathcal{L}}f\,+\,\mathcal{O}(\varepsilon^2)\,. \tag{6}$$

We construct Φ as the time–ε map of a Hamiltonian flow; this is the so called Lie method, which for our purposes has the important property of having an intrinsically global character. The problem then reduces to the determination of the Hamilton function χ of Φ. Since $g\circ\Phi=g+\varepsilon\{\chi,g\}+\mathcal{O}(\varepsilon^2)$ for any function g, the function χ has to satisfy the 'homological' equation

$$\{h,\chi\} \;=\; f\;-\;\Pi_{\mathcal{L}}f\,. \tag{7}$$

The key observation is now that χ solves (7) if and only if its local representatives χ_λ solve the equations $\{h,\chi_\lambda\}=f_\lambda\;-\;\Pi_{\mathcal{L}}f_\lambda$, where f_λ are the local representatives of f (those of h are all equal). These 'local' homological equations are solved by well known Fourier series expansions:

$$\chi_\lambda(I,p_\lambda,q_\lambda,\varphi_\lambda) \;=\; \sum_{k\notin\mathcal{L}}\frac{\langle f_\lambda\rangle_k(I_\lambda,p_\lambda,q_\lambda)}{i\,\omega(I)\cdot k}\,e^{ik\cdot\varphi_\lambda}\,. \tag{8}$$

Using (5), it is easy to verify that (provided the series (8) converge, which can be controlled chart by chart), the functions χ_λ defined in this way are the local representatives of a function χ defined on the toral bundle. Consequently, the Hamiltonian flow corresponding to χ is defined globally in $\pi^{-1}(R)$, and so is the normal form (6).

Nekhoroshev's Theorem, and an Application

The described method can be used to prove a 'global' formulation of the well known theorem by Nekhoroshev (1977) on the stability of the actions for exponentially long times. Such a generalization is meaningful and useful since, at variance with KAM, Nekhoroshev's theorem applies quite naturally to degenerate systems. A rather standard iteration of the construction above provides the so called 'analytic part' of the proof of this theorem, which deals with the construction of a normal form with an exponentially small remainder in a set $\pi^{-1}(R)$ characterized by definite resonance properties. The 'geometric part' of the proof, which consists in choosing suitably the resonant regions R so to cover the action space and in providing a confinement mechanism for the actions, goes just as in the local case. For details, see Fassò, 1993.

As an application of this result, consider the 'fast' motion of a symmetric rigid body with a fixed point, under the influence of external forces derived by an analytic potential. This system can be naturally regarded as a small perturbation of Euler–Poinsot. One then can prove that, for *all* motions with initial angular velocity of modulus Ω large enough, the actions G and L remain almost constant (their relative variations being of the order of Ω^{-1}) for times which increase as $\exp\sqrt{\Omega}$. For greater details see Fassò, 1991, and Benettin and Fassò, 1993.

REFERENCES

Andoyer, H., 1923, Cours de Mécanique Céleste, Gauthier–Villars, Paris.

Arnold, V.I., 1976, Méthodes Mathématiques de la Mécanique Classique, MIR, Moscow.

Benettin, G., and Fassò, F., 1993, Paper in preparation.

Coffey, S.L., Deprit, A., and Miller, B.R., 1986, The critical inclination in artificial satellite theory, *Cel. Mech.* 39:365.

Cushman, R., 1983, Geometry of the energy momentum mapping of the spherical pendulum, *Centrum voor Wiskunde en Informatica Newsletter* 1:4.

Cushman, R., 1984, Normal form for hamiltonian vectorfields with periodic flow, in Differential Geometric Methods in Mathematical Physics, S. Sternberg ed., Reidel, Dordrecht.

Cushman, R., and Knörrer, H., 1985, The energy–momentum mapping of the Lagrange top, in Differential Geometric Methods in Mathematical Physics, H.D. Doebner and J.D. Henning eds., *Lect. Notes Math.* No. 1139, Springer Verlag, Berlin.

Dazord P., and Delzant T., 1987, Le probleme general des variables actions-angles, *J. Diff. Geom.* 26:223.

Deprit, A., 1967, Free rotation of a rigid body studied in phase plane, *Am. J. Phys.* 55:424.

Duistermaat, J.J., 1980, On global action–angle coordinates, *Commun. Pure Appl. Math.* 33:687.

Fassò, F., 1991, Fast Rotations of the Rigid Body and Hamiltonian Perturbation Theory. Ph.D. Thesis, SISSA, Trieste.

Fassò, F., 1993, Geometry of the symmetric Euler–Poinsot system and Hamiltonian perturbation theory on a manifold. Preprint, University of Trento.

Nekhoroshev, N.N., 1972, Action–angle variables and their generalizations. *Trudy Moskov. Mat. Obsc.* 26:181.

Nekhoroshev, N.N., 1977, An exponential estimate of the time of stability of nearly integrable Hamiltonian systems. *Usp. Math. Nauk.* 32:5.

A NON-INTEGRABILITY TEST FOR PERTURBED HAMILTONIAN SYSTEMS OF TWO DEGREES OF FREEDOM

E. Meletlidou and S. Ichtiaroglou

Department of Physics
University of Thessaloniki
54006 Greece
CAAZ07@GRTHEUN1.BITNET

INTRODUCTION

The non-integrability of Hamiltonian systems and especially nearly integrable ones has been extensively investigated in the past and criteria for establishing this property in given Hamiltonians have appeared. Although these criteria do not always supply the answer to the question of non-integrability, they can often be applied to physical problems and may be divided into different categories, each one exploiting certain characteristic properties peculiar to integrable or non-integrable systems.

One of the methods relies on proving the existence of transverse homoclinic points and is based on what is known as Melnikov's homoclinic theorem. One can apply this method to perturbed Hamiltonians if the unperturbed, integrable part possesses a homoclinic orbit (Poincaré, 1899, Vol.III, p.397; Melnikov, 1963; Arnold, 1964; Kozlov, 1983 and references therein).

Ziglin (1983) has also introduced an important theorem on non-integrability by studying the branching of solutions of an integrable system in the complex domain. Applications of Ziglin's theorem have been made in several systems (Yoshida, 1987, 1988; Ito, 1985, 1987).

At the end of the previous century, Poincaré (1899, Vol.I, p.233), in his pioneering work on the three-body problem, supplied a theorem for proving non-integrability of a two degree

Hamiltonian Mechanics, Edited by J. Seimenis,
Plenum Press, New York, 1994

of freedom Hamiltonian of the form $H=H_0+\varepsilon H_1$ where H_0 is a non-degenerate integrable part and ε a small parameter. The criterion proved below is based on this theorem but makes use of a convenient parametrization of phase space along the orbits of the unperturbed Hamiltonian H_0. Thus one can prove that, if action-angle variables for H_0 can be defined in an open domain, H is non-integrable for values of $\varepsilon\neq 0$ in an open interval around zero, if, for a dense set of resonant tori of H_0, the average value of H_1, evaluated along the periodic orbits on each one of them, depends on the particular orbit, i.e. it is different in at least two of them.

We apply this criterion to prove non-integrability of the separable quartic oscillator with a quadratic perturbation. Hamiltonians of this form may appear in problems of galactic dynamics (e.g. Contopoulos, 1990), field theory (Matinyan et al, 1981) etc. Due to the homogeneity of H_0 and H_1 in our example, the result is valid for all values of $\varepsilon\neq 0$. By a suitable point transformation we also show that a non-isotropic harmonic oscillator with any desired frequency ratio, perturbed by any member of a monoparametric family of quartic perturbations is non-integrable.

A CRITERION FOR NON-INTEGRABILITY

We consider a nearly integrable, autonomous, two degree of freedom Hamiltonian

$$H=H_0+\varepsilon H_1 \tag{1}$$

with H_0 integrable. We assume that H is integrable, with an integral Φ, independent of H and analytic in ε around $\varepsilon=0$, i.e.

$$\Phi=\Phi_0+\varepsilon\Phi_1+O(\varepsilon^2). \tag{2}$$

By using the involution property of H and Φ we obtain at zeroeth order

$$[\Phi_0,H_0]=0, \tag{3}$$

that is, Φ_0 is an integral of H_0, and at first order

$$[\Phi_0,H_1]+[\Phi_1,H_0]=0. \tag{4}$$

Equations (3) and (4) hold identically in phase space. It is known (Poincaré, 1899) that Φ can always be selected such that Φ_0 does not depend only on H_0. This is an important property we are going to use later on.

Let I_0 be a known independent integral of H_0. Then we can define action-angle variables J_i, w_i, in the domain of bounded motion of H_0, where the gradients of H_0, I_0 are linearly independent. It is then known (Poincaré, 1899) that Φ_0 does not depend on the angles w_i in this open domain of the phase space, if the non-degeneracy condition

$$\det\left(\frac{\partial^2 H_0}{\partial J_i \partial J_j}\right)\neq 0 \tag{5}$$

holds.

At this point, Poincaré (1899) in order to prove his well-known non-integrability theorem expands both H_1, Φ_1 as doubly periodic Fourier series in w_1, w_2 and poses necessary conditions for integability on the Fourier coefficients of H_1.

We deviate from the above method by using a suitable parametrization, in which equation (4) takes the form

$$\frac{d\Phi_1}{dt}=[H_1,\Phi_0] \tag{6}$$

where both sides of (6) are calculated on a particular solution of H_0. On a resonant torus $\omega_1/\omega_2=n/m$ of H_0, where $m,n\in Z\setminus\{0\}$ and are relative primes, every solution is periodic with period

$$T=\frac{2\pi n}{\omega_1}=\frac{2\pi m}{\omega_2} \tag{7}$$

where $\omega_i = \partial H_0/\partial J_i$. Therefore Φ_1, calculated on such a periodic solution, satisfies

$$\Phi_1(T)=\Phi_1(0). \tag{8}$$

By integrating (6) along a periodic solution, we obtain

$$\sum_{i=1}^{2}\frac{\partial\Phi_0}{\partial J_i}\int_0^T\frac{\partial H_1}{\partial w_i}dt = 0. \tag{9}$$

The unperturbed solution on the particular torus is given by

$$\begin{aligned} w_1&=\omega_1 t+\vartheta_1\\ w_2&=\omega_2 t+\vartheta_2 \end{aligned} \tag{10}$$

where w_i are mod(2π) and ϑ_i are arbitrary constants. We define now

$$\vartheta=m\vartheta_1-n\vartheta_2 \tag{11}$$

and note that on the certain resonant torus (n/m) there is a one-to-one correspondence between a particular orbit and ϑ (mod 2π). Equation (9) takes the form

$$\sum_{i=1}^{2}\frac{\partial\Phi_0}{\partial J_i}\frac{\partial}{\partial\vartheta_i}\langle H_1\rangle = 0 \tag{12}$$

where $\langle H_1\rangle$ is the average value of H_1 along a periodic orbit on this torus. Since $\langle H_1\rangle$ is evaluated on a particular orbit, it depends on ϑ_i only through ϑ. By considering (11), equation (12) yields

$$(m\frac{\partial\Phi_0}{\partial J_1}-n\frac{\partial\Phi_0}{\partial J_2})\frac{\partial}{\partial\vartheta}\langle H_1\rangle = 0. \tag{13}$$

On this torus $\omega_1/\omega_2=n/m$, so equation (13) obtains the form

$$\frac{D(\Phi_0,H_0)}{D(J_1,J_2)}\frac{\partial}{\partial\vartheta}\langle H_1\rangle = 0. \tag{14}$$

If at least on one orbit on this torus

$$\frac{\partial}{\partial\vartheta}\langle H_1\rangle \neq 0, \tag{15}$$

then

$$\frac{D(\Phi_0,H_0)}{D(J_1,J_2)} = 0 \tag{16}$$

on the whole torus, since it depends only on the actions. If in addition relation (15) is valid on a dense set of resonant tori, at least on one orbit on each torus, equation (16) holds identically in the corresponding open domain of phase space, i.e. functions H_0, Φ_0 must be dependent in this domain and thus we have arrived at a contradiction. So at this point we may state the following:

Proposition: Consider the two-degree of freedom autonomous Hamiltonian $H=H_0+\varepsilon H_1$ where H_0 is a non-degenerate integrable part for which we can define action-angle variables at least in an open domain of phase space. If there exists a set of resonant tori of H_0, which is dense in the action space, such that on each one of them the average value of the perturbative function H_1, evaluated on the corresponding periodic orbits, depends on the particular orbit, there exists an interval $(-\varepsilon_0,\varepsilon_0)\backslash\{0\}$ of ε in which the perturbed Hamiltonian H is non-integrable.

AN APPLICATION TO THE QUARTIC OSCILLATOR

We will apply the results of the preceding section to the Hamiltonian

$$H=\frac{1}{2}(p_x^2+p_y^2)+\frac{1}{4}(x^4+y^4)+\varepsilon(\alpha x^2+\beta y^2+2xy) \tag{17}$$

where

$$p_x=\dot{x}, \quad p_y=\dot{y}$$

and α, β are constant parameters. The integrable part

$$H_0=\frac{1}{2}(p_x^2+p_y^2)+\frac{1}{4}(x^4+y^4) \tag{18}$$

corresponds to a separable unharmonic oscillator and the general solution of the equations of motion for H_0 is given by (Davis, 1962, p.207)

$$\begin{aligned} x&=\lambda cn(\lambda t-\varphi_1,\frac{1}{\sqrt{2}}) \\ y&=\mu cn(\mu t-\varphi_2,\frac{1}{\sqrt{2}}) \end{aligned} \tag{19}$$

where λ, μ, φ_1, φ_2 are arbitrary constants which depend on the initial conditions. Two independent integrals of motion for H_0 are the following

$$\begin{aligned} I_1&=2p_x^2+x^4=\lambda^4 \\ I_2&=2p_y^2+y^4=\mu^4 \end{aligned} \tag{20}$$

and action-angle variables may be defined everywhere in the phase space except where

$$rank\left|\frac{D(I_1,I_2)}{D(x,y,p_x,p_y)}\right|<2,$$

that is except the points where λ or μ are zero, which correspond to the straight-line solutions along the two axes. The invariant tori of the system are defined by pairs of non-zero values of λ, μ. The frequencies of x and y are respectively

$$\omega_1=2\pi\lambda/4K, \qquad \omega_2=2\pi\mu/4K$$

where $K=K(1/\sqrt{2})$ is the complete elliptic integral of the first kind with modulus $k=1/\sqrt{2}$.

On the orbits of the resonant torus $\omega_1/\omega_2=\lambda/\mu=n/m$, the common period is

$$T=\frac{4K}{g} \tag{21}$$

where g is defined by

$$g=\lambda/n=\mu/m.$$

The parameter g labels a specific torus of the family of resonant tori with frequency ratio (n/m). Solution (19) acquires the form

$$x=gn\ cn(gnt-\varphi_1,\frac{1}{\sqrt{2}})$$
$$y=gm\ cn(gmt-\varphi_2,\frac{1}{\sqrt{2}})$$

and the parameter which defines a particular orbit on the resonant torus is found to be

$$\varphi=m\varphi_1-n\varphi_2 \qquad \mathrm{mod}(4K).$$

The perturbation H_1 is the perturbative potential

$$V_1=\alpha x^2+\beta y^2+2xy$$

If one keeps only the separable part of V_1, the resulting Hamiltonian is integrable and, as one can easily verify, the integral

$$\int_0^T(\alpha x^2+\beta y^2)dt$$

assumes the same value along every solution on a specific resonant torus. On the other hand, the average value of 2xy is

$$c_0(\varphi_1,\varphi_2) = \frac{g^3mn}{2K}\int_0^{4K/g} cn(gnt-\varphi_1)\ cn(gmt-\varphi_2)\ dt. \tag{22}$$

For $\varphi_1=K$, $\varphi_2=0$ which correspond to $\varphi=mK$ (mod 4K), the integral (22) becomes

$$c_0(K,0) = \frac{g^2mnk'}{2K}\int_0^{4K}\frac{sn(n\tau)cn(m\tau)}{dn(n\tau)}d\tau \tag{23}$$

where $k'=(1-k^2)^{1/2}=1/\sqrt{2}$ and $\tau=gt$. Since the integrand is an odd function of τ,

$$c_0(K,0)=0$$

on the whole set of resonant tori, i.e. $\forall$ m,n. Sellecting now $\varphi_1=\varphi_2=0$, we form the integral

$$c_0(0,0)=\frac{g^2mn}{2K}\int_0^{4K}cn(n\tau)cn(m\tau)d\tau. \tag{24}$$

This choice of φ_1,φ_2 corresponds to $\varphi=0$, therefore, if m is a multiple of 4 we are actually dealing with the same orbit. In any other case, the two integrals (23) and (24) are evaluated on different orbits of the resonant torus defined by a particular non-zero value of g.

The constant coefficient of the complex Fourier expansion of the product of two even functions is given by

$$c_0=\frac{1}{4}a_0a'_0+\frac{1}{2}\sum_{p=1}^{\infty}a_pa'_p \tag{25}$$

where a_p, a'_p are the Fourier coefficients of the cosine series of these two functions. By considering the Fourier expansion of the Jacobi elliptic cosine (Davis, 1962, p.168), we may write

$$\begin{aligned} cn(n\tau) &= \frac{2\pi}{kK}\sum_{s=0}^{\infty}\frac{q^{s+1/2}}{1+q^{2s+1}}\cos\left[n(2s+1)\frac{\pi\tau}{2K}\right] \\ cn(m\tau) &= \frac{2\pi}{kK}\sum_{r=0}^{\infty}\frac{q^{r+1/2}}{1+q^{2r+1}}\cos\left[m(2r+1)\frac{\pi\tau}{2K}\right] \end{aligned} \tag{26}$$

where q is the nome, which in our case is

$$q=e^{-\pi}.$$

In order to apply (25), we notice from (26) that the coefficients a_p,a'_p multiplied together correspond to

$$p = |n|(2s+1) = |m|(2r+1). \tag{27}$$

Since m,n are relative primes, they cannot be both even. When one is odd and the other even, there are no integers p satisfying equation (27) and $c_0=0$ in this case. We sellect therefore both m,n to be odd and relative primes and we see from (27) that for every r such that 2r+1 is a multiple of |n| there exists only one s which satisfies (27), such that 2s+1 is a multiple of |m|. We define an integer j by the relations

$$\begin{aligned} 2r+1 &= |n|(2j+1) \\ 2s+1 &= |m|(2j+1) \end{aligned}$$

and expression (25), after using (26), yields

$$c_0(0,0) = \frac{8\pi^2}{K^2}\,g^2mn\sum_{j=0}^{\infty}\frac{q^{(j+1/2)(|m|+|n|)}}{(1+q^{|m|(2j+1)})(1+q^{|n|(2j+1)})}\ . \tag{28}$$

The infinte series in (28) is greater than zero and converges to a positive number.

Relation (28) is valid for the whole family of tori with frequency ratio of the form (odd/odd). On the other hand, rationals of this form are dense in $\mathbb{R}$. Therefore we have found a dense set of tori on which the average value of H_1 is different on two orbits on the same torus. According to our proposition, this proves that the Hamiltonian (17) is non-integrable for $\varepsilon\neq0$ in an open interval of ε around zero. This however means that the Hamiltonian (17) is actually non-integrable for any $\varepsilon\neq0$, since the corresponding equations of motion remain invariant under the transformation

$$x\to cx\ ,\quad y\to cy\ ,\quad t\to c^{-1}t\ ,\quad \varepsilon\to c^2\varepsilon$$

with arbitrary c.

By diagonalizing the quadratic form in V_1, we also have shown non-integrability of any potential of the form

$$V = \frac{1}{2}(Ax^2+By^2) + C(x^4+y^4) - Dxy(x^2-y^2) + Ex^2y^2, \qquad A\neq B$$

with

$$C/D/E = (1+\gamma^4)/4\gamma(1-\gamma^2)/12\gamma^2$$

where the ratio A/B and the non-zero parameter γ are arbitrary. The above form of the potential may be applicable in problems of galactic dynamics. The criterion cannot be applied directly to this potential, considering the harmonic oscillator as the integrable part, since it is degenerate, that is relation (5) does not hold.

CONCLUSIONS

The proposition proved in this paper is a useful tool in proving non-integrability of a two degree of freedom perturbed Hamiltonian. Although this result stems from Poincaré's theorem, its viewpoint is different. In order to apply Poincaré's theorem for a specific perturbation, one needs to transform to action-angle variables and evaluate the coefficients of the double Fourier series of H_1 with respect to the angles in an open domain of phase space and then show that an infinite number of them is different from zero, each at certain points of phase space. On the other hand, in applying the present proposition one needs to evaluate the average value of H_1 with respect to time along the periodic orbits of H_0 using any canonical variables. This is an easier task and can be met with more applications, although the knowledge of the periodic solutions on a dense set of tori is more or less equivalent to obtaining a global transformation to action-angle variables.

There is also an apparent connection of our result to the theory of subharmonic bifurcations of the periodic orbits under the perturbation (Poincaré, 1899, Vol.I, p.109; see also Wiggins, 1990, p.117) and our quantity $\partial\langle H_1\rangle/\partial\vartheta$ is actually the Melnikov subharmonic function. Simple zeroes of this function on a resonant torus correspond to periodic orbits which can be continued analytically for $\varepsilon\neq 0$. Attempt to prove non-integrability by using the isolateness of these orbits for $\varepsilon\neq 0$ and the continuity of the Jacobian of H and Φ with respect to ε is made in Poincaré (1890) but is demolished by the fact that one cannot prove the existence of isolated orbits of arbitrarily long periods for small but fixed $\varepsilon\neq 0$ (Kozlov, 1983; Veerman & Holmes, 1985). The existence of simple zeroes of this function on a dense set of tori of H_0 guarantees the existence of orbits on these tori where this function is different from zero and, according to our result, this suffices for non-integrability of H for small but fixed $\varepsilon\neq 0$, without any consideration of the bifurcated orbits and their stability.

We applied our result to the separable quartic oscillator with a non-separable quadratic perturbation and proved non-integrability of this system for all non-zero values of ε by taking advantage of the special form of the potential, i.e. sum of two homogeneous terms. Yoshida's (1988) criterion for such potentials or Melnikov's homoclinic method cannot be applied in this case, since both terms are by themselves integrable and H_0 does not possess a homoclinic loop.

Acknowledgements

One of the authors (E.M.) is supported by the Hellenic Scholarship Foundation (I.K.Y.).

REFERENCES

Arnold, V.I., 1964, Instabilities of dynamical systems with several degrees of freedom, *Soviet Math. Dokl.* 5:581.

Contopoulos, G.,1990, Asymptotic curves and escapes in Hamiltonian systems, *Astron. Astrophys.* 231:41.

Davis, H.T., 1962, "Introduction to Nonlinear Differential and Integral Equations", Dover, New York.

Ito, H., 1985, Non-integrability of Hénon-Heiles system and a theorem of Ziglin, *Kodai Math. J.* 8:120.

Ito, H., 1987, A criterion for non-integrability of Hamiltonian systems with nonhomogeneous potentials, *ZAMP* 38:459.

Kozlov, V.V., 1983, Integrability and non-integrability in Hamiltonian mechanics, *Uspekhi Mat. Nauk.* 38:3.

Matinyan, S.G., Savvidi, G.K. and Ter-Arutyunyan-Savvidi, N.G., 1981, Classical Yang-Mills mechanics. Nonlinear color oscillations, *Sov. Phys. JETP* 53:421.

Melnikov, V.K., 1963, On the stability of the center for time periodic perturbations, *Trans. Moscow Math. Soc.* 12:1.

Poincaré, H., 1890, Sur le problème des trois corps et les équations de la dynamique, *Acta Math.* 13:1.

Poincaré, H., 1899, "Les Méthodes Nouvelles de la Mécanique Céleste", Vol. I-III, Gauthier-Villars, Paris; *idem.*, 1957, Dover, New York.

Veerman, P. and Holmes, P., 1985, The existence of arbitrarily many distinct periodic orbits in a two degree of freedom Hamiltonian system, *Physica D* 14:177.

Wiggins, S., 1990, "Introduction to Applied Nonlinear Dynamical Systems and Chaos", Springer-Verlag, New York.

Yoshida, H., 1987, A criterion for the non-existence of an additional integral in Hamiltonian systems with a homogeneous potential, *Physica D* 29:128.

Yoshida, H., 1988, Non-integrability of the truncated Toda lattice Hamiltonian at any order, *Commun. Math. Phys.* 116:529.

Ziglin, S.L., 1983, Branching of solutions and nonexistence of first integrals in Hamiltonian mechanics, *Func. Anal. Appl.* 16:181; 17:6.

LIBRATIONAL INVARIANT SURFACES IN THE SPIN–ORBIT PROBLEM

Alessandra Celletti

Dipartimento di Matematica Pura e Applicata
Universitá di L'Aquila
Via Vetoio - I–67010 Coppito (L'Aquila)
e-mail: Celletti@vxscaq.aquila.infn.it

Abstract. We device a method for constructing librational invariant curves confining the motion associated to the synchronous resonance in the spin–orbit problem. In particular we perform some symplectic changes of variables in order to reduce the initial Hamiltonian associated to our model to a form which describes small curves around the equilibrium position. The construction of librational surfaces is obtained applying perturbation techniques. More specifically we apply the KAM theorem after normalizing the Hamiltonian by a Birkhoff procedure. Two concrete applications (i.e., the Moon–Earth and the Rhea–Saturn systems) are considered. An introduction to spin–orbit resonances of the II kind is also presented.

Keywords: Spin–orbit resonances, Librational tori, normal form, KAM theorem.

1. A spin–orbit model

We consider the stability of spin–orbit resonances in Celestial Mechanics. Let S be a triaxial ellipsoidal satellite rotating about an internal spin–axis and at the same time revolving around a central body P. We denote by T_{rot} and T_{rev} the periods of rotation about the spin–axis and that of revolution about the primary body.

Def.: A spin–orbit resonance of the I kind is a commensurability relation between the periods of rotation and revolution. More precisely, one has a spin–orbit resonance of

Hamiltonian Mechanics, Edited by J. Seimenis,
Plenum Press, New York, 1994

type $p : q$, with $p, q \in \mathbf{Z}\backslash\{0\}$, when

$$pT_{rev} - qT_{rot} = 0 .$$

Remark: A 1:1 (or synchronous resonance) implies that the period of rotation T_{rot} is identically equal to the period of revolution T_{rev}. The closest example of synchronous resonance is given by our Moon. Most of the evolved satellites of the solar system are observed in the 1:1 resonance. The only exception is provided by the Mercury–Sun system which satisfies a 3:2 resonance (i.e., $T_{rev}/T_{rot} = 3/2$) up to an error term of 10^{-4}.

Let us introduce now the spin–orbit resonances of the II kind, which couple the spin of a planet with the orbit of another planet. We consider a three body system (P_0, P_1, P_2), where P_0 is the central body and P_1, P_2 revolve around P_0. We denote by $\omega_{rot}^{(1)}$, $\omega_{rev}^{(1)}$ the frequencies of rotation and revolution of P_1 and by $\omega_{rev}^{(2)}$ the frequency of revolution of P_2 around P_0.

Def.: One has a spin–orbit resonance of the II kind of the type $p : q : r$, with $p, q, r \in \mathbf{Z}\backslash\{0\}$, when the frequencies of P_1 and P_2 are related by

$$p\omega_{rot}^{(1)} + q\omega_{rev}^{(1)} + r\omega_{rev}^{(2)} = 0 .$$

Remark: The only known example of a spin–orbit resonance of the II kind is provided by the Sun (P_0), Venus (P_1), Earth (P_2) system. In particular, up to a high degree of approximation, the following relation holds:

$$\omega_{rot}^{(1)} - 4\omega_{rev}^{(1)} - 5\omega_{rev}^{(2)} = 0 .$$

The above relation implies that at any inferior conjunction of Venus with the Earth (i.e. when the three bodies are aligned and their distance is minimum), Venus always presents the same face to the Earth.

The study of spin–orbit resonances of I and II kind is very similar. We start by introducing a simplified mathematical model for spin–orbit resonances of the I kind. In particular, due to the particular interest of the 1:1 resonance, we shall study its stability by applying perturbation techniques. The assumptions we make on the model are the following: we assume that

i) the satellite S moves on a *fixed* Keplerian ellipse around the central body;

ii) the spin–axis is orthogonal to the orbit plane and coincides with the shortest physical axis of the satellite;

iii) dissipative forces as well as gravitational perturbations due to other bodies of the solar system are neglected.

Denoting by $A < B < C$ the principal moments of inertia of the satellite, the equation of motion is derived from Euler's equations for a rigid body (see, e.g., Danby[5]) as

$$\ddot{x} + \frac{3}{2}\,\frac{B - A}{C}\,\left(\frac{a}{r}\right)^3 \sin(2x - 2f) = 0 , \tag{1}$$

where the mean motion $2\pi/T_{rev}$ has been set to one, a is the major semiaxis of the Keplerian orbit, r is the instantaneous orbital radius, f the true anomaly and x is the angle between the longest axis of the ellipsoid and the periapsis line.

Since the orbit is Keplerian, r and f are known functions, 2π–periodic in time. Therefore using standard Keplerian expansions one can use in place of (1) the equation

$$\ddot{x} + \varepsilon \sum_{m\neq 0, m=-\infty}^{\infty} W(\frac{m}{2}, e) \sin(2x - mt) = 0 , \tag{2}$$

where $\varepsilon \equiv \frac{3}{2}\frac{B-A}{C}$ (proportional to the *dynamical ellipticity* of the satellite) and the coefficients W's decay as powers of the orbital eccentricity e as $W(\frac{m}{2}, e) \propto e^{|m-2|}$ (see Goldreich and Peale[6]). We simplify further equation (2) truncating the series expansion to some finite order. In particular, the most relevant term we have neglected in (1) (or (2)) is the contribution which comes from the internal non–rigid structure of the satellite. Such dissipative force is in general small compared with the gravitational term and consistently we ignore in the series of (2) those terms whose size is of the same order of magnitude of the neglected dissipative torque. For small values of the eccentricity, the coefficients $W(\frac{m}{2}, e)$ become quite small and one can retain only a few number of terms in (2) obtaining an equation of the form

$$\ddot{x} + \varepsilon \sum_{m\neq 0, m=N_1}^{N_2} W(\frac{m}{2}, e) \sin(2x - mt) = 0 , \tag{3}$$

for some integers N_1, N_2 depending on the structure of the satellite.

Let us look now at the structure of the phase space associated to (3). Equation (3) is derived from Hamilton's equations associated to the nearly integrable Hamiltonian

$$H(y, x, t) = h(y) + \varepsilon V(x, t) , \qquad y \in \mathbf{R}, \quad (x, t) \in (\mathbf{R}/2\pi\mathbf{Z})^2 \equiv T^2 , \tag{4}$$

where

$$h(y) \equiv \frac{y^2}{2} , \qquad V(x, t) \equiv -\frac{1}{2} \sum_{m\neq 0, m=N_1}^{N_2} W(\frac{m}{2}, e) \cos(2x - mt) . \tag{5}$$

We will refer to the quantity $\Omega \equiv \frac{dh(y_0)}{dy} = y_0$ as the *rotation number*. Moreover, we select Ω in the class of the diophantine numbers (i.e., Ω has to satisfy the condition $|\Omega - \frac{p}{q}|^{-1} \leq Cq^2$, $\forall p, q \in \mathbf{Z}$, $q \neq 0$, for some $C > 0$). In the unperturbed case ($\varepsilon = 0$) the surface $\mathcal{T}_0(\Omega) \equiv \{y_0\} \times T^2$ is invariant. When $\varepsilon \neq 0$, but sufficiently small, the Kolmogorov–Arnold–Moser theory[7,1,8] ensures the existence of an invariant surface for the perturbed system, say $\mathcal{T}_\varepsilon(\Omega)$, on which a quasi–periodic motion, with frequency Ω, takes place. One of the outcomes of KAM theory is to provide a constructive algorithm which produces a lower bound on the perturbing parameter, say $\varepsilon_r \equiv \varepsilon_r(\Omega)$, ensuring the persistence of the invariant surface $\mathcal{T}_\varepsilon(\Omega)$ for any $\varepsilon \leq \varepsilon_r(\Omega)$.

We remark that the two–dimensional invariant surfaces separate into invariant regions the three–dimensional phase space associated to (4)–(5). We make use of this property in order to investigate the stability (in the sense of confinement of motions) of the synchronous resonance. More specifically, the Poincaré map associated to (3) shows a pendulum–like structure, in which *librational* and *rotational* regimes are defined (see fig. 1). We know from astronomical observations that the Moon deviates from the synchronous resonance just for *dynamical librations* whose amplitude is very small. Therefore (in the Poincaré map) we can locate our satellite close to the elliptic fixed point associated to the 1:1 resonance. Due to the above remark, we can establish the stability of the Moon, trapping its motion between two invariant *rotational* surfaces. This result has been obtained in Celletti[2], where the trapping invariant surfaces have been constructed for the *realistic* value of the perturbing parameter.

A more powerful constraint can be obtained constructing a *librational* invariant surface, enclosing the Moon's motion. To this end we compute a suitable normal form for (4)–(5) and apply the KAM theorem as sketched in the following paragraph. The details of the proof can be found in Celletti[3].

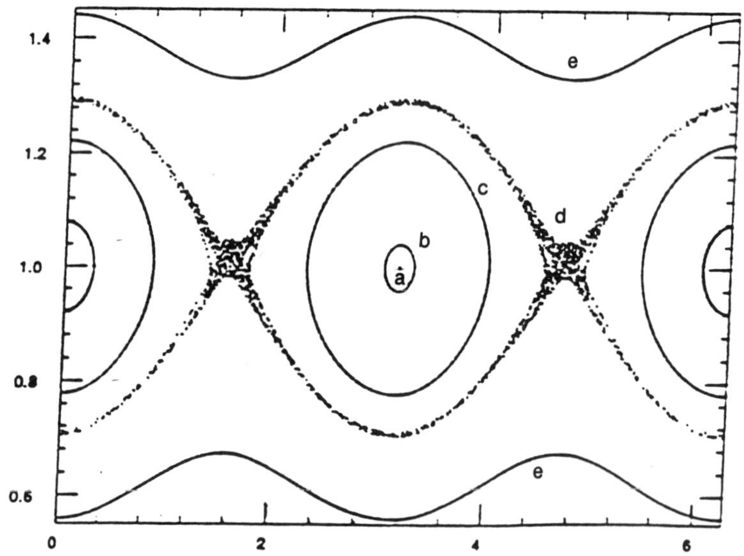

Fig. 1. Poincarè map of (5) associated to the 1:1 resonance. a) Periodic 1:1 orbit, b) librational curve, c) close–to–separatrix librational curve, d) chaotic separatrix, e) rotational invariant curves. The values of the parameters are $\varepsilon = 0.04$, $e = 0.0044$.

2. Construction of librational invariant surfaces

We transform the Hamiltonian (4)–(5), centering it on the synchronous resonance and placing the resonance on the (0,0) position by performing the symplectic changes of variables ($x' = 2x - 2t$, $y' = \frac{1}{2}(y-1)$), ($\overline{x} = x'$, $\overline{y} = y' - \frac{1}{2}$), so that one obtains

$$(6)\ H(\overline{y},\overline{x},t) = 2\,\overline{y}^2 - \varepsilon a \cos\overline{x} - \frac{\mu}{2} \sum_{m\neq 0,-2} \tilde{W}(\frac{m+2}{2}, e)\,[\cos\overline{x}\,\cos mt + \sin\overline{x}\sin mt]\,,$$

where $a \equiv \frac{1}{2}W(1,e)$, $\mu \equiv \varepsilon e$ and $\tilde{W}(\frac{m+2}{2}, e) = \frac{1}{e}W(\frac{m+2}{2}, e)$. Librational motions occurs in the unperturbed case (i.e. $\mu = 0$) as $|E| < \varepsilon a$. We fix a level energy $\overline{\delta}$ as $2\,\overline{y}^2 - \varepsilon a \cos\overline{x} = \overline{\delta}$ with $|\overline{\delta}| < \varepsilon a$. Next we expand (6) in Taylor series around the point (0,0) and then we transform to a harmonic oscillator form, performing the change of variables ($p = \alpha\,\overline{y}$, $q = \beta\,\overline{x}$), where $\alpha = \frac{\sqrt{2}}{(\varepsilon a)^{1/4}}$, $\beta = \frac{(\varepsilon a)^{1/4}}{\sqrt{2}}$. Therefore, denoting by $\omega \equiv 2\sqrt{\varepsilon a}$ the frequency of the harmonic oscillations, we obtain the following Hamiltonian

$$
\begin{aligned}
H(p,q,t) &= \frac{\omega}{2}(p^2+q^2) - \varepsilon a\left(\frac{q^4}{4!\beta^4} - \frac{q^6}{6!\beta^6} + \ldots\right) - \frac{\mu}{2}\sum_{m\neq 0,-2}\tilde{W}\left(\frac{m+2}{2}, e\right)\cdot \\
&\cdot\left[\cos(mt)\left(1 - \frac{q^2}{2\beta^2} + \frac{q^4}{4!\beta^4} + \ldots\right) + \sin(mt)\left(\frac{q}{\beta} - \frac{q^3}{3!\beta^3} + \frac{q^5}{5!\beta^5} + \ldots\right)\right].
\end{aligned} \tag{7}
$$

We next introduce action–angle variables (I, φ) for the harmonic oscillator, i.e. $p = \sqrt{2I}\,\cos\varphi$, $q = \sqrt{2I}\,\sin\varphi$. After some algebra the new Hamiltonian corresponding to (7) takes the form

$$
\begin{aligned}
H(I,\varphi,t) &= \omega I - \varepsilon a\left(\frac{I^2}{16\beta^4} - \frac{5I^3}{2\cdot 6!\beta^6} + \ldots\right) - \varepsilon a\left[-\frac{I^2}{12\beta^4}\cos 2\varphi + \frac{I^2}{48\beta^4}\cos 4\varphi\right. \\
&+ \left.\frac{I^3}{4\cdot 6!\beta^6}\cdot(15\cos 2\varphi - 6\cos 4\varphi + \cos 6\varphi) + \ldots\right] - \frac{\varepsilon e}{2}\sum_{m\neq 0,-2}\tilde{W}\left(\frac{m+2}{2}, e\right) \\
&\left\{\cos(mt)\left[1 - \frac{I}{2!\beta^2}(1-\cos 2\varphi) + \frac{I^2}{8\cdot 3!\beta^4}(3 - 4\cos 2\varphi + \cos 4\varphi) - \frac{I^3}{4\cdot 6!\beta^6}\right.\right. \\
&\left.(10 - 15\cos 2\varphi + 6\cos 4\varphi - \cos 6\varphi) + \ldots\right] + \sin(mt)\left[\frac{\sqrt{2I}}{\beta}\sin\varphi - \frac{\sqrt{2}\,I^{3/2}}{12\beta^3}\right. \\
&\left.\left.(3\sin\varphi - \sin 3\varphi) + \frac{\sqrt{2}I^{5/2}}{4\cdot 5!\beta^5}\,(10\sin\varphi - 5\sin 3\varphi + \sin 5\varphi) + \ldots\right]\right\},
\end{aligned} \tag{8}
$$

where we did not explicitate terms of order greater than three in the action variable. Notice that $|I|$ is the mean distance from the origin and that $|\overline{I}| \equiv \frac{\delta+\varepsilon a}{2\sqrt{\varepsilon a}}$. The series expansion in (8) converges provided $|I|/\beta^2 << 1$. For simplicity of reference we rewrite (8) in the form

$$
H(I,\varphi,t) = \omega\, I + \varepsilon\overline{h}(I) + \varepsilon h(I,\varphi) + \varepsilon e\; f(I,\varphi,t)\,, \tag{9}
$$

with an obvious identification of the functions $\overline{h}$, h and f.

The existence of librational surfaces is obtained applying the KAM theorem as developed in Celletti and Chierchia[4]. However, before applying KAM theorem we reduce the size of the perturbation in (8) by a normal form transformation, so to obtain a Hamiltonian of the form

$$
H_k(I',\varphi',t) \equiv h_k(I';\varepsilon) + \varepsilon^{k+1}\, R_k(I',\varphi',t)\,, \tag{10}
$$

where (I', φ') are the new set of variables. The normal form is attained performing the close–to–identity transformation

$$(11)\qquad \begin{cases} I = I' + \frac{\partial \Psi_k(I',\varphi,t)}{\partial \varphi} & \equiv I' + \sum_{j=1}^{k} \varepsilon^j \frac{\partial \Phi_j(I',\varphi,t)}{\partial \varphi} \\ \varphi' = \varphi + \frac{\partial \Psi_k(I',\varphi,t)}{\partial I'} & \equiv \varphi + \sum_{j=1}^{k} \varepsilon^j \frac{\partial \Phi_j(I',\varphi,t)}{\partial I'} \\ t' = t \ , \end{cases}$$

where the function Ψ_k is obtained imposing that the expression

$$(12)\qquad \begin{aligned} &\omega\,(I' + \frac{\partial \Psi_k}{\partial \varphi}) \ + \ \varepsilon\,\overline{h}(I' + \frac{\partial \Psi_k}{\partial \varphi}) \ + \ \varepsilon\,h(I' + \frac{\partial \Psi_k}{\partial \varphi}, \varphi) + \\ &+ \ \varepsilon\,e\,f(I' + \frac{\partial \Psi_k}{\partial \varphi}, \varphi, t) \ + \ \frac{\partial \Psi_k}{\partial t} \end{aligned}$$

does not depend on the angle variables (φ, t) up to $O(\varepsilon^{k+1})$.

In order to establish the existence of a librational surface we apply the computer–assisted KAM algorithm as developed in Celletti and Chierchia[4]. We remark that there exists an optimal order, say $k = k_*$, of the Birkhoff–KAM procedure, which is in our cases $k_* = 5$ for the Moon and $k_* = 4$ for Rhea.

Let us summarize the results as follows. Consider the Moon–Earth system described by the Hamiltonian function (5) with $N_1 = -1$, $N_2 = 5$ and let $e = 0.0549$, $\varepsilon_{obs} = 3.45 \cdot 10^{-4}$ (i.e., the observed value). Then for any $\varepsilon \leq \varepsilon_{obs}/5.26$, there exists an invariant trapping surface corresponding to a libration of about $8^o.79$. We notice that it was not possible to construct trapping surfaces for values of the parameter which are consistent with the astronomical observations. On the other hand, one gets interesting results in the Rhea–Saturn system, which is described again by (5) with $N_1 = 1$, $N_2 = 5$. Let $e = 0.00098$ and $\varepsilon_{obs} = 3.45 \cdot 10^{-3}$; then for any $\varepsilon \leq \varepsilon_{obs}$, there exists an invariant trapping surface (around the synchronous orbit) corresponding to a libration of about $1^o.95$.

The details of the proof can bc found in Celletti[3].

3. Numerical results on spin–orbit resonances of the II kind

The study of spin–orbit resonances of the II kind is analogous to that of I kind resonances. We consider a three body system (e.g., Sun–Venus–Earth), where Venus plays the role of the satellite, i.e. it is a triaxial ellipsoidal satellite satisfying the assumptions *i*), *ii*), *iii*) of I kind resonances. Moreover we assume that also the Earth moves on a Keplerian ellipse. Denoting by x the angle between the longest axis of Venus and the perihelium line and by ϱ, r the distances of Venus by the Sun and the Earth, the equation of motion is given by (see e.g. Goldreich and Peale[6]):

$$\ddot{x} + \varepsilon\,\frac{GM_E}{r^3}\sin 2\delta + \varepsilon\frac{GM_S}{\varrho^3}\sin 2\beta \ = \ 0 \ ,$$

where G is the gravitational constant, M_E and M_S are the masses of the Earth and the Sun, respectively. We can assume, in a first approximation, that Venus and the Earth revolve around the Sun with their respective Keplerian eccentricities $e_V = 0.0067$, $e_E = 0.0167$. However, we stress that the gravitational perturbation, due to the Earth, on the figure of Venus is comparable to that exerted on the orbital elements of Venus. Therefore we need to couple the above equation with the variation of semimajor axis and eccentricity of Venus, according to a three–body interaction.

As mentioned in the introduction, Venus and the Earth have rotational and revolutional frequencies very close to the resonant relation

$$\omega_{rot}^{V} = 4\omega_{rev}^{V} - 5\omega_{rev}^{E} .$$

Such resonance corresponds to a rotational motion in the phase space. A numerical integration of the above model shows a stability of the system (in the sense of computer evidence of the existence of trapping invariant rotational surfaces) for values of the perturbing parameter ε greater than the realistic value. Perturbation theory results on this model can be obtained following closely the approach presented in §2.

References

1. V.I. Arnold, *Proof of a Theorem by A.N. Kolmogorov on the invariance of quasi-periodic motions under small perturbations of the Hamiltonian,* Russ. Math. Surveys **18**, 9 (1963)
2. A. Celletti, *Analysis of resonances in the spin–orbit problem in Celestial Mechanics: The synchronous resonance (Part I),* J. of Appl. Math. and Phys. (ZAMP) **41**, 174 (1990)
3. A. Celletti, *Construction of librational invariant tori in the spin–orbit problem,* preprint CARR 1992, to appear in J. of Appl. Math. and Phys. (ZAMP)
4. A. Celletti, L. Chierchia, *Rigorous estimates for a computer-assisted KAM theory,* J. Math. Phys. **28**, 2078 (1987)
5. J.M.A. Danby, *Fundamentals of Celestial Mechanics,* Macmillan, New York (1962)
6. P. Goldreich, S. Peale, *The dynamics of planetary rotations,* Ann. Rev. Astron. Astroph. **6**, 287 (1970)
7. A.N. Kolmogorov, *On the conservation of conditionally periodic motions under small perturbation of the Hamiltonian,* Dokl. Akad. Nauk. SSR **98**, 469 (1954)
8. J. Moser, *On invariant curves of area-preserving mappings of an annulus,* Nach. Akad. Wiss. Göttingen, Math. Phys. Kl. II **1**, 1 (1962)

NORMALIZATION OF RESONANT HAMILTONIANS

David Farrelly,* Jeffrey Humpherys,*† and T. Uzer‡

*Department of Chemistry and Biochemistry
Utah State University, Logan, UT 84322-0300, U.S.A.

†Department of Mathematics and Statistics
Utah State University, Logan, UT 84322-3900, U.S.A.

‡School of Physics, Georgia Institute of Technology,
Atlanta, GA 30332-0430, U.S.A.

INTRODUCTION

The definition of integrability is simple to state; an autonomous N degree of freedom Hamiltonian is integrable if N independent global invariants exist and these are in involution with each other.[1] However, a failure to find such a set of global invariants does not exclude the possibility that the Hamiltonian in question is integrable. The detection of integrability is thus a critical issue in non-linear dynamics and a variety of analytical and numerical procedures has been developed to determine if a Hamiltonian is integrable. The most obvious approach is to try to establish if the Hamiltonian is separable, possibly using the Stäckel conditions to guide one to appropriate coordinate system. It should be noted, however, that finding coordinates that separate a particular problem can itself be a difficult task. More general and systematic approaches than simply seeking separability are therefore in order, e.g., the Whittaker program. An alternative method is the Painlevé test[2] in which the analytic structure of the equations of motion in the complex time plane is examined. This approach has been used to uncover integrability but must be applied gingerly because it cannot be guaranteed to succeed in every case. Probably the simplest numerical method is to generate Poincaré surfaces of section and determine by eye whether or not the motion is integrable. Of course, no numerical method by itself can definitively determine integrability.

Deprit and co-workers[3–7] in their studies of normal forms, and particularly in an application to the Toda Lattice[3] have discovered what might constitute a new symptom of integrability in Hamiltonian systems. They note a correlation between the persistence of degeneracy in the normal form to high order and integrability of the pre-normalized Hamiltonian, and conjecture that this might be a symptom of integrability.[6] However, tests of this conjecture have ultimately uncovered integrable systems that are also separable. In a recent study of a problem in atomic physics we have encountered a Hamiltonian possessing a non-separable, integrable limit.[7] Surprisingly, in this case the conjecture of Deprit and Miller seems not to hold which

Hamiltonian Mechanics, Edited by J. Seimenis,
Plenum Press, New York, 1994

led us to question whether normalization might sense separability rather than integrability *per se.* After all, it is well known quantum mechanically that separability leads to degeneracies and degenerate equilibria are the symptom that the normal form is supposed to exhibit in the case of integrability. In this paper we examine normalization of, *(i)* a class of perturbed isotropic oscillators to high order which admit various integrable limits, and, *(ii)* the hydrogen atom in a generalized van der Waals (GVDW) potential. We conclude that under certain circumstances normalization can detect separability, but that it may also overlook integrable cases, whether separable or not.

NORMALIZATION OF ELLIPTIC OSCILLATORS

The normalization of perturbed elliptic oscillators has been studied thoroughly by Deprit and co-workers.[3–7] In fact, our exhibition problem is the same as Miller's,[6] i.e., a perturbed elliptic oscillator of the form $\mathcal{H} = \mathcal{H}_0 + \mathcal{H}_1$ where

$$\mathcal{H}_0 = \frac{1}{2}(P_x^2 + P_y^2) + \frac{1}{2}(x^2 + y^2) \tag{1}$$

and

$$\mathcal{H}_1 = \epsilon(\alpha x^3 + \beta x y^2) \tag{2}$$

High order normalizations of (i.e., 1:1 resonant) elliptic oscillators perturbed by quartic and sextic polynomials have also been investigated recently: for these systems the integrable limits also turn out to be globally separable.[4–8] To start it is useful to consider the properties of the isotropic oscillator $\mathcal{H}_0$, in particular its integrals of motion. Including the energy there are four invariants

$$\pi_0 = E_0/2 = \frac{1}{4}(P_x^2 + x^2) + \frac{1}{4}(P_y^2 + y^2)$$

$$\pi_1 = D/2 = \frac{1}{4}(P_x^2 + x^2) - \frac{1}{4}(P_y^2 + y^2)$$

$$\pi_2 = L/2 = \frac{1}{2}(yP_x - xP_y)$$

$$\pi_3 = K/2 = \frac{1}{2}(P_x P_y + xy). \tag{3}$$

The $\{\pi_i, i = 1, 2, 3\}$ are called the *Hopf* variables and are related in a simple fashion to D, L, K which correspond to the more usual definitions of the invariants of the isotropic oscillator. The Hopf variables have the useful property of satisfying the same Poisson bracket relations as angular momentum, namely,

$$\{\pi_j, \pi_k\} = \epsilon_{jkl}\pi_l \tag{4}$$

where ϵ_{jkl} is equal to 1 (-1) for even (odd) permutations of its subscripts and to 0 otherwise. Together with the relation

$$\pi_0^2 = \pi_1^2 + \pi_2^2 + \pi_3^2 \tag{5}$$

the $\{\pi_i, i = 1, 2, 3\}$ generate the Lie algebra of the group $SU(2)$. Equation (5) is that of a sphere, sometimes nominated the Poincaré sphere, and on whose surface the phase flow of the reduced system can be portrayed.

Further geometrical insight can be obtained by recognizing that the Hopf variables may be transformed to action-angle variables j_x, j_y, ϕ_x, ϕ_y as follows,

$$x = \sqrt{2j_x} \sin \phi_x, \quad y = \sqrt{2j_y} \sin \phi_y$$

$$P_x = \sqrt{2j_x} \cos \phi_x, \quad P_y = \sqrt{2j_y} \cos \phi_y \tag{6}$$

After a further transformation

$$J_1 = \frac{(j_x + j_y)}{2}, \quad J_2 = \frac{(j_x - j_y)}{2}$$

$$\phi_1 = \phi_x + \phi_y, \quad \phi_2 = \phi_x - \phi_y \tag{7}$$

the $\{\pi_i\}$ become,

$$\pi_0 = J_1$$

$$\pi_1 = J_2$$

$$\pi_2 = \sqrt{J_1^2 - J_2^2} \cos \phi_2$$

$$\pi_3 = \sqrt{J_1^2 - J_2^2} \sin \phi_2 \tag{8}$$

which makes clear that the reduced phase space defined by π_0 = constant is a two-dimensional sphere. The action-angle variables just introduced are closely related to the coordinates of Hopf: transformations between the Hopf variables may be effected by a simple rotation on the Poincaré sphere.

Our intuition, which springs from molecular spectroscopy, causes us to think of the unperturbed Hamiltonian $\mathcal{H}_0$ as describing two modes (e.g., of a molecule) whose normal coordinates are taken to be x and y.[8] Working for now with the unperturbed Hamiltonian it is possible to characterize the invariant tori of the system in terms of the $\pi's$ or the quantities D, K, L. We label D-type dynamics *normal* mode, L-type dynamics *precessional* mode, and K-type dynamics *local* mode in analogy with the terminology used in molecular spectroscopy. In light of the Poisson bracket relations of eq. (4), π_0 and any *one* of D, K, and L (or linear combinations) may be used to parameterize the invariant tori of $\mathcal{H}_0$. The degeneracy of the unperturbed system means that any of the various possible representations are acceptable. Under the influence of a perturbation, however, it becomes critical to select the correct parametrization of the unperturbed tori. This must be done in such a way that, as one causes the perturbation to diminish, eventually to zero, the KAM tori of $\mathcal{H}$ transmute smoothly into the invariant tori of $\mathcal{H}_0$. In fact the three Hopf variables are associated with separability of the isotropic oscillator in Cartesian (π_1), polar (π_2) and rotated Cartesian coordinates (π_3). Importantly, rotations in the phase space coordinates transform the Hopf variables into each other, as can be proven by explicit calculation or by $SU(2)$ rotations on the Poincaré sphere.

Normalization of an elliptic oscillator perturbed by a real polynomial in the coordinates (x, y) about the center at the origin is guaranteed to produce an expansion in which each term is a polynomial in the $\pi's$. In the event that this polynomial is a function of a single Hopf variable the implication is that the dynamics of the reduced system is of purely local, normal or precessional mode nature. Does this imply anything regarding separability or integrability? It is time to normalize the exhibition problem.

Normal Form

Normalization of $\mathcal{H}$ produces the following expression through order ϵ^4 in terms of the Hopf variables

$$\mathcal{H}_{NF} = 2\,\pi_0 + \epsilon^2\left(\frac{-15\,\alpha^2}{4} - \frac{5\alpha\beta}{2} - \frac{25\,\beta^2}{12}\right)\pi_0{}^2$$

$$+\epsilon^2\left(\frac{-15\,\alpha^2}{2} + \frac{5\,\beta^2}{6}\right)\pi_0\,\pi_1 + \epsilon^2\left(\frac{-15\,\alpha^2}{4} + \frac{5\alpha\beta}{2} + \frac{5\,\beta^2}{4}\right)\pi_1{}^2$$

$$+\epsilon^2(-\alpha\beta + 2\,\beta^2)\,\pi_2{}^2$$

$$+\epsilon^4\left(\frac{-705\,\alpha^4}{16} - \frac{319\,\alpha^3\,\beta}{8} - \frac{151\,\alpha^2\,\beta^2}{8} - \frac{1427\,\alpha\,\beta^3}{72} - \frac{2887\,\beta^4}{432}\right)\pi_0{}^3$$

$$+\epsilon^4\left(\frac{-2115\,\alpha^4}{16} - \frac{319\,\alpha^3\,\beta}{8} + \frac{59\,\alpha^2\,\beta^2}{12} - \frac{689\,\alpha\,\beta^3}{72} + \frac{2485\,\beta^4}{432}\right)\pi_0{}^2\,\pi_1$$

$$+\epsilon^4\left(\frac{-2115\,\alpha^4}{16} + \frac{319\,\alpha^3\,\beta}{8} + \frac{151\,\alpha^2\,\beta^2}{8} + \frac{1091\,\alpha\,\beta^3}{72} + \frac{97\,\beta^4}{16}\right)\pi_0\,\pi_1{}^2$$

$$+\epsilon^4\left(\frac{-705\,\alpha^4}{16} + \frac{319\,\alpha^3\,\beta}{8} - \frac{59\,\alpha^2\,\beta^2}{12} + \frac{1025\,\alpha\,\beta^3}{72} - \frac{739\,\beta^4}{144}\right)\pi_1{}^3$$

$$+\epsilon^4\left(\frac{-41\,\alpha^3\,\beta}{4} - \frac{3\,\alpha^2\,\beta^2}{2} + \frac{767\,a\,\beta^3}{36} + \frac{13\,\beta^4}{2}\right)\pi_0\,\pi_2{}^2$$

$$+\epsilon^4\left(\frac{-41\,\alpha^3\,\beta}{4} + \frac{17\,\alpha^2\,\beta^2}{2} + \frac{545\,\alpha\,\beta^3}{36} - \frac{101\,\beta^4}{18}\right)\pi_1\,\pi_2{}^2 \qquad (9)$$

We now examine the normal form in the three known integrable limits.

(i) $\beta = 0$

$\mathcal{H}$ is clearly separable in the original Cartesian variables and the normal form in this limit, through 4th order, and using eq. (5) becomes

$$\mathcal{H}_{NF} = 2\,\pi_0 + \epsilon^2\alpha^2\left(\frac{-15\;\pi_0{}^2}{4} - \frac{15\;\pi_0\,\pi_1}{2} - \frac{15\;\pi_1{}^2}{4}\right)$$

$$+\epsilon^4\alpha^4\left(\frac{-705\;\pi_0{}^3}{16} - \frac{2115\;\pi_0{}^2\,\pi_1}{16} - \frac{2115\;\pi_0\,\pi_1{}^2}{16} - \frac{705\;\pi_1{}^3}{16}\right) \qquad (10)$$

Notice that the normal form depends on the *single* Hopf variable π_1. We have verified that this pattern persists through 20th order and we label it the normal mode limit. Importantly, π_1 is separable in the original coordinates and it is possible to express the normal form completely in terms of the two actions J_1 and J_2 of eq. (10) with no angle appearing. To us, dependence of the normal form on π_1 is the hallmark of separability in the original Cartesian coordinate system.

(ii) $\beta = 3\alpha$

In this case the normal form depends on a single Hopf generator, this time π_3, through at least 20th order. Through 4th order we obtain

$$\mathcal{H}_{NF} = 2\,\pi_0 - 15\,\alpha^2\,\epsilon^2\,\left(\pi_0{}^2 + \pi_3{}^2\right) - \frac{705}{2}\,\alpha^4\,\epsilon^4\,\pi_0\,\left(\pi_0{}^2 + 3\,\pi_3{}^2\right) \qquad (11)$$

In this limit $\mathcal{H}$ is also separable after a rotation

$$x = (x' - y')/\sqrt{2}, \qquad P_x = (P'_x - P'_y)/\sqrt{2}$$
$$y = (x' + y')/\sqrt{2}, \qquad P_y = (P'_x + P'_y)/\sqrt{2} \tag{12}$$

which, incidentally, has the effect of interchanging π_1 and π_3.

Taken together with results from previous studies of elliptic oscillators perturbed by quartic and sextic perturbations[4,5,7,8] it might seem that in the various integrable limits the normal form collapses into an expansion that involves a single Hopf variable. We have, however, discovered a counter example to this trend: $\mathcal{H}$ admits a third, and less well known integrable limit that corresponds to global separability in parabolic coordinates.

(iii) $\alpha = 2\beta$

Integrability in this limit was suggested by Chang, Tabor and Weiss[9] based on Painlevé analysis and numerical evidence. Motivated by this, J. Greene subsequently determined an analytic expression for the second invariant. This case is a particular limit of the more general Hamiltonian

$$\mathcal{H} = \frac{1}{2}(P_x^2 + P_y^2) + \frac{1}{2}(Ax^2 + By^2) + D(2x^3 + xy^2) \tag{13}$$

for which the second integral of the motion found by Greene is

$$I = D(yP_xP_y - xP_y^2) + y^2(D^2x^2 + D^2y^2/4 + DBx) + (B - A/4)(P_y^2 + By^2) \tag{14}$$

The Hamilton-Jacobi equation corresponding to eq. (14) is separable in shifted parabolic coordinates; for A = B = 1, and D = ϵ these coordinates are, $x = \sqrt{\xi\eta}$, $y = (\xi - \eta)/2 + 3/(4\epsilon)$.[10]. Examination of the 2nd order normal form suggests that $\alpha = 2\beta$ might correspond to an integrable limit: substitution reduces the normal form at that order to an expression in terms of π_1. However, to higher order the normal form is the following

$$\mathcal{H}_{NF} = 2\,\pi_0 - \frac{5\,\epsilon^2\,\left(53\,{\pi_0}^2 + 70\,\pi_0\,\pi_1 + 21\,{\pi_1}^2\right)}{12}$$
$$\frac{-5\,\epsilon^4\,\left(98999\,{\pi_0}^3 + 209755\,{\pi_0}^2\,\pi_1 + 145509\,\pi_0\,{\pi_1}^2\right)}{432}$$
$$-\frac{5\,\epsilon^4\,\left(33033\,{\pi_1}^3 + 3360\,\pi_0\,{\pi_2}^2 + 2016\,\pi_1\,{\pi_2}^2\right)}{432} \tag{15}$$

which does not fit the pattern that might be expected, i.e., dependence on a single Hopf variable. It thus appears that the normal form is only sensitive to separability in the original Cartesian or polar coordinates in which case the associated separation constants are directly related to the Hopf generators. We now provide a second counter example from atomic physics in which normalization fails to uncover a non-separable integrable limit, but does detect all separable cases.

THE GENERALIZED VAN DER WAALS POTENTIAL

In Cartesian coordinates and atomic units ($m_e = e = \hbar = 1$) the Hamiltonian for the hydrogen atom in a GVDW potential is,[7,8,11]

$$H = \frac{1}{2}(P_x^2 + P_y^2 + P_z^2) + \frac{\gamma}{2}(x^2 + y^2 + \beta^2 z^2) - \frac{1}{r} \tag{16}$$

where E is energy and γ and β are dimensionless physical parameters. Scaling the coordinates by $\gamma^{-1/3}$ and the momenta by $\gamma^{1/6}$ and converting to cylindrical coordinates ($x = \rho\cos\phi, y = \rho\sin\phi, z$) gives the Hamiltonian,

$$\mathcal{H} = \gamma^{-1/3}H = \frac{1}{2}(P_\rho^2 + P_z^2) + \frac{1}{2}(\rho^2 + \beta^2 z^2) - \frac{1}{r} + \frac{m^2}{2\rho^2}. \tag{17}$$

where $r = \sqrt{\rho^2 + z^2}$. In (17) m is the z-component of the angular momentum vector and is conserved due to axial symmetry.

The integrability of eq. (16) has been analyzed by a variety of methods, including Lie group and Painlevé methods. In particular, it has been shown that the equations of motion derived from (17) in semiparabolic coordinates with $m = 0$ have the Painlevé property for $\beta = \pm 1, \pm 2$, and $\pm 1/2$.[11] Global invariants were obtained for these cases using Noether's theorem, but only for $m = 0$. A simpler and more direct resolution of the issue of the integrability of eq. (16) has recently been proposed; the invariants when $m = 0$ have been shown to be connected to separability of eq. (16) in appropriate coordinates.[12] However, the case $\beta = \pm 1/2$ and arbitrary m is integrable but non-separable.

In order to normalize this problem we first pass over to the coordinates of Kustaanheimo and Stiefel (KS). The KS transformation has been described elsewhere in this connection[7] and results in a 4-dimensional 1:1:1:1 resonant Hamiltonian (after several scalings and rotations in phase space)

$$H_{KS} = \frac{4}{\omega} = \frac{1}{2}(\mathbf{P}_u^2 + |\mathbf{u}|^2) + \frac{8\epsilon^4(2-\beta^2)}{\omega^4}|\mathbf{u}|^2(u_1^2 + u_4^2)(u_2^2 + u_3^2)$$

$$+\frac{4\epsilon^4\beta^2}{\omega^4}|\mathbf{u}|^2\left[(u_1^2 + u_4^2)^2 + (u_2^2 + u_3^2)^2\right] \tag{17}$$

where $\omega = \sqrt{-8E}$ which restricts eq. (17) to bound states with $E < 0$. The normal form was obtained in these coordinates using Mathematica as implemented on an IBM RISC 6000.[13] The normal form is most simply expressed in terms of the components of the orbital angular momenta, $\mathbf{L}(L_x, L_y, L_z)$ and the Runge-Lenz vector $\mathbf{A}(A_x, A_y, A_z)$ which generate the Lie algebra of the group $SO(4)$ [isomorphic to $SU(2) \otimes SU(2)$]. Using the usual notation for the generators of $SO(4)$, we write, $\mathbf{L} = \mathbf{L}(S_{23}, S_{13}, S_{12})$ and $\mathbf{A} = \mathbf{A}(S_{14}, S_{24}, S_{34})$. The generators of $SU(2) \otimes SU(2)$ are two angular momenta $\mathbf{J}$ and $\mathbf{K}$ which are related to the $SO(4)$ generators by,

$$\mathbf{J} = \frac{(\mathbf{L}+\mathbf{A})}{2}, \quad \mathbf{K} = \frac{(\mathbf{L}-\mathbf{A})}{2} \tag{18}$$

where,

$$\{J_j, J_k\} = \epsilon_{jkl}J_l, \quad \{K_j, K_k\} = \epsilon_{jkl}K_l \tag{19}$$

and $(i, j, k) = (x, y, z)$. The normal form through 4th order is (where $n = \frac{1}{2}\mathcal{H}_0$)

$$\mathcal{H}_{NF} = 2n + 16\,\frac{\epsilon^4 n}{\omega^4}\left(n^2 + \beta^2 n^2 + {S_{12}}^2 - \beta^2\,{S_{12}}^2 + 4\,{S_{14}}^2 - \beta^2\,{S_{14}}^2\right)$$

$$+16\,\frac{\epsilon^4 n}{\omega^4}\left(4\,{S_{24}}^2 - \beta^2\,{S_{24}}^2 - {S_{34}}^2 + 4\beta^2\,{S_{34}}^2\right) \tag{20}$$

and the resulting equations of motion for the generators are

$$\dot{S}_{12} = 0$$

$$\dot{S}_{13} = \left(1 - \beta^2\right) S_{12} S_{23} - \left(4 - \beta^2\right) S_{14} S_{34} + \left(4\beta^2 - 1\right) S_{14} S_{34}$$
$$\dot{S}_{14} = \left(1 - \beta^2\right) S_{12} S_{24} - \left(4 - \beta^2\right) S_{12} S_{24} - \left(4\beta^2 - 1\right) S_{13} S_{34}$$
$$\dot{S}_{23} = -\left(1 - \beta^2\right) S_{12} S_{13} - \left(4 - \beta^2\right) S_{24} S_{34} + \left(4\beta^2 - 1\right) S_{24} S_{34}$$
$$\dot{S}_{24} = -\left(1 - \beta^2\right) S_{12} S_{14} + \left(4 - \beta^2\right) S_{12} S_{14} - \left(4\beta^2 - 1\right) S_{23} S_{34}$$
$$\dot{S}_{34} = \left(4 - \beta^2\right) S_{13} S_{14} + \left(4 - \beta^2\right) S_{23} S_{24} \tag{21}$$

Degenerate equilibria arise whenever the right hand sides of two or more of these equations vanish simultaneously. In fact, degenerate equilibria exist only in the three integrable limits. Further, this is true for all m except if $\beta = \pm 1/2$ when degenerate equilibria exist only if $m = 0$. Hence, the cases producing degenerate equilibria coincide with the separable limits. Evidently, normalization has failed to detect the integrable limit $\beta = \pm 1/2$, $|m| > 0$. Additional light is shed on the matter by examining the normal form itself in each integrable limit through higher order.

(i) $\beta = \pm 2$

$$\mathcal{H}_{NF} = 2n + \frac{\epsilon^4 \left(80\, n^3 - 48\, n\, {S_{12}}^2 + 240\, n\, {S_{34}}^2\right)}{\omega^4}$$
$$+ \frac{\epsilon^8}{\omega^8}(-12576\, n^5 + 12480\, n^3\, {S_{12}}^2 - 2208\, n\, {S_{12}}^4$$
$$+ \frac{\epsilon^8}{\omega^8}(-125760\, n^3\, {S_{34}}^2 + 37440\, n\, {S_{12}}^2\, {S_{34}}^2 - 62880\, n\, {S_{34}}^4) \tag{22}$$

In this limit, through all computed orders, the normal form is found to depend only on the squares of the two generators $\{S_{12}, S_{34}\}$ *alias* $\{L_z, A_z\}$ whose Poisson brackets mutually vanish.

(ii) $\beta = \pm 1$

This limit corresponds to rotational symmetry and the normal form becomes,

$$\mathcal{H}_{NF} = 2n + \frac{32\epsilon^4 n^3}{\omega^4} + \frac{48\epsilon^4 n}{\omega^4}(S_{14}^2 + S_{24}^2 + S_{34}^2) -$$
$$\frac{\epsilon^8 n^3}{\omega^8}\left({S_{14}}^2 + {S_{24}}^2 + {S_{34}}^2\right)\left(23\, {S_{14}}^2 + 23\, {S_{24}}^2 + 23\, {S_{34}}^2 + 84\, n^2\right) \tag{23}$$

In view of the constraint $\mathbf{L}.\mathbf{A} = 0$, pure dependence on $\mathbf{A^2}$ is equivalent to pure dependence on $\mathbf{L^2}$. This is expected, given the initial rotational invariance. Thus the reduced phase space is the sphere $SU(2)$ and the energies are degenerate with respect to m.

(iii) $\beta = \pm 1/2$

$$\mathcal{H}_{NF} = 2n + \frac{20\epsilon^4 n^3}{\omega^4} + \frac{12\epsilon^4 n}{\omega^4}(m^2 + 5S_{14}^2 + 5S_{24}^2) -$$
$$6\, n\, \frac{\epsilon^8}{\omega^8}\left(23\, m^4 + 230\, m^2\, n^2 + 131\, n^4\right) +$$

$$\frac{30\epsilon^8 n}{\omega^8}\left(-62\,{S_{14}}^2-62{S_{24}}^2+16{S_{34}}^2 m^2-131\left({S_{14}}^2+{S_{24}}^2\right)\left({S_{14}}^2+{S_{24}}^2+2n^2\right)\right) \tag{24}$$

Significantly, (S_{14}, S_{24}, S_{12}) generate an $SU(2)$ Lie algebra. Thus, through order ϵ^4 eq. (24) represents a dynamical symmetry limit of the original Hamiltonian because the normal form can be expressed entirely in terms of the dynamical quantities Λ^2 and Λ_3.. Therefore, the reduced phase space is the sphere $SU(2)$ defined by the components of Λ and a degenerate equilibrium exists. At orders higher than 4th, terms in S_{34}^2 appear, but vanish in the limit $m = 0$. It is these terms that destroy any hope of finding degenerate equilibria when $m \neq 0$. Unlike the previous two examples, when $\beta = \pm 1/2$, for non-zero m the original Hamiltonian (16), while integrable, is non-separable and this correlates with the appearance of terms that break the dynamical symmetry. Incidentally, this result differs from Alhassid[14] who indicated that $\beta = 1/2$ is a dynamical symmetry limit for all m.

Normalization has a role to play in the detection of integrability. However, we have found two examples that clearly violate the conjecture that the normal form exhibits particular patterns that signify integrability or separability. We conclude that further study of the relationship of normal form theory to integrability would be profitable.

ACKNOWLEDGEMENTS

Support of this research by NSF and the American Chemical Society is gratefully acknowledged. TU is grateful to the NATO Special Program Panel on Chaos, Order, and Patterns for a Research, Study, and Lecture Visit Award.

REFERENCES

1. V.I. Arnold, *"Dynamical Systems III,"* Springer-Verlag, New York, NY, (1985).
2. M. Tabor, *Annals N.Y. Academy of Sciences*, 536:43 (1988).
3. A. Deprit and B.R. Miller, *Annals N.Y. Academy of Sciences*, 536:101 (1988).
4. A. Deprit and S. Ferrer, *Phys. Lett. A*, 148:412 (1991).
5. A. Deprit and A. Elipe, *Celest. Mech. and Dynam. Astron.*, 51:227 (1991).
6. B.R. Miller, *Celest. Mech. and Dynam. Astron.*, 51:361 (1991).
7. D. Farrelly, J.P. Skelton and T. Uzer, *Celest. Mech. and Dynam. Astron.*, (submitted); M.J. Gourlay, T. Uzer and D. Farrelly, *Phys. Rev. A*, 47:3113 (1993).
8. J.A. Milligan and D. Farrelly, *Phys. Rev. A*, 47:3137 (1993).
9. Y.F. Chang, M. Tabor and J. Weiss, *J. Math. Phys.*, 23:531 (1982)
10. S. Wojciechowski, *Phys. Lett. A*, 100:277 (1984).
11. K. Ganesan and M. Lakshmanan, *Phys. Rev. Lett.*, 62:232 (1989); *Phys. Rev. A.*, 42: 2940 (1990); *ibid.* 45:1548 (1992).
12. D. Farrelly and J.E. Howard, *Phys. Rev. A*, (in press).
13. P.E. Raines and T. Uzer, *Comput. Phys. Commun.*, 70:569 (1992).
14. Y. Alhassid, E.A. Hinds and D. Meschede, *Phys. Rev. Lett.*, 59:945 (1987).

EFFECTIVE STABILITY FOR PERIODICALLY PERTURBED HAMILTONIAN SYSTEMS

Àngel Jorba,[1] and Carles Simó[2]

[1]Departament de Matemàtica Aplicada I, ETSEIB
Universitat Politècnica de Catalunya
Diagonal 647, 08028 Barcelona (Spain)
[2]Departament de Matemàtica Aplicada i Anàlisi
Universitat de Barcelona
Gran Via 585, 08007 Barcelona (Spain)

Abstract

In this work we present a method to bound the diffusion near an elliptic equilibrium point of a periodically time-dependent Hamiltonian system. The method is based on the computation of the normal form (up to a certain degree) of that Hamiltonian, in order to obtain an adequate number of (approximate) first integrals of the motion. Then, bounding the variation of those integrals with respect to time provides estimates of the diffusion of the motion.

The example used to illustrate the method is the Elliptic Spatial Restricted Three Body Problem, in a neighbourhood of the points $L_{4,5}$. The mass parameter and the eccentricity are the ones corresponding to the Sun-Jupiter case.

1 Introduction

The study of the nonlinear stability of an elliptic equilibrium point of a Hamiltonian system is a classical and difficult topic. There are mainly two kind of results concerning this: results of KAM type (perpetual stability on a Cantor set of initial conditions) and results of Nekhoroshev type (stability for an exponentially long time span, on an open set of initial conditions). A survey of both kind of methods can be found in Arnol'd.[1]

In this work we are going to focus on the results of Nekhoroshev type. Our purpose will be to bound the diffusion of the motion near an elliptic equilibrium point of a Hamiltonian system. The kind of methods we are going to use is very similar to the ones used by Giorgilli et al.[2] and Simó[3] for autonomous Hamiltonians.

Hamiltonian Mechanics, Edited by J. Seimenis,
Plenum Press, New York, 1994

Let us recall the definition of effective stability. Without loss of generality, let us assume that the elliptic equilibrium point is located at the origin. Let B_ρ a ball of radius ρ centered at the origin. Let η and T be positive numbers, with $\eta < 1$.

Definition 1 *The origin is (η, T)-stable iff for all initial condition z_0 in $B_{\eta\rho}$, the corresponding solution is in the ball B_ρ during a time span of lenght T.*

The values of η and T depend on the application we are interested in. For the Sun-Jupiter case, the usual values are $\eta = 0.9$ (we allow a diffusion of 10%) and T of the order of the age of the Solar system.

In the next sections we are going to see an adapted version of the method used by Simó,[3] in order to bound the diffusion (near an elliptic equilibrium point) of a time-periodic Hamiltonian system. For shortness, the methodology will be directly presented on a concrete example. The one used is the study of the nonlinear stability of the L_5 point of the Elliptic Spatial RTBP (from now on, ESRTBP), taking as eccentricity and mass parameter the ones corresponding to the Sun-Jupiter case, that is, $e = 0.048498458$ and $\mu = 0.95387536 \times 10^{-3}$.

2 Expansion of the Hamiltonian

The first step of the process is to obtain a power expansion of the Hamiltonian around the equilibrium point. The way to obtain that expansion depends strongly on the form of the Hamiltonian we are dealing with. In our example, we have used a recurrence based on the one of the Legendre polynomials.

Now, let us start from the equations of motion of the ESRTBP. They are (see Szebehely[4])

$$\begin{aligned} x'' - 2y' &= \frac{\partial \omega}{\partial x}, \\ y'' + 2x' &= \frac{\partial \omega}{\partial y}, \\ z'' + z' &= \frac{\partial \omega}{\partial z}, \end{aligned}$$

where

$$\omega = \frac{\Omega}{1 + e\cos f}, \qquad \Omega = \frac{1}{2}(x^2 + y^2 + z^2) + \frac{1-\mu}{r_1} + \frac{\mu}{r_2} + \frac{\mu(1-\mu)}{2},$$

and $r_1^2 = (x-\mu)^2 + y^2 + z^2$, $r_2^2 = (x - \mu + 1)^2 + y^2 + z^2$. Moreover, e is the eccentricity of the two primaries and $'$ stands for the derivative with respect to the true anomaly f.

Defining the momenta $p_x = x' - y$, $p_y = y' + x$, $p_z = z'$ and making the (canonic) change of variables $p_x \to p_x - \beta$, $p_y \to p_y + \alpha$, $p_z \to p_z$, $x \to x + \alpha$, $y \to y + \beta$ and $z \to z$, where (α, β) are the (x, y) coordinates of an equilibrium point, the Hamiltonian of the system is

$$\begin{aligned} H &= p_f + \frac{1}{2}(p_x^2 + p_y^2 + p_z^2) + \frac{1}{2}(x^2 + y^2 + z^2) + yp_x - xp_y - \\ &\quad - \frac{1}{1 + e\cos f}\left[\frac{1}{2}(x^2 + y^2 + z^2) + \alpha x + \beta y + \frac{1-\mu}{r_{PS}} + \frac{\mu}{r_{PJ}}\right], \end{aligned} \tag{1}$$

where $r_{PS}^2 = (x - x_S)^2 + (y - y_S)^2 + z^2$, $r_{PJ}^2 = (x - x_J)^2 + (y - y_J)^2 + z^2$, being $(x_S, y_S) = (\mu - \alpha, -\beta)$, $(x_J, y_J) = (\mu - 1 - \alpha, -\beta)$. The term p_f is the momentum corresponding to f, and has been added to obtain an autonomous Hamiltonian.

As we are interested in the points $L_{4,5}$ we have $(\alpha, \beta) = (\mu - 1/2, \pm\sqrt{3}/2)$, that is, $x_S = 1/2$, $y_S = \mp\sqrt{3}/2$, $x_J = -x_S$, $y_J = y_S$.

From now on, we are going to focus on the L_5 case (the same results will hold for L_4, due to the symmetry): $x_S = 1/2$, $y_S = \sqrt{3}/2$.

Our purpose now is to expand this Hamiltonian in a power series of x, p_x, y, p_y, z and p_z:

$$H(x, p_x, y, p_y, z, p_z, f, p_f) = p_f + \sum_{j\geq 2} H_j(x, p_x, y, p_y, z, p_z, f),$$

where H_j is an homogeneouos polynomial of degree j, whose coefficients are Fourier series with respect to f.

To reduce the global computational effort, we will perform the change of variables that puts H_2 in normal form, at the same time that the Hamiltonian is expanded. For this reason, the first point we are going to deal with is the normal form of H_2.

2.1 Normal Form of H_2

This corresponds to a linear differential equation with periodic coefficients, that can be reduced to constant coefficients by means of the classical Floquet Theorem. We have obtained the change of variables numerically, and we have performed a Fourier analysis in order to obtain an (approximate) analytical expression for the change. At the same time, we want that H_2 be of the form:

$$H_2 = \frac{\omega_1}{2}(x^2 + p_x^2) + \frac{\omega_2}{2}(y^2 + p_y^2) + \frac{\omega_3}{2}(z^2 + p_z^2),$$

in order to simplify the next steps. This is done by a linear change of variables, that can be composed with the Floquet one to obtain a single change making both things.

Finally, in order to obtain an adequate expression for the calculus of the normal form it is convenient to complexify the expansion. The reason is that, in suitable complex coordinates, H_2 takes the form

$$H_2 = \sum_{j=1}^{3} \sqrt{-1}\omega_j q_j p_j,$$

that will be very useful (it will save time, memory and the process will not be ill conditioned) during the computations. For that purpose we perform the change of variables

$$x = \frac{q_1 + \sqrt{-1}p_1}{\sqrt{2}}, \qquad p_x = \frac{\sqrt{-1}q_1 + p_1}{\sqrt{2}}, \tag{2}$$

and similar expressions for q_2, p_2, q_3 and p_3. The composition of this change with the previous ones provides the final change we were looking for.

2.2 Expansion

In order to expand (1), we are going to focus on the expansion of r_{PS}^{-1} (r_{PJ}^{-1} is obtained with the same procedure). It is known that

$$\frac{1}{r_{PS}} = \frac{1}{\sqrt{1 - 2\rho\cos\psi + \rho^2}} = \sum_{n=0}^{\infty} \rho^n P_n(\cos\psi),$$

where ψ is the angle between $(x_S, y_S, 0)$ and (x, y, z), $\rho^2 = x^2 + y^2 + z^2$ and P_n is the Legendre polynomial of degree n. Let us define A_n as $\rho^n P_n(\cos\psi)$ (note that A_n is an homogeneous polynomial of degree n). Then, from the recurrence of the Legendre polynomials is not difficult to obtain a recurrence for the A_n:

$$A_{n+1} = \frac{2n+1}{n+1}(x x_S + y y_S)A_n - \frac{n}{n+1}(x^2+y^2+z^2)A_{n-1}, \tag{3}$$

being $A_0 = 1$ and $A_1 = x x_S + y y_S$. We have used this recurrence to obtain the expansion of the Hamiltonian, using for x and y the expressions provided by the final "Floquet" change. Finally, the remaining terms of (1) are computed directly.

3 Normal form

In what follows, k will be a multiindex, splitted as $k = (k^1, k^2)$, where k^1 and k^2 correspond to positions and momenta respectively.

The computation of the normal form is based on the following proposition:

Proposition 1 *Let us consider the Hamiltonian*

$$H = p_f + H_2(q,p) + \sum_{i=3}^{r-1} H_i(q,p) + H_r(q,p,f) + H_{r+1}(q,p,f) + \cdots,$$

where $r > 2$, $H_2(q,p) = \sum \omega_i q_i p_i$ $(\omega_i \in \mathbb{C})$, $H_i(q,p) = \sum_{|k|=i} h_i^k q^{k^1} p^{k^2}$, $H_r(q,p,f) = \sum_{|k|=r} h_r^k(f) q^{k^1} p^{k^2}$ and $h_r^k(f) = \sum_j h_{r,j}^k \exp(jf\sqrt{-1})$. Let us define $G_r = G_r(q,p,f) = \sum_{|k|=r} g_r^k(f) q^{k^1} p^{k^2}$ as follows:

1. if $k^1 \neq k^2$:

$$g_r^k(f) = \frac{c_k - h_{r,0}^k}{< \omega, k^2 - k^1 >} + \sum_{j\neq 0} \frac{h_{r,j}^k}{j\sqrt{-1} - < \omega, k^2 - k^1 >} \exp(jf\sqrt{-1}).$$

2. if $k^1 = k^2$:

$$g_r^k(f) = \sum_{j\neq 0} \frac{h_{r,j}^k}{j\sqrt{-1}} \exp(jf\sqrt{-1}).$$

Then, the new Hamiltonian H' obtained from H by means of the change of variables given by the generating function G_r,

$$H' = H + \{H, G_r\} + \frac{1}{2!}\{\{H, G_r\}, G_r\} + \cdots,$$

satisfies that

$$H' = p_f + H_2(q,p) + \sum_{i=3}^{r-1} H_i(q,p) + H_r'(q,p) + H_{r+1}'(q,p,f) + \cdots,$$

where $H_r'(q,p) = \sum (h')_r^k q^{k^1} p^{k^2}$ and

$$(h')_r^k = \begin{cases} c_k & \text{if } k^1 \neq k^2 \\ h_{r,0}^k & \text{if } k^1 = k^2 \end{cases}$$

Remark: The value c_k that appears in the function G can be selected according to several criteria. In our case, as we want to obtain an integrable normal form, we have chosen that value equal to 0. It is also possible to chose $c_k = h_{r,0}^k$ for the "quasiresonant" values of k, to alleviate the effect of the small divisors, but what we obtain in this case is a (in general non-integrable) seminormal form.

3.1 The resonant case

In the elliptical problem, we have a (exact) resonance between the frequency of the true anomaly and the frequency of the vertical mode of the RTBP. Due to that resonance, it is not possible to obtain an autonomous normal form, because some of the divisors $j\sqrt{-1} - < \omega, k^2 - k^1 >$ appearing in the generating function are zero. In that case, we select the corresponding $g^k_{r,j}$ equal to zero.

3.2 Realification

The last step of the procedure is to return to real coordinates and to introduce action variables, in order to study the dynamics of the final Hamiltonian. The first part is done by using the inverse of the map given in (2). We denote again by x, p_x, etc. the real coordinates. The second one by using the Poincaré change $x = \sqrt{2I_1}\cos\varphi_1$, $p_x = -\sqrt{2I_1}\sin\varphi_1$, and similar for the other variables. In this way, the global transformation preserves the real character of variables and Hamiltonian.

With these changes, we obtain a final Hamiltonian of the form:

$$H = N + R^{(m+1)} + I_f,$$

where m is the degree of the normal form and

$$N = N(I_1, I_2, I_3, \varphi_3 - f),$$

that depends on $\varphi_3 - f$ because of the $1:1$ resonance between the true anomaly f and the angle φ_3 (corresponding to the action I_3) of the vertical mode of the RTBP. Here I_f is the actual momentum conjugated to f.

All the previous computations have been effectively carried out up to order 20. The Fourier coefficients which appear in all the process with amplitude less than 10^{-15} have been dropped.

3.3 An approximation to the dynamics

To study the flow of this Hamiltonian, let us perform the following change of variables

$$\begin{array}{llll} I_f = J_0 - J_3, & I_1 = J_1, & I_2 = J_2, & I_3 = J_3, \\ f = \psi_0, & \varphi_1 = \psi_1, & \varphi_2 = \psi_2, & \varphi_3 = \psi_3 + \psi_0, \end{array}$$

to obtain

$$H = N + R^{(m+1)} + J_0,$$

where

$$N = N(J_1, J_2, J_3, \psi_3) = \omega_1 J_1 + \omega_2 J_2 + O_2(J_1, J_2, J_3),$$

is an integrable Hamiltonian and $R^{(m+1)} = \sum_{j>m} R_j$. More concretely, up to second order in the J_j variables, N is given by

$$\begin{aligned} N_{(2)} &= \omega_1 J_1 + \omega_2 J_2 + \alpha_{1,1} J_1^2 + \alpha_{1,2} J_1 J_2 + \alpha_{2,2} J_2^2 + \\ &\quad + J_3[a J_1 + d J_2 + (b J_1 + e J_2)\cos 2\psi_3 + (c J_1 + f J_2)\sin 2\psi_3] + g J_3^2, \end{aligned}$$

where the numerical values obtained for the constants in the current example are given in Table 1.

The dynamics described by this Hamiltonian is very simple: as J_1 and J_2 are first integrals, we can take $J_1 = C_1$, $J_2 = C_2$ ($C_{1,2}$ are constants) to obtain an one-degree

Table 1

ω_1	=	$-.8080513430831042E-01$	ω_2	=	$.9967588604945699E+00$
$\alpha_{1,1}$	=	$.5842667892951852E+00$	$\alpha_{1,2}$	=	$-.1551131477053751E+00$
$\alpha_{2,2}$	=	$.5641749211325000E-02$	a	=	$.5439031817085827E-01$
b	=	$.8735695861611382E-04$	c	=	$.2162237916664492E-03$
d	=	$.5793623535537272E-02$	e	=	$-.5265749230961082E-05$
f	=	$-.7753646646134309E-05$	g	=	$-.1787604308798145E-03$

of freedom Hamiltonian. In our case this is, essentially, a pendulum depending on the parameters J_1 and J_2. Using $N_{(2)}$ as an approximation, the fixed points are located at

$$\cos 2\psi_3 = \pm n/\sqrt{n^2+p^2}, \quad \sin 2\psi_3 = \pm p/\sqrt{n^2+p^2}, \quad J_3 = (m \pm \sqrt{n^2+p^2})/(-2g),$$

where $m = aJ_1 + dJ_2$, $n = bJ_1 + eJ_2$ and $p = cJ_1 + fJ_2$. The points with the + sign are hyperbolic and the ones with the $-$ sign are elliptic. If $J_1 = J_2 = 0$ the angle ψ_3 is moving quite slowly, with limit frequency equal to zero when J_3 goes to zero. When J_1 and/or J_2 move away from zero a bifurcation occurs. This is related to the 1 to 1 resonance between the vertical mode and the frequency of the elliptic motion. In particular the existence of normally hyperbolic tori of dimensions 2 and 3 follows. However, this will not affect the existence of effective stability, as we shall show.

4 Bounding the diffusion

To bound the diffusion speed it is enough to bound J_1', J_2' and N', because N is also an approximate first integral of the motion. To do it, we recall that

$$J_j' = \{J_j, H\} = \{J_j, R^{(m+1)}\}, \; j = 1, 2, \quad N' = \{N, R^{(m+1)}\}, \tag{4}$$

and this can be bounded easily if we have bounds for $\|R_{m+1}\|$ and for the Poisson bracket.

4.1 Norms

The norm used to produce the bounds mentioned above is

$$\|U_r\| = \sum_{|k|=r} \|u_r^k(f)\| = \sum_{|k|=r} \sum_j |u_{r,j}^k|,$$

where U is an homogeneous polynomial of degree r, u_r^k are its coefficients, and $u_{r,j}^k$ are the corresponding Fourier coefficients.

This norm has two interesting properties:

1. Let us denote by $\|U_r\|_{B_\rho}$ the sup norm of the homogeneous polynomial U_r of degree r over the ball centered in the origin and with radius ρ, and for all (real) time f. Then
$$\|U_r\|_{B_\rho} \leq \|U_r\|\rho^r.$$
2. Let U_r and V_s be homogeneous polynomials of degree r and s. Then
$$\|\{U_r, V_s\}\| \leq rs\|U_r\|\|V_s\|.$$

Using the norms defined above it is possible to estimate the remainder $R^{(m+1)}$ and also to bound $\|J_j'\|_{B_\rho}$ and $\|N'\|_{B_\rho}$ over a ball of radius ρ. More accurate results can be obtained using different norms (see Simó[3]).

4.2 Final results

From (4), and using the norms mentioned above, it is not difficult to obtain

$$|J_j'| \le \sum_{l>m} l\|R_l\|\rho^l, \quad j = 1, 2,$$

where the values $\|R_l\|$ can be estimated of the following form: From the recurrence (3), we obtain bounds for the norms of the homogeneous polynomials H_r that appear in the Hamiltonian expansion. Then, we bound the norms of the homogeneous components of the successive Hamiltonians obtained (using G_3 to G_{20}). In particular we get $\|R_l\|$. We have used the recurrent relations given by the Lie transforms to obtain the bounds up to order 800. Note that, with this, we have bounded the diffusion with respect to the planar variables.

Now we would like to bound the diffusion in the vertical direction. To do that, instead of bounding $|N'|$, let us define first $\widetilde{N}$ as N minus the part of N depending only on J_1, J_2 (obviously, $\widetilde{N}$ is also an approximate first integral).

Now, we bound $\widetilde{N}'$ (as it was done with $J_{1,2}'$ by means of (4)). From this bound we need to derive a bound for the diffusion in the vertical direction. At this point we want to remark a difficulty in this approach: J_3 is not an approximate first integral but a fast variable of the system. The J_3 variable can oscillate around the elliptic equilibrium point mentioned in Section 3.3.

We have used the following approach: to ensure that if we start at a ball of radius $\eta\rho$ then we end, at most, at the boundary of a ball of radius ρ we should have

$$\min_{(z_i,z_f)\in\partial B_{\eta\rho}\times\partial B_\rho} |\widetilde{N}(z_f) - \widetilde{N}(z_i)| \ge T|\widetilde{N}'|_{B_\rho}.$$

We have used $\eta = 0.5$ (see later) and T the approximate age of the Solar system ($T = 4.5 \times 10^9$ years $\approx 2.4 \times 10^9$ adimensional units). The maximum value of ρ satisfying the previous condition is then obtained.

Notice the following fact about this method: even if the diffusion for $\widetilde{N}$ were exactly 0, we would have "diffusion" in the vertical direction, due to the (possible) libration of J_3. Obviously, this diffusion is not a real one. Otherwise, this "false" diffusion is a fast phenomenon that can be observed, and it has a physical meaning.

To have an idea of the amount of this diffusion, we can compute from the normal form a bound A of the maximum size of the above-mentioned libration motion. This size depends on the value J_3^* for which we have the elliptic equilibrium point (see Section 3.3). It is not difficult to obtain that $A/J_3^* < 0.27$ (this follows, essentially, from a careful analysis of $N_{(2)}$). This is the reason why we have taken a small value of η (0.5 instead of something close to 1).

Finally, using $\eta = 0.5$ and $T = 2.4 \times 10^9$, as stated before, the radius of the ball of (η, T)-stability is found to be $\rho = 0.571 \times 10^{-3}$.

Now we need to send that ball back to the original coordinates. We are going to split this process in two steps: the normal form change (the nonlinear part of the change obtained in Section 3) and the Floquet change.

The deformation produced by the normal form change can be bounded from the norm of the coefficients of this change of variables. This shows that this ball could be reduced in a factor of 5/6, that is, to the value $\widehat{\rho} = 0.476 \times 10^{-3}$.

As the Floquet change does not modify the vertical coordinates, we are going to focus only on the planar ones. Note that the final region produced by this change depends on the true anomaly f in a periodic way. To obtain a region of effective stability independent of f we have intersected (for all f) all these regions. This produces a domain with a "banana" shape, having a minimum diameter of $0.11\widehat{\rho}$ and a maximum one of $9.8\widehat{\rho}$. This implies that the largest ball contained in that domain and centered at the origin ($L_{4,5}$) has a radius of 0.52×10^{-4} adimensional units. On the other hand, in some "good" direction, the distance of stability is multiplied by a factor of $9.8/0.11 \approx 89$.

4.3 Remarks

Numerical simulations of the motion close to $L_{4,5}$ in the same problem show that there exist points close to $L_{4,5}$ (but outside the region mentioned before) that go away in a very short time (see Gómez et al.[5]; Simó[6]). This is due to the effect of the 1:1 resonance between the true anomaly and the vertical mode of the RTBP (this phenomenon is now being studied (see Jorba and Simó[7])). This effect is more evident when the value of μ is bigger (i.e. Earth-Moon system), because the (angle of) splitting between the stable and unstable manifolds of the hyperbolic tori mentioned above grows very fast when μ increases.

There are also other regions of stability that are even larger, allowing for large values for the inclinations (see Gómez et al.[5]). These regions are the ones containing the Trojan asteroids.

Acknowledgements. The authors have been partially supported by the CICYT grant ESP91-0403.

References

1. V.I. Arnol'd: "Dynamical Systems III", Springer-Verlag (1988).

2. A. Giorgilli, A. Delshams, E. Fontich, L. Galgani and C. Simó: Effective stability for a Hamiltonian system near an elliptic equilibrium point, with an application to the Restricted Three Body Problem, *Journal of Differential Equations* 77:1 (1989).

3. C. Simó: Estabilitat de sistemes Hamiltonians, Mem. de la Real Acad. de Cienc. i Art. de Barcelona, Vol. XLVIII, no. 7 (1989).

4. V. Szebehely: "Theory of Orbits", Academic Press (1967).

5. G. Gómez, A. Jorba, J. Masdemont and C. Simó: *Study of Poincaré maps for Orbits near Lagrangian Points*, ESOC contract 9711/91/D/IM(SC), Second Progress Report (1992).

6. C. Simó: Stability regions for the elliptic RTBP near the triangular points (in progress).

7. A. Jorba and C. Simó: Hyperbolic tori close to the equilateral points of the elliptic RTBP (in progress).

BIHAMILTONIAN SYSTEMS AND LAX REPRESENTATION

J. J. Morales and R. Ramirez

Departament de Matemàtica Aplicada II
c/ Pau Gargallo,5
E-08028 Barcelona Spain

INTRODUCTION

Between the methods for obtaining first integrals of hamiltonian systems there are two that have acquired more and more relevance in the study of integrable hamiltonian systems, they are the following:
(1). The representation of the system as a Lax Pair

$$\dot{L} = [L, A],$$

where L, A belong to a Lie algebra.
(2). The bihamiltonian method, which consist in the existence of an alternative Poisson bracket, such that the linear combinations of this bracket with the original one are a Poisson bracket too and the hamiltonian equation of the second bracket, with respect to some hamiltonian, is the same that the original one.

The first method is applied to finite dimensional as well as to infinite dimensional hamiltonian systems (partial differential equations) and practically the totality of the known integrable systems have been putting in Lax Pair form. The second method is mainly applied to partial differential equations and his systematic use in the majority of systems which however admits the first approaches is still open. It is clear that one possibility to obtain bihamiltonian systems is to relate the two above approaches in such a way that, under some restrictions, a hamiltonian system in Lax Pair form admit a bihamiltonian description. This is the point of view adopted by us in this work and it could be considered as an intent to apply several remarkable papers of Reyman[9,10] and Semenov-Tyan-Shanskii[9,11] to concrete finite dimensional examples.

Two family of examples covered are the Toda (non-periodic) type systems and the Neumann type systems.

The Toda type systems are the Toda system with the hamiltonian

$$H = \frac{1}{2}\sum_{i=1}^{i=n} y_i^2 + \sum_{i=0}^{i=n} \exp(x_i - x_{i+1}), \quad x_{n+1} = x_0 = 0,$$

and its generalizations[1].

The Neumann type systems are the Neumann system

$$H = \frac{1}{2}((\sum_{i=1}^{i=n} y_i^2)(\sum_{i=1}^{i=n} x_i^2) - \sum_{i=1}^{i=n} x_i y_i + \sum_{i=1}^{i=n} \alpha_i x_i y_i),$$

the geodesic flow on an ellipsoid,the especial Euler-Arnold equations and a central force problem on an ellipsoid[2].

BIHAMILTONIAN STRUCTURES

The adequate framework for the study the Lax Pair Theory from the Lie algebra point of view is that of (classical) R-matrices. We recall the main facts about this theory that is in fact a generalization of the Kostant-Symes-Adler Theory[1,2,6].

Let ($\mathfrak{g}$,[,]) a Lie algebra. We say that a linear operator R on $\mathfrak{g}$ is an R-matrix if the so called R-bracket

$$[X,Y]_R = [RX,Y] + [X,RY]$$

is such that $(\mathfrak{g},[,]_R)$ is a Lie algebra.

If a R-matrix exist in $\mathfrak{g}$ then we have two Poisson Structures on $\mathfrak{g}^*$

$$\{f,g\}_0(L) = < L,[df(L),dg(L)] >, \tag{1}$$

$$\{f,g\}_R(L) = < L,[df(L),dg(L)]_R >, \tag{2}$$

where $<,>$ is the duality pairing between $\mathfrak{g}$ and $\mathfrak{g}^*$.

Furthermore we assume the existence on $\mathfrak{g}$ of an invariant and non degenerate scalar product $(|)$. The condition of invariance means

$$([X,Y]|Z) = ([Y,Z]|X),$$

for all X, Y, Z, in $\mathfrak{g}$. This product permit us to identify $\mathfrak{g}$ and $\mathfrak{g}^*$. Then as shown by Semenov-Tyan-Shanskii[11] we have the following 1

Theorem 1 . *Let be R an R-matrix on the Lie algebra $\mathfrak{g}$ and also there exist a nondegenerate, invariant scalar product in $\mathfrak{g}$. Then*
(i) The Casimir functions $I(\mathfrak{g}^*)$ of $(\mathfrak{g},[,])$ are in involution with respect to $\{,\}_R$.
(ii) If H is a Casimir function, then the Hamiltonian equations of H with respect to $\{,\}_R$ has the Lax form

$$\dot{L} = [L, R(dH(L))]. \tag{3}$$

We remember that the Casimir functions of $\mathfrak{g}$ are the coadjoint invariant ones.

The best known class of R-matrices arise when the Lie algebra $\mathfrak{g}$ split in a direct sum (as vector spaces) of two subalgebras

$$\mathfrak{g} = \mathfrak{g}_+ + \mathfrak{g}_-.$$

Now, we define

$$R = \frac{1}{2}(P_+ - P_-),$$

where P_+, P_- are the projectors on $\mathfrak{g}_+$, $\mathfrak{g}_-$ along $\mathfrak{g}_-$, $\mathfrak{g}_+$ respectively. Then we have the following alternative Lax representation for a Casimir function H with respect to $\{,\}_R$ as shown by Reyman[10]

$$\dot{L} = [L, R(dH(L))] = [L, dH(L)_+] = -[L, dH(L)_-], \tag{4}$$

with $X_+ = P_+(X)$, $X_- = P_-(X)$.

Let us give two examples.

(1) We consider the Lie algebra $\mathfrak{g} = \mathfrak{sl}(n, \mathbb{C})$ of matrices with trace zero. This algebra split

$$\mathfrak{g} = \mathfrak{g}_+ + \mathfrak{g}_-,$$

being $\mathfrak{g}_+$, $\mathfrak{g}_-$ the upper triangular matrices, and $\mathfrak{g}_-$ the lower triangular matrices with zero diagonal. The scalar product in $\mathfrak{g}$ is the Killing form

$$(X|Y) = Tr(XY).$$

This product is nondegenerate ($\mathfrak{g}$ is semisimple) and invariant.

(2) Let $\mathfrak{g} = Mat_n(\mathbb{C}) \otimes \mathbb{C}[h, h^{-1}]$ the Lie algebra of Laurent polynomials in the formal variables h with coefficients in the algebra of complex matrices $n \times n$. An element $X = X(h)$ of $\mathfrak{g}$ is of the form

$$X = \sum_{i=-p}^{q} X_i h^i, \quad p, \quad q \in \mathbb{N}.$$

The nondegenerate, invariant scalar product is

$$(X|Y) = Tr \sum_{i+j=-1} X_i Y_j.$$

Let us suppose that $\mathfrak{g}$ is a Lie algebra obtained from an associative algebra. If as before we have a nondegenerate scalar product $(|)$ on $\mathfrak{g}$ such that:

(i) The multiplication in $\mathfrak{g}$ is symmetric with respect to $(|)$

$$(XY|Z) = (X|YZ)$$

.

(ii) The operator R is skewsymmetric and satisfy the so-called Yang-Baxter modified equation

$$[RX, RY] - R([X, Y]_R) = [Y, X].$$

Then as shown by Semenov-Tyan-Shanskii[11] R is an R-matrix and we can define another Poisson bracket on $\mathfrak{g}^* \approx \mathfrak{g}$

$$\{f, g\}(L) = (L|[X, R(LY)] + (L|[R(XL), Y]), \qquad X = df(L), \quad Y = dg(L) \tag{5}$$

Furthermore the two brackets $\{,\}_R$, $\{,\}$ are compatible (i.e. their linear combinations are also Poisson brackets), and the Casimir functions of $\mathfrak{g}$ are in involution with respect to both brackets. In the other hand, the Hamilton equations of a Casimir function H with respect to $\{,\}$ are in Lax pair form too

$$\dot{L} = [R(Ldh(L)), L]. \tag{6}$$

We remark that (5) is a generalization of the Adler's second Poisson bracket[1] for KdV type equations.

Now we can introduce the concept of bihamiltonian or, in general, multihamiltonian system on a manifold M. If there are m compatible Poisson brackets $\{,\}_i$, $i = 1, ..., m$ (their linear combinations are Poisson brackets again), and Γ is a hamiltonian vector field with respect to all of them with hamiltonians H_i

$$\Gamma = J_i dH_i, \quad i = 1, ..., m \tag{7}$$

J_i being the hamiltonian mappings from $\{,\}_i$,then we say that the field Γ is a multihamiltonian system. If $m = 2$ the system is bihamiltonian.

We are interested in analysing the bihamiltonian systems on a finite dimensional Poisson subspace M with respect to $\{,\}_R$. For example the subspace

$$\mathfrak{g}_{-p,q} = \{\sum_{i=-p}^{q} A_i h^i | \quad A_i \in Mat_n(\mathbb{C})\} \tag{8}$$

is a Poisson subspace with respect to $\{,\}_R$.

Theorem 2. *Let R be a skewsymmetric linear operator on $\mathfrak{g}$, M is a Poisson finite dimensional subspace of $\{,\}_R$ and H_0 is a Casimir function on M. Then exist a sequence of functions H_n on M, such that*
(i) $J(L)dH_{n-1}(L) = J_R(L)dH_n(L)$, $\quad L \in M$, $\quad n \geqslant 1$
(ii) $\{H_i, H_j\} = \{H_i, H_j\}_R = 0$, $i, j \geqslant 0$.

Sketch of proof. We inductively define

$$dH_k(L) = LdH_{k-1}(L)$$

and we obtain

$$[L, dH_{k-1}(L)] = 0,$$

since, by the induction hypothesis, H_{k-1} is a Casimir function. This implies that the 1 form $dH_k(L) = LdH_{k-1}(L)$ is closed. Hence H_k exist locally and is a Casimir function. But all the Casimir functions on $\mathfrak{g}$ are in involution and we have obtained (ii).

The hamiltonian mappings of $\{,\}$, $\{,\}_R$ are respectively

$$J(L)X = [R(LX), L] - LR([X, L]),$$

$$J(L)X = [R(X), L] - R([X, L]).$$

Then

$$J(L)dH_{k-1}(L) = [R(LdH_{k-1}(L)), L] = [R(dH_k(L)), L] = J_R(L)dH_k(L),$$

that is (i).

We observe that, although this theorem is a consequence of that at Semenov-Tyan-Shanskii[11], we start here with only a hamiltonian H_0 and then obtain the hierarchy of bihamiltonian systems

$$\Gamma_k = J_R dH_k = JdH_{k-1},$$

whose flows commute.

If $\mathfrak{g} = Mat_n(\mathbb{C}) \otimes \mathbb{C}[h, h^{-1}]$ and

$$R = \frac{1}{2}(P_+ - P_-)$$

then the subspace $\mathfrak{g}_{-p,q}$ is a Poisson one with respect to $\{,\}_R$. In this situation as shown by Reyman and Semenov-Tyan-Shanskii[9] $R_k = Rh^k$ (h^k multiplicative linear operator on

$\mathfrak{g}$) is also an R-matrix, $-p \le k \le q+1$. Futhermore the family of Lie-Poisson brackets $\{,\}_{R_k}$ are compatible and the Casimir functions are in involution with respect to any of them.

Hence we obtain the multihamiltonian system on M

$$\dot{L} = [L, R(dH(L))] = [L, R_{-p}(dH^{-p}(L))] = \cdots = [L, R_k(dH^k(L))] = \cdots$$

where $H = H(L)$ is a Casimir function and

$$H^k(L) = h^{-k} H(L).$$

Now we can apply the theorem 2 to each R_k since this operator is skewsymmetric and we have the following diagram associated to the same hamiltonian vector field $\Gamma(L) = J(L)dH_0(L)$ (we maintain the theorem's notations)

$$\begin{array}{ccc} dH_0(L) & \xrightarrow{L} & dH_1(L) = LdH_0(L) \\ {\scriptstyle h^{-k}}\downarrow & & \downarrow{\scriptstyle h^{-k}} \\ dH_0^k(L) & \xrightarrow{L} & dH_1^k(L) = LdH_0^k(L) \end{array}$$

Analogously, we obtain similar diagrams for each field

$$\Gamma_n(L) = J(L)dH_{n-1}(L) = J_R(L)dH_n(L)$$

of the hierarchy obtained in the theorem 2.

Then from H_0 we deduce the hamiltonians H_i^k, $i = 1, 2$, such that any of them generate the same hamiltonian field Γ, but with respect to a different Poisson bracket. This is valid also for every element H_n of the hierarchy.

EXAMPLES

For the Toda lattice $\mathfrak{g} = \mathfrak{sl}(n, \mathbb{C})$ and by the application of theorem 2 we recover the bihamiltonian structure obtained by Adler[1]. The second hamiltonian is the total moment.

In order to apply the theorem 2 to Neumann system we shall work with

$$\mathfrak{g} = Mat_n(\mathbb{C}) \otimes \mathbb{C}[h, h^{-1}],$$

$$L = \alpha h^2 + \Gamma_{xy} h - \Gamma_{xx},$$

where,

$$\alpha = \text{diag}(\alpha_1, \alpha_2, ..., \alpha_n), \quad \alpha_i \in \mathbb{R},$$

$$(\Gamma_{xx})_{ij} = x_i x_j, \quad (\Gamma_{xy})_{ij} = x_i y_j - x_j y_i$$

(for the splitting and the scalar product in $\mathfrak{g}$ see the examples in the above parts).

We observe that

$$L \in \mathfrak{g}_{0,2}.$$

For the initial hamiltonian we take

$$H_0(L) = (Lh^{-1}|I),$$

where I is the identity matrix in $\mathfrak{g}$. It is the Casimir function

$$H_0 = -(x, x),$$

being (,) the usual scalar product in $\mathbb{R}^n$.

Then aplying the theorem 2 and its consequence, we obtain for $k = 2$ the following diagram

$$\begin{array}{ccc} dH_0(L) = Ih^{-1} & \xrightarrow{L} & dH_1(L) = Lh^{-1} \\ {\scriptstyle h^{-2}}\downarrow & & \downarrow{\scriptstyle h^{-2}} \\ dH_0^2(L) = Ih^{-3} & \xrightarrow{L} & dH_1^2(L) = Lh^{-3} \end{array}$$

In consequence

$$H_1 = -(x, x)^2, \quad H_0^2 = \sum \alpha_i,$$

$$H_1^2 = (x, x)(y, y) - (x, y)^2 + 2(\alpha x, x).$$

In such a way, we have obtained the bihamiltonian system of Adler and Van Moerbeke for the Neumann system with hamiltonians H_1, H_0^2.

FINAL REMARKS

(1). There are several difficulties to extending this method to certain hamiltonian systems which admit a Lax Pair representation. For example, as shown Leo and others[7] the periodic Toda system is a bihamiltonian system, but the application of the multihamiltonian approach of Reiman and Semenov-Tyan-Shanskii only gives trivial hamiltonians . Furthermore, we have problems in order to apply the generalized Adler Poisson bracket $\{,\}_R$, since the linear operator R is not skewsymmetric if we select $R = 1/2(P_+ - P_-)$ (we take here the formulation of Adler and Van Moerbeke on the periodic Toda system). But this system is bihamiltonian as shown by Leo et al.[7].

(2). Some authors[3,8] studied the reciprocal problem: admit a bihamiltonian system a Lax Pair representation?. There are some difficulties in order to answer to this question.

(3). From a lagrangian point of view the analogous to bihamiltonian systems are the bilagrangian systems[3,5,8].

REFERENCES

1. M. Adler, On a Trace Functional for Formal Pseudo-Diferential Operators andThe Symplectic Structure of the Korteweg-DeVries Type Equations. *Inv. Math.* 50(1979),219.
2. M. Adler, P. van Moerbeke, Completely Integrable Systems, Euclidean Lie Algebras and Curves. *Adv. in Math.* 38(1980),267.
3. M. Crampin, A Note on non-Noether constants of motion, *Phys. Lett.* 95A(1983), 209.
4. L. Fadeev, L.A. Takhtajan," Hamiltonian Methods in the Theory of Solitons". Springer-Verlag 1985.
5. S. Hojman, H. Harleston, Equivalent Lagrangians: Multidimensional case, *J. Math. Phys.* 22 (1981), 1414.
6. B. Kostant, The Solutions to a Generalized Toda Lattice and Representation Theory, *Adv.Math.* 34(1979), 195.
7. M. Leo, R. A. Leo, G. Soliani, L. Solombino, G. Mancarella, Symmetry Properties and Bi-Hamiltonian Structure of Toda Lattice, *Lett. Math. Phys.* 8(1984), 267.
8. M. Lutzki, Non canonical Symmetries and Isoespectral Representation of Hamiltonian Systems, *Phys. Lett.* 87A(1982), 274.

9. A. G. Reyman, M. A. Semenov-Tyan-Shanskii, Compatible Poisson Structures for Lax Equations: an r-matrix approach, *Phys. Lett.* A130(1988), 456.
10. A. G. Reyman, Integrable Systems conected with graded Lie algebra, *J. Sov. Math.* 19 (1982),1507.
11. M. A. Semenov-Tyan-Shanskii, What is a Classical R-matrix ?, *Fuct. Anal. Appl.* 17 (1983),259.

AN EFFICIENT METHOD FOR COMPUTING PERIODIC ORBITS OF CONSERVATIVE DYNAMICAL SYSTEMS

M. N. VRAHATIS AND T. C. BOUNTIS

Department of Mathematics, University of Patras
GR–261.10 Patras, Greece

ABSTRACT

The accurate computation of periodic orbits and the precise knowledge of their properties are very important for studying the behavior of many dynamical systems of physical interest. In this paper, we present an efficient numerical method for computing to any desired accuracy periodic orbits (stable, unstable and complex) of any period. This method always converges rapidly to a periodic orbit independently of the initial guess, which is important when the mapping has many periodic orbits, stable and unstable close to each other, as is the case with conservative systems. We illustrate this method first, on the 2–Dimensional quadratic Hénon's mapping, by computing rapidly and accurately several periodic orbits of high period. We also apply our method here to a 3–D conservative mapping as well as a 4–D complex version of Hénon's map.

1. INTRODUCTION

It is well known that two dimensional (2–D) mappings of the (x_1, x_2) plane onto itself defined by :

$$T \; : \; \begin{cases} \hat{x}_1 = \varphi_1(x_1, x_2), \\ \hat{x}_2 = \varphi_2(x_1, x_2), \end{cases} \tag{1.1}$$

can be used to study dynamical systems with two degrees of freedom. Such mappings can model conservative dynamical systems, if the determinant of the Jacobian of the map, i.e. $|det\, J_T| = 1$, or dissipative ones if $|det\, J_T| < 1$, [2, 3, 10, 11]. We say that $X = (x_1, x_2)$ is a fixed point of T if $T(X) = X$ and a fixed point of *order* p, (or a periodic orbit of *period* p), if :

Hamiltonian Mechanics, Edited by J. Seimenis,
Plenum Press, New York, 1994

$$X = T^p(X) \equiv \underbrace{T\Big(T(\cdots(T(X))\cdots)\Big)}_{p \text{ times}}. \tag{1.2}$$

In general, it is difficult to find in the literature efficient methods for computing orbits of high period if the mapping is not decomposable into involutions [6, 20]. Also, traditional iterative schemes, such as Newton's method and related classes of algorithms [5, 16], often fail since they converge to a fixed point almost independently of the initial guess, while there may exist several different fixed points, close to each other, which are all desirable for the applications. Moreover, these methods are affected by the mapping evaluations taking large values in the neighborhood of *unstable* or *saddle–hyperbolic* periodic orbits, or may fail due to the nonexistence of derivatives or poorly behaved partial derivatives near fixed points [5, 16].

In this paper, we describe an efficient numerical method for rapidly computing periodic orbits (stable or unstable) of any period and to any desired accuracy. This method exploits topological degree theory to provide a criterion for the existence of a periodic orbit of an iterate of the mapping within a given region. More specifically, the method begins by constructing a polyhedron in such a way that the value of the topological degree of an iterate of the mapping relative to this polyhedron is ± 1, which means that there exists a periodic orbit within this polyhedron. Then it repeatedly subdivides its edges (and diagonals) so that the new polyhedron also contains a periodic orbit within its interior, avoiding any computation of the topological degree. These subdivisions continue iteratively until the periodic orbit is computed to the desired accuracy.

This method is especially useful for the computation of high period orbits (stable or unstable) and is quite efficient, since the only computable information required is the algebraic signs of the components of the mapping. Thus it is not affected by the mapping evaluations taking large values in neighborhoods of unstable periodic orbits.

In the next section we start by giving a criterion for the existence of a periodic orbit inside a converging sequence of smaller and smaller characteristic polyhedra (the so–called CP–criterion). Also, in that section, we describe a generalized bisection method used in combination with the CP–criterion to compute the desired periodic orbit to any accuracy.

In Section 3, this procedure is applied to the calculation of stable and unstable periodic orbits of the 2–D Hénon's mapping for periods which reach up to the thousands. Then in Section 4, we apply our methods to periodic orbits of a 3–D conservative mapping and a 4–D complex version of the 2–D Hénon's map.

Finally we end, in Section 5, with some concluding remarks and a discussion of ongoing work on the application of these methods to symplectic mappings, which are known to model the behavior of Hamiltonian systems.

2. THE CP–CRITERION

In this section, we implement topological degree theory to give a criterion for the existence of a periodic orbit within a given region of the phase space of the system. This criterion is based on the construction of a sequence of "characteristic polyhedra" within a scaled translation of the unit cube. The concept of a characteristic polyhedron will be reviewed and a procedure for its construction will be presented. The theoretical development of the concepts employed here can be found in [23, 28].

As we said previously, the problem of finding periodic orbits of nonlinear mappings $T = (\varphi_1, \varphi_2, \ldots, \varphi_n)\colon \mathcal{D} \subset \mathbb{R}^n \to \mathbb{R}^n$ of period p amounts to finding points $X^\star = (x_1^\star, x_2^\star, \ldots, x_n^\star)^\top \in \mathcal{D}$ which satisfy the following equation :

$$T^p(X^\star) = X^\star. \tag{2.1}$$

Obviously, the problem of finding such a periodic orbit is equivalent to solving the following system :

$$F(X) = \mathcal{O}, \tag{2.2}$$

with $F = (f_1, f_2, \ldots, f_n) = T^p - I_n$, where I_n indicates the identity mapping and $\mathcal{O} = (0, 0, \ldots, 0)^\top$ is the origin of $\mathbb{R}^n$.

Many problems require the solution of systems of nonlinear equations for which Newton's method and related classes of algorithms [5, 16] fail due to nonexistence of derivatives or poorly behaved partial derivatives. Also, Newton's method, (as well as Newton–like methods), often converge to a solution X^* of $F(X) = \mathcal{O}$ almost independently of the initial guess, while there may exist several solutions nearby all of which are desired for the application. Because of these reasons, various approaches based upon topological degree theory and generalized bisection methods have been investigated in recent years [7, 10, 22, 23, 24, 28].

Bisection methods for finding solutions of systems of equations depend on a criterion, which guarantees that a solution lie within a given region. Then this region is subdivided in such a way that the criterion can again be applied to the new refined region.

In one dimension, this criterion consists of the product of the signs of the function evaluations at the endpoints of a given interval. Specifically, if one desires to locate a solution of a equation $f(x) = 0$ in the interval (a, b) where $f\colon (a, b) \subset \mathbb{R} \to \mathbb{R}$ one examines whether the following relation is fulfilled :

$$sgn\, f(a) \cdot sgn\, f(b) = -1, \tag{2.3}$$

where sgn is the sign function with values :

$$sgn\, \psi = \begin{cases} -1, & \text{if} \quad \psi < 0\ ; \\ 0, & \text{if} \quad \psi = 0\ ; \\ 1, & \text{if} \quad \psi > 0\ . \end{cases} \tag{2.4}$$

If (2.3) holds, then we know that there is at least one solution within (a, b). This is known as ***Bolzano's existence criterion*** and it can be generalized to higher dimensions [15, 25].

Instead of Bolzano's criterion however one may also use condition :

$$deg[f, (a, b), 0] = \frac{1}{2}\{sgn\, f(b) - sgn\, f(a)\}, \tag{2.5}$$

where $deg[f, (a, b), 0]$ is the ***topological degree of f at the origin relative to*** (a, b). Now, if the value of $deg[f, (a, b), 0]$ is not zero, we know with certainty that there is at least one solution in (a, b), since, in that case, Bolzano's criterion is fulfilled. The value of $deg[f, (a, b), 0]$ gives additional information concerning the behavior of the solutions of $f(x) = 0$ in (a, b) relative to the slopes of f [10]. For example, if $deg[f, (a, b), 0] = 1$ (which means that $f(b) > 0$ and $f(a) < 0$), then the number of solutions at points where $f(x)$ has a positive slope exceeds by one the number of solutions at points at which $f(x)$ has a negative slope.

The topological degree as well as Bolzano's criterion transfer all information regarding the roots to the boundary of the given region. Now, using the value of the topological degree, (or Bolzano's criterion), one can calculate a solution of $f(x) = 0$ by bisecting the interval (a, b). So we subdivide (a, b) into two subintervals $(a, c]$, $[c, b)$, where $c = (a + b)/2$ is the midpoint of (a, b), and keep the subinterval for which the value of the topological degree is not zero relative to itself, by checking the information on the boundaries. In this way, we keep at least one solution within a smaller interval. We can continue this procedure until the endpoints of the final subinterval differ from each other by less than a fixed amount. This method is called ***bisection method*** and can be expressed as follows [12, 23, 24, 27, 29] :

$$x_{n+1} = x_n + sgn\, f(a) \cdot sgn\, f(x_n) \cdot (b-a)/2^{n+1}, \quad x_0 = a, \quad n = 0, 1, \ldots . \tag{2.6}$$

Of course, it converges to a solution x^* in (a, b) if for some x_n, $n = 1, 2, \ldots$, we have :

$$sgn\, f(x_0) \cdot sgn\, f(x_n) = -1. \tag{2.7}$$

Also, the minimum number of iterations γ, which are required to obtain an approximate solution x' such that $|x' - x^*| \leq \varepsilon$ for some $\epsilon \in (0, 1)$ is given by :

$$\gamma = \left\lceil \log_2\left((b-a) \cdot \varepsilon^{-1}\right)\right\rceil , \tag{2.8}$$

where the notation $\lceil \cdot \rceil$ refers to the smallest integer which is not less than the real number quoted.

Based on the relation (2.8), it has been proved in [21] that the bisection method is *optimal*, i.e. that it possesses asymptotically the best rate of convergence. Also, it is worth mentioning that the only computable information required by the bisection method is the signs of various function evaluations, and, as it always converges within the given interval (a, b) independently of its length, it is a global convergence method. Using then the relation (2.8), one can easily find out the number of iterations needed for the attainment of an approximate solution to a predetermined accuracy.

It would be very desirable, of course, to generalize the above bisection method to higher dimensions. To do this we extend Bolzano's criterion in the following way : Let us define a characteristic n–polyhedron by constructing the $2^n \times n$ matrices $\mathcal{M}_n$ whose rows are formed by all possible combinations of -1, 1. For example for $n = 1, 2, 3$ we have :

$$\mathcal{M}_1 = \begin{bmatrix} -1 \\ 1 \end{bmatrix}, \qquad \mathcal{M}_2 = \begin{bmatrix} -1 & -1 \\ -1 & 1 \\ 1 & -1 \\ 1 & 1 \end{bmatrix}, \qquad \mathcal{M}_3 = \begin{bmatrix} -1 & -1 & -1 \\ -1 & -1 & 1 \\ -1 & 1 & -1 \\ -1 & 1 & 1 \\ 1 & -1 & -1 \\ 1 & -1 & 1 \\ 1 & 1 & -1 \\ 1 & 1 & 1 \end{bmatrix}. \tag{2.9}$$

Now, for $n = 1$ we consider the segment $[x_1, x_2]$ and evaluate the sign of $f(x)$ at the endpoints. Then, if the matrix :

$$\mathcal{S}(f;\, [x_1, x_2]) = \begin{bmatrix} sgn\, f(x_1) \\ sgn\, f(x_2) \end{bmatrix}, \tag{2.10}$$

agrees with $\mathcal{M}_1$, up to a permutation of the rows, then we say that $[x_1, x_2]$ is a *characteristic polyhedron*. Suppose now that $\Pi^n = \langle \Upsilon_1, \Upsilon_2, \ldots, \Upsilon_{2^n} \rangle$ is an oriented n–dimensional polyhedron with 2^n vertices, $\Upsilon_k \in \mathbb{R}^n$, (i.e. an orientation has been assigned to its vertices), and let $F = (f_1, f_2, \ldots, f_n)\colon \Pi^n \subset \mathbb{R}^n \to \mathbb{R}^n$ be a nonlinear mapping from Π^n into $\mathbb{R}^n$. Then we call a *matrix of signs associated with F and Π^n*, and denote it by $\mathcal{S}(F;\, \Pi^n)$, the $2^n \times n$ matrix whose entries in the kth row are the corresponding coordinates of the vector :

$$sgn\, F(\Upsilon_k) = \left(sgn\, f_1(\Upsilon_k), sgn\, f_2(\Upsilon_k), \ldots, sgn\, f_n(\Upsilon_k)\right)^{\top}. \tag{2.11}$$

The n-polyhedron $\Pi^n = \langle \Upsilon_1, \Upsilon_2, \ldots, \Upsilon_{2^n} \rangle$ in $\mathbb{R}^n$ is called a *characteristic n-polyhedron relative to* $F = (f_1, f_2, \ldots, f_n)\colon \Pi^n \subset \mathbb{R}^n \to \mathbb{R}^n$, if the matrix of signs associated with F and Π^n, $\mathcal{S}(F;\, \Pi^n)$, is identical with the n–complete matrix $\mathcal{M}_n$.

Suppose now that Π^n is a characteristic n–polyhedron and that $F = (f_1, f_2, \ldots, f_n)\colon \Pi^n \subset \mathbb{R}^n \to \mathbb{R}^n$ is continuous. Then, under suitable assumptions on the boundary of Π^n the value of the topological degree of F at $\mathcal{O}$ relative to Π^n is given by :

$$deg[F, \Pi^n, \mathcal{O}] = \sum_{X \in F^{-1}(\mathcal{O})} sgn\, det\, J_F(X) = \pm 1 \neq 0, \tag{2.12}$$

(see [28, Theorem 2.9]), which implies the existence of a periodic orbit inside Π^n. For a detailed description of how to construct a characteristic n-polyhedron to locate a desired periodic orbit see [23, 24, 26].

Next, we turn to the description of a generalized bisection method, used in combination with the CP–criterion outlined above, for computing periodic orbits of any period and any desired accuracy. This method has all the advantages of one–dimensional bisection and is particularly useful in cases where the period of the orbit is high since it always converges within the initial specified region. Moreover, it is very efficient, since the only information it requires is the algebraic signs of the components of the mapping.

This method is based on the refinement of a characteristic n–polyhedron and may be called *characteristic bisection*. In the literature several bisection methods are available [7, 10, 22] that require the computation of the topological degree in order to secure its nonzero value. In the bisection method outlined here, however, the computation of the topological degree is avoided by making sure that it retains a nonzero value at every iteration.

The method bisects a characteristic n–polyhedron Π^n in such a way that the new refined n–polyhedron is also a characteristic one. To do this it computes the midpoint of a proper 1–simplex (edge) of Π^n and uses it to replace that vertex of Π^n for which the vectors of their signs are identical (see [23, 24, 26, 28] for details).

The number of characteristic bisections of the proper 1–simplexes of a Π^n required to obtain a new refined characteristic n–polyhedron $\Pi^n_\star$ such that the length of its longest edge $\Delta(\Pi^n_\star)$ satisfies $\Delta(\Pi^n_\star) \leq \varepsilon$, for some $\varepsilon \in (0, 1)$, is given by :

$$\zeta = \left\lceil \log_2 \left(\Delta(\Pi^n) \cdot \varepsilon^{-1} \right) \right\rceil , \tag{2.13}$$

(see [28] for a proof). Based on the above formula we conjecture that our generalized bisection method is "optimal". We plan, however, to address the complete proof of this in a future publication.

3. An Application to Hénon's 2–D Mapping

We shall first illustrate the method of Section 2 on a quadratic area–preserving 2–D mapping originally due to Hénon [11] :

$$T \ : \ \begin{cases} \hat{x}_1 = x_1 \cos a - (x_2 - x_1^2) \sin a, \\ \hat{x}_2 = x_1 \sin a + (x_2 - x_1^2) \cos a. \end{cases} \tag{3.1}$$

For a point $X^\star$ which has period p this means that $T^p(X^\star) = X^\star$ so we consider the following mapping :

$$F = (f_1, f_2) = T^p - I_2, \tag{3.2}$$

where I_2 is the identity mapping, and solve, for any period p, the following system of 2 equations in 2 unknowns :

$$F(X) = \mathcal{O} = (0, 0)^\top. \tag{3.3}$$

To do this we choose a starting point :

$$X^0 = (x_1^0, x_2^0)^\top, \tag{3.4}$$

and two stepsizes in each coordinate direction :

$$H = (h_1, h_2)^\top, \tag{3.5}$$

in such a way that the box thus constructed forms a domain within which we will attempt to locatc and compute a solution of the system $F(X) = \mathcal{O}$, which is a periodic orbit of the mapping T^p.

Suppose now that a periodic point $X_1^\star$ has been computed within a predetermined accuracy ε such that :

$$\|T^p(X_1^\star) - X_1^\star\| \leq \varepsilon. \tag{3.6}$$

Then, in order to compute all the other points $X_i^\star$, $i = 2, \ldots, p$ with the same accuracy ε we iterate the mapping T as follows : First we obtain an approximation $\hat{X}_2$ of the next point of the orbit $X_2^\star$ by the following relation :

$$\hat{X}_2 = T(X_1^\star), \tag{3.7}$$

and check if the following relation is fulfilled :

$$\|T^p(\hat{X}_2) - \hat{X}_2\| \leq \varepsilon. \tag{3.8}$$

Let us illustrate the above procedure on the mapping (3.1). In general, a visualization of the orbits of the mapping is very helpful for choosing the starting point X^0 and the stepsizes H. In any case, if such a visualization is not available, one can search within various boxes taking a suitable grid for the domain of interest. The phase plots (Figures 1, 2) which are shown here are drawn using the new software package GIOTTO of [29].

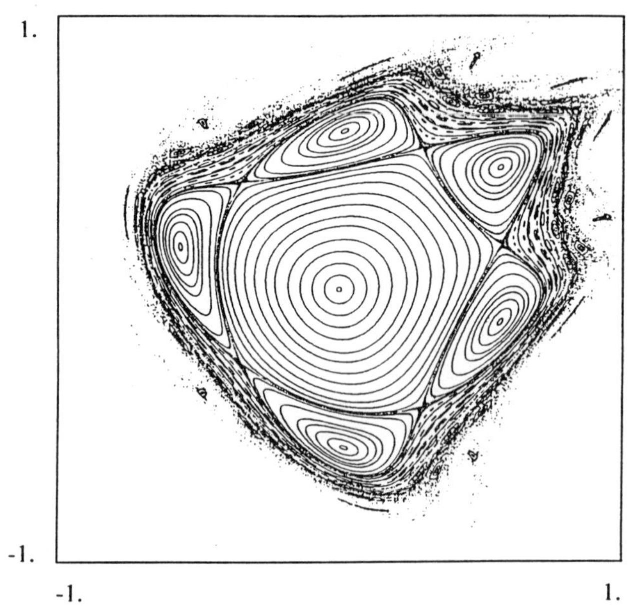

Figure 1. Hénon's mapping for $\cos a = 0.24$.

Taking for example $a = \cos^{-1}(0.24)$, we can see in Figure 1 that there is a chain of five big islands around the origin. So, in this case, we shall search for five elliptic and five hyperbolic fixed points of period $p = 5$. The reason for this choice of a is that the corresponding phase plot has a large region of stability around the origin (see [3, p.1868], [11, p.298]) and may be of interest to applications in beam dynamics.

Now in order to find the elliptic periodic orbit of this period we choose one island, include it into a box by taking appropriate values for X^0 and H, e.g. :

$$X^0 = \begin{pmatrix} x_1^0 \\ x_2^0 \end{pmatrix} = \begin{pmatrix} 0.3 \\ -0.3 \end{pmatrix}, \qquad H = \begin{pmatrix} h_1 \\ h_2 \end{pmatrix} = \begin{pmatrix} 0.4 \\ 0.4 \end{pmatrix},$$

and apply our method with accuracy $\varepsilon = 10^{-16}$. Thus, we compute the following stable periodic orbit :

$$\begin{aligned}
X_1^5 &= (\ \ 0.5672405470221847,\ -0.1223202134278941)^\top, \\
X_2^5 &= (\ \ 0.5672405470221847,\ \ \ 0.4440820516139216)^\top, \\
X_3^5 &= (\ \ 0.0173925844399303,\ \ \ 0.5800185952239573)^\top, \\
X_4^5 &= (-0.5585984457571741,\ \ \ 0.1560161118011652)^\top, \\
X_5^5 &= (\ \ 0.0173925844399305,\ -0.5797160932304572)^\top,
\end{aligned}$$

utilizing about 6.5 *m*secs of CPU time on the CERN VAX 9000–410 system.

It is well known that periodic orbits are identified by their *rotation number* σ :

$$\sigma = \frac{\nu}{2\pi} = \frac{m_1}{m_2},$$

where ν is the *frequency* of the orbit and m_1, m_2, are two positive integers (see [4, 9]). So from the sequence with which the above points are created on the x_1, x_2 plane, we can infer the rotation number of this orbit $\sigma = m_1/m_2 = 1/5$, indicating that it has produced $m_2 = 5$ points by rotating around the origin $m_1 = 1$ times.

Now to compute the unstable periodic orbit of period $p = 5$ we may choose the initial values :

$$X^0 = \begin{pmatrix} x_1^0 \\ x_2^0 \end{pmatrix} = \begin{pmatrix} 0.1 \\ -0.7 \end{pmatrix}, \qquad H = \begin{pmatrix} h_1 \\ h_2 \end{pmatrix} = \begin{pmatrix} 0.4 \\ 0.4 \end{pmatrix},$$

and apply our method with the same accuracy $\varepsilon = 10^{-16}$. Thus, the following orbit is computed :

$$\begin{aligned}
X_1^5 &= (\ \ 0.2942106885737921,\ -0.4274862418615337)^\top, \\
X_2^5 &= (\ \ 0.5696326513533602,\ \ \ 0.1622406787439296)^\top, \\
X_3^5 &= (\ \ 0.2942106885737916,\ \ \ 0.5140461711325987)^\top, \\
X_4^5 &= (-0.3443814883177751,\ \ \ 0.3882084578625210)^\top, \\
X_5^5 &= (-0.3443814883177746,\ -0.2696098483665559)^\top.
\end{aligned}$$

also utilizing around 6.5 *m*secs of CPU time.

Similarly we can use this procedure to compute periodic orbits of higher periods. Looking at the phase plot (Figure 1) of Hénon's mapping for the same value of a as before, we are able to distinguish 16 islands across the "boundary" of the mapping. Hence, in this case, we search for 16 stable and 16 unstable periodic orbits of period $p = 16$. To do this we choose one island and enclose it into a box by choosing appropriate values for X^0 and H :

$$X^0 = \begin{pmatrix} x_1^0 \\ x_2^0 \end{pmatrix} = \begin{pmatrix} 0.8 \\ 0.1 \end{pmatrix}, \qquad H = \begin{pmatrix} h_1 \\ h_2 \end{pmatrix} = \begin{pmatrix} 0.1 \\ 0.1 \end{pmatrix},$$

and apply our method (with accuracy of $\varepsilon = 10^{-16}$). Note that, from the sequence with which these points are created on the x_1, x_2 plane, we can infer the rotation number of this orbit :

$$\sigma = m_1/m_2 = 3/16,$$

indicating that it has produced $m_2 = 16$ points by rotating around the origin $m_1 = 3$ times.

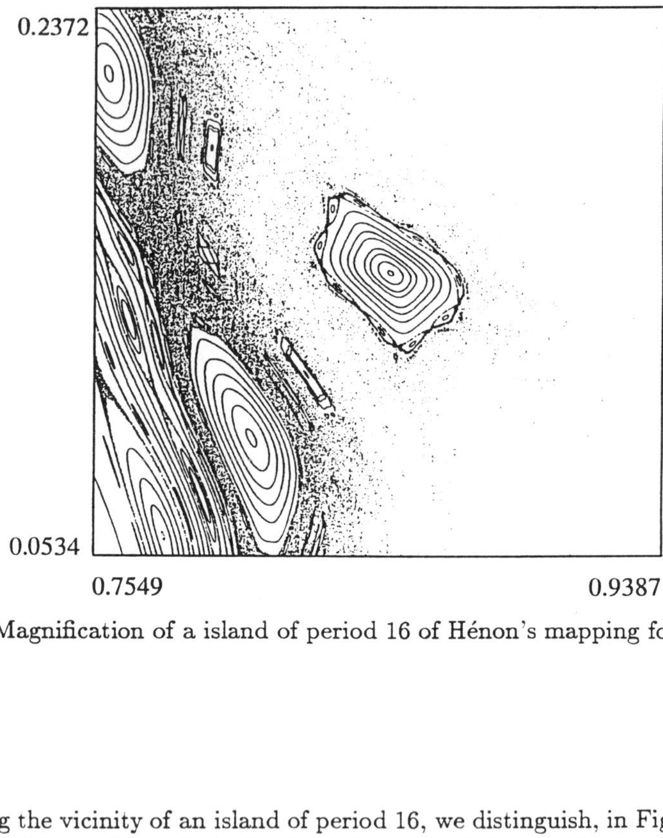

Figure 2. Magnification of a island of period 16 of Hénon's mapping for $\cos a = 0.24$.

Enlarging the vicinity of an island of period 16, we distinguish, in Figure 2, a chain of 8 islands around it, so the period of each of these islands is $p = 16 \times 9 = 144$. To compute all the points of this period, we start again with the computation of one which is stable, taking, e.g.

$$X^0 = \begin{pmatrix} x_1^0 \\ x_2^0 \end{pmatrix} = \begin{pmatrix} 0.865 \\ 0.131 \end{pmatrix}, \qquad H = \begin{pmatrix} h_1 \\ h_2 \end{pmatrix} = \begin{pmatrix} 0.005 \\ 0.005 \end{pmatrix}.$$

Enlarging further the vicinity of an island of period 144 a chain of 9 islands appears around it, with period $p = 144 \times 9 = 1296$. Finally, by enlarging the vicinity of an island of this period we distinguish a chain of 8 islands around it, with period $p = 1296 \times 8 = 10368$, where each one has a chain of 6 islands around it, with period $p = 10368 \times 6 = 62208$, etc. The rotation numbers of these orbits, respectively, are $\sigma = 243/1296$, 1944/10368, 11664/62208 and the coordinates of some of their points are listed in Table 1. The calculation of the corresponding unstable orbits requires CPU times of the same, if not smaller magnitude.

It is now clear, from the above, that one can similarly proceed to calculate orbits of higher and higher period in CPU times which do not grow so rapidly as to make the calculation of these orbits impractical (see Table 1).

Table 1 . Rotation Numbers and Periodic Orbits of Hénon's 2–D Mapping

Rotation number σ	x_1	x_2	CPU time
1/5	0.5672405470221847	0.4440820516139216	6.5 *msec*
3/16	0.8504309709743801	0.1490801034942473	11.1 *msec*
27/144	0.8685006897387088	0.1341772208865609	46 *msec*
246/1296	0.8656539452140570	0.1320418594308142	2.2 sec
1944/10368	0.8655111610927791	0.1321457880746314	5.6 sec

4. Periodic Orbits of Higher–Dimensional Maps

Recently, we have begun to study the sequence of period–doubling bifurcations in a 3–D mapping given by the equations :

$$T : \begin{cases} \hat{x}_1 = x_2, \\ \hat{x}_2 = x_3, \\ \hat{x}_3 = -x_1 + x_2^2 + x_3^2 + \mu\left(x_2^2 x_3 + x_2 x_3^2\right), \end{cases} \tag{4.1}$$

where μ is a free parameter. Such mappings are known to arise as special solutions of certain discretized lattice equations [17, 18], as well as in some problems of fluid dynamics [8]. The system (4.1) has the properties of being volume–preserving and reversible [20].

We have followed the bifurcations of an orbit of period 2 to an orbit of period 4 and then to orbits of period 8, 16, 32 etc., for several values of μ, and have observed that they occur in a very similar way as in the case of 2–D area–preserving mappings : In other words, periodic orbits of period 2^n are "born" stable on a symmetry plane $x_1 = x_3$ of initial conditions x_1, x_2 at points where the "mother" orbit of period 2^{n-1} destabilizes.

This destabilization occurs with two of the eigenvalues of the period 2^{n-1} orbit "colliding" at -1 while the 3d eigenvalue is always $+1$. In Table 2 we list some of these bifurcation points computed by our method to an accuracy of 10^{-16}.

Table 2 . Bifurcation Points for $\mu = 4$

Bifurcation	x_1	x_2	x_3
$2 \leftrightarrow 4$	0.9276675561404863	-0.5446097058809972	0.9276675561404863
$4 \leftrightarrow 8$	1.1388838325396091	-0.5565872257732393	1.1388838325396091
$8 \leftrightarrow 16$	1.0758404065131902	-0.5017191586916615	1.0758404065131902
$16 \leftrightarrow 32$	1.0912618737438293	-0.5163084493440770	1.0912618737438293

These results lead us to believe that (4.1) may have one analytic integral for all μ and thus be reducible to a 2–D mapping [19]. Such an integral can be easily found for $\mu = 0$:

$$\mathcal{I}_0 = \left(x_2 - x_1 - x_3 + x_2^2\right)^2 = \text{const.}, \tag{4.2}$$

and $\mu = \infty$,

$$\mathcal{I}_\infty = x_1^2 + x_2^2 + x_3^2 = \text{const.}, \tag{4.3}$$

but is as yet unknown for general values of μ.

Furthermore, the reversibility and perhaps also integrability of (4.1) may be destroyed by multiplying the cubic terms in (4.1c) by two different parameters $\mu_1 \neq \mu_2$, thus breaking the symmetry of the equations under $x_1 \leftrightarrow x_3$. All this can be studied by following the bifurcation properties of period 2^n orbits, using the methods described in this paper and is currently under investigation [12].

Finally, let us apply our method to the computation of periodic orbits of Hénon's mapping (1.3) with complex components, which is an example of a real 4–D conservative mapping. In order to compute such complex periodic orbits we expand a given 2–D mapping to 4 dimensions by separating real and imaginary parts. Thus we construct a 4–D real map T_4 and solve the following system of 4 equations :

$$F = T^p - I_4 = \mathcal{O} = (0,0,0,0)^\top . \tag{4.4}$$

where I_4 is the identity mapping.

Let us illustrate this procedure on Hénon's mapping (3.1) : Replacing x_1, x_2 by their complex form, $x_j + i\, y_j$, $j = 1, 2$, $x_j, y_j \in \mathbb{R}$, yields the following system of equations :

$$\begin{pmatrix} \hat{x}_1 \\ \hat{x}_2 \\ \hat{y}_1 \\ \hat{y}_2 \end{pmatrix} = \begin{pmatrix} \cos a & -\sin a & 0 & 0 \\ \sin a & \cos a & 0 & 0 \\ 0 & 0 & \cos a & -\sin a \\ 0 & 0 & \sin a & \cos a \end{pmatrix} \cdot \begin{pmatrix} x_1 \\ x_2 - x_1^2 + y_1^2 \\ y_1 \\ y_2 - 2x_1 y_1 \end{pmatrix}, \tag{4.5}$$

cf. (4.4), for a given value of the frequency a.

In this case, of course, a visualization of these four dimensional orbits is quite complicated. For our purposes, it is convenient to consider two dimensional projections of these orbits and use them to obtain good initial values of X^0 and H. If that is not easily achieved, we may also take an appropriate grid of initial conditions in 4–D space to start the computation.

Let us apply our method to mapping (4.5) to compute some of its periodic orbits for $a = \cos^{-1}(0.24)$ and $p = 2$. Selecting a proper grid we thus calculate within an accuracy of $\varepsilon = 10^{-16}$ the following point of period 2 :

$$X_1^2 = \begin{pmatrix} -1.2773327473170111 \\ -1.0000000000000000 \\ 1.9056702094980712 \\ -2.4341749641783542 \end{pmatrix}.$$

Proceeding in the same way as before, we have been able to compute periodic orbits of various periods. For example, with $a = \cos^{-1}(0.24)$ and $p = 3$ we have computed within an accuracy of $\varepsilon = 10^{-16}$ the period–3 point :

$$X_1^3 = \begin{pmatrix} -1.2773327473170111 \\ 0.0611205432937182 \\ 1.2285511493682861 \\ -1.5692685842514041 \end{pmatrix}.$$

Note that the mapping (4.5), while conservative, it is not symplectic [13, 14]. Our method, however, can also be easily applied to periodic orbits of $2N$–dimensional symplectic mappings (with $N > 1$) which are analogous to N–degree of freedom Hamiltonian systems. In fact, we have already started such a generalization and computed some of the low order periodic orbits of a 4–dimensional quadratic mapping, which has two frequencies a_1 and a_2 and is closely related to (4.5) :

$$\begin{pmatrix} \hat{x}_1 \\ \hat{x}_2 \\ \hat{x}_3 \\ \hat{x}_4 \end{pmatrix} = \begin{pmatrix} \cos a_1 & -\sin a_1 & 0 & 0 \\ \sin a_1 & \cos a_1 & 0 & 0 \\ 0 & 0 & \cos a_2 & -\sin a_2 \\ 0 & 0 & \sin a_2 & \cos a_2 \end{pmatrix} \cdot \begin{pmatrix} x_1 \\ x_2 + x_1^2 - x_3^2 \\ x_3 \\ x_4 - 2x_1 x_3 \end{pmatrix}. \tag{4.6}$$

This is an interesting model, since mappings of this type are of direct relevance to the dynamics of particle beams (in a 4–Dimensional phase space) passing repeatedly through FODO cells of magnetic focusing elements [4]. In this context, the computation of high period orbits can be quite useful in helping us study the existence and structure of nearby invariant surfaces, by investigating e.g. the validity of Greene's criterion in these higher–dimensional systems [9]. More detailed results on this question, however, are expected to appear in a future publication [30].

5. Concluding Remarks

An efficient method for rapidly and accurately computing periodic orbits of nonlinear mappings has been described in this paper. This method exploits topological degree theory to construct a characteristic polyhedron and locate the periodic orbit within a given region, without making any computation of the topological degree. Then it repeatedly subdivides this polyhedron to compute the periodic orbit rapidly and to any accuracy.

The method is very efficient, since the only computable information that is required is the algebraic signs of the components of the mapping. Thus it is not affected by the mapping evaluations taking large or imprecise values. Moreover, it always converges rapidly to a periodic orbit within the initial specified region independently of the initial guess, which is particularly useful in cases where the period of the periodic orbit is very high and the mapping has many periodic orbits, close to each other.

It is also a globally convergent method, it can be applied to nondifferentiable continuous functions and does not involve derivatives or approximations of such derivatives. Furthermore, using this method the number of iterations needed to find out a periodic orbit to a predetermined accuracy is known. Based on this we have conjectured that the method of the present paper is optimal (i.e. that it possesses asymptotically the best rate of convergence).

We have illustrated this method first to Hénon's 2–Dimensional mapping, used in the study of beam dynamics in particle accelerators and have succeeded in overcoming the difficulties of other schemes in the computation of periodic orbits which are strongly unstable and/or of very high period.

Moreover, we have applied our method to higher dimensions, by using it to calculate periodic orbits of 3–D and 4–D mappings of conservative dynamical systems.

Since the method is especially suited for the calculation of orbits of high period, it might be used to approximate quasiperiodic orbits which lie on invariant tori of nonlinear mappings. Thus, we may be able to check in the case of 4 (and higher) dimensions the validity of the famous Greene's conjecture concerning the break–up of invariant surfaces and the onset of large scale chaos [9]. Such topics will be addressed, however, in a future publication.

Acknowledgments

We gratefully acknowledge many stimulating and useful discussions with Drs J.A.G. Roberts and G.R.W. Quispel as well as with Prof. G. Turchetti.

References

1. A. Bazzani, P. Mazzanti, G. Servizi, and G. Turchetti, Normal forms for Hamiltonian maps and nonlinear effects in a particle accelerator, *Il Nuovo Cimento* **102,** 51–80 (1988).
2. G. D. Birkhoff, Dynamical systems with two degrees of freedom, *Trans. Amer. Math. Soc.* **18,** 199–300 (1917).
3. T. C. Bountis, and R. H. G. Helleman, On the stability of periodic orbits of two–dimensional mappings, *J. Math. Phys.* **22,** 1867–1877 (1981).
4. T. C. Bountis, and S. Tompaidis, Strong and weak instabilities in a 4D mapping model of FODO cell dynamics, in *Future Problems in Nonlinear Particle Accelerators*, edited by G. Turchetti and W. Scandale (World Scientific, 1991).

5. J. E. DENNIS Jr., and R. B. SCHNABEL, *Numerical methods for unconstrained optimization and nonlinear equations* (Prentice–Hall, Inc., Englewood Cliffs, New Jersey, 1983).
6. R. DEVOGELAERE, On the structure of symmetric periodic solutions of conservative systems with applications, in *Contributions to the theory of nonlinear oscillations, pp. 53–84*, edited by S. Lefschetz (Princeton U.P., Princeton, New Jersey, 1958).
7. A. EIGER, K. SIKORSKI, and F. STENGER, A bisection method for systems of nonlinear equations, *ACM Trans. Math. Softw.* **10,** 367–377 (1984).
8. M. FEINGOLD, L. P. KADANOFF, and O. PIRO, Passive scalars, three–dimensional volume–preserving, and chaos, *J. Stat. Phys.* **50,** 529–565 (1988).
9. J. M. GREENE, A method for determining a stochastic transition, *J. Math. Phys.* **20,** 1183–1201 (1979).
10. J. M. GREENE, Locating three–dimensional roots by a bisection method, *J. Comp. Phys.* **98,** 194–198 (1992).
11. M. HÉNON, Numerical study of quadratic area–preserving mappings, *Quart. Appl. Math.* **27,** 291–312 (1969).
12. S. KOMINEAS, M. N. VRAHATIS, and T. C. BOUNTIS, A study of period–doubling bifurcations in a 3–D conservative mapping, preprint, University of Patras, (1993).
13. H. T. KOOK, and J. D. MEISS, Periodic orbits for reversible, symplectic mappings, *Physica D* **35,** 65–86 (1989).
14. R. S. MACKAY, J. D. MEISS, and I. C. PERCIVAL, Transport in Hamiltonian systems, *Physica D* **13,** 55–81 (1985).
15. C. MIRANDA, Un' osservazione su un teorema di Brouwer, *Bol. Un. Mat. Ital.* **3,** 5–7 (1940).
16. J. M. ORTEGA, and W. C. RHEINBOLDT, *Iterative solution of nonlinear equations in several variables* (Academic Press, New York, 1970).
17. V. G. PAPAGEORGIOU, F.W. NIJHOFF, and H.W. CAPEL, Integrable mappings and nonlinear integrable lattice equations, *Phys. Lett. A* **147,** 106–114, (1990).
18. G. R. W. QUISPEL, H.W. CAPEL, V.G. PAPAGEORGIOU, and F.W. NIJHOFF, Integrable mappings derived from soliton equations, *Physica A* **173,** 243–266, (1991).
19. J. A. G. ROBERTS, The dynamics of trace maps, in this volume.
20. J. A. G. ROBERTS, and G. R. W. QUISPEL, Chaos and time–reversal symmetry. Order and Chaos in reversible dynamical systems, *Physics Reports* **216,** No 2&3, 63–117 (1992).
21. K. SIKORSKI, Bisection is optimal, *Numer. Math.* **40,** 111–117 (1982).
22. M. N. VRAHATIS, An error estimation for the method of bisection in $\mathbb{R}^n$, *Bull. Soc Math. Grèce* **27,** 161–174 (1986).
23. M. N. VRAHATIS, Solving systems of nonlinear equations using the nonzero value of the topological degree, *ACM Trans. Math. Software* **14,** 312–329 (1988).
24. M. N. VRAHATIS, CHABIS: A mathematical software package for locating and evaluating roots of systems of non–linear equations, *ACM Trans. Math. Software* **14,** 330–336 (1988).
25. M. N. VRAHATIS, A short proof and a generalization of Miranda's existence theorem, *Proc. Amer. Math. Soc.* **107,** 701–703 (1989).
26. M. N. VRAHATIS, An efficient method for locating and computing periodic orbits of nonlinear mappings, submitted.
27. M. N. VRAHATIS, T. C. BOUNTIS, and N. BUDINSKY, A convergence–improving iterative method for computing periodic orbits near bifurcation points, *J. Comp. Phys.* **88,** 1–14 (1990).
28. M. N. VRAHATIS, and K. I. IORDANIDIS, A rapid generalized method of bisection for solving systems of nonlinear equations, *Numer. Math.* **49,** 123–138 (1986).
29. M. N. VRAHATIS, G. SERVIZI, G. TURCHETTI, and T. C. BOUNTIS, A procedure to compute the periodic orbits and visualize the orbits of a 2D map, CERN SL/93 (AP) (1993).

30. M. N. Vrahatis, G. Turchetti, G. Servizi, and T. C. Bountis, Periodic orbits and invariant surfaces in 4D mapping models of particle beams, in preparation.

THE DYNAMICS OF TRACE MAPS

John A. G. Roberts and Michael Baake

Department of Mathematics, University of Melbourne
Parkville, VIC 3052, Australia

INTRODUCTION

Trace maps are 3-dimensional (3D) mappings derived from transfer matrix approaches to various physical processes that display quasiperiodicity in space or time. A subset of trace maps, to be considered here, possess one (and the same) constant of motion and so induce dynamics on the 2D level sets of this integral. When motivated by a particular physical problem, these trace maps are often studied in a particular regime of the phase space and with particular initial conditions. Here we take a global interest in them, motivated more by them being interesting dynamical systems. We show that, particularly with regard to the structure and location of their periodic orbits, these mappings are closely related to 2D Hamiltonian (i.e., area-preserving) mappings. We highlight a particular level set where the dynamics is exactly solvable and closely related to that of a hyperbolic toral automorphism or 'cat map'. Finally, we show how time-reversal symmetry in such trace maps can be identified from a related property of matrices belonging to $PGl(2,\mathbb{Z})$.

DERIVATION OF TRACE MAPS

Trace maps get their name from the fact that they relate the traces of successive pairs of matrices that are built up in a systematic way. Let A and B be unimodular 2×2 matrices, i.e. A, $B \in Sl(2,\mathbb{C})$ (actually, we restrict our discussion to such matrices with real traces). Now generate a sequence of pairs of matrices $\{(A_n, B_n)\}$ using the matrix map

$$A_n \mapsto A_{n+1} = \varrho(A_n), \quad B_n \mapsto B_{n+1} = \varrho(B_n). \tag{1}$$

Here ϱ is a given 2-letter *substitution rule*

$$\varrho : \begin{array}{lcl} A & \rightarrow & \varrho(A) = w_A(A, B, A^{-1}, B^{-1}) \\ B & \rightarrow & \varrho(B) = w_B(A, B, A^{-1}, B^{-1}), \end{array} \tag{2}$$

Hamiltonian Mechanics, Edited by J. Seimenis,
Plenum Press, New York, 1994

with w_A and w_B being given 'words' in the matrices $\{A, B, A^{-1}, B^{-1}\}$ and their powers, and ϱ obeys the homomorphism relation $\varrho(x_1 x_2 \dots x_r) = \varrho(x_1)\varrho(x_2)\dots\varrho(x_r)$, where x_i is A, B, A^{-1}, or B^{-1}. Eqn. (2), together with the homomorphism relation, just embodies in a formal way a simple iterative process to generate the pair (A_n, B_n). Namely, given the initial pair (A_0, B_0), each of which is also a word in A, B and their inverses, and the rule ϱ, then $(A_1, B_1) = (\varrho(A_0), \varrho(B_0))$ follows by replacing every occurrence of A and B in (A_0, B_0) using the right-hand side of (2). Also, A^{-1} is replaced by $(w_A)^{-1}$ and B^{-1} by $(w_B)^{-1}$. Similarly, (A_2, B_2) follows by replacing every occurrence of A and B in (A_1, B_1) again using (2), etc.

As an example, consider, for $\ell \in \mathbb{Z}$, the substitution rule

$$\varrho_\ell : \begin{array}{lcl} A & \to & \varrho_\ell(A) = B \\ B & \to & \varrho_\ell(B) = B^\ell A. \end{array} \tag{3}$$

Employing the above scheme with $(A_0, B_0) = (A, B)$, we find $(A_1, B_1) = (B, B^\ell A)$, $(A_2, B_2) = (B^\ell A, (B^\ell A)^\ell B)$, etc.

The replacement rule ϱ is often called an inflation/deflation rule for obvious reasons. It is called *invertible* if there exists another rule, denoted $\varrho^{-1} : A \to \varrho^{-1}(A), B \to \varrho^{-1}(B)$, such that successive application to any matrix word w_0 of ϱ^{-1} and ϱ leaves the word unchanged, i.e. $\varrho(\varrho^{-1}(w_0)) = \varrho^{-1}(\varrho(w_0)) = w_0$. The example (3) is invertible with $\varrho_\ell^{-1} : A \to A^{-\ell}B,\ B \to A$. More generally, the product of two substitution rules, written $\sigma\varrho$, is defined by first applying σ and then ϱ so that $\sigma\varrho(w_0) = \varrho(\sigma(w_0))$.

The *substitution matrix* R_ϱ corresponding to the substitution rule ϱ in (2) is the integer matrix defined by

$$R_\varrho = \begin{pmatrix} \#_A(w_A) & \#_B(w_A) \\ \#_A(w_B) & \#_B(w_B) \end{pmatrix} =: \begin{pmatrix} \alpha & \beta \\ \gamma & \delta \end{pmatrix}. \tag{4}$$

In (4), $\#_A(w_A)$ is the number of A's in $w_A(A, B, A^{-1}, B^{-1})$ of (2), (adding $+1$ for every occurrence of A, and -1 for every occurrence of A^{-1}) and $\#_B(w_A)$ is the number of B's etc. If the integer row vector $(\#_A(w_0), \#_B(w_0))$ gives the number of A's and B's in a word w_0, then multiplication of this vector on the right by ${R_\varrho}^n$ gives a new row vector $(\#_A(w_n), \#_B(w_n))$ whose components are the number of A's and B's in $w_n = \varrho^n(w_0)$. For our example (3) above

$$R_\ell = \begin{pmatrix} 0 & 1 \\ 1 & \ell \end{pmatrix}. \tag{5}$$

Matrix recursions like (1) arise in the study of various properties of 1D nonperiodic structures in space or time (Kohmoto et al., 1983; Ostlund et al., 1983; Sutherland, 1986). The matrices A and B are transfer matrices over two constituent length or time scales of the structure, and A_n and B_n relate to the transfer matrices acting over longer and longer approximants to the infinite nonperiodic structure. Although the matrix recursion (1) is certainly an interesting dynamical system in its own right (Kramer, 1993), a lower-dimensional dynamical system is obtained by considering the relation between the traces of the successive matrix pairs. Indeed, it can be shown (Horowitz, 1975; Allouche and Peyrière, 1986) that for *any* matrix substitution rule (2), there exists a polynomial mapping $F_\varrho : \mathbb{R}^3 \mapsto \mathbb{R}^3$ with integer coefficients that expresses the coordinate triple $(\mathrm{tr}(\varrho(A)), \mathrm{tr}(\varrho(B)), \mathrm{tr}(\varrho(AB)))$ in terms of $(\mathrm{tr}(A), \mathrm{tr}(B), \mathrm{tr}(AB))$. If $(A_0, B_0) = (A, B)$ in (1), then $A_n = \varrho^n(A)$ and $B_n = \varrho^n(B)$, and $\mathrm{tr}(A_n)$ and $\mathrm{tr}(B_n)$ are given, respectively, by the first and second components of ${F_\varrho}^n(\mathrm{tr}(A), \mathrm{tr}(B), \mathrm{tr}(AB))$. Because many physical properties depend simply on the traces of the transfer matrices

A_n and B_n, this explains the interest in the dynamics of the *trace map* F_ϱ. This connection was recognised in the pioneering papers cited above, and a substantial literature now exists (e.g., see Baake et al., 1993, and refs. therein).

For our example substitution rule (3), the corresponding trace map is

$$F_\ell : \begin{pmatrix} x \\ y \\ z \end{pmatrix} \mapsto \begin{pmatrix} y \\ zU_{\ell-1}(y) - xU_{\ell-2}(y) \\ zU_\ell(y) - xU_{\ell-1}(y) \end{pmatrix} . \tag{6}$$

Note that in (6) we have used the convention of taking $\frac{1}{2}\text{tr}(A)$, etc., for the trace map coordinates. Here, $U_\ell(y)$ are Chebyshev's polynomials of the second kind. They can be defined by the recurrence relation $U_{\ell+1}(y) = 2yU_\ell(y) - U_{\ell-1}(y)$ together with the initial conditions $U_{-1} \equiv 0$ and $U_0 \equiv 1$. They appear because for $B \in Sl(2,\mathbb{C})$, we find by induction and the Cayley-Hamilton theorem that $B^\ell = U_{\ell-1}(y)B - U_{\ell-2}(y)\mathbb{1}$ with $y = \frac{1}{2}\text{tr}(B)$. For all ℓ, the trace map F_ℓ is volume-preserving, but orientation-reversing, so that the determinant of the Jacobian matrix of (6) equals -1. Note that for $\ell = 1$, we obtain

$$F_1 : \begin{pmatrix} x \\ y \\ z \end{pmatrix} \mapsto \begin{pmatrix} y \\ z \\ 2yz - x \end{pmatrix} , \tag{7}$$

which is the much-studied Fibonacci trace map on which the pioneering studies in the field were made (other members of the family F_ℓ are often called generalized Fibonacci trace maps, cf. Kalugin et al., 1986).

We will focus here exclusively on invertible substitution rules, such as (3). We use the term *Nielsen trace maps* to denote the trace maps derived from such rules. There is a rich and powerful algebraic structure connecting invertible substitution rules, and their associated trace maps and substitution matrices (Horowitz 1975, Iguchi 1991, Peyrière 1991, Peyrière et al., 1992, Kramer 1993, Roberts and Baake 1992). For example, the substitution matrix and trace map of a product $\sigma\varrho$ of two substitution rules are given by $R_{\sigma\varrho} = R_\sigma \cdot R_\varrho$, $F_{\sigma\varrho} = F_\sigma \circ F_\varrho$, where $\cdot$ is matrix multiplication and $\circ$ is functional composition. Taking these relations with $\sigma = \varrho^{-1}$, it follows that the inverse of a Nielsen trace map exists and is a Nielsen trace map, and that the corresponding (integer) substitution matrix has an inverse which is also an (integer) substitution matrix. Consequently, the set of substitution matrices of invertible substitution rules form the group $Gl(2,\mathbb{Z})$ (so that in (4), the integers satisfy $\alpha\delta - \beta\gamma = \pm 1$), and the set of Nielsen trace maps form a group too, denoted $\mathcal{G}$.

It turns out that the substitution matrix associated with a Nielsen trace map plays an important role in characterising the latter. This is because invertible substitution rules ϱ and σ lead to the same trace map if and only if $R_\varrho = R_\sigma$ or $R_\varrho = -R_\sigma$ (Peyrière et al., 1992). Another way of saying this is that one can uniquely associate to a substitution matrix R_F a Nielsen trace map F (the matrix $-R_F$ also corresponds to F). In fact, there is an isomorphism between the group $\mathcal{G}$ of Nielsen trace maps and the matrix group $PGl(2,\mathbb{Z}) \simeq Gl(2,\mathbb{Z})/\{\pm\mathbb{1}\}$ (cf. Roberts and Baake, 1992). A significant consequence of this is that one can calculate a Nielsen trace map directly from its corresponding substitution matrix, effectively bypassing the substitution rule. To do this, one notes that every matrix in $Gl(2,\mathbb{Z})$ can be written as a string of products of 3 generator matrices

$$R_u = \begin{pmatrix} 1 & 1 \\ 0 & 1 \end{pmatrix} , \quad R_p = \begin{pmatrix} 0 & 1 \\ 1 & 0 \end{pmatrix} , \quad R_s = \begin{pmatrix} -1 & 0 \\ 0 & 1 \end{pmatrix} . \tag{8}$$

We find the generator Nielsen trace maps corresponding to these simple matrix generators are, respectively,

$$u : \begin{pmatrix} x \\ y \\ z \end{pmatrix} \mapsto \begin{pmatrix} z \\ y \\ 2yz - x \end{pmatrix}, \; p : \begin{pmatrix} x \\ y \\ z \end{pmatrix} \mapsto \begin{pmatrix} y \\ x \\ z \end{pmatrix}, \; s : \begin{pmatrix} x \\ y \\ z \end{pmatrix} \mapsto \begin{pmatrix} x \\ y \\ 2xy - z \end{pmatrix}. \tag{9}$$

Every Nielsen trace map can hence be written as a word in the generators u, p and s (and their powers), and the form of this word follows the word in R_u, R_p and R_s which describes its substitution matrix (e.g. in (5), $R_\ell = R_p \cdot {R_u}^\ell$ and so $F_\ell = p \circ u^\ell$, which agrees with (6)). Obviously properties common to the generator trace maps in (9) that are retained under functional composition will hold for all Nielsen trace maps. For instance, all Nielsen trace maps are volume-preserving and orientation-preserving or orientation-reversing according to the determinant of their substitution matrix.

INTEGRAL OF MOTION AND LEVEL SETS

It was shown already in Kohmoto et al. (1983) and Ostlund et al. (1983) that the trace map (7) is only nominally a 3D mapping, in that it has a constant of motion $\hat{I}$. That is, $\hat{I}(F_1\boldsymbol{x}) = \hat{I}(\boldsymbol{x})$, for all $\boldsymbol{x} = (x, y, z) \in \mathbb{R}^3$, where

$$\hat{I}(x, y, z) = x^2 + y^2 + z^2 - 2xyz - 1. \tag{10}$$

This invariance can easily be checked by direct substitution. Note the (unnecessary) constant term in (10) is kept for historical reasons. It is now known that the invariance of (10) is a property common to all Nielsen trace maps because it is shared by the generators (9), cf. also Horowitz (1975), Iguchi (1991), Peyrière (1991). Consequently, the level sets of $\hat{I}$ are invariant under all $F \in \mathcal{G}$ and F defines an orientation-reversing or orientation-preserving map of each level set to itself. We denote the level sets by

$$\mathcal{M}_\mu := \{(x, y, z) \in \mathbb{R}^3 \mid \hat{I}(x, y, z) = \mu\} = \mathcal{M}_\mu^c \cup \mathcal{M}_\mu^{nc}. \tag{11}$$

The last part of (11) indicates the division of $\mathcal{M}_\mu$ into its compact (c) and noncompact (nc) parts. The topology of $\mathcal{M}_\mu$ is one of four basic types. For $\mu < -1$, $\mathcal{M}_\mu^c$ is empty and $\mathcal{M}_\mu^{nc}$ comprises 4 disconnected non-compact cones going to infinity and confined to 4 of the 8 octants of $\mathbb{R}^3$. For $-1 < \mu < 0$ (cf. Figure 1(a)), we additionally have a non-empty $\mathcal{M}_\mu^c$ which is equal to a compact and invariant set around the origin diffeomorphic to a sphere (when $\mu = -1$, $\mathcal{M}_\mu^c$ degenerates to being the origin). For $\mu = 0$ (cf. Figure 1(b)), $\mathcal{M}_0^c$ is only homeomorphic to a sphere and reaches the 4 cones of $\mathcal{M}_0^{nc}$ in a set $\mathcal{P}$ of 4 'pinches':

$$\mathcal{P} := \mathcal{M}_0^c \cap \mathcal{M}_0^{nc} = \{(1,1,1)\} \cup \hat{\mathcal{P}}, \;\; \hat{\mathcal{P}} := \{(1,-1,-1), (-1,1,-1), (-1,-1,1)\}. \tag{12}$$

Finally, for $\mu > 0$, the 4 pinches open into necks, so that $\mathcal{M}_\mu$ is always a connected, non-compact C^∞-manifold. We remark that the different physical applications of trace maps lead one to consider their dynamics in the various different topological regimes of $\mathcal{M}_\mu$ (cf. Baake et al., 1993, and refs. therein).

The set $\mathcal{P}$ of (12) is an important fixed set of any Nielsen trace map. Via the generators above, it can be seen that $(1,1,1)$ is always a fixed point. The remaining 3 points belonging to $\hat{\mathcal{P}}$ form a 3-cycle, 3 fixed points or a 2-cycle and a fixed point, depending on a symmetry property of the trace map. This symmetry property also controls the dynamics on the cones of $\mathcal{M}_\mu^{nc}$ for $\mu \leq 0$. Furthermore, every Nielsen trace map also fixes the origin.

Possession of a smooth integral or invariant $I = I(x,y,z)$ by a 3D mapping F has important dynamical implications, particularly for its n-cycles. Let $\boldsymbol{p}_0 = (x_0, y_0, z_0)$ be a point of an n-cycle of F, and call the set of eigenvalues of the linearization $dF^n(\boldsymbol{p}_0)$ the *eigenvalue spectrum* of the cycle containing $\boldsymbol{p}_0$. Then if $I(\boldsymbol{p}_0) = \mu_0$ and $\nabla I(\boldsymbol{p}_0) \neq \mathbf{0}$, the implicit function theorem gives: [P1] the eigenvalue spectrum of the cycle includes $\{+1\}$; [P2] if $\{+1\}$ is an eigenvalue of multiplicity 1, the point $\boldsymbol{p}_0$ is embedded locally into a unique curve $\boldsymbol{p}(\mu) = (x(\mu), y(\mu), z(\mu))$ that crosses the level set $I(\boldsymbol{x}) = \mu_0$, and whose points are points of n-cycles of F on adjacent level sets $I(\boldsymbol{x}) = \mu$; and [P3] the curve $\boldsymbol{p}(\mu)$ can intersect a curve of points of a $(j \cdot n)$-cycle at $\boldsymbol{p}_0$ only if $dF^n(\boldsymbol{p}_0)$ has an eigenvalue equal to a j-th root of unity.

For the Nielsen trace maps with invariant $I = \hat{I}$, one finds $\nabla \hat{I} \neq \mathbf{0}$ for all points except the origin and the set $\mathcal{P}$ of (12). Consequently, barring these 5 points, all other n-cycles of a Nielsen trace map satisfy [P1] above. Furthermore, since all Nielsen trace maps are volume-preserving, the eigenvalue spectrum is of the form $\{1, \lambda, \lambda^{-1}\}$ when F is orientation-preserving, or orientation-reversing and n is even, and it is $\{1, \lambda, -\lambda^{-1}\}$ when F is orientation-reversing and n is odd. The eigenvalue equal to $+1$ corresponds to motion normal to the level set, so that the linearized motion on the surface is described by the remaining two eigenvalues. It follows that the periodic orbit on the surface is typically *hyperbolic* with real eigenvalues $\{\lambda, \lambda^{-1}\}$ or $\{\lambda, -\lambda^{-1}\}$ with $\lambda \neq \pm 1$, or *elliptic* with complex eigenvalues $\{e^{i\theta}, e^{-i\theta}\}$ (in the orientation-reversing case, elliptic cycles are necessarily of even period).

We illustrate [P2] which says that the periodic orbits typically form curves of points with the same period, with the example (7). The 2-cycles of this mapping can be written $\{(a, \frac{a}{2a-1}, a), (\frac{a}{2a-1}, a, \frac{a}{2a-1})\}$, where a is an arbitrary variable, so that both points lie on the single curve $\{(x, \frac{x}{2x-1}, x) | x \in \mathbb{R}\}$. The value, μ, of $\hat{I}$ on which each 2-cycle lies follows from substituting one of its coordinates into (10) , and the eigenvalue spectrum of each 2-cycle is given by $\{1, \lambda_2, \lambda_2^{-1}\}$ with $\lambda_2 + \lambda_2^{-1} = \frac{8a^2-2a+1}{2a-1}$. Similarly, the 4-cycles of (7) each have 2 points on the curve $\{(-\frac{1}{2}, x, -\frac{1}{2}) | x \in \mathbb{R}\}$ and 2 points on the curve $\{(x, -\frac{1}{2}, \frac{1}{2} - x) | x \in \mathbb{R}\}$. We observe that the curve of 2-cycles intersects the curves of 4-cycles in the points $(-\frac{1}{2}, \frac{1}{4}, -\frac{1}{2})$ and $(\frac{1}{4}, -\frac{1}{2}, \frac{1}{4})$. From (10) these points of intersection lie on the level set $\mathcal{M}^c_{\mu_1}$ of (11) with $\mu_1 = -\frac{9}{16} = -0.5625$. At the point of intersection, $\lambda_2 + \lambda_2^{-1} = -2$, implying $\lambda_2 = -1$, which agrees with [P3] above. We can continue to look for the intersection of the 4-cycles of (7) with its 8-cycles, etc. Some of the results are illustrated in Figure 2. In this picture, we show the projection of the plane $\{x = z\}$ onto a part of the plane $\{z = 0\}$. The reason we look in the plane $\{x = z\}$ is that each of the even-cycles involved is found to have two points always on this plane (this is a consequence of a time-reversal symmetry in (7), see below). It is seen that the point of intersection with curves of double the period occurs precisely at the changeover point in the cycle's stability from elliptic to hyperbolic, when its spectrum is $\{1, -1, -1\}$. In Figure 2 are also shown the intersections of some of the compact surfaces $\mathcal{M}^c_\mu$ with the plane $\{x = z\}$, for $\mu \in (-1, 0]$. The sequence of values $\{\mu_n\}$ labelling the compact level sets on which intersections between 2^n-cycles and 2^{n+1}-cycles occur, converges asymptotically geometrically to $\mu_\infty = -0.27775\ldots$ with the period-doubling exponent

$$\delta = 8.721\ldots. \tag{13}$$

This is the familiar value for 2D area-preserving or 2D reversible mappings (cf. Roberts and Quispel, 1992, and refs. therein).

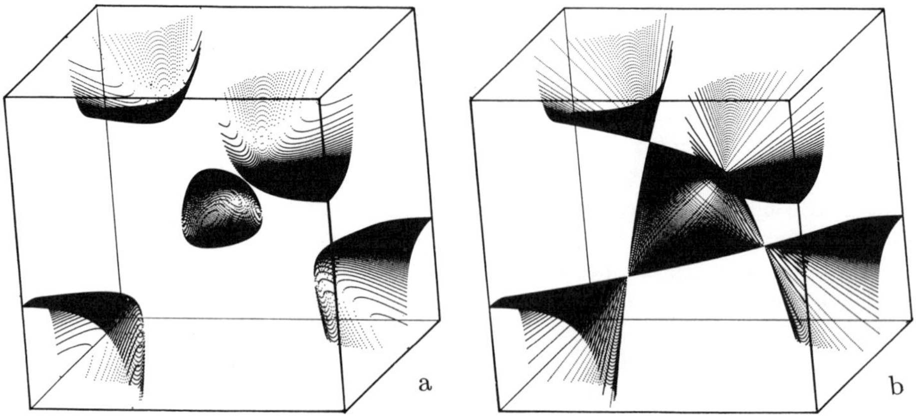

Figure 1. Topology of the level sets $\hat{I} = \mu$ of the Nielsen trace maps when (a) $\mu = -0.5$; and (b) $\mu = 0$.

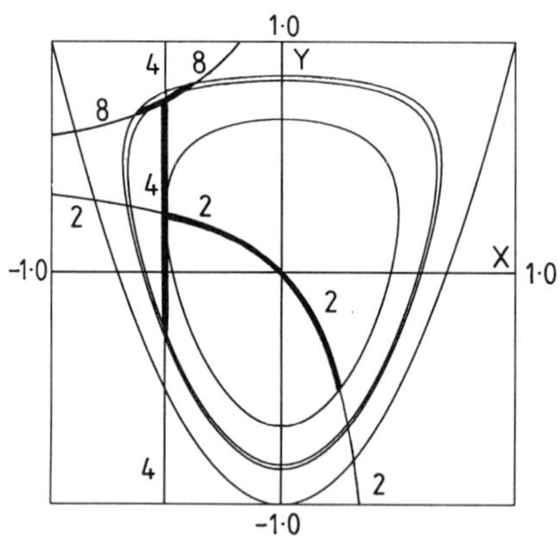

Figure 2. Curves of points of 2-,4- and 8-cycles of F_1 lying on the plane $\{x = z\}$. The thick part of each curve indicates where the corresponding cycles are elliptic. The closed curves (inner to outer) are the intersections of $\mathcal{M}^c_\mu$ with $\{x = z\}$ when $\mu \in \{-0.5625, -0.3125, -0.281835\ldots\}$. The outer parabolic curve and the line $\{y = 1\}$ are the intersection of $\mathcal{M}^c_0$ with $\{x = z\}$.

In Roberts and Baake (1992) this is explained by the fact that, in a region near periodic orbits, the 3D mapping F_1 is related to a 2D conservative mapping. Indeed, for any Nielsen trace map F, $\hat{I}(x,y,z) = \mu$ with $\hat{I}$ given by (10) defines, say, two z values for given x, y and μ, namely $z_\pm = J_\pm(x,y,\mu) = xy \pm \sqrt{(x^2-1)(y^2-1)+\mu}$. This divides the level sets $\mathcal{M}_\mu$ into an *upper* '+' branch and a *lower* '−' branch with respect to the value of z. Suppose there is a region of the 3D phase space in which the dynamics of F^n is confined to one set of branches such as the '−' branches (e.g. in the neighbourhood of a point $\boldsymbol{p}_0 = (x_0, y_0, z_0)$ of an n-cycle of F). We can then switch coordinates unambiguously between (x, y, z_-) and (x, y, μ), and study the transformed mapping

$$M : \begin{pmatrix} x \\ y \\ \mu \end{pmatrix} \mapsto \begin{pmatrix} f(x,y,J_-(x,y,\mu)) \\ g(x,y,J_-(x,y,\mu)) \\ \mu \end{pmatrix} =: \begin{pmatrix} F(x,y,\mu) \\ G(x,y,\mu) \\ \mu \end{pmatrix}, \tag{14}$$

where f and g are the x and y parts of F^n. The mapping M is effectively a 2D mapping dependent on the parameter μ; because F is volume-preserving, M is measure-preserving or conservative. This shows that the intersections of curves of periodic orbits are governed by the 2D bifurcation theory of conservative mappings, and shows why a cascade of intersections (bifurcations) belongs to the universality class of 2D conservative mappings. Given a good initial condition (x_0, y_0) from $\boldsymbol{p}_0 = (x_0, y_0, z_0)$, and initial parameter value $\mu_0 = \hat{I}(x_0, y_0, z_0)$, the mapping M can be used for a 2D 'tracking' of an n-cycle of F onto neighbouring level sets, subject to the self-consistency of the nearby cycles still lying on one set of branches of $\hat{I} = \mu$.

A LEVEL SET WITH SOLVABLE DYNAMICS

In a typical nonlinear dynamical system one can analytically only determine low-period orbits. It turns out for the Nielsen trace maps that on the surface $\mathcal{M}_0^c$, the 'tetrahedral ball' in Figure 1(b), we can explicitly find periodic orbits of every period. In a one-parameter 2D measure-preserving mapping like (14), this is analogous to knowing all the periodic orbits at one parameter value.

This comes about because on $\mathcal{M}_0^c$ we have

$$F\,|_{\mathcal{M}_0^c} \circ S = S \circ R_F, \tag{15}$$

where F is the trace map and $R_F \in Gl(2, \mathbb{Z})$ is its associated substitution matrix. In (15), $S : \mathbb{R}^2 \to \mathbb{R}^3$ is a parametrization of $\mathcal{M}_0^c$ given by

$$S : \begin{pmatrix} \vartheta_1 \\ \vartheta_2 \end{pmatrix} \mapsto \begin{pmatrix} x \\ y \\ z \end{pmatrix} = \begin{pmatrix} \cos 2\pi\vartheta_1 \\ \cos 2\pi\vartheta_2 \\ \cos 2\pi(\vartheta_1 + \vartheta_2) \end{pmatrix}. \tag{16}$$

One can check that with (x, y, z) of this form, $\hat{I}(x,y,z)$ of (10) is identically zero. Because the function $\cos 2\pi\vartheta$ identifies ϑ and $\vartheta + m$, $m \in \mathbb{Z}$, the action of the integer matrix R_F on the right-hand side of (15) can be considered to be acting on the torus (i.e. take ϑ mod 1). In the typical case that R_F is a hyperbolic matrix, it follows that the dynamics of a Nielsen trace map on $\mathcal{M}_0^c$ is semi-conjugate to that of the hyperbolic toral automorphism induced by its associated substitution matrix. Hyperbolic toral automorphisms are the canonical examples of *Anosov* systems; popularly, they are known as 'cat maps' because of their strong mixing properties (e.g. on the face of a cat, cf. Arnol'd and Av-z, 1968). The periodic orbits of hyperbolic toral automorphisms are dense in the phase space and the motion is homogeneously chaotic, e.g. cf. Devaney (1989). However, it is worth stressing that (15) is a semi-conjugacy and not a conjugacy (or equivalence). The transformation S identifies the points of the torus $\boldsymbol{\vartheta} = (\vartheta_1, \vartheta_2)$ and $-\boldsymbol{\vartheta} = (-\vartheta_1, -\vartheta_2)$. Consequently, the action of F on $\mathcal{M}_0^c$ is described by the quotient of a toral automorphism by reflection through the origin $(0,0)$, and it is therefore a *pseudo − Anosov* mapping. Pseudo-Anosov mappings play an important role in topology via the Nielsen-Thurston classification, cf. Casson and Bleiler (1988).

The semi-conjugacy (15) makes the dynamics on $\mathcal{M}_0^c$ solvable. The fixed points of F^n on $\mathcal{M}_0^c$, i.e. $\{\boldsymbol{x} \in \mathcal{M}_0^c \mid F^n\boldsymbol{x} = \boldsymbol{x}\}$, are given by $\boldsymbol{x} = S\boldsymbol{\vartheta}$, where

$$R_F{}^n\boldsymbol{\vartheta} = \pm\boldsymbol{\vartheta} \quad (\text{mod } 1). \tag{17}$$

The solutions to (17) can be explicitly calculated. Firstly, because $R_F \in Gl(2,\mathbb{Z})$, $R_F{}^n$ is known. When $R_F \in Sl(2,\mathbb{Z})$, we gave the formula below eq. (6). When R_F has determinant -1, we use $R_F{}^n = V_{n-1}(y)R_F + V_{n-2}(y)\mathbb{1}$ with $y = \frac{1}{2}\text{tr}(R_F)$, and $V_{n+1}(y) = 2yV_n(y) + V_{n-1}(y)$ with $V_{-1} \equiv 0$ and $V_0 \equiv 1$. The solution of $N\boldsymbol{\vartheta} = \epsilon\boldsymbol{\vartheta}$ (mod 1), $N \in Gl(2,\mathbb{Z})$ and $\epsilon = \pm 1$, depends on the entries of $N := \left(\begin{smallmatrix} a & b \\ c & d \end{smallmatrix}\right)$. For example, if $d - \epsilon$ and b are coprime, then $\boldsymbol{\vartheta} = (i/D, [(a-\epsilon)q_0 - cp_0]i/D)$ (mod 1), where $D = \det(N - \epsilon\mathbb{1})$, (p_0, q_0) are a pair of integers satisfying $(d-\epsilon)p_0 - bq_0 = 1$, and i ranges over $|D|$ consecutive values. The total number of solutions to (17), after identifying $\boldsymbol{\vartheta}$ and $-\boldsymbol{\vartheta}$, is $|\text{tr}(R_F{}^n)|$, so this is the number, ℓ_n, of fixed points of F^n on $\mathcal{M}_0^c$. A fixed point of F^n need not be a point of an n-cycle; it may belong to any k-cycle with $k|n$. To get p_n, the number of n-cycles on $\mathcal{M}_0^c$, we can invert the formula $\ell_n = \sum_{k|n} kp_k$ using the Möbius inversion formula. We obtain

$$p_n = \frac{1}{n}\sum_{k|n}\mu(\frac{n}{k})\ell_k = \frac{1}{n}\sum_{k|n}\mu(\frac{n}{k})|\text{tr}(R_F{}^k)|, \tag{18}$$

where $\mu(d)$ is the Möbius function. It follows from Llibre and MacKay (1991) that p_n is always non-zero, i.e. $F|_{\mathcal{M}_0^c}$ possesses a genuine n-cycle for every $n \in \mathbb{N}$, so that F has periodic orbits of all periods. As an example, the first numbers (starting with p_1) for the Fibonacci trace map (7) explicitly read 1,1,1,1,2,2,4,5,8,11,18,25,... .

From (15) also follow the two remaining eigenvalues (apart from $+1$) of the eigenvalue spectrum of n-cycles on $\mathcal{M}_0^c$ disjoint from the pinches (12). The eigenvalue spectrum of such an n-cycle is $\{1, \pm\lambda^n, \pm\lambda^{-n}\}$ ($\{1, \pm\lambda^n, \mp\lambda^{-n}\}$), where λ is an eigenvalue of R_F, and F is orientation-preserving or orientation-reversing with n even (orientation-reversing with n odd). Consequently, every n-cycle 'off' the pinches on $\mathcal{M}_0^c$ is hyperbolic if R_F is hyperbolic.

Studies of the energy spectrum of the tight binding problem with various quasiperiodic potentials suggest that on the surfaces $\mathcal{M}_\mu$ with $\mu > 0$, many (if not all) periodic orbits of F_ℓ of (6) are also hyperbolic. If this is true, then no periodic orbits can be created or destroyed in this regime and they all lie on curves which cut across adjacent level sets and cannot intersect (cf. [P2], [P3] above). It follows that $\mathcal{M}_0^c$ acts as the 'gateway' to this regime, as all periodic orbits for $\mu > 0$ are then the continuation of the periodic orbits that can be explicitly calculated on this surface (note that there are no periodic orbits of Nielsen trace maps on $\mathcal{M}_0^{nc}$ except for the pinches, cf. Roberts and Baake 1992). Using the mapping M of (14) it is possible to follow n-cycles off $\mathcal{M}_0^c$ into the region with $\mu > 0$ and test whether they remain hyperbolic (one can use a 2D tracking algorithm which is 'immune' to the growing hyperbolicity of the initial n-cycles, e.g. cf. Vrahatis and Bountis, 1993). Certainly, for $\mu < 0$, periodic orbits can be elliptic, cf. Figure 2. In this regime, they can be created and destroyed, and we can see period-doubling and period-multifurcations happening as a function of the level set value μ. Nevertheless, $\mathcal{M}_0^c$ acts a 'barrier' to such cascades continuing through $\mu = 0$ because of the hyperbolicity of the cycles on this surface.

TIME-REVERSAL SYMMETRY

It is well-known that symmetries of a dynamical system, if present, can relate and impose constraints on the periodic orbits of the system. An important example is generalized time-reversal symmetry, or *reversibility* (cf. Roberts and Quispel, 1992, and references within). This symmetry is common to many Hamiltonian systems, and we now ask how common is it among the Nielsen trace maps.

A mapping L (which we assume to be a diffeomorphism) is said to be *weakly reversible* if there exists another diffeomorphism G satisfying

$$L \circ G \circ L = G \iff L^{-1} = G \circ L \circ G^{-1}. \tag{19}$$

In many applications G, which is called a reversing symmetry of L, is found to be an involution ($G \circ G = Id$), and this case is usually called reversibility. In this case, L can be written as a composition of two involutions $H := L \circ G$ and G. Many of the familiar 2D area-preserving mappings have this property, e.g. the standard map, the area-preserving Hénon map (but note that not all area-preserving mappings are reversible, cf. Roberts and Capel (1992)). Weak reversibility or reversibility guarantees that if $\boldsymbol{x}$ is a point of a j-cycle whose eigenvalue spectrum includes λ, then $G\boldsymbol{x}$ also belongs to a j-cycle and this cycle's spectrum includes λ^{-1}. Consequently, a periodic orbit of a reversible mapping can be invariant under a reversing symmetry G, in which case it is called *symmetric*, and its eigenvalue spectrum comes in reciprocal pairs. Alternatively, the periodic orbit and its image under G form an *asymmetric* pair, whose eigenvalue spectra are reciprocals of one another. An advantage of reversibility is that it helps to locate symmetric periodic orbits. These can be found by taking initial conditions in $Fix(G) := \{\boldsymbol{x} \mid G\boldsymbol{x} = \boldsymbol{x}\}$ and/or $Fix(L \circ G)$. These fixed sets are typically of lower dimension than the phase space, e.g. for 2D reversible mappings they are 1D curves.

Casdagli (1986) observed that the Fibonacci trace map (7) is reversible because $F_1 = H_1 \circ G_1$, where

$$H_1 : \begin{pmatrix} x \\ y \\ z \end{pmatrix} \mapsto \begin{pmatrix} y \\ x \\ 2xy - z \end{pmatrix}, \quad G_1 : \begin{pmatrix} x \\ y \\ z \end{pmatrix} \mapsto \begin{pmatrix} z \\ y \\ x \end{pmatrix}, \tag{20}$$

are involutions. Observe that $Fix(G_1)$ is a 2-dimensional set, the plane $\{(x, y, x) \mid x, y \in \mathbb{R}\}$, whereas $Fix(H_1)$ is a 1-dimensional set, the curve $\{(y, y, y^2) \mid y \in \mathbb{R}\}$. Reversibility, and the different dimensions of these two fixed sets, is sufficient to explain why one sees isolated symmetric odd-cycles and curves of non-isolated symmetric even-cycles for F_1. The curves of symmetric even-cycles lie along the intersections of iterates of the plane $Fix(G_1)$ under F_1, and an orbit of F_1 is a symmetric $2j$-cycle if and only if it contains precisely two points $\boldsymbol{x}$ and $F_1^j\boldsymbol{x}$ on the plane $Fix(G_1)$. This explains why in Figure 2 we choose the plane $\{x = z\}$ to find symmetric $2j$-cycles and calculate the intersections between the curves they form. In these respects, F_1 is typical of 3D orientation-reversing reversible mappings. Of course, F_1, and indeed all F_ℓ, are examples of such mappings with the added (and atypical) feature of an integral of motion. Interestingly, however, it appears that even without an integral of motion, a 3D reversible mapping can have geometric self-similarity in the period-doubling branching of symmetric even-cycles, described by familiar 2D scalings like (13), cf. Roberts and Lamb (1993). The effect of an integral on symmetric odd-cycles of F_ℓ, and any other reversible orientation-reversing Nielsen trace map, is that their eigenvalue spectrum is restricted to be $\{-1, 1, 1\}$. This is not true if no integral is present, cf. Turner and Quispel (1993) who show that period-multifurcations of odd symmetric cycles in 3D orientation-reversing reversible mappings occur with 2D scalings.

Although the generator trace maps in (9) are reversible, reversibility is not (in general) a property retained under mapping composition, unlike some of the other properties considered above. As a first step to asking about the prevalence of (weak) reversibility in Nielsen trace maps, we can ask for (weak) reversibility with respect to reversing symmetries that are also Nielsen trace maps. Then we can use the above-mentioned isomorphism between the group $\mathcal{G}$ of Nielsen trace maps and $PGl(2,\mathbb{Z})$. From (19), we deduce that a Nielsen trace map F with substitution matrix $R_F \in Gl(2,\mathbb{Z})$ is weakly reversible in $\mathcal{G}$ if and only if there exists $R_G \in Gl(2,\mathbb{Z})$ such that

$$R_F \circ R_G \circ R_F = \pm R_G \quad \Longleftrightarrow \quad R_F^{-1} = \pm R_G \circ R_F \circ R_G^{-1}. \tag{21}$$

If a matrix R_G can be found to satisfy (21) then the corresponding trace map G can be calculated using the associations between the respective matrix and trace map generators (8) and (9). Note that systematic methods have been developed to test a diffeomorphism for reversibility locally, in the neighbourhood of a possible symmetric periodic orbit (Quispel and Capel, 1989; Lamb et al., 1993). Here we have a way of testing a diffeomorphism for reversibility globally, at least with respect to a particular class of reversing symmetry. This is interesting because apart from trying for a linear G in (19), the problem of finding or excluding nonlinear G's is *a priori* arduous.

In Baake and Roberts (1993), we show the problem of finding R_G satisfying (21) is tractable. There exists a closed form characterisation of the solution. Alternatively, there is a constructive way of explicitly finding R_G for given R_F, or for indicating that no such R_G exists. This method is related to the number theory problem of determining when two integer quadratic forms are equivalent, an algorithm for which was developed by Gauss (also cf. Percival and Vivaldi, 1987). Using the concept of class number, it can be deduced that for some values of the trace of a substitution matrix all corresponding Nielsen trace maps are reversible. Furthermore, for every possible value of $\mathrm{tr}(R_F)$, there exists a reversible Nielsen trace map whose substitution matrix has this trace value. This can be seen from the following special cases. When R_F satisfies $\beta = \gamma$ (cf. (4)), there always exists the solution to (21) of $R_G = \pm\left(\begin{smallmatrix} 0 & -1 \\ 1 & 0 \end{smallmatrix}\right)$. Using (8) and (9), we find that this matrix corresponds to the Nielsen trace map H_1 of (20). Because R_ℓ of (5) has $\beta = \gamma = 1$, we conclude that the Nielsen trace maps F_ℓ of (6) are all reversible with this involution. Furthermore, by changing ℓ in (5) we obtain an orientation-reversing substitution matrix of any given trace. To obtain an orientation-preserving substitution matrix of any given trace, we can replace one of the off-diagonal 1's in (5) with -1. A substitution matrix R_F with $\beta = -\gamma$ always satisfies (21) with $R_G = \pm\left(\begin{smallmatrix} 0 & 1 \\ 1 & 0 \end{smallmatrix}\right)$, whose corresponding trace map is given by p of (9).

Finally, we can ask about matrices R_F in (21) for which there exists no R_G. An example is $R_F = \pm\left(\begin{smallmatrix} 1 & 5 \\ 3 & 14 \end{smallmatrix}\right)$. This matrix is the word $\mp R_p \cdot {R_u}^3 \cdot R_p \cdot R_s \cdot {R_u}^5$ in the matrix generators (8), and so the corresponding trace map is $F = p \circ u^3 \circ p \circ s \circ u^5$. This trace map is consequently not weakly reversible in the Nielsen class. A numerical check suggests that it may indeed have *no* differentiable reversing symmetry (Baake and Roberts, 1993).

CONCLUDING REMARKS

In this paper we have considered trace maps derived from invertible substitution rules (Nielsen trace maps). These are 3D mappings with one constant of motion, and we have highlighted the resulting connections to 2D Hamiltonian maps. We have also shown how the corresponding substitution matrix, contained in $Gl(2,\mathbb{Z})$, features in the dynamics of the trace map and the investigation of time-reversal symmetry. Finally, investigation of the full matrix map (1) reveals a 6D dynamical system with up to 3 constants of motion, so there is only one more effective dimension than in the Nielsen trace map. We hope to explore these in the future, cf. also Kramer (1993).

REFERENCES

Allouche, J.-P., and Peyrière, J., 1986, Sur une formule de récurrence sur les traces de produits de matrices associés à certaines substitutions, *C. R. Acad. Sci. Paris* 302(II):1135.

Arnol'd, V.I., and Avez, A., 1968, "Ergodic Problems of Classical Mechanics," Benjamin, New York.

Baake, M., Grimm, U., and Joseph, D., 1993, Trace maps, invariants, and some of their applications, *Int. J. Mod. Phys. B* 7:1527.

Baake, M., and Roberts, J.A.G., 1993, in preparation.

Casdagli, M., 1986, Symbolic dynamics for the renormalization map of a quasiperiodic Schrödinger equation, *Commun. Math. Phys.* 107:295.

Casson, A.J., and Bleiler, S.A., 1988, "Automorphisms of Surfaces after Nielsen and Thurston," Cambridge Univ. Press, Cambridge.

Devaney, R.L., 1989, "An Introduction to Chaotic Dynamical Systems (Second Edition)," Addison-Wesley, Redwood City.

Horowitz, R.D., 1975, Induced automorphisms on Fricke characters of free groups, *Trans. Am. Math. Soc.* 208:41.

Iguchi, K., 1991, Theory of quasiperiodic lattices II : generic trace map and invariant surface, *Phys. Rev. B* 43:5919.

Kalugin, P.A., Kitaev, A.Yu., and Levitov, L.S., 1986, Electron spectrum of a one-dimensional quasicrystal, *Sov. Phys. JETP* 64:410.

Kohmoto, M., Kadanoff, L.P., and Tang, C., 1983, Localization problem in one dimension: mapping and escape, *Phys. Rev. Lett.* 50:1870.

Kramer, P., 1993, Algebraic structures for one-dimensional quasiperiodic systems, *J. Phys. A* 26:213.

Lamb, J.S.W., Roberts, J.A.G., and Capel, H.W., 1993, Conditions for local reversing symmetries in dynamical systems, *Physica A* 197:379.

Llibre, J., and MacKay, R.S., 1991, Pseudo-Anosov homeomorphisms on a sphere with four punctures have all periods, *Warwick Preprint* 70.

Ostlund, S., Pandit, R., Rand, D., Schellnhuber, H.J., and Siggia, E.D., 1983, One-dimensional Schrödinger equation with an almost periodic potential, *Phys. Rev. Lett.* 50:1873.

Percival, I., and Vivaldi, F., 1987, Arithmetical properties of strongly chaotic motions, *Physica D* 25:105.

Peyrière, J., 1991, On the trace map for products of matrices associated with substitutive sequences, *J. Stat. Phys.* 62:411.

Peyrière, J., Wen Zhi-xiong, and Wen Zhi-ying, 1992, Algebraic properties of trace mappings associated with substitutive sequences, *Modern Mathem. (China)* to appear.

Quispel, G.R.W., and Capel, H.W., 1989, Local reversibility in dynamical systems, *Phys. Lett. A* 142:112.

Roberts, J.A.G., and Baake, M., 1992, Trace maps as 3D reversible dynamical systems with an invariant, *Melbourne Preprint* 9, to appear *J. Stat. Phys.* (Feb. 1994).

Roberts, J.A.G., and Capel, H.W., 1992, Area-preserving mappings that are not reversible, *Phys. Lett. A* 162:243.

Roberts, J.A.G., and Lamb, J.S.W., 1993, in preparation.

Roberts, J.A.G., and Quispel, G.R.W., 1992, Chaos and time-reversal symmetry. Order and chaos in reversible dynamical systems, *Phys. Rep.* 216:63.

Sutherland, B., 1986, Simple system with quasiperiodic dynamics : a spin in a magnetic field, *Phys. Rev. Lett.* 57:770.

Turner, G.S., and Quispel, G.R.W., 1993, Tupling in three-dimensional reversible mappings, *Latrobe Preprint*.

Vrahatis, M.N., and Bountis, T., 1993, An efficient method for computing periodic orbits of symplectic mappings, this volume.

SCARS IN GROUPS OF EIGENFUNCTIONS

F. Borondo[1], G.G. de Polavieja[1] and R.M. Benito[2]

[1]Dep. de Química, C-XIV. Universidad Autónoma de Madrid
CANTOBLANCO - 28049 Madrid (Spain)
[2]Dep. de Física Aplicada. ETSI de Telecomunicación
Universidad Politécnica de Madrid. 28040 Madrid (Spain)

INTRODUCTION

A great deal of interest has been devoted recently to the study of quantum systems whose classical counterparts are chaotic[1-3]. Many authors have considered the distribution of energy levels, and demonstrated the validity of random matrix theory[4]. Also, the long range correlations in chaotic spectra have been related to classical periodic orbits (PO)[5] using a PO summation formulae due to Gutzwiller[1,6]. In a series of papers, Gutzwiller constructed a semiclassical version of the quantum mechanical Green's function in terms of classical orbits, and applied it to the calculation of eigenvalues of classically chaotic systems[6,7].

The topological characteristics of the wavefunctions of classically chaotic systems (nodal patterns[8], localization properties, etc.) have also been considered. Berry[9] and Voros[10], based on the theorem of Schnirelman[11], formulated the conjecture that the quantum expectation value of a smooth operator is the classical microcanonical average, for almost all states. Accordingly, high lying wavefunctions would be very delocalized, corresponding to the projection of the classical microcanonical distribution on the coordinate space. This early conjecture was contradicted by the numerical studies of McDonald and Kauffman[12] on the stadium; some eigenfunctions appeared strongly "scarred" by classical PO's. This phenomenon was qualitatively explained by Heller[13] in terms of the relationship existing between spectra and wavepacket dynamics; time recurrences of a packet in the vicinity of a PO being responsible for the localization on that PO. Also a time independent Feshbach-type explanation for scars has been given in the literature[14]: loosely speaking, the region around a PO is classically approximately decoupled from the surroundings, where the motion is more chaotic. Similarly, one can, quantum mechanically, formulate this idea in terms of projection operators, **P** and **Q**, for the PO subspace and the remaining of the phase space.

The localization effects can also be observed in the quasiprobability density distributions in phase space, such as Wigner or Husimi's distributions[15,16].

Using Gutzwiller's summation formulae, Bogolmony[17] obtained an expression for the smoothed quantum probability density over small intervals in energy and in space, which also exhibits localization on the PO's.

Hamiltonian Mechanics, Edited by J. Seimenis,
Plenum Press, New York, 1994

In this paper we will consider the localization properties of linear combinations of groups of eigenstates.

SYSTEM

In the following we will consider the quartic oscillator hamiltonian

$$H = \frac{1}{2}\left(P_x^2 + P_y^2\right) + \frac{1}{2}x^2y^2 + \frac{\alpha}{4}\left(x^4 + y^4\right) \tag{1}$$

with α=0.01. We have included the x^4 and y^4 terms to make the system completely bounded, and then the quantum calculations easier, although eigenvalues can also be obtained for α=0[18]. The classical dynamics corresponding to this hamiltonian is highly chaotic. Also, the potential function used in (1) is homogeneous and then H shows mechanical similarity. This makes the classical study much simpler, since trajectories at any energy can be obtained by scaling the results corresponding to E=1.

Quantum mechanically hamiltonian (1) has been the subject of numerous studies. For example, Waterland et al.[19] calculated quantum surfaces of section based on Husimi functions, with the result that they were strongly influenced by the manifolds emanating from hyperbolic fixed points of the classical Poincaré map. Also, Eckhardt et al. [20] presented a comprehensive study of the eigenfunctions of (1) showing that many of them were scarred by PO's.

The quantum states of hamiltonian (1) can be classified according to the five irreducible representations: A_1, B_1, A_2, B_2 and E, of the C_{4v} symmetry group. We have calculated the eigenvalues and eigenfunctions of the lowest 600 states, using a basis set of 2011 symmetry adapted linear combinations of distributed gaussian functions[21] centered on the knots of a certain net. To optimize the basis set we customize it to the shape of the potential; starting with a square net of equally spaced knots we eliminated those centers for which the potential energy was greater than a predetermined cut-off (65 in our case). The gaussian exponents were chosen accordingly to the semiclassical prescription of Hamilton and Light[22]. The resulting eigenvalues obtained in this way were converged to five significant figures.

RESULTS

In Fig. 1 we present the probability densities for some representative wavefunctions of hamiltonian (1). In agreement with the results of Eckhardt et al.[20] we found that around half of the calculated eigenfunctions are localized around a few PO's. The other half show a much more complicated pattern, which may be the result of the interference of many PO's or true delocalization. The first eigenfunction presented in Fig. 1, (state number 67 of A_1 symmetry) corresponds to that kind of states. On the other hand, the rest of the states presented in this figure are strongly scarred by PO's: the 100-th B_1 state by orbits running along the axis, the 27-th B_2 state by orbits localized on the diagonals, the 31-th B_2 state by a box-type PO, the 71-th B_2 state by a more complicated trajectory, and finally the 86-th A_1 state by a combinations of the diagonals and box-type PO's simultaneously. This localization properties of single eigenstates can also clearly be seen in the corresponding quasiprobability distributions in phase space, such as Wigner or Husimi distributions[21].

Let us consider now a more general version of the problem and study the scarring effects of PO's on groups of eigenstates. This can be done by following the dynamics of wavepackets that move in the surroundings of a given PO.

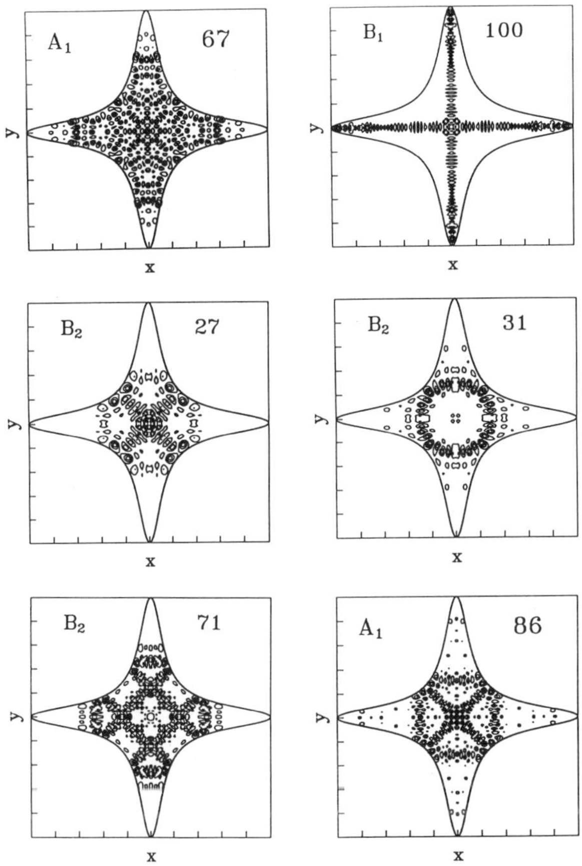

Figure 1. Quantum probability density, $|\Psi|^2$, for some representative states of the quartic hamiltonian (1). The eigenstate number and symmetry class is indicated in each panel. Some of these states appear strongly localized along a few PO of the system.

For example, let us consider a symmetrized wavepacket whose centers are initially placed at the turning points of the diagonal PO at an energy of 25; this corresponds to

$$\Phi = [\phi(x,y;x_0,y_0)+\phi(x,y;-x_0,-y_0)] \tag{2}$$

with

$$\phi(x,y;x_0,y_0) = (2\alpha/\pi)^{1/2}\,\exp[-\alpha(x-x_0)^2-\alpha(y-y_0)^2] \tag{3}$$

and $x_0 = y_0 = 2.655834$.

Since the system is bounded, instead of solving the corresponding time-dependent Schrödinger equation, the corresponding dynamics can be studied by simply calculating the infinite resolution spectrum

$$I_\infty = \frac{1}{2\pi}\int_{-\infty}^{\infty} dt <\Phi(0)|\Phi(t)> e^{-iEt} = \sum_n |<\Phi(0)|n>|^2\, \delta(E-E_n) \tag{4}$$

where $|n\rangle$ is the n-th eigenstate of the system. With the choice made in eq. (2) only states of symmetry A_1 and B_2 will appear in the spectrum.

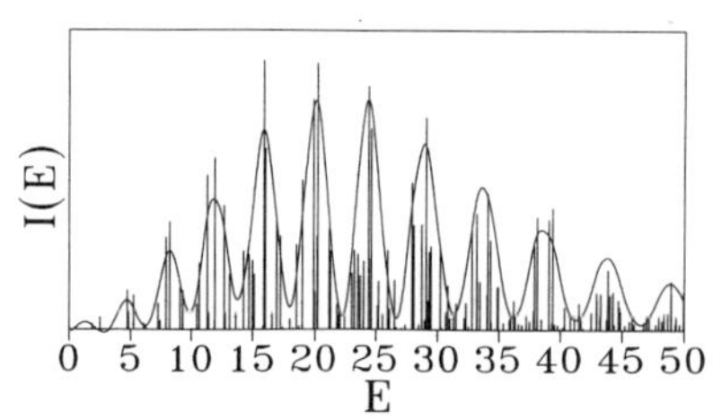

Figure 2. Infinite resolution stick spectrum, $I_\infty(E)$, and its low resolution (smooth curve) version, $I_T(E)$, for a wavepacket initially centered on the PO along the diagonal.

The corresponding calculated infinite resolution spectrum in presented in Fig. 2 in the form of stick spectrum. It shows all characteristics of chaotic, unassignable spectra: randim fluctuation of intensities, level repulsion, etc. This result is expected since it contains information on the long term dynamics of the wavepacket. Since we are only interested in the effect of PO's, which have relatively small periods, we can consider instead a low resolution/finite time version of the spectrum,

$$I_T = \frac{1}{2\pi}\int_{-T}^{T} dt <\Phi(0)|\Phi(t)> e^{-iEt} \tag{5}$$

where T is the recurrence time of the correlation function $\langle\Phi(0)|\Phi\rangle$ (which should be

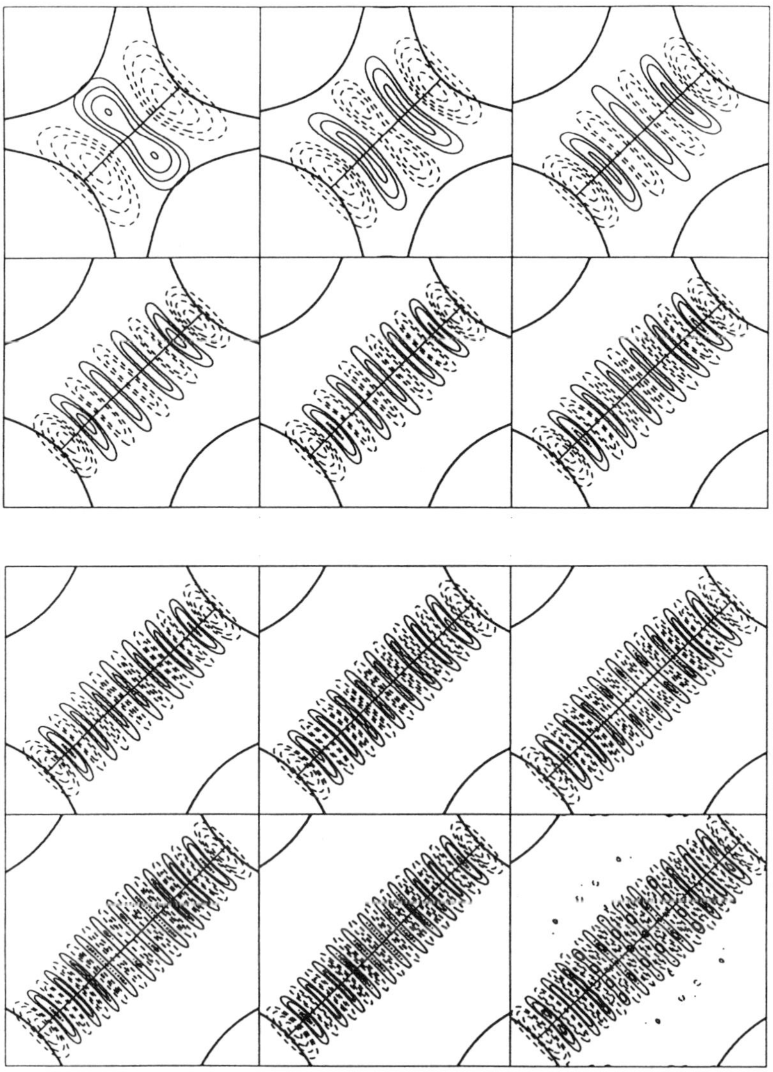

Figure 3. Squared band-wavefunctions corresponding to the twelve bands appearing in the low resolution version of the spectrum in Figure 2. The PO and the equipotential calculated at the energy of the center of the band is also shown. Notice that all of them are strongly localized along the "diagonal" PO and that they appears with an increasing number of nodes.

approximately equal to one half of the period of the diagonal PO). The result for $I_T(E)$ is also shown in Fig. 2 superimposed in continuous line to the original stick spectrum. It shows a much more simpler estructure consisting of a series of bands roughly equally spaced. The wavefunctions associated to these bands can be calculated in a straightforward manner by Fourier transform

$$\Psi^{BAND} = \frac{1}{2\pi} \int_{-\tau}^{\tau} dt \; |\Phi(t)\rangle \; e^{-iE_0 t} \tag{6}$$

where E_0 is the energy corresponding to the center of each band and τ the time associated with its width. An approximate expression of eq. (6) has been used in Ref. 23 to study the transition state corresponding to the isomerization LiCN↔LiNC, and in Ref. 24 to analyze certain features of molecular spectra.

The results for the band-wavefunctions associated to the twelve peaks in Fig. 2 are presented in Fig. 3. The corresponding "diagonal" PO and the equipotentials calculated at the energy of the center of the bands are also shown. It can be seen that these band-wavefunctions appear very localized along the PO, with an increasing number of nodes along it.

The localization mechanism can be studied in more detail if we consider the topology of the eigenfunctions that contribute to each band. In the case of the spectrum of Fig. 2 there appears two possibilities. In the first one, corresponding for example to the eleventh band, quite a lot eigenstates contribute to a single band. In this case, none of the eigenfunctions show a preferred localization on the classical PO, and it is the combined effect of the coefficients $\langle n|\Phi(0)\rangle$, weighting the eigenfunctions in eq. (6), which eliminates density in the regions far from the vicinity of the PO, giving rise to the locali-

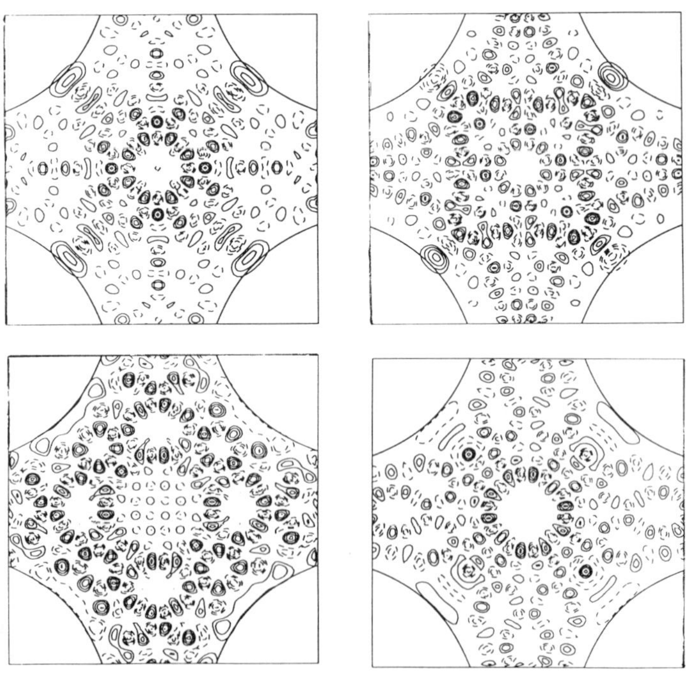

Figure 4. Wavefunctions (squared) contributing the most to the eleventh band of the low resolution spectrum in Fig.2. Notice that none of them show a preferred localization on the "diagonal" PO.

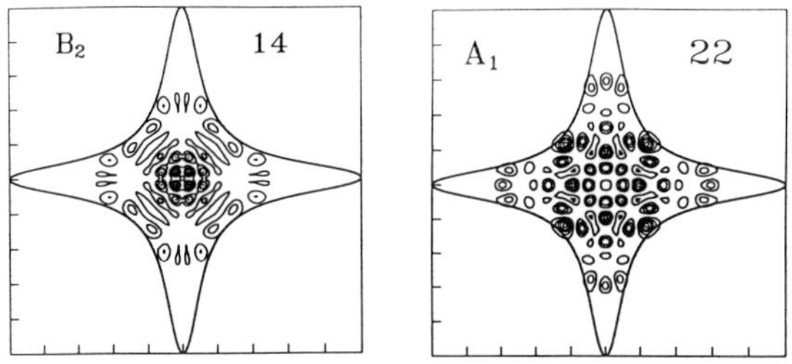

Figure 5. The only two wavefunctions (squared) that contribute significantly to the fifth band in the low resolution spectrum of Fig. 2. Note that both show are importantly scarred along the "diagonal" PO.

zation effect. This is apparent in Fig. 4 where we present the four most important contributions to the eleventh band of the spectrum. On the other hand, we have bands, like the fifth one, in which only very few eigenstates represent important contributions to the band-wavefunctions. In this case, the localization effect is more or less obvious in the individual eigenstates (see Fig. 5).

CONCLUSIONS

In this paper we have presented a general method to construct wavefunctions highly localized around a given PO in classical chaotic systems. We have applied it to the quartic oscillator and construct a series of wavefunctions with incresing excitation on a mode corresponding to a classical PO running along the diagonal. We have found situations in which the mixing of eigenstates is particularly complicated, and then the results are particularly striking. On the other hand, when the individual eigenstates show a clear scar, very few eigenstates are needed to construct the localized wavefunctions.

Future developments should include a study of the limitations of the method.

ACKNOWLEDGEMENTS

This work has been supported in part by the Dirección General de Investigación Científica y Técnica (Spain) under project PB89-117. GGP acknowledges a fellowship from Ministerio de Educación y Ciencia (Spain).

REFERENCES

1. M.C. Gutzwiller. "Chaos in Classical and Quantum Mechanics", Springer Verlag, New York (1990).
2. M.J. Giannoni, A. Voros and J. Zinn-Justin, eds., "Chaos and Quantum Physics", North- Holland, Amsterdam (1991).

3. K. Nakamura, "Quantum Chaos: A New Paradigm of Nonlinear Dynamics", Cambridge University Press, Cambridge (1993).
4. S.W. McDonald and A.N. Kauffman, Phys. Rev. Lett. 42:1189 (1979); O. Bohigas, M.J. Giannoni and C. Smit, Phys. Rev. Lett. 52:1 (1984).
5. J. Main, G. Wiebush, A. Holle and K.H. Welge, Phys. Rev. Lett. 57:2789 (1986); D. Wintgen, Phys. Rev. Lett. 58:1589 (1987); G.S. Ezra, K. Richter, G. Tanner and D. Wintgen, J. Phys. B 24:L413 (1991).
6. M.C. Gutzwiller, J. Math. Phys. 12:343 (1971).
7. M.C. Gutzwiller, Phys. Rev. Lett. 45:150 (1980).
8. R.M. Stratt, N.C. Handy and W.H. Miller, J. Chem. Phys. 71:3311 (1979).
9. M.V. Berry, J. Phys. A 10:2083 (1977).
10. A. Voros, in: "Stochastic Behaviour in Classical and Quantum Hamiltonian Systems", G. Casati and J. Ford, eds., Springer Verlag, New York (1979).
11. A.I. Schnirelman, Ups. Mat. Nauk. 29:181 (1974).
12. S.W. MacDonald and Kauffman, Phys. Rev. Lett. 42:1189 (1979).
13. E.J. Heller, Phys. Rev. Lett. 53:1515 (1984); S. Tomsovic and E.J. Heller, Phys. Rev. Lett. 70:1405 (1993).
14. H.S. Taylor, J. Zakrzewski and S. Saini, Chem. Phys. Lett. 145:555 (1988).
15. E. Wigner, Phys. Rev. 40:749 (1932).
16. K. Husimi, Proc. Phys. Soc. Japan, 22:264 (1940).
17. E.B. Bogomolny, Physica D 31:169 (1988).
18. B. Simon, Ann. Phys. 146:209 (1983).
19. R.L. Waterland, J.M. Yuan, C.C.. Martens, R.E. Gillilan and W.P. Reinhardt, Phys. Rev. Lett. 61:2733 (1988).
20. B. Eckhardt, G. Hose and E. Pollak, Phys. Rev. A 39:3776 (1989).
21. G.G. de Polavieja, F. Borondo and R.M. Benito, Quantum Localization on the quartic Potential, An. Fis. (in press).
22. I.P. Hamilton and J.C. Light, J. Chem. Phys. 84:306 (1986).
23. J.M. Gómez-Llorente, F. Borondo, N. Berenguer and R.M. Benito, Chem. Phys. Lett. 192:430 (1992).
24. M.J. Davis, Chem. Phys. Lett. 192:479 (1992).

A MODEL OF POINCARÉ AND RIGOROUS PROOF OF THE SECOND ELEMENT OF THERMODYNAMICS FROM MECHANICS

L. D. Pustyl'nikov

Research Institute of Electrical Energy
Moscow, Russia

INTRODUCTION

The main aim of this work is to point out the physical reason, which explains how irreversible processes causing a growth of the Gibbs entropy occur in reversible systems of classical mechanics. To do so we consider the models of three-dimensional or one-dimensional gas consisting of a finite number of particles moving in a container of fixed volume.

This problem interested scientists for long time and after the work of Klausius, who had introduced thermodynamic entropy, the numerous attempts have been undertaken to solve it (J. Gibbs, L. Boltzman, M. Plank, H. Poincaré, T. Ehrenfest and P. Ehrenfest, N.N. Bogolubov and others). However, L.D. Landau and E.M. Lifshitz point in their book[1], that the solution of this problem is not found and it is not known what physical reasons cause irreversibility. To solve this problem H. Poincaré in his work[2] suggested to consider one-dimensional and three-dimensional gas consisting of large number of particles $p_1, \ldots, p_N$, moving in a container which has a form of parallelepiped.

The three-dimensional model is as follows. In the three-dimensional space with orthogonal coordinates q_1, q_2, q, we consider a parallelepiped $a_1 < q_1 < b_1$, $a_2 < q_2 < b_2$, $a < q < b$, which has along the vertical direction q the edge of length $l = b - a$. In this parallelepiped there are N identical point particles $p_1, \ldots, p_N$, having each mass equal 1, that move in it interacting in accordance to the law of elastic collision when they encounter each other. In the one-dimensional model the container is the vertical interval $a < q < b$. Poincaré considered the situation when the particles are under influence of the forces produced by an external body, that at first approaches to the container then moves away from it. Poincaré shows that in that model the delicate entropy (that is the Gibbs entropy) is always constant while the coarse entropy, determined by the average of the statistical function along the cells of the phase space, can increase. However, the substitution of the delicate entropy by the coarse entropy can not explain the reason

Hamiltonian Mechanics, Edited by J. Seimenis,
Plenum Press, New York, 1994

why the reversible mechanical system becomes irreversible because as the result of this substitution the system does not change at all and the law of increase of the entropy can not depend on the averaging.

THE MODEL AND DESCRIPTION OF RESULTS

In this work we consider models which comprise described above Poincaré models. The peculiarity of our models is that the effect of external body is modelled by the action of the walls of the container on the particles coming into contact with them. In the three-dimensional model it is assumed that the following boundary conditions hold: after reflection at time t from boundaries of the container described by the equation $q = a$ or $q = b$ the particle moves as if it had been reflected at that time from a lower horizontal wall that moves in the direction q in accordance with the periodic law $q = f_1(t)$ or from an upper horizontal wall that moves in the direction q in accordance with the periodic law $q = f_2(t)$; after reflection at time t from boundaries of the container described by the equation $q_1 = a_1$ or $q_1 = b_1$ the particle moves as if it had been reflected at that time from a wall that moves in the direction q_1 in accordance with the law $q_1 = g_1(t)$ or from a wall that moves in the direction q_1 in accordance with the law $q_1 = g_2(t)$; after reflection at time t from boundaries of the container described by the equation $q_2 = a_2$ or $q_2 = b_2$ the particle moves as if it had been reflected at that time from a wall that moves in the direction q_2 in accordance with the law $q_2 = h_1(t)$ or from a wall that moves in the direction q_2 in accordance with the law $q_2 = g_2(t)$. In the one-dimensional model there are only two boundary conditions: from the boundaries of the container described by the equations $q = a$ and $q = b$ in the three-dimensional model.

It is assumed that walls are infinitely heavy, the functions $f_1(t)$ and $f_2(t)$ are twice continuously differentiable having period 1 with respect to t, i.e., $f_i(t + 1) = f(t)$, $i = 1, 2$. The functions $g_i(t)$, $h_i(t)$, $i = 1, 2$, are differentiable and the modules of their derivatives decrease sufficiently quickly with the time. The walls introduced above act at the particle only if it is in the boundaries of the container and in this time change the velocity of the particle. It is assumed that at the initial time the modules of the velocity vectors of all the particles are greater than some constant, and therefore at all the subsequent times they will never be so small that after reflection from the boundary the velocity component of the particle perpendicular to the boundary at the time t has the same direction as at the time t before reflection from that boundary. The assumptions made above are physically justified—the functions $f_i(t)$, $g_i(t)$, $h_i(t)$, $i = 1, 2$ model fields that have small oscillation amplitudes produced by the atoms in walls of the container, and the velocities of particles in the container are appreciably greater that the rates of changes of these fields.

We consider two cases.

1. The motion of the particles takes place in accordance with Newtonian mechanics.
2. The motion of the particles takes place in the framework of the special theory of relativity.

The results obtained here for these two cases are completely opposite. Namely, in the case 1 the Gibbs entropy for all t is constant independent of the time, whereas in the case 2 the Gibbs entropy increases with increasing t to infinity if fairly general conditions are satisfied by the statistical distribution functions at the initial time and by the functions $f_1(t)$ and $f_2(t)$ and if the interval $[a, b]$ has irrational length l.

Thus, we show that the presence of the relativistic effects creates a mechanism for the occurrence of irreversibility in a mechanical system consisting of a finite number of particles. This is a fundamentally new phenomenon since hitherto irreversibility has been established in connection with coarse graining of real physical systems and a passage to the limit in which the number of particles and the volume of the container tend to infinity. An important feature of the process is that in the relativistic case the energy of the particles tends exponentially to infinity, whereas in the Newtonian case it is always bounded. Therefore, one could attempt to use this relativistic model to construct small accelerators and generators of high energy particles.

This work is based on the works of the author in which he studied the Fermi-Ulam model[3,4], relativistic analog of the Fermi-Ulam model[5,6], and a model of the same type as here but with one particle[6]. This work demonstrated the possibility of obtaining arbitrary high energies in systems with bounded configuration space. In particular, in the Fermi-Ulam model the particle energy is always bounded, and the proof given by the author[4] can be extended to the model studied in this work in the Newtonian case, whereas in the generalized relativistic model studied by the author[6] the particle energy tends to infinity, this phenomena being possible moreover in an arbitrary small region of space (for arbitrary small separation between the walls), and the proof can be extended to the model studied in this work in the relativistic case.

The physical significance of the condition satisfied by the functions $f_1(t)$ and $f_2(t)$ under which irreversibility occurs is that the energy transferred from the walls of the container to the particles after a sufficiently long interval of time will be positive in the relativistic case. Regarding the irrationality of the container length l, from the physical point of view this restriction is unimportant, since the rational numbers form a set of measure zero on R^1.

The considered model is an ideal form of a thermodynamic system, but it reveals the reasons for the occurrence of irreversibility in statistical physics, since even a fraction of the particles with high velocities is capable of causing a growth of the entropy as a result of the mechanism described here.

For the model of one-dimensional gas these results were obtained by the author[7].

REFERENCES

1. L.D. Landau and E.M. Lifshitz , " Statistical Physics", Publisher Pergamon Press, Oxford, (1969).
2. H. Poincaré , 1906, Reflextions sur la theorie cinetique des gaz, *J.Phys.thoret. et appl.*, 4° ser 5.
3. L.D. Pustyl'nikov, 1983, On the Ulam problem, *Teor.Mat.Fiz*, 57, 1.
4. L.D. Pustyl'nikov, 1987, On the Fermi-Ulam problem, *Dokl.Akad.Nauk SSSR*, 292, 3.
5. L.D. Pustyl'nikov, 1988, A new mechanism of particle accerelation and a relativistic analog of the Fermi-Ulam model, *Teor.Mat.Fiz*, 77, 1.
6. L.D. Pustyl'nikov, 1990, A new mechanism of particle accerelation and rotation number, *Teor.Mat.Fiz*, 82, 2.
7. L.D. Pustyl'nikov, 1990, A mechanism of irreversibility and unbounded growth of the energy in a model of statistical mechanics, *Teor.Mat.Fiz*, 82, 2.

NEKHOROSHEV AND KAM THEOREMS REVISITED VIA A UNIFIED APPROACH*

Amadeu Delshams† Pere Gutiérrez‡

Abstract

We consider an n-degrees of freedom nearly-integrable Hamiltonian system. The system is put in normal form and estimates are obtained for the size of the remainder. The transformation to normal form is constructed by means of an iterative process. This normal form provides the analytic part required for both proofs of KAM and Nekhoroshev theorems.

Assuming non-degeneracy conditions on the unperturbed Hamiltonian, the geometric part can be fulfilled. Our estimates rely on bounds of the associated vectorfields, and provide, in Nekhoroshev theorem, the "optimal" stability exponent: $1/2n$ for quasiconvex Hamiltonians.

1 Introduction

We consider a *nearly-integrable Hamiltonian* written in *action–angle* variables:

$$H(\phi, I) = h(I) + f(\phi, I), \tag{1}$$

where $\phi = (\phi_1, \ldots, \phi_n) \in \mathbf{T}^n$ and $I = (I_1, \ldots, I_n) \in \mathcal{G} \subset \mathbf{R}^n$ are, respectively, the *angular* and *action* variables, and f is a small perturbation ($f = \mathcal{O}(\varepsilon)$) of the integrable Hamiltonian h. In general, the dynamics of (1) can be very complicated, and even Arnold diffusion can take place. Concerning stability, the main results are provided by Nekhoroshev and KAM theorems.

Nekhoroshev theorem, first proved in [Nek77], leads to the concept of *effective stability*. It states, under *steepness* conditions on h, an estimate of the type

$$|I(t) - I(0)| < R_0\, \varepsilon^b \qquad \text{for} \quad |t| \le T_0 \cdot e^{\left(\frac{A}{\varepsilon}\right)^a},$$

which holds for *all* initial conditions $(\phi(0), I(0)) \in \mathbf{T}^n \times \mathcal{G}$. The *stability exponents* a and b are positive constants that have been successively improved. Thus, for a quasiconvex unperturbed Hamiltonian, the exponent $a = 2/(n^2+n)$ was found in [BGG85], and $a < 1/(2n+1)$ in [Loc90]. Recently, the exponents

*This work has been partially supported by the spanish grant DGCYT PB90–0580.

†Dep. de Matemàtica Aplicada I, Univ. Pol. de Catalunya. Diagonal 647, 08071 Barcelona.

‡Dep. de Matemàtica Aplicada II, Univ. Pol. de Catalunya. Pau Gargallo 5, 08071 Barcelona.

Hamiltonian Mechanics, Edited by J. Seimenis,
Plenum Press, New York, 1994

$$a = b = \frac{1}{2n} ,$$

have been stated in papers [Pos92], [LN92]. It has been conjectured ([Chi79]) that the exponent $a = 1/2n$ is optimal. Related optimal estimates, for the case of a perturbation of a system of harmonic oscillators, also appear in [GD*89], [Fas90] and [DG93].

KAM theorem ([Kol78], [Arn63], [Mos62], [Pos82]) states that, for ε small enough, most of the n-dimensional invariant tori of the unperturbed system are preserved in the perturbed system (1). Such preserved tori fill a Cantorian set parameterized by frequencies satisfying a Diophantine condition. In this way, one gets *perpetual stability*, but only for initial conditions in a set which does not contain any open set.

This talk is concerned about the proof of both theorems via a unified approach that allows to stress the close relationship between the different types of stability. The method we follow, which is standard in classical perturbation theory, consists of seeking for a suitable canonical transformation Ψ bringing our Hamiltonian H into a normal form $H^* = H \circ \Psi$ depending on less angles, none if possible.

It is well-known that an obstruction for the construction of such Ψ is found on the resonances or near them. A *resonance* is characterized by a given module $\mathcal{M} \subset \mathbf{Z}^n$:

$$V_{\mathcal{M}} := \{I \in \mathcal{G} \;:\; k \cdot \omega(I) = 0 \;\; \forall k \in \mathcal{M}\},$$

where $\omega(I) := \frac{\partial h}{\partial I}(I)$ is the vector of the unperturbed frequencies. The obstruction comes from the presence of the small divisors $k \cdot \omega(I)$, with $k \in \mathcal{M}$, and forces the normal form to accept dependence on combinations of angles $k \cdot \phi$, with $k \in \mathcal{M}$. The union of all such resonances is dense in the set of frequencies, but one only needs to consider resonances up to a given finite order K. For an integer vector $k = (k_1, \ldots, k_n) \in \mathbf{Z}^n$, its *order* is defined as $|k| := |k|_1 = \sum_{j=1}^n |k_j|$, and a module $\mathcal{M} \subset \mathbf{Z}^n$ is said to be a *K-module* if it is generated by vectors of order less or equal than K.

It turns out that the effect of the resonances of higher orders is exponentially small in K. In fact, in the first part of this talk, we will restrict ourselves to a subset of the domain $\mathcal{G}$, where the frequency vector $\omega(I)$ is allowed to satisfy resonance relations corresponding to a fixed K-module $\mathcal{M}$, but a neighbourhood of all other resonances of order less or equal than K is excluded. Thus, for given $\mathcal{M}$, K, α, a subset $G \subset \mathcal{G}$ is said to be *α, K-nonresonant modulo $\mathcal{M}$*, with respect to h, if

$$|k \cdot \omega(I)| \geq \alpha \qquad \forall k \in \mathbf{Z}^n \setminus \mathcal{M}, |k| \leq K, \quad \forall I \in G. \tag{2}$$

On the set G we construct a transformation $(\phi, I) = \Psi(\phi^*, I^*)$ and the transformed Hamiltonian is written as $H^*(\phi^*, I^*) = h(I^*) + Z^*(\phi^*, I^*) + R^*(\phi^*, I^*)$, where the Fourier series expansion of Z^* in the angular variables is restricted to the form

$$Z^*(\phi^*, I^*) = \sum_{\substack{k \in \mathcal{M} \\ |k| \leq K}} Z_k^*(I^*)\, e^{i\, k \cdot \phi^*}.$$

A function with such an expansion is said to be in *normal form with respect to $\mathcal{M}$ of degree K*. The set of such functions will be denoted by $\mathbf{R}(\mathcal{M}, K)$. The remainder R^* does not need to be in normal form, but it is required to be as small as possible.

In the first part of the proofs of both theorems, the transformation Ψ is constructed by means of an iterative process, which is finite for the proof of Nekhoroshev theorem and infinite for KAM theorem. The Iterative Lemma, common to both proofs, provides estimates for one given step of this process. We make use of a *vectorfield norm* which allows us to optimize this lemma with respect to other related results.

From successive application of the Iterative Lemma, with a fixed K, we get the Normal Form Theorem, which gives a remainder that is exponentially small in K. This theorem completes the analytic part of the proof of Nekhoroshev theorem. In the geometric part of the proof, the whole domain $\mathcal{G}$ is covered by a family of suitable sets $G = G_{\mathcal{M}}$, associated to the different modules $\mathcal{M}$, where the Normal Form Theorem can be applied. Under a quasi-convexity condition, one gets Nekhoroshev theorem, with the optimal exponent.

Concerning the proof of KAM theorem, its analytic part consists of applying the Iterative Lemma successively with the trivial module $\mathcal{M} = 0$ and with K increasing to infinity, in order to exclude all resonances. Then, the remainder tends to zero but the set of validity of the process shrinks to a Cantorian set, filled with n-dimensional invariant tori. In the geometric part, the measure of this set is easily estimated under some non-degeneracy condition on the unperturbed Hamiltonian h.

2 The common part

2.1 A norm for Hamiltonian vectorfields

We work on complex neighbourhoods of the domain $\mathbf{T}^n \times G$. Given $\rho = (\rho_1, \rho_2) > 0$ (i.e., $\rho_j > 0,\ j = 1, 2$), we first introduce complex neighbourhoods for $\mathbf{T}^n$ and G:

$$\mathcal{W}_{\rho_1}(\mathbf{T}^n) := \{\phi \ :\ \operatorname{Re}\phi \in \mathbf{T}^n,\ |\operatorname{Im}\phi|_\infty < \rho_1\},$$
$$\mathcal{V}_{\rho_2}(G) := \{I \in \mathbf{C}^n \ :\ |I - I'|_2 < \rho_2 \text{ for some } I' \in G\},$$

and define

$$\mathcal{D}_\rho = \mathcal{D}_\rho(G) := \mathcal{W}_{\rho_1}(\mathbf{T}^n) \times \mathcal{V}_{\rho_2}(G)$$

as a complex neighbourhood of $\mathbf{T}^n \times G$. Moreover we introduce $\mathcal{U}_{\rho_2}(G) := \mathcal{V}_{\rho_2}(G) \cap \mathbf{R}^n$ as a real neighbourhood of G.

For a bounded analytic function f on $\mathcal{D}_\rho(G)$ (2π-periodic in ϕ), let us consider its Fourier series $f(\phi, I) = \sum_{k \in \mathbf{Z}^n} f_k(I)\, e^{i\, k \cdot \phi}$, and define the norm

$$\|f\|_\rho := \sum_{k \in \mathbf{Z}^n} |f_k|_{\rho_2} \cdot e^{|k|\rho_1} \tag{3}$$

where $|f_k|_{\rho_2} := \sup_{I \in \mathcal{V}_{\rho_2}(G)} |f_k(I)|$. This norm (3), analogous to the one used in [Pos92], allows to carry out easily a separate control of harmonics.

An important remark is that the Hamiltonian function H does not take part directly in the Hamiltonian equations, but rather its derivative

$$DH = \left(\frac{\partial H}{\partial \phi}, \frac{\partial H}{\partial I}\right) = \left(\frac{\partial H}{\partial \phi_1}, \ldots, \frac{\partial H}{\partial \phi_n}, \frac{\partial H}{\partial I_1}, \ldots, \frac{\partial H}{\partial I_n}\right).$$

In order to avoid unnecessary uses of the Cauchy inequalities, we will work with a vectorfield norm. This idea was suggested to us by A. I. Neishtadt, but in fact it goes back to the work [Fas90] of F. Fassò, where it is fully developed. So, we introduce

$$\|Df\|_\rho = \|Df\|_{\rho,c} := \max\left(\frac{1}{c_2}\sum_{j=1}^{n}\left\|\frac{\partial f}{\partial \phi_j}\right\|_\rho, \frac{1}{c_1}\max_{j=1,\ldots,n}\left\|\frac{\partial f}{\partial I_j}\right\|_\rho\right), \tag{4}$$

where c_1, c_2 are positive numbers that we will fix in advance, satisfying $\max(c_1, c_2) = 1$. We will write $c = c_1 \cdot c_2 = \min(c_1, c_2)$.

For $\delta = (\delta_1, \delta_2) < \rho$ we write $|\delta| := |\delta|_\infty = \max(\delta_1, \delta_2)$, and take $c_j := \delta_j / |\delta|$, $j = 1, 2$. For analytic functions f and g on $\mathcal{D}_\rho$, we can easily check the following properties of the norm (3–4):

P1 $\|Df\|_{\rho-\delta,c} \le \frac{1}{c|\delta|} \|f\|_\rho$.

P2 $\|\{f,g\}\|_\rho \le 2c \|Df\|_{\rho,c} \cdot \|Dg\|_{\rho,c}$.

P3 $\|f_{>K}\|_{(\rho_1-\delta_1,\rho_2)} \le e^{-(K+1)\delta_1} \cdot \|f\|_\rho$.

By properties P1 and P2, we see that the vectorfield norm (4) provides a compact writing, avoiding to carry out different estimates for the ϕ-derivatives and the I-derivatives. But its main feature is that a good choice of the numbers c_1, c_2 in the definition (4) makes easier to obtain better estimates, even in the case of very different δ_1, δ_2. In fact, δ_2 will often be chosen much smaller than δ_1, i.e., $c_1 = 1$, $c = c_2 << 1$.

2.2 The Iterative Lemma

Given $\mathcal{M}$, K, α, we restrict our Hamiltonian (1) to a subset $G \subset \mathcal{G}$, which is assumed to be α, K-nonresonant modulo $\mathcal{M}$. On the set G, we consider H written in the form $H(\phi, I) = h(I) + Z(\phi, I) + R(\phi, I)$, with $Z \in \mathbf{R}(\mathcal{M}, K)$.

We assume that H is real analytic on $\mathcal{D}_\rho = \mathcal{D}_\rho(G)$, and let us introduce

$$M := \left|\frac{\partial^2 h}{\partial I^2}\right|_{\rho_2} = \sup_{I \in \mathcal{V}_{\rho_2}(G)} \left|\frac{\partial^2 h}{\partial I^2}(I)\right|_2 , \tag{5}$$

where $|\cdot|_2$ stands for the matrix norm induced by the 2-norm for vectors. If

$$\rho_2 \le \frac{\alpha}{2MK} , \tag{6}$$

and G is α, K-nonresonant modulo $\mathcal{M}$, it is easy to see that its neighbourhood $\mathcal{V}_{\rho_2}(G)$ is still $\frac{\alpha}{2}, K$-nonresonant modulo $\mathcal{M}$. We shall see in section 3.2 that condition (6) imposes an important restriction on the domain $\mathcal{V}_{\rho_2}(G)$. An exception is the very special case of a system of harmonic oscillators, where $M = 0$ (see section 3.3).

We will construct the canonical transformation Ψ as a product of canonical transformations $\Phi^{(1)}, \Phi^{(2)}, \ldots$ At the step q we will get

$$H^{(q)} = H \circ \Phi^{(1)} \circ \cdots \circ \Phi^{(q)} = h + Z^{(q)} + R^{(q)},$$

where $Z^{(q)} \in \mathbf{R}(\mathcal{M}, K)$. Obviously we start with $Z^{(0)} = Z$ and $R^{(0)} = R$.

Now we write H, Z, R, $\tilde{Z}$, $\tilde{R}$, Φ, instead of $H^{(q-1)}$, $Z^{(q-1)}$, $R^{(q-1)}$, $Z^{(q)}$, $R^{(q)}$, $\Phi^{(q)}$, respectively. The transformation Φ is constructed as the flow at time 1 associated to a Hamiltonian determined through the Lie series method, as described in [DG93].

Theorem 1 (Iterative Lemma) *Let $H = h + Z + R$ real analytic on $\mathcal{D}_\rho = \mathcal{D}_\rho(G)$, and assume that G is α, K-nonresonant modulo $\mathcal{M}$, and that $Z \in \mathbf{R}(\mathcal{M}, K)$. Let $\delta = (c_1, c_2)\,|\delta| < \rho$, and $L := 1 + \frac{2M}{\alpha} \cdot \frac{c_2}{c_1}$. Assume:*

$$\rho_2 \le \frac{\alpha}{2MK} , \qquad \|DR\|_\rho \le \frac{\alpha|\delta|}{74L} .$$

Then there exists a real canonical transformation $\Phi : \mathcal{D}_{\rho-\delta} \longrightarrow \mathcal{D}_\rho$ such that $H \circ \Phi = h + \tilde{Z} + \tilde{R}$, with $\tilde{Z} \in \mathbf{R}(\mathcal{M}, K)$, and the following bound holds:

$$\left\|D\tilde{R}\right\|_{\rho-\delta} \le e^{-(K+1)\delta_1} \cdot \|DR\|_\rho + \frac{14L}{\alpha|\delta|} \left(\|DZ\|_\rho + \|DR\|_\rho\right) \cdot \|DR\|_\rho . \tag{7}$$

3 Nekhoroshev theorem

3.1 The analytic part

Now, starting with $H = h + Z + R$ on $\mathcal{D}_\rho$, we apply Q times the Iterative Lemma and obtain an estimate for the remainder. Given $\delta = (c_1, c_2)\,|\delta| < \rho$, we define

$$\rho^{(0)} = \rho, \qquad \rho^{(q)} = \rho^{(q-1)} - \hat{\delta} = \rho - q\hat{\delta}, \;\; 1 \leq q \leq Q,$$

with $\hat{\delta} = \delta/Q$. At the step q, we obtain $\Phi^{(q)} : \mathcal{D}_{\rho^{(q)}} \longrightarrow \mathcal{D}_{\rho^{(q-1)}}$. Starting with $H^{(q-1)}$ on $\mathcal{D}_{\rho^{(q-1)}}$, we get an estimate for $H^{(q)} = H^{(q-1)} \circ \Phi^{(q)}$ on $\mathcal{D}_{\rho^{(q)}}$. At the last step $q = Q$, we estimate $H^* = H^{(Q)}$ on $\mathcal{D}_{\rho-\delta}$. By choosing Q as $[(K+1)\delta_1/2]$ we get an exponentially small remainder, as shown in the next theorem.

Theorem 2 (Normal Form Theorem) *Let $H = h + Z + R$ real analytic on $\mathcal{D}_\rho = \mathcal{D}_\rho(G)$, and assume that G is α, K-nonresonant modulo $\mathcal{M}$, and that $Z \in \mathbf{R}(\mathcal{M}, K)$. Let $\delta = (c_1, c_2)\,|\delta| < \rho$. Assume:*

$$\rho_2 \leq \frac{\alpha}{2MK}\,, \qquad \|DZ\|_\rho + \|DR\|_\rho \leq \frac{\alpha}{61 c_1 L (K+1)}\,.$$

Then there exists a real canonical transformation $\Psi : \mathcal{D}_{\rho-\delta} \longrightarrow \mathcal{D}_\rho$ such that $H \circ \Psi = h + Z^* + R^*$, with $Z^* \in \mathbf{R}(\mathcal{M}, K)$, and the following bound holds:

$$\|DR^*\|_{\rho-\delta} \;\leq\; 3\, e^{-\frac{(K+1)\delta_1}{2}} \cdot \|DR\|_\rho\,.$$

3.2 The geometric part

From the Normal Form Theorem, one can obtain *stability estimates* on the set G where we apply this theorem. If G is α, K-nonresonant modulo 0 (a nonresonant region), the normal form $Z^*(I^*)$ only depends on the action variables, and therefore they can be bounded up to an exponentially large time.

In order to obtain stability estimates for the orbits starting in a resonant region, it is necessary to impose some geometric condition on h. In the original Nekhoroshev's proof ([Nek77]), a general *steepness* condition was imposed. But the main geometric ideas of the proof are contained in the simpler quasiconvex case, which is considered in [BGG85], [Loc90] and other papers. Following [Loc90], we say that h is *m-quasiconvex* on U if, at every point $I \in U$, and for all $v \in \mathbf{R}^n$ such that $\omega(I) \cdot v = 0$, we have

$$\left| \frac{\partial^2 h}{\partial I^2}(I)\, v \cdot v \right| \geq m\, |v|_2^2\,.$$

Under this condition, stability estimates on a resonant region $G = G_\mathcal{M}$ are obtained following [Loc90] and [Pos92]. The basic point is that, for a Hamiltonian in normal form with respect to $\mathcal{M}$ with an exponentially small remainder, the variation of the action variables along the $\mathcal{M}^\perp$-direction is exponentially small. Then the quasi-convexity condition and the energy conservation allow to bound the variation along the $\mathcal{M}$-direction, giving rise to the stability estimate.

In order to obtain stability estimates for all orbits starting in $\mathbf{T}^n \times \mathcal{G}$, the whole domain $\mathcal{G}$ is covered by a family of sets $G_\mathcal{M}$ (the trivial module $\mathcal{M} = 0$ is also considered). For each module $\mathcal{M}$, the set $G_\mathcal{M}$ has to be close to $\mathcal{M}$-resonances, and to satisfy the non-resonance condition (2) up to a fixed order K of the form $[(A/\varepsilon)^{1/2n}]$.

The construction of such a covering is carried out in [Nek77] and [BGG85]. The quantitative aspect was improved in the Geometric Lemma stated in [Pos92]. From this lemma and the stability estimates we obtain our main result on effective stability.

Theorem 3 (Nekhoroshev theorem) *Let* $H(\phi, I) = h(I) + Z(I) + R(\phi, I)$ *real analytic on* $\mathcal{D}_\rho(\mathcal{G})$*, and assume that* h *is* m*-quasiconvex on* $\mathcal{U}_{\rho_2}(\mathcal{G})$*. Define:*

$$\Omega := \sup_{I \in \mathcal{U}_{\rho_2}(G)} |\omega(I)|_2 , \qquad F := \tfrac{2882 M^2}{m^2} , \qquad \hat{\rho} := \min(\rho_1, \rho_2) .$$

Let $0 < \lambda < \frac{23 M^2 \hat{\rho}}{m}$ *given, and assume:*

$$\varepsilon := \|Z\|_\rho + \|R\|_\rho \ \leq \ A := \tfrac{16\lambda^2}{m\,F^{2n}} \cdot \min\left(\tfrac{\rho_1}{6} , \tfrac{1}{\sqrt{n}}\right) .$$

Then, for every orbit $(\phi(t), I(t))$ *of* H*, with* $(\phi(0), I(0)) \in \mathbf{T}^n \times \mathcal{G}$ *and satisfying* $|\omega(I(0))|_2 \geq \lambda$ *, one has*

$$|I(t) - I(0)|_2 < \hat{\rho} \cdot \left(\tfrac{\varepsilon}{A}\right)^{1/2n} \qquad \textit{for} \quad |t| \leq \tfrac{F}{\Omega}\, e^{\frac{m\rho_1}{12M}\left(\frac{A}{\varepsilon}\right)^{1/2n}} .$$

Remark. The stability exponents proclaimed in the introduction: $a = b = 1/2n$, have been obtained in the proof of theorem 3 by taking the worst possible case in all situations. However, the stability exponents can be improved by means of a particular analysis, if one is only interested in a given region. In particular, in a neighbourhood of a given resonant manifold $V_{\mathcal{M}_0}$, one can get the exponents $a = b = 1/2\nu_0$, where ν_0 is the codimension of $\mathcal{M}_0$. We also notice that, in the case of a nonresonant set, the stability exponents are $a = 1/2n$, $b = 1/2$ (see these results in [Pos92] and [DG93]).

3.3 A perturbation of a system of harmonic oscillators

As a simple application of the stability estimates in the nonresonant case, one can also consider the case $h(I) = \omega \cdot I$, i.e., H is a perturbation of a system of n harmonic oscillators. The frequencies vector ω is assumed to satisfy standard Diophantine condition (8). In this case, where no geometric part is required, we obtain (like in papers [GD*89], [Fas90], [Pos92] and [DG93]) the "optimal" stability exponent $1/(\tau+1)$, where τ is the exponent in the Diophantine condition for ω. We notice that, since $M = 0$, condition (6) does not impose any restriction on ρ_2, and K can be chosen of the form $[(A/\varepsilon)^{1/(\tau+1)}]$.

Theorem 4 *Let* $H(\phi, I) = \omega \cdot I + Z(I) + R(\phi, I)$ *real analytic on* $\mathcal{D}_\rho(\mathcal{G})$*, and assume that* ω *satisfies the Diophantine condition:*

$$|k \cdot \omega| \geq \frac{\alpha}{|k|^\tau} \qquad \forall k \in \mathbf{Z}^n \setminus \{0\}, \tag{8}$$

for some $\alpha > 0$*,* $\tau > n - 1$*. Assume* $\varepsilon := \|Z\|_\rho + \|R\|_\rho \ \leq \ A := \frac{\alpha\rho_2}{732}$ *. Then, for every orbit* $(\phi(t), I(t))$ *of* H*, with* $(\phi(0), I(0)) \in \mathbf{T}^n \times \mathcal{G}$*, one has*

$$|I(t) - I(0)|_2 < \tfrac{\rho_2}{2} \qquad \textit{for} \quad |t| \leq \tfrac{\rho_1 \rho_2}{24\,\varepsilon}\, e^{\frac{\rho_1}{12}\left(\frac{A}{\varepsilon}\right)^{1/(\tau+1)}} .$$

4 KAM theorem

In order to prove KAM theorem, it is necessary to apply infinitely many times the Iterative Lemma in such a way that the canonical transformation Ψ takes the remainder off and the transformed Hamiltonian becomes integrable: $H^*(\phi^*, I^*) = h(I^*) + Z^*(I^*)$.

The domain where the transformation Ψ holds is then filled with invariant tori. This domain will be a Cantorian set corresponding to Diophantine frequencies and the measure of its complement will be small.

Let us describe the scheme of the iterations. We start with $H = h + R = h^{(0)} + R^{(0)}$ on $\mathcal{D}_\rho(\mathcal{G})$. After we have applied $q-1$ times the Iterative Lemma, let us assume that we have constructed $H^{(q-1)} = h^{(q-1)} + R^{(q-1)}$, which is real analytic on a domain $\mathcal{D}_{\rho^{(q-1)}}(G_{q-1})$ to be specified below. The Iterative Lemma provides a canonical transformation $\Phi^{(q)} : \mathcal{D}_{\rho^{(q)}}(G_{q-1}) \longrightarrow \mathcal{D}_{\rho^{(q-1)}}(G_{q-1})$, and we obtain a new Hamiltonian

$$H^{(q)} = H^{(q-1)} \circ \Phi^{(q)} = h^{(q)} + R^{(q)},$$

with $h^{(q)} := h^{(q-1)} + Z^{(q)}$. The set $\mathcal{D}_{\rho^{(q)}}(G_{q-1})$, where $H^{(q)}$ is defined, contains $\mathcal{D}_{\rho^{(q)}}(G_q)$ and in this way the process may go on.

Next we sketch the quantitative aspect. For $\alpha > 0$ and $\tau > n-1$ given, several parameters are chosen in such a way that the tori corresponding to the frequencies satisfying Diophantine condition (8) are conserved. For $q \geq 0$, we consider sequences α_q (increasing from $\alpha/2$ to α) and K_q (increasing to ∞), and we define:

$$G_q := \left\{ I \in \mathcal{G} \ : \ \left|k \cdot \omega^{(q)}(I)\right| \geq \frac{\alpha_q}{|k|^\tau} \quad \forall\, 0 < |k| \leq K_q \right\},$$

where $\omega^{(q)} := \partial h^{(q)}/\partial I$. Clearly, this set is $\frac{\alpha_q}{K_q^\tau}, K_q$-nonresonant modulo 0, with respect to $h^{(q)}$. We write $\delta^{(q)} = \rho^{(q-1)} - \rho^{(q)} = (c_1^{(q)}, c_2^{(q)}) \left|\delta^{(q)}\right|$, which will be taken decreasing in a geometric way. For the angular component, we can choose the sequence $\delta_1^{(q)}$ in such a way that $\rho_1^{(\infty)} = \rho_1/4$. But, concerning the action component, condition (6) forces $\rho_2^{(q)}$ to be chosen tending to zero: $\rho_2^{(\infty)} = 0$. We also consider a sequence M_q of bounds for $\left|\frac{\partial^2 h^{(q)}}{\partial I^2}\right|_{\rho_2^{(q)}}$.

For the first iteration, the Iterative Lemma imposes a condition to be fulfilled by $\|R\|_\rho$ (see theorem 5 below), and we proceed by induction in the subsequent iterations. For $q \geq 1$, we assume that $H^{(q-1)} = h^{(q-1)} + R^{(q-1)}$ is real analytic on $\mathcal{D}_{\rho^{(q-1)}}(G_{q-1})$ and satisfies the hypothesis of the Iterative Lemma for $\mathcal{M} = 0$, with the constants α_{q-1}, K_{q-1}, M_{q-1}, $\rho^{(q-1)}$ and $\delta^{(q)}$. Then, if K_0 is large enough, the set G_{q-1} contains G_q, and $H^{(q)}$ is also real analytic on $\mathcal{D}_{\rho^{(q)}}(G_q)$ and satisfies the hypothesis of the Iterative Lemma for $\mathcal{M} = 0$, but now with the constants α_q, K_q, M_q, $\rho^{(q)}$ and $\delta^{(q+1)}$. As a consequence, the size of the remainder $R^{(q)}$ tends to zero for $q \to \infty$.

The composition of the successive transformations $\Phi^{(1)}, \Phi^{(2)}, \ldots$ provides a canonical transformation $\Psi : \mathcal{W}_{\rho_1/4}(\mathbf{T}^n) \times G_\infty \longrightarrow \mathcal{D}_\rho(\mathcal{G})$, where

$$G_\infty := \left\{ I \in \mathcal{G} \ : \ \left|k \cdot \omega^{(\infty)}(I)\right| \geq \frac{\alpha}{|k|^\tau} \quad \forall\, k \neq 0 \right\}.$$

Then one checks that $\mathcal{T}_{\alpha,\tau} := \Psi\left(\mathbf{T}^n \times G_\infty\right)$ is a real invariant set for the equations associated to the initial Hamiltonian H.

Finally, in order to find a lower bound for the measure of this invariant set, we need to introduce some non-degeneracy condition on the unperturbed Hamiltonian h. The simplest condition is to say that h is *m-nondegenerate* on $\mathcal{G}$ if

$$\left|\frac{\partial^2 h}{\partial I^2}(I) \cdot v\right|_2 \geq m\,|v|_2 \quad \forall\, v \in \mathbf{R}^n, \quad \forall\, I \in \mathcal{G},$$

but the measure estimates are also possible under different conditions like, for instance, the isoenergetic non-degeneracy.

Theorem 5 (KAM theorem) *Let $H(\phi, I) = h(I) + R(\phi, I)$ real analytic on $\mathcal{D}_\rho(\mathcal{G})$, and assume that h is m-nondegenerate on $\mathcal{G}$. Let $\alpha > 0$ and $\tau > n-1$ given. Then there exists a constant C_1 such that, if*

$$\|R\|_\rho \leq C_1 \alpha^2,$$

there is an invariant set $\mathcal{T}_{\alpha,\tau} \subset \mathbf{T}^n \times \mathcal{G}$ filled with invariant tori with associated frequencies satisfying the Diophantine condition (8). Moreover,

$$\text{mes}\,[(\mathbf{T}^n \times \mathcal{G}) \setminus \mathcal{T}_{\alpha,\tau}] \leq C_2\, \alpha.$$

The constants C_1, C_2 do not depend on α.

References

[Arn63] V. I. Arnold. Proof of a theorem of A. N. Kolmogorov on the preservation of quasi-periodic motions under small perturbations of the Hamiltonian. *Russian Math. Surveys 18:* 9–36, 1963.

[BGG85] G. Benettin, L. Galgani, A. Giorgilli. A proof of Nekhoroshev's theorem for the stability times in nearly integrable Hamiltonian systems. *Cel. Mech. 37:* 1–25, 1985.

[Chi79] B. V. Chirikov. A universal instability of many-dimensional oscillator systems. *Physics Reports 52:* 263–379, 1979.

[DG93] A. Delshams, P. Gutiérrez. Effective stability for nearly integrable Hamiltonian systems. In C. Perelló, C. Simó, J. Solà-Morales (editors), *Equadiff 91.* World Scientific, Singapore, 1993.

[Fas90] F. Fassò. Lie series method for vector fields and Hamiltonian perturbation theory. *J. Appl. Math. Phys. (ZAMP) 41:* 843–864, 1990.

[GD*89] A. Giorgilli, A. Delshams, E. Fontich, L. Galgani, C. Simó. Effective stability for a Hamiltonian system near an elliptic equilibrium point, with an application to the restricted three body problem. *J. Diff. Equations 77:* 167–198, 1989.

[Kol78] A. N. Kolmogorov. The general theory of dynamical systems and classical mechanics. In R. H. Abraham, J. E. Marsden (editors), *Foundations of Mechanics:* 741–757. The Benjamin/Cummings Publishing Company, 1978.

[Loc90] P. Lochak. Stabilité en temps exponentiels des systèmes hamiltoniens proches de systèmes integrables: résonances et orbites fermées. Preprint. Laboratoire de Mathématiques, École Normale Supérieure, 1990.

[LN92] P. Lochak, A. I. Neishtadt. Estimates of stability time for nearly integrable systems with a quasiconvex Hamiltonian. *Chaos 2:* 495–499, 1992.

[Mos62] J. Moser. On invariant curves of area-preserving mappings of an annulus. *Nachr. Akad. Wiss. Gottingen Math. Phys. Kl. 2:* 1–20, 1962.

[Nek77] N. N. Nekhoroshev. An exponential estimate of the time of stability of nearly-integrable Hamiltonian systems. *Russian Math. Surveys 32:* 1–65, 1977.

[Pos82] J. Pöschel. Integrability of Hamiltonian systems on Cantor sets. *Comm. Pure Appl. Math. 35:* 653–695, 1982.

[Pos92] J. Pöschel. On Nekhoroshev estimates for quasi-convex Hamiltonian systems. *Math. Z.* (to appear), 1992.

DYNAMICS AND K-SYMMETRIES

G.R.W. Quispel[1] and J.S.W. Lamb[2]

[1] Department of Mathematics, La Trobe University
Bundoora, Melbourne 3083, Australia
[2] Institute for Theoretical Physics, University of Amsterdam
Valckenierstraat 65, 1018 XE Amsterdam, The Netherlands

1 INTRODUCTION

The presence of symmetries is common in many dynamical systems of interest, endowing the system with certain simplifying features. A lot of attention has been given to continuous symmetries, since they give rise to conservation laws allowing one to reduce the number of degrees of freedom. During the past two decades attention has been directed to discrete symmetries as well, often in the context of bifurcation problems (Sattinger, 1979; Vanderbauwhede, 1982; Golubitsky et al., 1988).

There are two types of discrete symmetries. *Symmetries* are transformations in phase space that leave the equations of motion invariant. *Reversing symmetries* leave the equations of motion invariant if the direction of time is also reversed (Lamb 1992). It is the latter case which is of primary interest here. For a review of systems with time reversal symmetry see Roberts and Quispel (1992).

When a discrete symmetry has a simple geometric interpretation in phase space its presence is quite manifest in the phase portrait. It has been realised recently (Lamb, 1993a; D'Ariano et al., 1992; Hoveijn, 1992) that it is possible to see manifest discrete symmetry in the phase portrait in the absence of a symmetry. An example of this is shown in Figure 1. The map which gives rise to this phase portrait, Eqn. (3.1), possesses symmetry properties only if considered on a proper time scale: i.e. it is only certain iterates of the map which possess the symmetry. This phenomenon can also be observed in various well known dynamical systems that occur in the context of problems in physics, e.g. the standard map (Chirikov, 1979), web maps (Lamb, 1993a; Zaslavsky et al., 1991; Hoveijn, 1992), a class of trace maps (Roberts and Baake, 1993), and the kicked rotator (D'Ariano et al., 1992). This last example will be discussed in section 4.

The outline of the rest of this paper is as follows. In the next section we briefly review the concepts of symmetries and reversing symmetries of dynamical systems and

Hamiltonian Mechanics, Edited by J. Seimenis,
Plenum Press, New York, 1994

Figure 1. Phase portrait of (3.1) with $\lambda = -0.2$ and $\kappa = 0.34$ on $[-2,2] \times [-2,2]$

present some standard results. Section 3 contains our main results. There we consider dynamical systems that possess symmetry properties only if considered on certain time scales. We discuss a generalization of the method for obtaining periodic orbits to systems with these k-symmetries.

Throughout this paper we confine ourselves to systems with discrete time, i.e. mappings. These occur as Poincaré sections of dynamical systems with continuous time, as well as in problems of a truly discrete nature.

The reader who is interested in a more extended motivation and theoretical treatment of k-symmetries may wish to consult Lamb and Quispel (1993) of which the present paper is an abridged version.

2 SYMMETRIES AND REVERSING SYMMETRIES

Let us first review some familiar concepts and results both in order to make our presentation more self contained, but also so that we may make a comparison with results of the next section.

We consider a discrete dynamical system, i.e. a map on a state space Ω

$$L : \Omega \mapsto \Omega. \tag{2.1}$$

Let us first give some notation.

The map M is a symmetry of L if

$$M \circ L = L \circ M. \tag{2.2}$$

The map S is a reversing symmetry of L if

$$S \circ L = L^{-1} \circ S. \tag{2.3}$$

The set $\mathcal{G}$ of symmetries of L is called the *symmetry group* of L.[1] We denote the set comprising both the symmetries and the reversing symmetries the *reversing symmetry group* $\mathcal{E}$. Discussion of the group structure of $\mathcal{E}$ and $\mathcal{G}$ can be found in Lamb (1992).

We denote the orbit of a point $\mathbf{x}_0$ by

$$\Gamma(\mathbf{x}_0) = \{\mathbf{x} \in \Omega | \mathbf{x} = L^n \mathbf{x}_0, n \in \mathbb{Z}\} \tag{2.4}$$

and the set of fixed points of a map $A : \Omega \mapsto \Omega$ by

$$\text{Fix}(A) = \{\mathbf{x} \in \Omega \mid A\mathbf{x} = \mathbf{x}\}. \tag{2.5}$$

The presence of discrete symmetries naturally has consequences for the dynamics. For instance, given an invariant set $\Delta \subset \Omega$ it follows that $U\Delta$ is also an invariant set if U is a discrete symmetry of the equations of motion. If $\Delta = U\Delta$ also, then we say that Δ is a *symmetric* invariant set. A more fruitful consequence of reversing symmetries is that their presence permits one to readily locate symmetric periodic orbits via the Fixed Set Iteration (FSI) Method (DeVogelaere (1958), Piña and Jiménez Lara (1987), Piña and Cantoral (1989), Ichikawa et al. (1989), Richter et al. (1990), D'Ariano et al. (1992), Roberts and Quispel (1992), Lamb (1992)). Prior to stating the method let us first present some Lemmas which are of independent interest.

Lemma 1 *$\Gamma(\mathbf{x}_0)$ is symmetric with respect to the reversing symmetry S if and only if $\mathbf{x}_0 \in \text{Fix}(L^m \circ S)$ for some value of $m \in \mathbb{Z}$.*

Proof.
For all $n \in \mathbb{Z}$ we have

$$S\mathbf{x}_0 = L^{-m}\mathbf{x}_0 \Rightarrow L^n \circ S\mathbf{x}_0 = L^{n-m}\mathbf{x}_0 \Rightarrow S \circ L^{-n}\mathbf{x}_0 = L^{n-m}\mathbf{x}_0. \tag{2.6}$$

The converse follows by definition. □

Lemma 2 *Let S be a reversing symmetry of L and $m \in \mathbb{Z}$, then $\mathbf{x}_0 \in \text{Fix}(L^m \circ S) \Rightarrow \mathbf{x}_0 \in \text{Fix}(S^2)$.*

Proof.
$L^m \circ S\mathbf{x}_0 = \mathbf{x}_0 \Rightarrow L^m \circ S \circ L^m \circ S\mathbf{x}_0 = \mathbf{x}_0 \Leftrightarrow S^2\mathbf{x}_0 = \mathbf{x}_0$. □

Now we present the fixed set iteration (FSI) method.

Theorem 3 *$\Gamma(\mathbf{x}_0)$ is a symmetric periodic orbit with respect to a reversing symmetry S if and only if there are $n, m \in \mathbb{Z}$ ($n \neq m$) such that $\mathbf{x}_0 \in \text{Fix}(L^n \circ S) \cap \text{Fix}(L^m \circ S)$.*

Proof.

$$\begin{aligned} \mathbf{x}_0 &\in \text{Fix}(L^n \circ S) \cap \text{Fix}(L^m \circ S) && (2.7)\\ &\Rightarrow L^{n-m} \circ L^m \circ S \circ L^m \circ S\mathbf{x}_0 = \mathbf{x}_0 && (2.8)\\ &\Rightarrow L^{n-m} \circ S^2\mathbf{x}_0 = \mathbf{x}_0 \quad \text{as S is a symmetry} && (2.9)\\ &\Rightarrow L^{n-m}\mathbf{x}_0 = \mathbf{x}_0. \quad \text{via Lemma 2} && (2.10) \end{aligned}$$

[1] We assume that both the map and the (reversing) symmetries are invertible.

The symmetry of the orbit follows from the premise and Lemma 1. Conversely, if the orbit is symmetric with respect to a reversing symmetry S, then there is an n in $\mathbb{Z}$ such that $L^{-n}\mathbf{x}_0 = S\mathbf{x}_0$. Since the orbit is periodic it follows directly that $L^{p-n}\mathbf{x}_0 = S\mathbf{x}_0$. □

The use of Theorem (3) is made easier by the fact that

$$\mathrm{Fix}(L^{2n} \circ S) = L^n(\mathrm{Fix}(S)) \tag{2.11}$$

$$\mathrm{Fix}(L^{2n+1} \circ S) = L^n(\mathrm{Fix}(L \circ S)) \tag{2.12}$$

when S is a reversing symmetry.

In the case that a periodic orbit is symmetric with respect to more that one reversing symmetry other results exist (Lamb and Quispel, 1993).

3 K-SYMMETRIES AND REVERSING K-SYMMETRIES

We come now to the main concern of this paper: the concept of a *k-symmetry.* Let us motivate our discussion with an example.

Let us consider the map L of the plane given by

$$L: \begin{cases} x' = \lambda \sin(\pi x) - y + \kappa \sin\left[\pi x - \pi\kappa \sin\left[\pi y - \pi\lambda \sin(\pi x)\right]\right] \\ y' = -x + \kappa \sin[\pi y - \pi\lambda \sin(\pi x)] - \lambda \sin(\pi x') \end{cases} \tag{3.1}$$

where κ, $\lambda \in \mathbb{R}$ are parameters. Admittedly this map is contrived but it provides a convenient example with which to illustrate our discussion. An example from a physical application is discussed in the next section.

A cursory view of the phase portrait of the map L, Figure 1, shows clear evidence of discrete symmetry.[2] The phase portrait is symmetric under reflection in both the x and the y axes. One can readily show that these reflections are neither symmetries nor reversing symmetries. In fact one can readily show that the mirrors in the x-axis, M_x, and y-axis, M_y, satisfy

$$M_x = L^2 \circ M_x \circ L^2$$

$$M_y = L^2 \circ M_y \circ L^2$$

This study motivates us to look for maps M and S satifying either

$$M \circ L^k = L^k \circ M \tag{3.2}$$

or

$$S \circ L^k = L^{-k} \circ S \tag{3.3}$$

We denote the set of all symmetries of L^k by $\mathcal{G}_k$. Similarly the set of all symmetries of L^k *together with* its reversing symmetries is denoted by $\mathcal{E}_k$.

In order to make a precise definition of the concept of both k-symmetry and k-symmetric orbits we define a map ϕ_L as follows

$$\phi_L(M) = L \circ M \circ L^{-1} \tag{3.4}$$

$$\phi_L(S) = L \circ S \circ L$$

[2]There is also translational invariance in this example. In fact, the phase portrait has crystallographic symmetry and possesses a stochastic web. For some discussions of the last phenomenon see e.g. Zaslavsky et al. (1991), Hoveijn (1992), and Lamb (1993a, 1993b).

Note that if M is a symmetry of L^k then so is $\phi_L(M)$. Similarly if S is a reversing symmetry of L^k then so is $\phi_L(S)$. In the above example $\phi_L(M_x) = M_y$.

The map ϕ_L enjoys the following properties

$$\begin{array}{lrcl} \text{(i)} & \phi_L^n(M \circ \tilde{M}) & = & \phi_L^n(M) \circ \phi_L^n(\tilde{M}), \\ \text{(ii)} & \phi_L^n(S \circ \tilde{S}) & = & \phi_L^n(S) \circ \phi_L^{-n}(\tilde{S}), \\ \text{(iii)} & \phi_L^n(M \circ S) & = & \phi_L^n(M) \circ \phi_L^n(S), \\ \text{(iv)} & \phi_L^n(S \circ M) & = & \phi_L^n(S) \circ \phi_L^{-n}(M). \end{array} \tag{3.5}$$

$\forall n \in \mathbb{Z}$. Note that ϕ_L is a group automorphism of $\mathcal{G}_k$ but not of $\mathcal{E}_k$.

We first present the

Definition 4 *$U \in \mathcal{E}_{\tilde{k}}$ is called a (reversing) k-symmetry of L if k is the smallest positive integer such that*

$$\phi_L^k(U) = U. \tag{3.6}$$

We denote this as $\#_L(U) = k$.

If U is an element of $\mathcal{E}_k$ then it is certainly an element of $\mathcal{E}_{\tilde{k}}$ if $\tilde{k}$ is a multiple of k. This definition avoids the ambiguity which may possibly arise from this fact.

Let us now look at how k-symmetry manifests itself. Suppose Γ is an invariant set of L. If U was a (reversing) symmetry then it would follow that $U\Gamma$ would also be an invariant set. If $U\Gamma = \Gamma$ we would say that the invariant set was symmetric. The most natural definition of a k-symmetric invariant set would mirror this definition excepting that U would now be a k-symmetry. For the k-symmetric case however the situation is more complicated and this definition is inappropriate. As an example of what can happen suppose $\mathbf{x}_0$ is a fixed point of L, then if U is a (reversing) k-symmetry iof L it follows that $U\mathbf{x}_0$ in general belongs to a periodic orbit of period k. This orbit is given by $\{\phi_L^i(U)\mathbf{x}_0 : i = 1, \ldots, k-1\}$. More importantly suppose Γ is an invariant set and U is a (reversing) k-symmetry, then $U\Gamma$ need not be an invariant set. In the general case one can see that $\bigcup_{i=0}^{k-1} \phi_L^i(U)\Gamma$ is an invariant set but it is in general a union of orbits. Because of these difficulties the "natural" definition of a k-symmetric orbit is too strong. Of course a choice of definition is dictated by its utility. It turns out that useful results can be demonstrated for k-symmetric orbits defined in the following sense

Definition 5 *An orbit $\Gamma(\mathbf{x}_0)$ is k-symmetric with respect to a (reversing) k-symmetry U if for all $\mathbf{x}_i \in \Gamma(\mathbf{x}_0)$ there is an integer p such that $\phi_L^p(U)\mathbf{x}_i \in \Gamma(\mathbf{x}_0)$.*

Thus, for example, a fixed point of the map L is k-symmetric with respect to U if $\phi_L^i(U)\mathbf{x}_0 = \mathbf{x}_0$ for some i. In general it means that for each point $\mathbf{x}_i \in \Gamma$ one of the elements of the set $\{\phi_L^i(U)\mathbf{x}_i\}_{i=0}^{k-1}$ must be in the original orbit.

As mentioned in the introduction, the presence of k-symmetries is reflected in the symmetries observed in the global phase portrait of a dynamical system. In fact, if $\mathcal{P} \in \Omega$ is the set of all (points on) periodic orbits of L, and $\mathcal{C} \in \Omega$ is the set of all (points on) chaotic orbits of L, then for all $U \in \mathcal{E}_{\mathbb{N}}$,

$$U\mathcal{P} = \mathcal{P}, \tag{3.7}$$

$$U\mathcal{C} = \mathcal{C}. \tag{3.8}$$

In fact, in Lamb (1993a), symmetry properties of stochastic webs were understood in the spirit of (3.8).[3]

[3]For discussions on the way (reversing) symmetries manifest themselves in the global phase portrait, see e.g. also Chossat and Golubitsky (1988), Kimball and Dumas (1990), and Lamb (1993b).

Let us now present the analog of the FSI method. First we present the analog of the lemmas of the last section.

Lemma 6 $\Gamma(\mathbf{x}_0)$ *is k-symmetric with respect to the reversing k-symmetry S if and only if* $\exists m, n \in \mathbb{Z}$ *such that* $\mathbf{x}_0 \in \text{Fix}(L^m \circ \phi_L^n(S))$.

Proof.

$$\mathbf{x}_0 = L^m \circ \phi_L^n(S)\mathbf{x}_0 \Leftrightarrow \forall p \in \mathbb{Z} \quad \phi_L^{n-p}(S) \circ L^p \mathbf{x}_0 = L^{-m-p}\mathbf{x}_0 \Box \tag{3.9}$$

Once again the fixed sets of interest here can be obtained via iteration of a few fixed sets

$$\text{Fix}(L^{2n} \circ \phi_L^q(S)) = L^n(\text{Fix}(\phi_L^{q+n}(S)) \tag{3.10}$$
$$\text{Fix}(L^{2n+1} \circ \phi_L^q(S)) = L^n(\text{Fix}(L \circ \phi_L^{q+n}(S))) \tag{3.11}$$

Lemma 7 $\mathbf{x}_0 \in \text{Fix}(L^m \circ S) \;\Rightarrow\; \mathbf{x}_0 \in \text{Fix}(\phi_L^m(S) \circ S)$.

Proof.
$L^m \circ S\mathbf{x}_0 = \mathbf{x}_0 \;\Rightarrow\; L^m \circ S \circ L^m \circ S\mathbf{x}_0 = \mathbf{x}_0 \Box$

Now we state a generalisation of the FSI method for systems with reversing k-symmetries.

Theorem 8 *Suppose* Γ *is a periodic orbit of a map L that is k-symmetric with respect to a reversing k-symmetry S. Then all points of* Γ *lie on intersections of iterates of* $\text{Fix}(\phi_L^p(S))$ *and/or* $\text{Fix}(L \circ \phi_L^q(S))$ *for integer value of p and q.*

Proof. If Γ is k-symmetric then $\exists$ m, n such that

$$\phi_L^n(S)\mathbf{x}_0 = L^{-m}\mathbf{x}_0 \;\Rightarrow\; \mathbf{x}_0 \in \text{Fix}(L^m \circ \phi_L^n(S)) \tag{3.12}$$

Suppose the period of the k-symmetric orbit is $\tilde{p}$ then

$$\phi_L^n(S)\mathbf{x}_0 = L^{-m}\mathbf{x}_0 \;\Rightarrow\; \phi_L^n(S)\mathbf{x}_0 = L^{-m-\tilde{p}}\mathbf{x}_0 \;\Rightarrow\; \mathbf{x}_0 \in \text{Fix}(L^{m+\tilde{p}} \circ \phi_L^n(S)) \tag{3.13}$$

Thus

$$\mathbf{x}_0 \in \text{Fix}(L^m \circ \phi_L^n(S)) \cap \text{Fix}(L^{m+\tilde{p}} \circ \phi_L^n(S)) \tag{3.14}$$

Using (3.10) and (3.11) the result follows. □

The converse of Theorem 8, i.e. that all intersections of iterates of $\text{Fix}(\phi_L^p(S))$ and/or $\text{Fix}(\phi_L^q(S))$ for all $p, q \in \mathbb{Z}$ are periodic points, can be established with some mild technical conditions (see Lamb and Quispel (1993)).

4 EXAMPLE

There are a number of examples of dynamical systems, arising in physics, which possess (reversing) k-symmetries. Here we shall present the example of a kicked rotator. Other examples are discussed in Lamb and Quispel (1993). A detailed discussion of these examples is beyond the scope of the present paper.

The equations of motion (of the components of the angular momentum of the rotator) for a kicked rotator can be written as (D'Ariano et al., 1992)

$$L:\begin{cases} x' &= -z, \\ y' &= y\cos(kz) - x\sin(kz), \\ z' &= x\cos(kz) + y\sin(kz). \end{cases} \tag{4.1}$$

L possesses the reversing symmetry P_{xz}

$$P_{xz}:\begin{cases} x' &= z, \\ y' &= y, \\ z' &= x. \end{cases} \tag{4.2}$$

Consider the maps

$$I_{xy}:\begin{cases} x' &= -x, \\ y' &= -y, \\ z' &= z, \end{cases} \quad \text{and } I_{yz}:\begin{cases} x' &= x, \\ y' &= -y, \\ z' &= -z. \end{cases} \tag{4.3}$$

and $I_{xz} = I_{xy} \circ I_{yz}$.

L further possesses the symmetry I_{xz} . Moreover, L^2 possesses I_{xy} as a symmetry (D'Ariano et al. 1992) , implying that I_{xy} is a 2-symmetry of L. In fact,

$$\phi_L(I_{xy}) = I_{yz}. \tag{4.4}$$

We find that these symmetries satisfy

$$\phi_L(I_{xz}) = I_{xz}, \tag{4.5}$$

$$\phi_L^2(I_{xy}) = \phi_L(I_{yz}) = I_{xy}, \tag{4.6}$$

$$\phi_L(P_{xz}) = P_{xz}, \tag{4.7}$$

$$\phi_L(P_{xz} \circ I_{xz}) = P_{xz} \circ I_{xz}, \tag{4.8}$$

$$\phi_L^2(P_{xz} \circ I_{xy}) = \phi_L(P_{xz} \circ I_{yz}) = P_{xz} \circ I_{xy}. \tag{4.9}$$

It would be very interesting to study k-symmetric periodic orbits in the spirit of Theorem 8.

5 CONCLUDING REMARKS

In this paper we generalized the concepts of (reversing) symmetries to (reversing) k-symmetries.

In setting up the frame-work for (reversing) k-symmetry groups, we carefully insisted that $k \in \mathbb{Z}$ be finite. We have chosen not to adopt the notion of "commutation of a group $\mathcal{G}$ with a map L", as used by MacKay and others (MacKay, 1984; Llibre and MacKay, 1992; Roberts and Baake, 1993) which permits k to be (formally) infinite. A group $\mathcal{G}$ commutes with a map L if

$$\forall U \in \mathcal{G},\ \exists \tilde{U} \in \mathcal{G}, \quad \text{such that } U \circ L = L \circ \tilde{U}. \tag{5.1}$$

Although from a group theoretical point of view (5.1) is a nice equivalence relation, this equivalence does not automatically include important dynamical features. In particular, an $U \in \mathcal{G}$ satisfying (5.1) does not have to preserve the set of periodic orbits and the set of chaotic orbits, in contrast to (3.7) and (3.8). For instance, the group Aut(Ω), consisting of all invertible maps of the state space into itself, commutes with every $L \in$ Aut(Ω). Of course, this observation does not reveal much about the dynamics of L. The closure of the orbits under ϕ_L is essential for the preservation of dynamical features such as periodic and chaotic orbits.

ACKNOWLEDGMENTS

This paper was written with the assistance of C.Pisani. This investigation is part of the research program of the Australian Research Council.

REFERENCES

Chirikov, B.V., 1979, A universal instability of many-dimensional oscillator systems, *Phys.Rep.* 216:63

Chossat, P., and Golubitsky, M., 1988, Symmetry-increasing bifurcation of chaotic attractors, *Physica D* 32:423

D'Ariano, G.M., Evangelista, L.R., and Saraceno, M., 1992, Classical and quantum structures in the kicked-top model, *Phys.Rev.A* 45:3646

DeVogelaere, R., 1958, On the structure of symmetric periodic solutions of conservative systems, with applications, *in*: "Contributions to the Theory of Nonlinear Oscillations" Vol. 4, ed. S.Lefschetz, Princeton University Press, Princeton.

Golubitsky, M., Stewart, I., and Schaeffer, D.G., 1988, "Singularities and Groups in Bifurcation Theory", Vol. 2, Applied Mathematical Sciences, Vol. 69, Springer, New York.

Hoveijn, I., 1992, Symplectic reversible maps, tiles and chaos, *Chaos, Solitons and Fractals* 2:81

Kimball, J.C., and Dumas, H.S., 1990, Symmetry in regular motion: relation to chaos, *Phys.Lett.A* 144:201

Lamb, J.S.W., 1992, Reversing symmetries in dynamical systems, *J.Phys.A* 25:925

Lamb, J.S.W., 1993a, Crystallographic symmetries of stochastic webs, *J.Phys.A* 26 (in press)

Lamb, J.S.W., 1993b, Stochastic webs with fourfold rotation symmetry, this volume

Lamb, J.S.W., and Quispel, G.R.W., 1993, Reversing *k*-Symmetries in Dynamical Systems, preprint

Llibre, J. and MacKay, R.S., 1992, Pseudo Anosov homeomorphisms on a sphere with four punctures have all periods, preprint

MacKay, R.S., 1984, Equivariant universality classes, *Phys.Lett.A* 106:99

Piña, E. and Jiménez Lara, L., 1987, On the symmetry lines of the standard mapping, *Physica D* 26:369

Piña, E. and Cantoral, E., 1989, Symmetries of the quasi-crystal mapping, *Phys.Lett.A* 135:190

Ichikawa, Y.H., Kanimura, T., Hatori, T., and Kim, S.Y., 1989, Stochasticity and symmetry of the standard map, *Prog. Theor.Phys.Supp.* 98:1

Richter, P.H., Scholz, H., and Wittek, A., 1990, A breathing chaos, *Nonlinearity* 3:45

Roberts, J.A.G., and Baake, M., 1993, Trace maps as 3-d reversible dynamical systems with an invariant, preprint

Roberts, J.A.G., and Quispel, G.R.W., 1992, Chaos and time-reversal symmetry. Order and chaos in reversible dynamical systems, *Phys.Rep.* 216:63

Sattinger, D.H., 1979, "Group Theoretic Methods in Bifurcation Theory", Lecture Notes in Mathematics, Vol. 762, Springer, New York.

Vanderbauwhede, A., 1982, "Local Bifurcation and Symmetry", Research Notes in Mathematics, Vol. 75, Pitman, London.

Zaslavsky. G.M., Sagdeev, R.Z., Usikov, D.A., and Chernikov, A.A., 1991 "Weak Chaos and Quasiregular Patterns", Cambridge Nonlinear Science Series Vol.1, Cambridge University Press, Cambridge.

QUANTAL-CLASSICAL MIXED MODE DYNAMICS OF COUPLED OSCILLATORS

Ersin Yurtsever

Middle East Technical University
Chemistry Department
Ankara, Turkey

ABSTRACT

The dynamics of a coupled two-dimensional anharmonic oscillator system is studied by fully classical and mixed-mode approaches. Previous classical calculations have shown that the system behaved chaotically above a critical value. However, upon quantization of one of the modes, the dynamics becomes highly regular with autocorrelation functions and Fourier transforms pointing to a very periodic motion.

INTRODUCTION

One of the ultimate aims of chemical physics is to perform "controlled experiments" in which a specific bond is formed or broken by controlled excitation of the target molecule. The transferred energy is then restributed ("Intramolecular Vibrational Energy Redistribution", IVR) within the molecule to achieve the desired reaction . The mechanisms and the time scales of IVR are the crucial questions to be answered (1).

Even for the very simple molecules, the equations which govern energy transfer processes are of a nonlinear nature and the numerical solutions exhibit a great deal of variety including highly chaotic behaviour. The correct procedure calls for the solution of the time-dependent Schrödinger equation, yet even numerical solutions in more than three dimensions are not feasible. On the other hand, the dynamical behaviour can be obtained from classical mechanical considerations but these solutions are not realistic since they exclude quantum effects such as zero point energy. Still, the bulk of the effort in literature is based on classical mechanical solutions.

Another point of interest in these different approaches lies in the definition of the quantum analog of the classical chaos (2). Although the concepts of integrability and chaos are well defined in any form of the classical mechanical formulation, so far there is no widely accepted corresponding measure in quantum mechanics which describes the changes from regular to "irregular" behaviour.

Hamiltonian Mechanics, Edited by J. Seimenis,
Plenum Press, New York, 1994

In this work we apply a mixed-mode approach in which certain modes are treated quantum mechanically and the others by classical formulations (3,4). This way, the chemically interesting processes are treated correctly and the remaining slow modes are analyzed in a somewhat more approximate way. These approaches are well known in quantum mechanics starting with the Born-Oppenheimer separation of the electronic and nuclear motion. In the example described here dynamical behaviour of two modes are studied by solving Hamilton's and Schrödinger's equations iteratively by employing self-consistent-field (SCF) concepts.

MIXED-MODE SELF-CONSISTENT-FIELD-METHOD

In this section we give a brief formulation of the mixed-mode SCF equations which can be extended to multimode systems without loss of generality. Let us write a two-dimensional nonlinearly coupled Hamiltonian as:

$$H = H_x + H_y + \lambda V(x, y) \tag{1}$$

In SCF approximation this operator can be written as a sum of two operators. Each Hamitonian defines the motion along one coordinate which is coupled to the other coordinate through an average but constant (for small time δt) field. For the explicit coupling of $x^2 y^2$:

$$h_x = H_x + \lambda x^2 < y^2 > \tag{2-a}$$

and

$$h_y = H_y + \lambda y^2 < x^2 > \tag{2-b}$$

Here <...> denotes an average value, that is either an expectation value over a wavepacket, a coordinate from a trajectory or an average over a bundle of trajectories. In our example, x-mode remains the classical mode and its dynamics can be studied by the numerical integration of Hamilton's equations:

$$\frac{\partial h_x}{\partial x} = - \frac{\partial p_x}{\partial t} \tag{3-a}$$

and

$$\frac{\partial h_x}{\partial p_x} = \frac{\partial x}{\partial t} \tag{3-b}$$

On the other hand y-mode is quantized, that is its dynamics must be analyzed by the solution of the time-dependent Schrödinger equation:

$$i \frac{\partial \varphi(y, t)}{\partial t} = h_y \varphi(y, t) \tag{4}$$

The above equation can be solved in various ways. We have used the basis set expansion method in which the wavefunction φ is expressed as a linear combination of the eigefunctions of the integrable Hamiltonian H_y. Then the time dependent coefficients of the wavepacket can be obtained by a technique of successive diagonalizations employing the condition that the wavepacket must be continuous in time. The details of this procedure are given elsewhere (3,4).

RESULTS AND DISCUSSION

The Hamiltonian is of the form:

$$H = 0.5\,(p_x^2 + p_y^2 + x^2 + 1.44\,y^2) - 0.05\,x^3 + 0.00140625\,x^4$$

$$- 0.0864\,y^3 + 0.02916\,y^4 + 0.1\,x^2 y^2 \qquad (5)$$

We have carried out two sets of calculations. What we termed as fully classical approach (FC) consists of the Runge-Kutta integration of the Hamilton's equations for the two-mode system. Initial conditions are generally chosen from the four-dimensional phase space of x, y, p_x and p_y such that they are on the same constant energy surface. In the mixed-mode analysis, we have chosen a point on a two-dimensional x-p_x phase space and a wavepacket along the coordinate y such that the total energy is as same as fully classical calculations. Hamilton and Schrödinger equations for x and y modes are then integrated iteratively. All the reported values of coordinates, momenta, energy and time are in atomic units.

Previously we have shown that classical dynamics of the operator in eq.5 displays a transition from regular to chaotic behaviour as the energy increases (5,6). The critical value is around H=E=2.0 and after E=8 all the largest Lyapunov exponents become positive. We decided to compare both methods in the region where classical behaviour is completely chaotic. The energy along x-mode is 15 which corresponds to the inflection point of H_x. The quantum mode has relatively low energy (around 3.5) and the total energy comes out to be 18.53 (including the energy in the coupling). The initial quantum wavepacket is chosen as:

$$\varphi(y,0) = \frac{1}{\sqrt{2}}(\phi_2 - \phi_3)$$

where ϕ are the eigenfunctions of the anharmonic oscillator. The corresponding coordinate and momentum in the fully classical calculations are chosen such that the energy in the y-mode and in the coupling should be same in both classical and MM calculations.

We have carried out integrations up to 200 time units (with time steps of 0.01). The coordinates, momenta (expectation values in case of the quantum mode), and energy of each mode are recorded. The autocorrelation functions of these variables as well as their Fourier transforms are computed.

In fig.1 we present plots of energy of both modes and the coupling for classical and MM computations. The variation of the x-mode is highly nonperiodic for the FC calculations whereas the same function shows a very periodic behaviour when the other mode is quantized. This behaviour is also seen for the coupling energy. The fluctuations

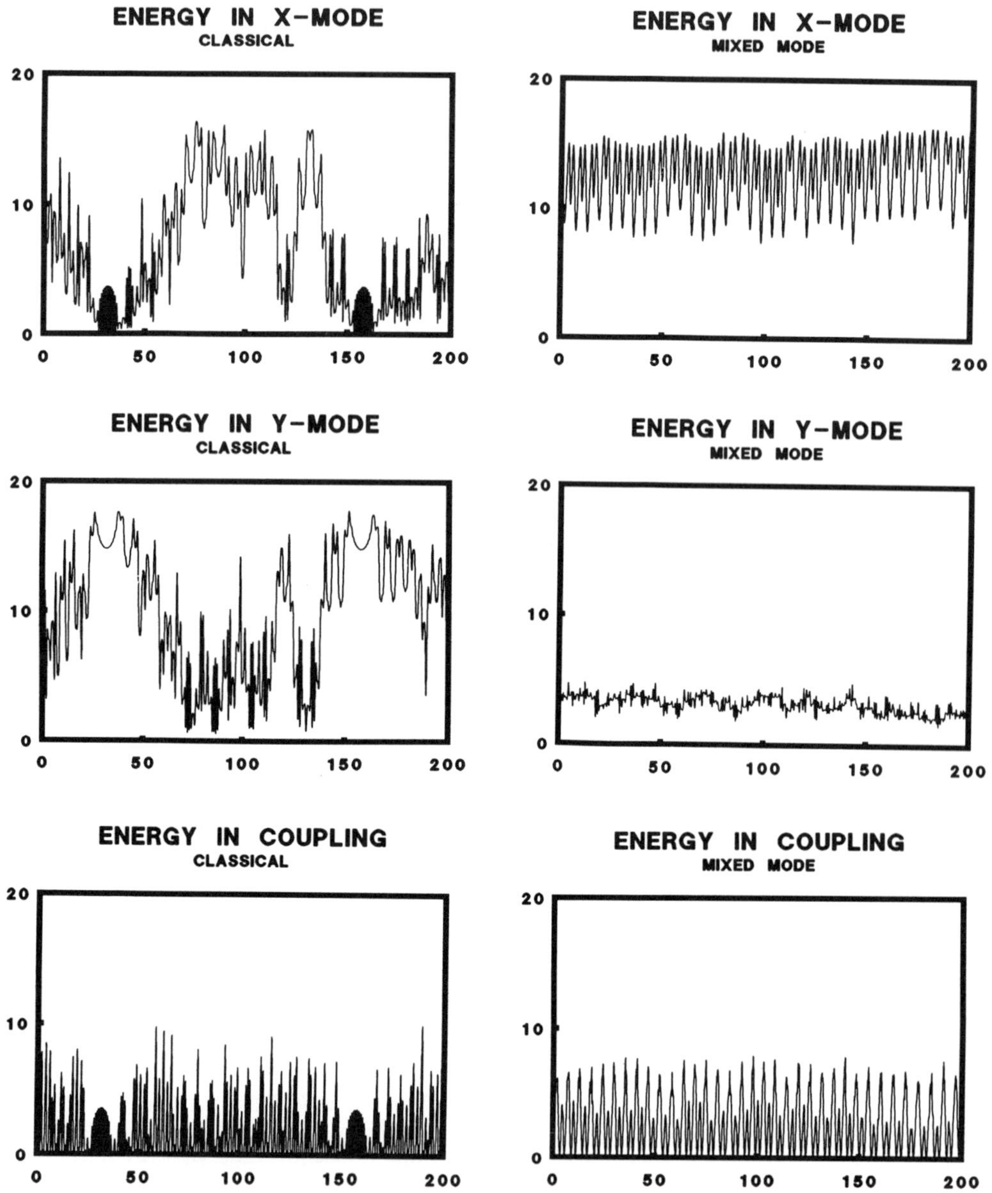

Fig. 1. Energy of both modes and the coupling from FC and MM calculations

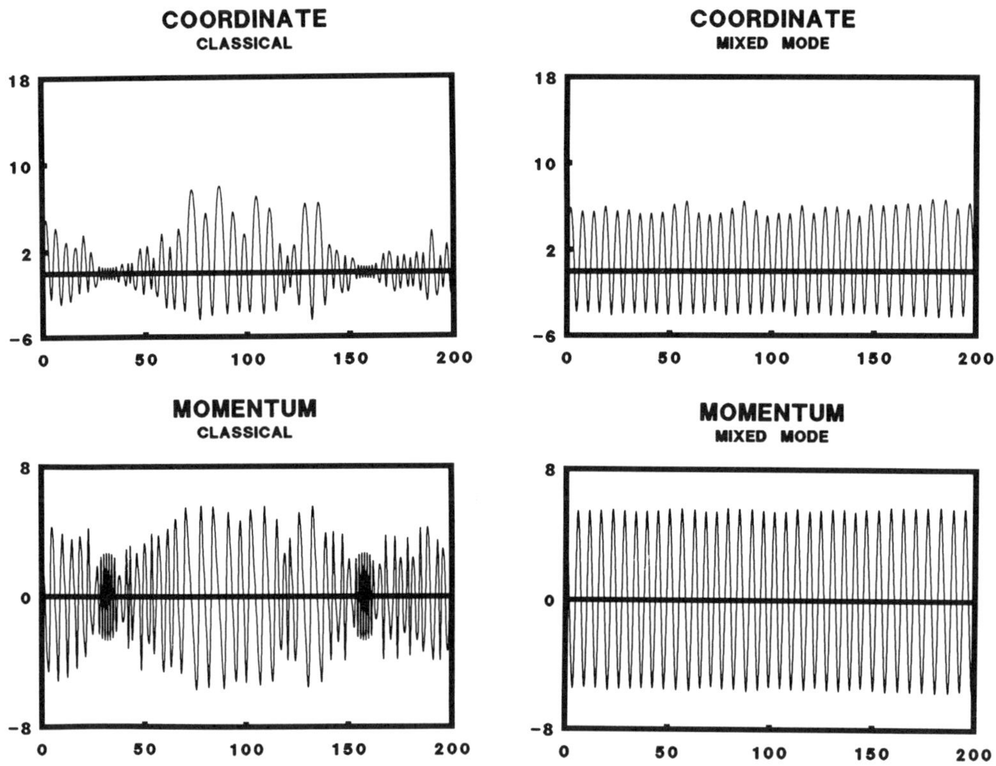

Fig.2. Coordinate and momentum of x-mode

of the quantum energy along the y-mode are very small and consequently it is difficult to make a visual judgment. In fig.2. the coordinates and momenta of the x-mode are displayed with the same characteristics, namely the chaotic behaviour of the particle along one coordinate is regularized upon quantization of the other mode. The autocorrelation functions which may be aptly termed as memory functions support this observation. In fig.3 we present autocorrelation functions and Fourier transforms of the energy in x-mode. In all cases we see a highly periodic structure when the perturbing mode is a quantum one. The Fourier transforms for MM computations show a set of well defined peaks where FC results display the signature of chaos with an array of peaks at different frequencies.

In this work we have tried to study the effects of the quantization on the occurence of chaos. There are various errors introduced by SCF approximation and the lack of an exact definition of the correspondence between a wavepacket and classical points. Among these two sources of errors, the former is thought to be very small especially when the dimensionality is very small. For the latter question, a better way would have been selection of the points by Wigner or Husimi transforms (7), although both of these approaches are more or less intiutive. We have tried to select

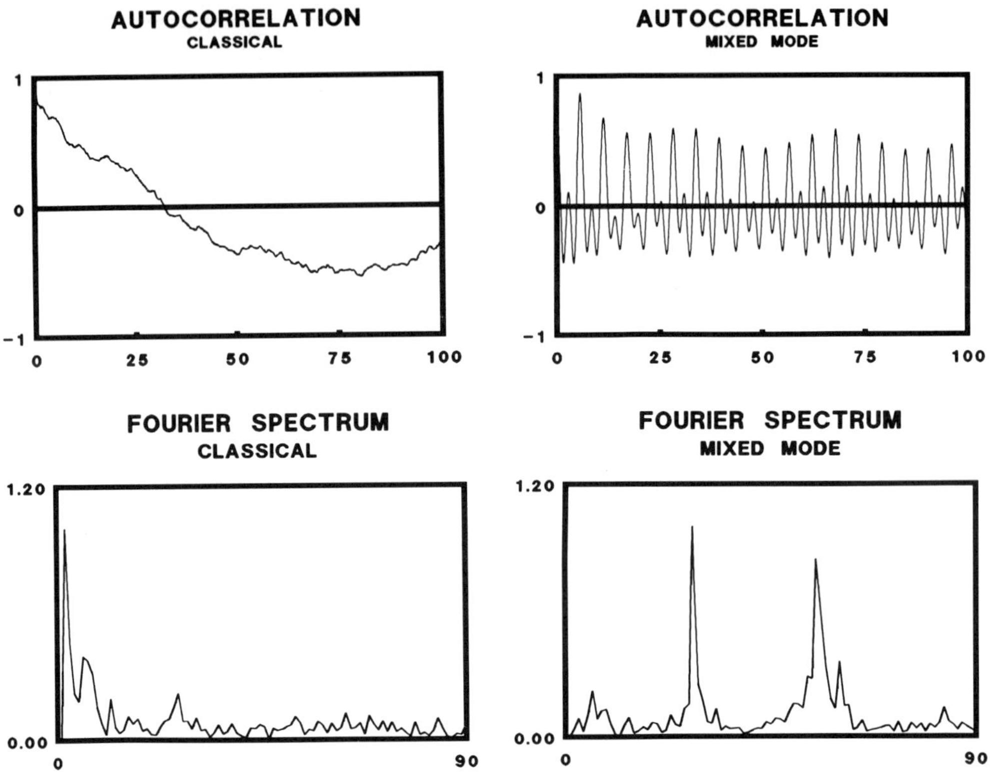

Fig.3. Autocorrelation functions and Fourier transforms of the energy in x-mode

a bundle of trajectories so that a statistical picture would develop.Due to lack of space, the results of other trajectories cannot be displayed here, but the qualitative behaviour is always the same, that is once the perturbing mode is a quantum mode, then the chaos along the classical mode seems to disappear. These results, in addition to our previous findings (3,5) imply that quantum mechanics has a strong smoothing effect on classical chaos.

REFERENCES

1) T.Uzer, Phys.Rep. **199**, 73 (1991)
2) F.Haake, " Quantum Signatures of Chaos", Springer Verlag, Berlin, (1991)
3) E.Yurtsever and J.Brickmann, Ber.Bunsenges.Phys.Chem. **96**, 142 (1992)
4) E.Yurtsever and T.Uzer, Ber.Bunsenges.Phys.Chem. **96**, 906 (1992)
5) E.Yurtsever and J.Brickmann, Phys.Rev.A **41**, 6688 (1990)
6) E.Yurtsever and J.Brickmann, Ber.Bunsenges.Phys.Chem. **94**, 804 (1990)
7) J.H.Harriman and M.E.Casida, Int.J.Quant.Chem. **45**, 263 (1993)

THE THREE-WAVE INTERACTION OF FOUR WAVES REVISITED: A LAX PAIR AND POSSIBLY GENERAL SOLUTION

Filipe J. Romeiras

Departamento de Matemática
e Centro de Electrodinâmica
Instituto Superior Técnico
1096 Lisboa Codex, Portugal

INTRODUCTION

The nonlinear resonant interaction of coherent waves is a fundamental process in the study of wave phenomena which has received a great deal of attention in its many aspects (Kaup et al.,1979; Craik, 1985; Ablowitz and Clarkson, 1991; and references therein).

Here we restrict our attention to a simplified model of the interaction which is described by a Hamiltonian system of autonomous ordinary differential equations. This model applies when the (slow) space variation of the amplitudes of the interacting waves can be neglected, the waves are marginally stable and exactly resonant.

The most elementary interaction involves a single triplet of waves and is exactly solvable in terms of elliptic functions (Craik, 1985; Davidson, 1972). The Hamiltonian system is integrable in the sense of Liouville (Arnold, 1978) and Lax representations (Lax, 1968), both with and without a spectral parameter, have been obtained for it (Wojciechowski et al., 1986). When waves of different energy sign are allowed to interact, the solutions can become singular in finite time: this is appropriately called the "explosive" instability (Davidson, 1972).

The interaction of waves constituting several coupled triplets has also been considered by several authors, mostly in the form of the so-called restricted multiple three-wave interaction, in which an arbitrary number of wave triplets interact through a shared wave (Verheest, 1988; and references therein), but also in the form of four waves in two triplets, with two of the waves common to both triplets (Sugihara, 1968; Aristov et al., 1972; Karplyuk et al., 1973; Walters and Lewak, 1977; Romeiras, 1983; Verheest, 1987).

Regarding this last interaction it was shown by Aristov et al. (1972), Karplyuk et al. (1973) and Walters and Lewak (1977) that the system had three integrals of motion and that it could be solved exactly in terms of elliptic functions when one of these integrals (the Hamiltonian of the system) took the constant value zero. It was also shown by Aristov et al. (1972) and Walters and Lewak (1977) that, when one allowed for negative energy waves, the presence of a second (stable) triplet could stabilize the explosive instability of the first one. Sugihara (1968) considered the case in which the interaction coefficients of the two triplets were equal and obtained a partially exact solution, again involving elliptic functions. Romeiras (1983) showed that the system was integrable by obtaining a fourth integral of motion. Verheest (1987) recovered this integral by a different method.

In the present paper we return to this interaction of four waves in two triplets with the purpose of addressing essentially three open questions: (1) obtain a Lax pair for the system; (2) obtain the general solution of the system (for arbitrary values of the parameters and the integrals of motion); (3) obtain a general stability criterion for the explosive instability. We summarize the most important aspects of our study of these questions, leaving the details to another publication (Romeiras, 1993).

BASIC EQUATIONS

The equations that describe the time evolution of the complex amplitudes of the interacting waves can be written as a Hamiltonian system

$$\frac{d\mathbf{a}}{dt} = \frac{\partial H}{\partial \mathbf{p}}, \qquad \frac{d\mathbf{p}}{dt} = -\frac{\partial H}{\partial \mathbf{a}}, \tag{1}$$

with Hamiltonian function $H : \mathbb{C}^4 \times \mathbb{C}^4 \to \mathbb{C}$, defined by

$$iH(\mathbf{a},\mathbf{p}) = V_{312}S_3a_1a_2p_3 + V_{312}^*S_1S_2a_3p_1p_2 + V_{124}S_1a_2a_4p_1 + V_{124}^*S_2S_4a_1p_2p_4. \tag{2}$$

Here $\mathbf{a} = (a_1, a_2, a_3, a_4)$, $\mathbf{p} = (p_1, p_2, p_3, p_4)$, a_n is the amplitude of wave n, $p_n = S_n a_n^*$ is the variable canonically conjugate to a_n and $S_n = \pm 1$ is the energy sign factor of wave n. V_{312} and V_{124} are the time independent interaction coefficients. The i is the imaginary unit and the asterisc denotes the complex conjugate.

Note that there is some freedom in the choice of the form of the equations that describe the interaction. Our own choice, Eqs. (1)–(2), is common in Plasma Physics (Davidson, 1972) and corresponds to the interaction of waves whose frequencies $\{\omega_n\}$, $n = 1,2,3,4$, satisfy the resonance conditions $\omega_3 = \omega_1 + \omega_2$, $\omega_4 = \omega_1 - \omega_2$, and with the amplitudes normalized in such way that the wave energy of wave n has the form $W_n = S_n\omega_n|a_n|^2$. Total wave energy $\sum_{n=1}^4 W_n$ is conserved by Eqs. (1)–(2) (see Eqs. (7)–(8) below).

Eq. (2) includes two complex parameters, V_{312}, V_{124}, and four integer parameters, S_1, S_2, S_3, S_4. A new Hamiltonian with a single (real non-negative) parameter

$$R = \frac{|V_{124}|}{|V_{312}|},$$

($V_{312} \neq 0$) can be obtained by introducing an extended canonical transformation (Goldstein, 1980). In fact, if we write

$$\begin{bmatrix} \mathbf{a} \\ \mathbf{p} \end{bmatrix} = \mathbf{T} \begin{bmatrix} \tilde{\mathbf{a}} \\ \tilde{\mathbf{p}} \end{bmatrix}, \tag{3}$$

where $\mathbf{T} = \{t_{jk}\}$ is an 8×8 matrix with non-zero elements

$$t_{jj} = \mu_j\alpha \ (j = 1,2,4), \qquad t_{33} = \mu_3, \qquad t_{12} = \mu_1\beta, \qquad t_{21} = \mu_2\beta, \qquad t_{48} = -i\mu_4\beta,$$

$$t_{jj} = \nu_{j-4}\alpha \ (j = 5,6,8), \qquad t_{77} = \nu_3, \qquad t_{56} = \nu_1\beta, \qquad t_{65} = \nu_2\beta, \qquad t_{84} = -i\nu_4\beta,$$

with

$$\mu_n \nu_n = \frac{S_1 S_2 S_3}{|V_{312}|^2} \quad (n = 1, 2, 3, 4), \qquad \mu_3 = S_3 V_{312} \mu_1 \mu_2, \qquad \mu_4 = S_1 \delta \frac{V_{124}^*}{|V_{312} V_{124}|} \frac{\mu_1}{\mu_2},$$

$$\delta = \exp\left\{ i \frac{\pi}{4} (1 - S_3 S_4) \right\}, \qquad \alpha = \tfrac{1}{2}(1 + \delta^2), \qquad \beta = \tfrac{1}{2}(1 - \delta^2),$$

and set

$$\tilde{H} = S_1 S_2 S_3 |V_{312}|^2 H,$$

we obtain the Hamiltonian system

$$\frac{d\tilde{\mathbf{a}}}{dt} = \frac{\partial \tilde{H}}{\partial \tilde{\mathbf{p}}}, \qquad \frac{d\tilde{\mathbf{p}}}{dt} = -\frac{\partial \tilde{H}}{\partial \tilde{\mathbf{a}}}, \tag{4}$$

with Hamiltonian function $\tilde{H}$

$$i\tilde{H}(\tilde{\mathbf{a}}, \tilde{\mathbf{p}}) = \tilde{a}_1 \tilde{a}_2 \tilde{p}_3 + \tilde{a}_3 \tilde{p}_1 \tilde{p}_2 + R(\tilde{a}_2 \tilde{a}_4 \tilde{p}_1 + \tilde{a}_1 \tilde{p}_2 \tilde{p}_4). \tag{5}$$

From now on, and except where explicitly indicated, we drop the tildes from all quantities. For future use we write in full the system of equations of motion (4)–(5):

$$\begin{aligned}
i\dot{a}_1 &= a_3 p_2 + R a_2 a_4, & -i\dot{p}_1 &= a_2 p_3 + R p_2 p_4, \\
i\dot{a}_2 &= a_3 p_1 + R a_1 p_4, & -i\dot{p}_2 &= a_1 p_3 + R a_4 p_1, \\
i\dot{a}_3 &= a_1 a_2, & -i\dot{p}_3 &= p_1 p_2, \\
i\dot{a}_4 &= R a_1 p_2, & -i\dot{p}_4 &= R a_2 p_1,
\end{aligned} \tag{6}$$

where the dot denotes derivative with respect to t.

A function $I : \mathbb{C}^4 \times \mathbb{C}^4 \to \mathbb{C}$ which is a constant on trajectories of the Hamiltonian equations is called an *integral of motion*. This is equivalent to requiring that

$$\{I, H\} = 0,$$

where the *Poisson bracket* $\{F, G\}$ of two differentiable functions $F, G : \mathbb{C}^4 \times \mathbb{C}^4 \to \mathbb{C}$ is defined by

$$\{F, G\} = \sum_{n=1}^{4} \left(\frac{\partial F}{\partial a_n} \frac{\partial G}{\partial p_n} - \frac{\partial F}{\partial p_n} \frac{\partial G}{\partial a_n} \right).$$

The Hamiltonian (5) is an integral of motion. Two other integrals of motion are very easily obtained from Eqs. (6):

$$C_1(\mathbf{a}, \mathbf{p}) = a_1 p_1 + a_3 p_3 + a_4 p_4, \tag{7}$$

$$C_2(\mathbf{a}, \mathbf{p}) = a_2 p_2 + a_3 p_3 - a_4 p_4. \tag{8}$$

A fourth integral of motion was obtained by Romeiras (1983):

$$\begin{aligned}
E(\mathbf{a}, \mathbf{p}) = {} & Q_1 Q_2 (a_1 p_1 - a_2 p_2)^2 - Q_2 (R a_1^2 - 2 a_3 a_4)(R p_1^2 - 2 p_3 p_4) \\
& + Q_1 (R a_2^2 + 2 a_3 p_4)(R p_2^2 + 2 a_4 p_3),
\end{aligned} \tag{9}$$

where $Q_1 = -\frac{1}{2}(1 - R^2)$, $Q_2 = -\frac{1}{2}(1 + R^2)$.

The four integrals of motion,

$$I_1 = C_1, \qquad I_2 = C_2, \qquad I_3 = E, \qquad I_4 = H,$$

are functionally independent and are in involution, that is, they satisfy, $\{I_n, I_m\} = 0$, for all $n, m = 1, 2, 3, 4, n \neq m$. The Hamiltonian system (4)–(5) is therefore completely integrable in the sense of Liouville (Arnold, 1978).

LAX REPRESENTATION

A powerful method for the study of integrable Hamiltonian systems is the isospectral deformation method. Introduced by Lax (1968) in the context of infinite dimensional dynamical systems and, in particular, the Korteweg-de Vries equation, it can also be applied to finite dimensional systems and, in particular, to nonlinear wave interaction systems of the type we are considering (see Rauch-Wojciechowski, 1990; and references therein).

In the context of finite-dimensional systems the Lax method consists in writing the dynamical system in the form

$$\dot{\mathbf{L}} = [\mathbf{A}, \mathbf{L}] \equiv \mathbf{AL} - \mathbf{LA}, \tag{10}$$

where $\mathbf{L}$ and $\mathbf{A}$ are $N \times N$ matrices which are functions of the canonical variables. It then follows that the matrix $\mathbf{L}$ is similarity invariant to a constant matrix and therefore that its eigenvalues, or, equivalently, the coefficients of its characteristic polynomial, are integrals of motion. The matrices $\mathbf{L}$ and $\mathbf{A}$ constitute the *Lax pair* for the dynamical system. Additional flexibility can be obtained by allowing the matrices to depend on a (spectral) parameter.

We obtained a Lax pair with a spectral parameter (μ) for our system (6) by assuming that matrices $\mathbf{L}$ and $\mathbf{A}$ are of the form

$$\mathbf{L}(\mu) = \begin{bmatrix} l_{11}\mu & l_{12}a_1 & l_{13}a_3 & l_{14}a_2 \\ l_{21}p_1 & l_{22}\mu & l_{23}a_2 & l_{24}p_4 \\ l_{31}p_3 & l_{32}p_2 & l_{33}\mu & l_{34}p_1 \\ l_{41}p_2 & l_{42}a_4 & l_{43}a_1 & l_{44}\mu \end{bmatrix}, \qquad \mathbf{A}(\mu) = \begin{bmatrix} a_{11}\mu & a_{12}a_1 & a_{13}a_3 & a_{14}a_2 \\ a_{21}p_1 & a_{22}\mu & a_{23}a_2 & a_{24}p_4 \\ a_{31}p_3 & a_{32}p_2 & a_{33}\mu & a_{34}p_1 \\ a_{41}p_2 & a_{42}a_4 & a_{43}a_1 & a_{44}\mu \end{bmatrix},$$

and further taking, without loss of generality,

$$l_{33} = a_{33} = a_{13} = a_{31} = 0.$$

Substituting into Eqs. (6) we arrived at the following relationships among the coefficients $\{l_{jk}\}$ and $\{a_{jk}\}$:

$$l_{22} = l_{11}\frac{z}{1+z}, \qquad l_{24} = \frac{l_{14}l_{23}}{l_{13}}2R(1+r), \qquad l_{34} = \frac{l_{14}l_{32}}{l_{12}}r,$$

$$l_{44} = l_{11}\frac{r^2 z}{1+r^2 z}, \qquad l_{42} = \frac{l_{41}l_{32}}{l_{31}}2R(1+r), \qquad l_{43} = \frac{l_{41}l_{23}}{l_{21}}r,$$

$$l_{12}l_{23}l_{31} = l_{21}l_{32}l_{13}, \qquad l_{13}l_{31} = 2l_{12}l_{21}(1+z), \qquad l_{14}l_{41} = l_{12}l_{21}\frac{1+z}{1+r^2 z},$$

$$\frac{a_{12}}{iwl_{12}} = \frac{a_{21}}{iwl_{21}} = -z, \qquad \frac{a_{14}}{iwl_{14}} = \frac{a_{41}}{iwl_{41}} = -rz, \qquad \frac{a_{23}}{iwl_{23}} = \frac{a_{32}}{iwl_{32}} = 1,$$

$$\frac{a_{24}}{iwl_{24}} = \frac{a_{42}}{iwl_{42}} = \frac{1-rz}{1+r}, \quad \frac{a_{34}}{iwl_{34}} = \frac{a_{43}}{iwl_{43}} = \frac{1}{r}, \quad a_{11} = 0, \quad \frac{a_{22}}{iwl_{22}} = 1, \quad \frac{a_{44}}{iwl_{44}} = \frac{1}{r},$$

where

$$r = \frac{Q_2}{Q_1} = \frac{1+R^2}{1-R^2}, \qquad z = \frac{l_{23}l_{32}}{l_{12}l_{21}}, \qquad w = \frac{l_{12}}{l_{13}l_{32}}.$$

The integrals of motion H, C_1, C_2, E appear in the coefficients of the characteristic polynomial of the matrix $\mathbf{L}$:

$$\det[\mathbf{L}(\mu) - \lambda \mathbf{I}] = \lambda^4 + D_3(\mu)\lambda^3 + D_2(\mu)\lambda^2 + D_1(\mu)\lambda + D_0(\mu),$$

where

$$D_3(\mu) = -\mu\Sigma_1, \qquad D_1(\mu) = -\mu^3\Sigma_3 + \mu D_{11} + D_{10},$$

$$D_2(\mu) = \mu^2\Sigma_2 + D_{20}, \qquad D_0(\mu) = \mu^4\Sigma_4 + \mu^2 D_{02} + \mu D_{01} + D_{00},$$

$$\Sigma_1 = l_{11} + l_{22} + l_{44}, \quad \Sigma_2 = l_{11}l_{22} + l_{22}l_{44} + l_{11}l_{44}, \quad \Sigma_3 = l_{11}l_{22}l_{44}, \quad \Sigma_4 = 0,$$

$$D_{20} = -(\Pi_{12} + \Pi_{34})C_1 - (\Pi_{14} + \Pi_{23})C_2, \qquad D_{02} = -l_{11}(l_{22}\Pi_{34}C_1 + l_{44}\Pi_{23}C_2),$$

$$D_{10} = -l_{31}(l_{12}l_{23} + l_{14}l_{43})(iH), \qquad D_{01} = l_{31}(l_{12}l_{23}l_{44} + l_{14}l_{43}l_{22})(iH),$$

$$D_{11} = [l_{44}\Pi_{12} + (l_{11} + l_{22})\Pi_{34}]C_1 + [l_{22}\Pi_{14} + (l_{11} + l_{44})\Pi_{23}]C_2, \qquad D_{00} = \frac{\Pi_{13}\Pi_{24}}{4R^2}E,$$

with $\Pi_{jk} = l_{jk}l_{kj}$.

A simple example of a Lax pair is obtained by setting

$$l_{13} = l_{31} = l_{11} = 2, \qquad l_{12} = l_{21} = l_{23} = l_{32} = 1, \qquad l_{14} = l_{41} = \sqrt{\frac{2}{1+r^2}}.$$

SOLUTIONS OF THE EQUATIONS OF MOTION

It is known that the system (6) can be solved exactly in terms of elliptic functions in two cases:
(I) $R = 0$ (see, for example, Davidson, 1972);
(II) $H = 0$ (Aristov et al., 1972; Karplyuk et al., 1973; Walters and Lewak, 1977).

We have been able to find two more classes of exact particular solutions, also in terms of elliptic functions, details of which are given elsewhere (Romeiras, 1993):
(III) $R = 1$;
(IV) $\hat{J} = 0$, where $\hat{J}$ is an auxiliary quantity.

So far we have not been able to obtain the exact solution of system (6) in the general case, that is, for arbitrary values of the parameter R and of the constants of motion.

What we have been able to obtain in the general case is a solution in the form of a Laurent series in the variable $\tau = t - t_\infty$, where t_∞ is an arbitrary constant, with a finite number of negative powers of τ; in other words, a solution with a movable pole at $\tau = 0$.

The existence of this solution was suggested by the studies of the relationship between integrability and the singularity structure of the solutions of ordinary differential equations (see Ramani et al. (1989) for a recent review); these studies imply that a necessary condition for a system to be integrable is that it has a solution of the referred (Laurent series) form with a full set of constants of integration. In our case the system (6) is integrable and should therefore have a solution of this form. Note that the exact particular solutions (I)-(IV) satisfy this condition: they are expressed in terms of elliptic functions whose singularities in the complex plane are poles.

For $R \neq 1$ the Laurent series solution has the form

$$\mathbf{X}(t) = \sum_{k=0}^{\infty} \tau^{k-1}\mathbf{X}^{(k)}. \tag{11}$$

where $\mathbf{X} = (a_3, p_3, a_4, p_4, a_1, p_1, a_2, p_2)$. The coefficients $\mathbf{X}^{(k)}$ are determined by substituting (11) into (6), reordering terms, and equating to zero the coefficients of distinct powers of τ . The equations so obtained can be solved successively with the introduction of seven arbitrary constants (Romeiras, 1993). If we add to these the constant t_∞ we conclude that solution (11) has the full set of arbitrary constants. The dominant ($k = 0$) term of the expansion is particulary relevant for the following discussion:

$$\mathbf{X}^{(0)} = (A_3, P_3, A_4, P_4, A_1, P_1, A_2, P_2),$$

where the constants (A_n, P_n) are such that

$$A_3 = iA_1A_2, \qquad P_3 = -iP_1P_2, \qquad A_4 = iRA_1P_2, \qquad P_4 = -iRA_2P_1,$$

$$A_1P_1 = \frac{1}{2Q_1} = -\frac{1}{1-R^2}, \qquad A_2P_2 = \frac{1}{2Q_2} = -\frac{1}{1+R^2},$$

with two of them, A_1 and A_2, say, arbitrary. Note that $\mathbf{X}(t) = \tau^{-1}\mathbf{X}^{(0)}$ is an exact elementary solution of (6).

For $R = 1$ the Laurent series solution has the form

$$\mathbf{Y}(t) = \sum_{k=0}^{\infty} \tau^{k-2}\mathbf{Y}^{(k)}, \qquad \mathbf{Z}(t) = \sum_{k=0}^{\infty} \tau^{k-1}\mathbf{Z}^{(k)}, \tag{12}$$

where $\mathbf{Y} = (a_3, p_3, a_4, p_4, a_1, p_1)$, $\mathbf{Z} = (a_2, p_2)$. The coefficients $\mathbf{Y}^{(k)}$ and $\mathbf{Z}^{(k)}$ are determined by a procedure similar to the described above. Again the solution involves eight arbitrary constants. In this case the dominant terms are

$$\mathbf{Y}^{(0)} = (\bar{A}_3, \bar{P}_3, \bar{A}_4, \bar{P}_4, \bar{A}_1, \bar{P}_1), \qquad \mathbf{Z}^{(0)} = (\bar{A}_2, \bar{P}_2),$$

where the constants $(\bar{A}_n, \bar{P}_n)$ are now such that

$$\bar{A}_3 = \frac{i}{2}\bar{A}_1\bar{A}_2, \quad \bar{P}_3 = -\frac{i}{2}\bar{P}_1\bar{P}_2, \quad \bar{A}_4 = \frac{i}{2}\bar{A}_1\bar{P}_2, \quad \bar{P}_4 = -\frac{i}{2}\bar{A}_2\bar{P}_1, \quad \bar{A}_2\bar{P}_2 = -2,$$

with three of them, $\bar{A}_1, \bar{P}_1$ and $\bar{A}_2$, say, arbitrary. Note that

$$\mathbf{Y}(t) = \tau^{-2}\mathbf{Y}^{(0)} + \frac{2}{\bar{A}_1\bar{P}_1}(\rho\bar{A}_3, -(1+\rho)\bar{P}_3, -\rho\bar{A}_4, (1+\rho)\bar{P}_4, 0, 0), \qquad \mathbf{Z}(t) = \tau^{-1}\mathbf{Z}^{(0)},$$

where ρ is an arbitrary constant, is an exact solution of system (6).

"EXPLOSIVE" INSTABILITY

The series solution of the previous Section allows us to obtain a stability criterion for the explosive instability in the resonant interaction of four waves in two triplets.

In fact if we use the extended canonical transformation (3) to return to the original (physical) variables and note that for these variables $p_n = S_n a_n^*$, the dominant terms of solutions (11) and (12) lead to:

(a) $R \neq 1$

$$\begin{bmatrix} |a_1|^2 \\ |a_2|^2 \\ |a_3|^2 \\ |a_4|^2 \end{bmatrix} \sim \frac{1}{|V_{312}|^2 \tau^2} \begin{bmatrix} -S_2 S_3 (1 - R^2 S_3 S_4)^{-1} \\ -S_1 S_3 (1 + R^2 S_3 S_4)^{-1} \\ S_1 S_2 (1 - R^4)^{-1} \\ S_1 S_2 R^2 (1 - R^4)^{-1} \end{bmatrix}, \qquad \text{as } \tau \to 0,$$

(b) $R = 1$,

$$\begin{bmatrix} |a_m|^2 \\ |a_3|^2 \\ |a_4|^2 \end{bmatrix} \sim \frac{|\mu_3 A_1 A_2|^2}{4|\tau|^4} \begin{bmatrix} -2S_m S_3 \\ 1 \\ 1 \end{bmatrix}, \qquad |a_{3-m}|^2 \sim \frac{-2S_m S_3}{|V_{312}|^2 \tau^2}, \qquad \text{as } \tau \to 0,$$

where $m = \frac{1}{2}(3 - S_3 S_4)$.

Requiring that the right hand sides be positive with t and t_∞ real we immediately obtain the following necessary conditions for the occurrence of the explosive instability:

(I) $\quad S_1 = S_2 = -S_3, \quad S_4 = \pm 1, \qquad R < 1;$

(II) $\quad S_2 = S_4 = -S_1, \quad S_3 = \pm 1, \qquad R > 1;$

(III) $\quad S_2 = S_3 = -S_1, \quad S_4 = +1, \quad$ or $\quad S_2 = S_4 = -S_3, \quad S_1 = +1, \qquad R = 1.$

Condition (I) shows that on its own ($R = 0$) the triplet $(1, 2, 3)$ can only be explosively unstable when $S_1 = S_2 = -S_3$; and that the coupled system can only be explosively unstable while $R < 1$, that is, while the interaction coefficient of the second triplet $(1, 2, 4)$ is smaller than that of the first one, irrespective of the energy sign factor of wave 4. Condition (II) has an identical interpretation, with the role of the two triplets reversed. Condition (III) shows that the coupled system can be explosively unstable whenever one of the triplets is unstable. Note that these are not sufficient conditions because, for given initial conditions, the parameter t_∞ that occurs in solution (11) may be non-real and therefore the solution is not singular in real time.

Reversing the argument we conclude that the explosive instability of one triplet will be eliminated by the presence of a second (necessarily) stable triplet with a larger interaction coefficient. This is the same conclusion as obtained previously (Aristov et al., 1972; Walters and Lewak, 1977), except that it has now been proved valid for arbitrary initial conditions (and not only for those that lead to $H = 0$).

CONCLUSIONS

Referring to the three questions posed in the introduction we can now say that:
(1) we have obtained a Lax pair with a spectral parameter for the system;
(2) we have obtained a solution in the form of a Laurent series around a movable pole with a full set of constants of integration; however we have not yet shown that this is the general solution of our system;
(3) we have obtained a stability criterion for the explosive instability.

ACKNOWLEDGEMENTS

This work was supported by the Junta Nacional de Investigação Científica e Tecnológica, under Project No. STRDA/P/CEN/528/92 .

REFERENCES

Ablowitz, M.A., and Clarkson, P.A., 1991, "Solitons, Nonlinear Evolution Equations and Inverse Scattering", Cambridge University Press, Cambridge.

Aristov, V.V., Karplyuk, K.S., and Pavlenko, V.P., 1972, Coupling effects of three-wave interaction on development of explosive instability, Ukrain. Phys. J. 17: 307.

Arnold, V.I., 1978, "Mathematical Methods of Classical Mechanics", Springer-Verlag, New York.

Craik, A.D.D., 1985, "Wave Interactions and Fluid Flows", Cambridge University Press, Cambridge.

Davidson, R.C., 1972, "Methods in Nonlinear Plasma Theory", Academic Press, New York.

Goldstein, H., 1980, "Classical Mechanics", Addison-Wesley, Reading, Massachusetts.

Karplyuk, K.S., Oraevskii, V.N., and Pavlenko, V.P., Dynamics of the nonlinear interaction of MHD waves, Plasma Phys. 15: 113.

Kaup, D.J., Reiman, A., and Bers, A., 1979, Space-time evolution of nonlinear three-wave interactions. I. Interaction in a homogeneous medium, Rev. Mod. Phys. 51: 275.

Lax, P.D., 1968, Integrals of nonlinear equations of evolution and solitary waves, Commun. Pure Appl. Math. 21: 467.

Ramani, A., Grammaticos, B., and Bountis, T., The Painlevé property and singularity analysis of integrable and non-integrable systems, Phys. Rep. 180: 159.

Rauch-Wojciechowski, S., 1990, Hamiltonian structures and complete integrability in analytical mechanics, in "Soliton Theory: A Survey of Results", A.P. Fordy, ed., Manchester University Press, Manchester.

Romeiras, F.J., 1983, Integrability of double three-wave interaction, Phys. Lett. A 93: 227.

Romeiras, F.J., 1993, to be published.

Sugihara, R., 1968, Interaction between an electromagnetic wave, plasma waves and an ion acoustic waves, Phys. Fluids 11: 178.

Verheest, F., 1987, Non-linear wave interactions in a complex Hamiltonian formalism, J. Phys. A: Math. Gen. 20: 103.

Verheest, F., 1988, Integrability of restricted multiple three-wave interactions. II., Coupling constants with ratios 1 and 2, J. Math. Phys. 29: 2197.

Walters, D., and Lewak, G.J., 1977, Dynamics of four coupled plasma waves to second order, J. Plasma Phys. 18: 525.

Wojciechowski, S., Jiang, Z., and Bullough, R.K., 1986, Integrable multiwave interaction systems of ODEs, Phys. Lett. A 117: 399.

CHAOTIC FRIEDMAN-ROBERTSON-WALKER COSMOLOGY COUPLED TO A REAL FREE MASSIVE SCALAR FIELD IN MAUPERTUIS PICTURE

Marek Szydlowski

Astronomical Observatory, Jagiellonian University
Orla 171, 30–244 Krakow, Poland

Abstract

The generalization of the Maupertuis principle in the relativistic cosmology context is presented. The connections between the sign of the Gauss curvature for a space with a Jacobi metric (which is indefinite in our case) and the chaotic behaviour is also discussed.

To present an example of chaos in FRW cosmology, let us consider the system with the Lagrange function

$$\mathcal{L}(\dot{q}, q) = \frac{1}{2} g_{\alpha\beta}(q)\dot{q}^\alpha \dot{q}^\beta - V(q)\,, \qquad \alpha, \beta = 1, 2, \ldots, N\,, \tag{1}$$

where $g_{\alpha\beta}$ is of the signature $(-, +, +, \cdots, +)$.

The Hamilton function for this system is

$$\mathcal{H}(p, q) = \frac{1}{2} g^{\alpha\beta}(q) p_\alpha p_\beta + V(q) \tag{2}$$

where

$$p_\alpha = g_{\alpha\beta}\dot{q}^\beta\,.$$

Because of cosmological applications we shall only discuss systems with the "Hamilton constraint" $\mathcal{H}(p, q) = 0$. It is clear that the value of the Hamiltonian along trajectories is determined by the choice of the initial conditions and remains constant. Let us consider the energy functions $\mathcal{E}(\dot{q}, q) = \frac{1}{2} g_{\alpha\beta}\dot{q}^\alpha \dot{q}^\beta + V(q)$. In such a case, the "Hamiltonian constraint" means that $\mathcal{E}(\dot{q}, q) = 0$, i.e., $g_{\alpha\beta}\dot{q}^\alpha \dot{q}^\beta = -2V(q)$. Therefore, trajectories of the system in the tangent space to $\mathbf{R}^N$ with coordinates $(\dot{q}^\alpha, q^\alpha)$ are situated in the region Ω given by

$$\Omega = \{(\dot{q}^\alpha, q^\alpha) \in \mathbf{R}^{2N}\,; \quad g_{\alpha\beta}\dot{q}^\alpha \dot{q}^\beta + 2V(q) = 0\}\,.$$

A vector $v \in T_q\mathbf{R}^N$, where $T_q\mathbf{R}^N$ denotes the tangent space to $\mathbf{R}^N$ at $q \in \mathbf{R}^N$, is timelike if $g_{\alpha\beta}(q)v^\alpha v^\beta < 0$, spacelike if $g_{\alpha\beta}(q)v^\alpha v^\beta > 0$ or null if $g_{\alpha\beta}(q)v^\alpha v^\beta = 0$.

Hamiltonian Mechanics, Edited by J. Seimenis,
Plenum Press, New York, 1994

An open subset of the configuration space $\mathbf{R}^N$ on which $V(q) > 0$ will be denoted by D_T; an open subset on which $V(q) < 0$ by D_S and a closed subset on which $V(q) = 0$ by ∂D. It is clear that $\partial D_T = \partial D_S = \partial D$ where ∂D_T and ∂D_S are the boundaries of D_T and D_S respectively.

On the strength of the "Hamiltonian constraint", the tangent vector to any trajectory in D_T is timelike and in D_S is spacelike. The tangent vector to a trajectory crossing ∂D at a point $q_0 \in \partial D$ is null, i.e., it is situated on the cone given by $g_{\alpha\beta}(q_0)\xi^\alpha\xi^\beta = 0$. It is evident that when a trajectory goes from a region to another, say from D_T to D_S, it must go through ∂D and that at points of ∂D the tangent vector is situated on the cone determined by the form of the "kinetic energy" (see Fig. 1).

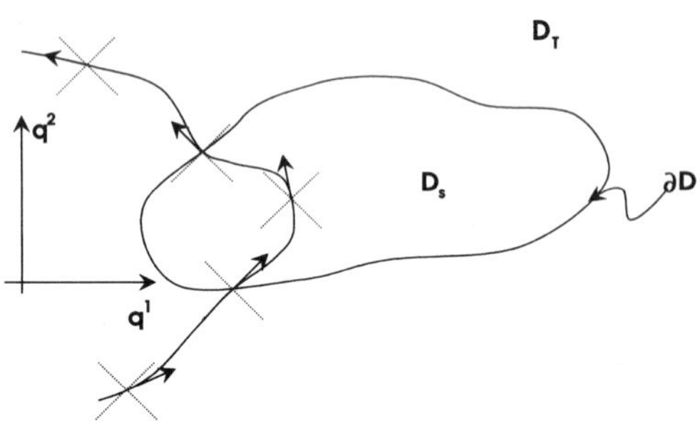

Fig. 1. The typical trajectory crossing the boundary.

The Euler-Lagrange equations of our system are

$$\ddot{q}^\alpha + \Gamma^\alpha_{\mu\nu}\dot{q}^\mu\dot{q}^\nu = -g^{\alpha\beta}\frac{\partial V}{\partial q^\beta} \tag{3}$$

where

$$\Gamma^\alpha_{\mu\nu} = \frac{1}{2}g^{\alpha\beta}[\partial_\mu g_{\beta\nu} + \partial_\nu g_{\beta\mu} - \partial_\beta g_{\mu\nu}]$$

are components of the metric connection describing the form of the "kinetic energy" $[g_{\alpha\beta}]$.

In the region D_T and D_S we define the Jacobi metric

$$\hat{g}_{\alpha\beta} := 2|V|g_{\alpha\beta}\,. \tag{4}$$

The "natural parameter" s ($s = s(t)$ or $t = t(s)$) is defined, along a trajectory in D_T or D_S, in the following way

$$\frac{ds}{dt}(t) = 2|V(q(t))|\,. \tag{5}$$

One can see that

$$\hat{g}_{\alpha\beta}\frac{dq^\alpha}{ds}\frac{dq^\beta}{ds} = -\mathrm{sgn}[V(q(s))] = \frac{2|V(q)|}{(2|V|)^2}g_{\alpha\beta}\frac{dq^\alpha}{dt}\frac{dq^\beta}{dt}\,. \tag{6}$$

Now, one can easily show that the Euler-Lagrange equations in D_T and D_S are equivalent to the equations of geodesics corresponding to the *pseudoriemannian geometry* with the metric $\hat{g}_{\alpha\beta} = 2|V(q)|g_{\alpha\beta}$, where the parameter $s = s(t)$ is such that $\frac{ds}{dt}(t) = 2|V(q(t))|$.

As an example, let us consider the geodesic equation in the region D_S

$$\frac{d^2q^\alpha}{ds^2} + \hat{\Gamma}^\alpha_{\mu\nu}\frac{dq^\mu}{ds}\frac{dq^\nu}{ds} = 0\,, \qquad \hat{g}_{\alpha\beta} = -2V g_{\alpha\beta} \tag{7}$$

and

$$\frac{ds}{dt} = -2V(q(t))\,.$$

After simple manipulation the above equations reduce to the Euler-Lagrange equations. We can perform similar calculations for the region D_T. Finally, we can see that the Euler-Lagrange equations in D_T and D_S are equivalent to the geodesic equations of the pseudoriemannian geometry with the metric $\hat{g}_{\alpha\beta} = 2|V(q)|g_{\alpha\beta}$ where $s(t)$ is defined such that $\frac{ds}{dt}(t) = 2|V(q(t))|$.

Let the Hamiltonian function of the system (discussed in [1]) with two degrees of freedom, has the form

$$\mathcal{H}(\pi, p, a, \phi) = \frac{1}{2}(-\pi^2 + p^2 - a^2 + m^2a^2\phi^2) \tag{8}$$

where $p_1 = \pi, p_2 = p$ are momenta corresponding the generalized coordinates $q^1 = a, q^2 = \phi$, respectively.

The subset ∂D of the configuration space is given by

$$(q^2)^2[1 + m^2(q^1)^2] = (q^1)^2 \tag{9}$$

In Fig. 2 the domains D_T, D_S and ∂D are presented as well as a fragment of the typical trajectory of the system.

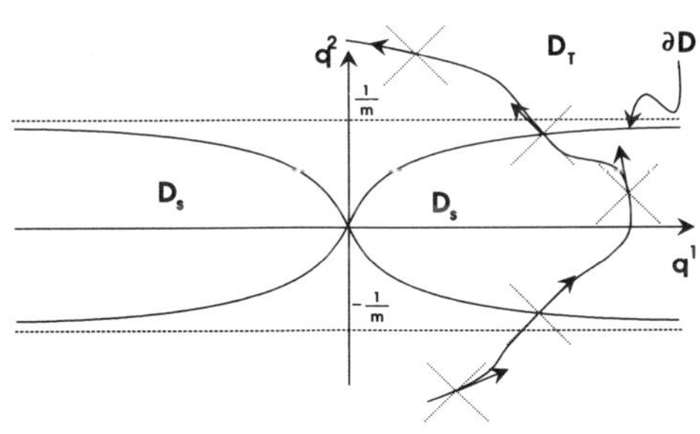

Fig. 2. The domains D_T, D_S and the boundary ∂D.

Let us consider the special case $m = 0$ which is integrable. The solution of the equations of motion have the following form

$$q^1(t) = q^1_0 \cos t + \dot{q}^1_0 \sin t$$
$$q^2(t) = q^2_0 \cos t + \dot{q}^2_0 \sin t$$

and they are represented, in the cofiguration space, by closed intervals and elipses. If $\dot{q}_0^1 = 0$ and $q_0^2 = 0$ then $(\dot{q}_0^2)^2 = (q_0^1)^2$, and the trajectory is a circle; it is shown on Fig. 3.

The dependence between the parameters s and t, for this particular case, is shown on Fig. 4.

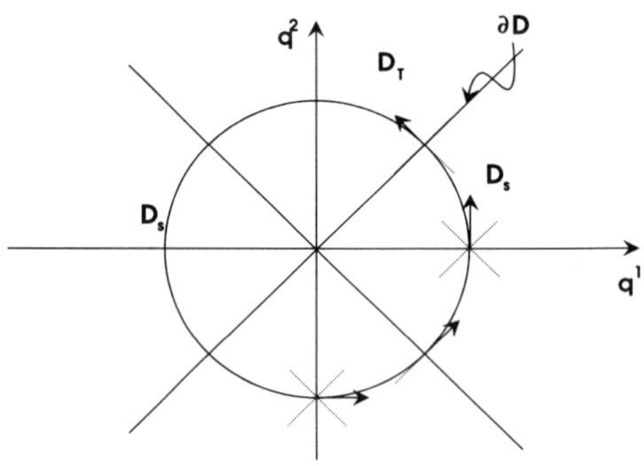

Fig. 3. The special trajectory is a circle if $\dot{q}_0^1 = 0$ i $\dot{q}_0^2 = 0$.

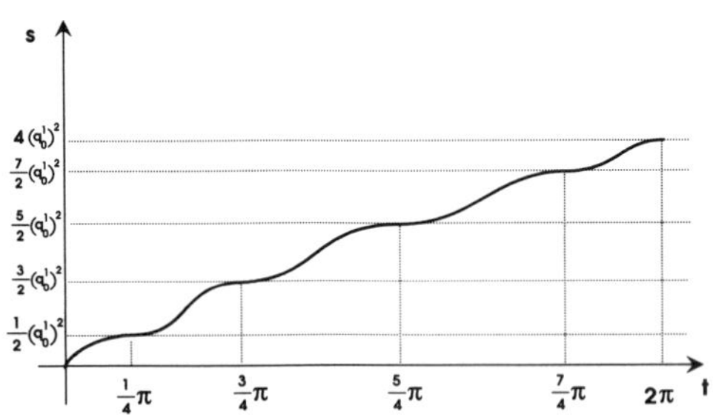

Fig. 4. The dependence between the parameters s and t for particular integrable case.

This example illustrates the general principle that the natural parameter $s = s(t)$ is a strictly increasing and differentiable function with respect to t such that its diagram has inflexion points for the values of t corresponding to the intersection of the trajectory with the boundary ∂D.

From the geodesic deviation equation [2] it follows that the Gauss curvature contains qualitative information about the divergence of nearby geodesics.

The Gauss curvature $\hat{K}$ for the Jacobi metric in the domains D_T and D_S, respectively, is given by

$$\hat{K}(q^1,q^2) = \frac{m^2}{F^3}[(q^1)^4 + (q^2)^4 + 4(q^1)^2(q^2)^2 + \\ + m^4(q^1)^4(q^2)^4 - m^2(q^1)^2(q^2)^2 F(q^1,q^2)] \quad (10)$$

and

$$\hat{K}(q^1,q^2) = -\frac{m^2}{F^3}[(q^1)^4 + (q^2)^4 + 4(q^1)^2(q^2)^2 + \\ + m^4(q^1)^4(q^2)^4 - m^2(q^1)^2(q^2)^2 F(q^1,q^2)] \quad (11)$$

where

$$F(q^1,q^2) = 2V(q^1,q^2) \qquad \text{and} \qquad V(q^1,q^2) = \frac{1}{2}[-(q^1)^2 + (q^2)^2 + m^2(q^1)^2(q^2)^2].$$

In the integrable case, $m = 0$, we obtain

$$\hat{K}(q^1,q^2) = 0$$

whereas for $m \approx 0$ the Gaussian curvature $\hat{K}$ is positive in both regions D_T and D_S. This means that nearby geodesics diverge in both these domains (because in our case $\hat{g}(n,n)\,\hat{g}(u,u) < 0$, where u and n are the tangent and normal geodesic separation vectors respectively) [2]. When a trajectory approaches the boundary, its behaviour is determined by the kinetic energy form. The points at which trajectories cross the boundary define a discrete mapping of the boundary points onto themselves. It seems straightforward that this mapping is a counterpart of the Poincare mapping in standard dynamics.

Conclusions

1. There is natural motivation to study the Hamiltonian dynamics with non-definite kinetic energy form, it is connected with applications in General Relativity and cosmology.

2. The generalized Maupertuis principle gives us a geometric picture of the Hamiltonian dynamics which in turn may be used in searching invariant characterization of chaos in General Relativity.

Acknowledgements

This work was supported by KBN (Poland) under project No. 2 2108 91 02.
I am grateful A. Krawiec, M. Maciejewski and J. Strelcyn.

REFERENCES

[1] E. Calcetta and C.E. Hasi, Chaotic FRW cosmology, preprint (1993).

[2] M. Szydlowski, *Phys. Lett.* **B215**, 711 (1988); M. Szydlowski and M. Biesiada, *Phys. Lett.* **B220**, 33 (1989); M. Szydlowski and A. Lapeta, *Phys. Lett.* **A160**, 123 (1991); M. Szydlowski and A. Lapeta, *Gen. Relat. Grav.* **23**, 151 (1991); M. Szydlowski and M. Biesiada, *Phys. Rev.* **D44**, 2369 (1991); M. Szydlowski, J. Szczesny and M. Biesiada, *Chaos, Fractals and Solitons* **1**, 233 (1991).

THE METHOD OF MODULAR SMOOTHING

Nikola Buric[1] and Ian C. Percival[1,2]

School of Mathematical Sciences [1]
Department of Physics [2]
Queen Mary & Westfield College
University of London

1. INTRODUCTION

Since the KAM theorem on the existence of regular motion was proved over 30 years ago, the great advances in the rigorous theory and the deeper understanding of Hamiltonian systems have not been reflected adequately in the practical applications. The primary aim of the method of modular smoothing is to help to remedy this situation by developing practical methods based on these advances.

Consider a one parameter family of Hamiltonian systems $H(p, q; k)$ such that $H(p, q; 0)$ is integrable. The regular motion of the system for small k is confined to invariant tori, which may be divided into classes, such as the class of rotational tori which support motion smoothly conjugate to the rotation with a frequency ν on the unperturbed tori with $k = 0$. Such tori are called KAM tori. Typically, for a given frequency ν a rotational torus exists only for a finite range of k, bounded by critical values of k, for which the torus can be destroyed by arbitrarily small perturbation. The dependence of such a critical parameter on the frequency is a critical function, denoted by $K(\nu)$ and illustrated in figure 1.

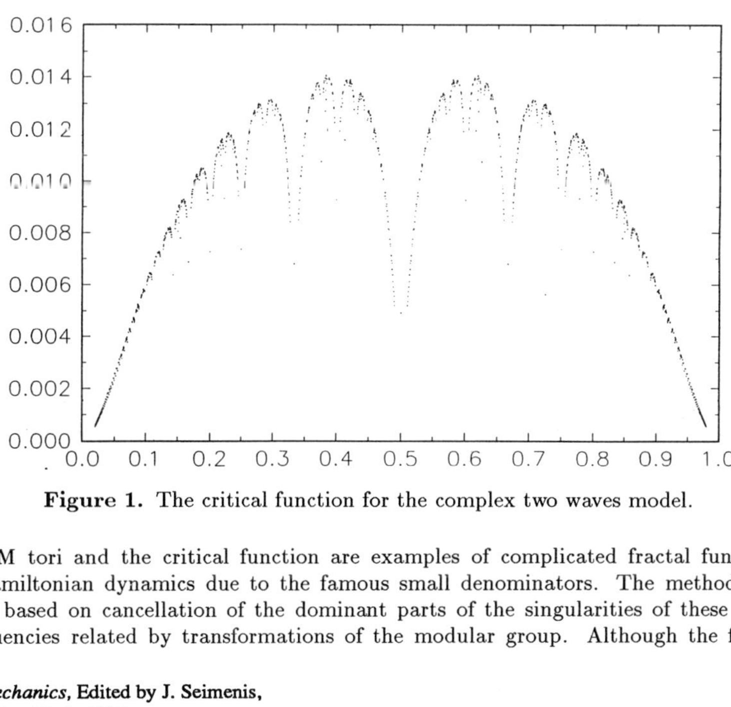

Figure 1. The critical function for the complex two waves model.

The KAM tori and the critical function are examples of complicated fractal functions which appear in Hamiltonian dynamics due to the famous small denominators. The method of modular smoothing is based on cancellation of the dominant parts of the singularities of these functions at rational frequencies related by transformations of the modular group. Although the functions are

Hamiltonian Mechanics, Edited by J. Seimenis,
Plenum Press, New York, 1994

complicated their transformation properties under the action of the modular group on frequencies are described by much simpler functions, which can be relatively easily approximated by piecewise smooth functions. This provides us with a relatively efficient and accurate method for calculation of these fractal functions.

The main idea of the method of modular smoothing was put forward for the first time in a reference[1], were it was applied for calculating the critical function of a complex area-preserving map. The method was developed further in a reference [2], were it was applied on a real area-preserving map and a continuous time system. It was shown that the method requires only a finite, low order perturbation expansion for efficient and quite accurate calculations of the critical functions. Some exact results were reported in a reference[3]. In a reference[4] the idea of the modular smoothing was applied for calculations of the KAM tori instead of just the critical functions.

Let us describe the main idea. The transformation properties of the invariant tori of an integrable system with two degrees of freedom under the action of the modular group on frequencies are trivial, because the tori $q_i(\theta_1, \theta_2; \nu)$ are almost everywhere smooth functions of ν. Thus a transformation from the point $q_1(\theta_1, \theta_2; \nu), q_2(\theta_1, \theta_2; \nu)$ on a ν torus into the point $q_1(\theta_1', \theta_2'; \nu'), q_2(\theta_1', \theta_2'; \nu')$ on a ν' torus, where $\nu' = M(\nu)$, $\theta' = M(\theta)$ and M is a modular transformation, is a smooth function of the frequency. The knowledge of one invariant torus with the frequency ν and the transformation properties gives all invariant tori with frequencies related to ν by modular transformations.

In the case of nonintegrable Hamiltonian systems with two degrees of freedom, the canonical transformation $q(\theta; k; \nu)$ which conjugates the KAM torus with a frequency ν to the corresponding torus of the integrable system is not a smooth function of the frequency, as it is for the integrable case. It has singularities at all resonant values of the frequencies. However, the singularities of $q(\theta; k; \nu)$ and of $q(\theta'; k'; \nu')$ are of the same type. This suggests that that there might be a function $k'(k, \nu)$ such that a nonlinear transformation from a point $q(\theta; k; \nu)$ on the ν-torus to the point $q(\theta'; k'(k, \nu); \nu')$ on the ν' torus could be made a smooth function of the frequency. In order to find such functions we compare perturbation expansions of the KAM tori at frequencies ν and ν' in the double limit $\nu \to m/n$ and $k \to 0$, and fix the relation between k and k' so that the singularities cancel in this limit. It turns out that the relation between k and k' in this limit is a continuous function of the frequency, whose values at the rationals are given exactly by a finite perturbation expansion. Corrections for frequencies away from rationals are given also by continuous functions.

To illustrate the idea we consider here an example of a complex Hamiltonian systems. Such systems have all typical features of real Hamiltonian systems but the analysis is much simpler.

2. MODULAR SMOOTHING OF THE KAM TORI

Consider a Hamiltonian system with one-and-a-half degrees of freedom, given by the following Hamiltonian:

$$H = p^2/2 - k(\exp iq + \exp i(q - t)), \tag{2.1}$$

which we call the complex two waves model. It is a complex version of the corresponding real two waves model which has two sine-waves instead of the exponentials.

The KAM torus with frequency ν is a function $q(k; \nu; \theta, t) = \theta + \tau(k; \nu; \theta, t)$ such that $\tau(k; \nu; \theta, t)$ satisfies the following equation:

$$D^2\tau = ik \exp i(\theta + \tau) + ik \exp i(\theta + \tau - t) \tag{2.2}$$

where $D = \nu\partial/\partial\theta + \partial/\partial t$.

Introducing the following change of the independent variables θ and t, and the function τ:

$$u_1 = k \exp i\theta, \quad u_2 = \exp -it \qquad i\tau = \sigma, \tag{2.3}$$

a Taylor expansion in k and a Fourier expansion in θ and t of the solution τ become the Taylor expansions in u_1 and u_2. of σ

$$\sigma = \sum_{l_1=0, l_2=-\infty}^{\infty} b(l_1, l_2; \nu) u^{l_1} u^{l_2}, \tag{2.4}$$

where the coefficients $b(l_1, l_2; \nu)$ can be calculated recursively. Due to the following initial condition: $b(0, l_2; \nu) = 0$, we have $b(l_1, l_2; \nu) = 0$ whenever $l_2 > l_1$ or $l_2 < 0$. The details can be found in reference[4].

In the case of the real two waves model the perturbation coefficients $b(l, l_1, l_2; \nu)$ have three indexes (l, l_1, l_2) where l corresponds to the Taylor expansion in k and (l_1, l_2) to the Fourier expansions in θ and t(reference[2]). In what follows we shall consider the KAM tori in a small neighbourhood of an m/n

resonance. Such resonance appears for the first time at the n-th order of the perturbation expansion, that is with the coefficient $b(n,n,m;\nu)$ (reference[2]). All other terms which contain the m/n resonance are multiplied by higher powers of k, and are of lower orders in the neighbourhood of the m/n resonance.

Let us concentrate on the KAM tori of the complex two waves model at frequencies ν and $\nu' = 1/\nu - [1/\nu]$, where $[1/\nu]$ indicates the integer part of $1/\nu$, in the following double limit: $\nu \to m/n$ and $k < K(\nu) \to 0$. We are guided by the following observation. In the limit the dominant terms in the expansion of the ν-torus correspond to the m/n resonance, and for the ν'-torus to the m'/n' resonance. The corresponding coefficients $b(n,m;\nu)$ and $b(n',m';\nu')$ are dominated by a pole of the same, second order. Thus if the ratio $k^n/k'^{n'}$ is properly determined, the ratio:

$$\frac{b(n,m;\nu)k^n}{b(n',m';\nu')k'^{n'}} \tag{2.5}$$

will in the limit approach unity. We then have to verify that this choice of k and k' makes equal all terms in the expansions of the ν and ν' tori, which are important in this limit.

The double limit $\nu \to m/n$ and $k \to 0$ allows us to replace the full expansion of the invariant torus by a much simpler one in which we neglect all nonresonant terms as follows:

$$\sigma = \sum_{l_1=1,l_2=0}^{\infty} b(l_1,l_2;\nu)k^{l_1} \exp i(l_1\theta - l_2 t) \tag{2.6}$$

$$= \sum_{d=1}^{\infty} b(dn,dm;\nu) \exp id(n\theta - mt) + \text{nonresonant terms.} \tag{2.7}$$

$$\approx f_{m/n}(k;\nu;\theta,t) = \sum_{d=1}^{\infty} b_a(dn,dm;\nu)k^{dn} \exp id(n\theta - mt). \tag{2.8}$$

The symbol $\approx$ means that the ratio of the left and the right hand side tends to one as $\nu \to m/n$ and $k < K(\nu) \to 0$.

Comparing the expansion (2.8) of $f_{m/n}(k;\nu;\theta,t)$ with the expansion of $\frac{m}{n} f_{m'/n'}(k';\nu';\theta',t')$ where

$$\nu' = 1/\nu - [n/m], \quad m' = n - m[n/m], \quad n' = m,$$

$$\theta' = [n/m]\theta - t, \quad t' = -\theta,$$

and

$$k' \approx \left[\frac{nb(n,m;\nu)}{n'b(n',m';\nu')}\right]^{1/m} k^{n/m} \equiv \chi(n/m;\nu)k^{n/m}. \tag{2.9}$$

we see that if

$$\frac{nb(dn,dm;\nu)}{n'b(dn',dm';\nu')} \approx \left[\frac{nb(n,m;\nu)}{n'b(n',m';\nu')}\right]^d \tag{2.10}$$

then

$$\tau(k;\nu;\theta,t) \approx \tau(k';1/\nu - [1/\nu]; -[1/\nu]\theta + t, -\theta), \qquad k < K(\nu) \to 0,\ \nu \to m/n, \tag{2.11}$$

where

$$\frac{k}{k'} \approx \frac{K(\nu)}{K(1/\nu - [1/\nu])} \approx K(1/\nu - [1/\nu])^{\nu-1} \exp -L_1(m/n), \quad \nu \to m/n, \tag{2.12}$$

and

$$L_1(m/n) = \frac{1}{n} \lim_{\nu \to m/n} \ln \frac{nb(n,m;\nu)}{n'b(n',m';\nu')}. \tag{2.13}$$

The condition (2.10) is checked by direct calculation using the recursion relations for the b-coefficients. The relations (2.11) with (2.12), (2.13) and (2.14) represent the relation between the KAM tori close to a resonance.

Provided that the function $L_1(m/n)$, given by the formula (2.13), is continuous at rationals it can be extended to a function $L_1(\nu) = \lim_{m/n \to \nu} L_1(m/n)$ defined for all $\nu \in [0,1]$.

In figure 2. we present $L_1(m/n)$, which shows that $L_1(\nu)$ is continuous and can be extrapolated from $L_1(m/n)$.

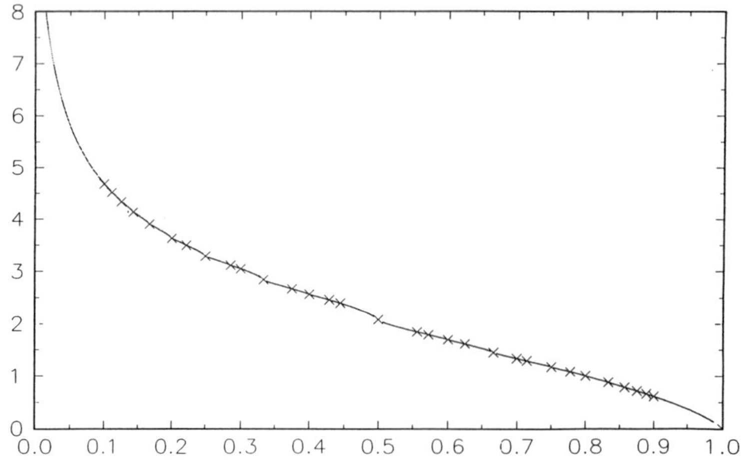

Figure 2. The function $L_1(m/n)$ for all $n < 80$. The values at $n < 11$ are presented by X.

The function $L_1(\nu)$ gives us a continuous relation between $K(\nu)$ and $K(\nu')$(reference [1]). Indeed, from the equation (2.13) we see that the expression $-\ln K(\nu')^{m/n} + \ln K(\nu) \approx L_1(m/n)$ gives: $\ln K(\nu) - \nu \ln K(\nu') = L_1(\nu)$, or:

$$\frac{K(\nu)}{K(\nu')^{\nu}} = \exp -L_1(\nu) \approx \exp -L_1(m/n), \quad m/n \approx \nu. \tag{2.14}$$

Similar formulae for $L_1(m/n)$ have been obtained for the standard map and the real two waves model[2].

The values of the function L_1 at rationals $\nu = 1/n$ are given by the following explicit formula:

$$L_1(1/n) = \frac{1}{n} \ln[\frac{n^{2n+3}}{2^{n-1}}]. \tag{2.15}$$

A similar formula was obtained for the real two waves model[3].

If the frequency ν is a noble number which is transformed into the golden mean γ by d_ν applications of the Gauss transformation, the exact relation in (2.14) can be iterated $d_\nu - 1$ times until $\nu' = \gamma$. In this way we obtain:

$$K(\nu) = K(\gamma)^{(\nu'\nu''\ldots\nu^{d_\nu - 1})} \exp - \left[L_1(\nu) + L_1(\nu')\nu + L_1(\nu'')\nu\nu' \ldots + L_1(\nu^{d_\nu - 1})\nu^{d_\nu - 2} \ldots \nu\right] \tag{2.16}$$

We have thus expressed the fractal function $K(\nu)$ by the continuous function $L_1(\nu)$. A good approximation of the critical function can now be obtained by calculating the function $L_1(m/n)$ at only few m/n with small n, and interpolating the function $L_1(\nu)$ by a piecewise linear approximation through the points $L_1(m/n)$.

The simple description of the transformation properties of the KAM tori in the limit of a rational frequency is made possible by cancellation of the singularity at $\nu = m/n$ in $b(n, m, \nu)$ by the corresponding singularity in $b(n', m', \nu')$, and because in this limit only the resonance m/n is important. In order to obtain a relation between the KAM tori for arbitrary frequency we need to know the ratios $b(n, m, \nu)/b(n', m', \nu')$ as functions of ν. An analysis of these ratios shows that the following functions

$$L_1(m, n; \nu) = \frac{1}{n} \ln \left[\frac{nb(n, m, \nu)}{n'b(n', m', \nu')}\right]. \tag{2.17}$$

have no singularities in a small neighbourhood around $\nu = m/n$. Thus, they can be approximated by the first few terms in their Taylor expansions. We expect that the coefficients in these expansions will give functions of m/n which can be interpolated easily. Indeed, the first derivatives $L'_1(m,n;\nu)|_{\nu=m/n}$ form a fractal discontinuous function of m/n, but the following function

$$R_1(m/n) = \frac{\frac{m}{n} L'_1(m,n;\nu)|_{\nu=m/n}}{\frac{m'}{n'} L'_1(m_1,n_1;\nu)|_{\nu=m_1/n_1}}, \tag{2.18}$$

where $m_1 = m'$ and $n_1 = n'$ if $m'/n' > 0.5$ or $m_1 = n' - m'$ and $n_1 = n'$ if $m'/n' < 0.5$, and m'/n' is the Gauss transform of m/n, looks like a smooth function of m/n on the interval $(0.5, 1)$, except at $2/3$. This is illustrated in figure 3.

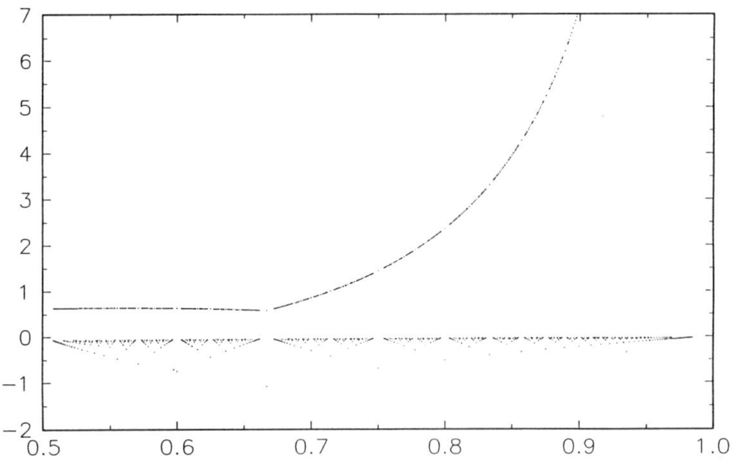

Figure 3 The functions $L'_1(m,n;m/n)$ and $R_1(m,n;m/n)$

Notice that the continuous function $L_1(m/n)$ represents a relation between two resonances, while the smooth function $R_1(m/n)$ gives the relation between three resonances. One can interpolate the ratio (2.18) by the piecewise linear interpolation through few m/n, for example $(n-1)/n$, $n = 2,3,4,5,6$, and use this approximation to calculate the first order corrections to the ratio of the b-coefficients for any m/n. The results of this approximation are reported in the reference [4], where it is shown that with negligible additional calculations the results of the ordinary perturbation theory are improved by at least one third.

3. SUMMARY

We shall now summarise the main results so far of the modular smoothing programme. The basic idea is the following one: The transformation properties of important but complicated fractal functions in Hamiltonian dynamics, like the critical functions and the KAM tori, under the action of the modular group on frequencies can be expressed via functions which are much simpler than the original objects. This provides us with an efficient method of calculating them.

In more details the main results are summarised as follows:

1) The function $L_1(\nu) = \ln K(\nu) - \nu \ln K(\nu')$ is a continuous function. This was shown first for the semi-standard map and then for the standard map, two-waves model and the complex two waves model.

2) Values of the function L_1 at rationals are expressed by coefficients of a finite perturbation expansion which involves only computations with rationals.

3) The values $L_1(1/n)$ fit an explicit formula.

4) There is sufficient numerical evidence that a function $L_2(\nu) = (\nu+1)L_1(\nu+1) - \nu L_1(\nu)$ for the semi-standard map and the complex two-waves model is differentiable. Numerical evidence for the standard map and the two-waves model is not as convincing as for the complex models but the same result is indicated.

5) Transformation properties of the KAM tori for all perturbations up to the critical are expressed via continuous functions. The zero order in $(\nu - m/n)$ is given by the function $L_1(m/n)$, and the first order by the function $R_1(m/n)$ which looks smooth.

At the end we would like to outline what we consider to be the main problems for future research.

1) More complete transformation theory for the KAM tori and a more accurate method of approximating them, related to applications in practical problems.

2) Smoothing of other fractals originating in Hamiltonian and other dynamical systems with small denominators.

3) Extension on more than two degrees of freedom.

REFERENCES

1) N. Buric, I.C. Percival and F. Vivaldi, Critical function and modular smoothing, *Nonlinearity* 3: 21 (1990).

2) N. Buric and I.C. Percival, Modular smoothing and finite perturbation theory, *Nonlinearity* 4: 981 (1991).

3) O. Piro, N. Buric and I.C. Percival, Exact results on the critical function for the motion of an electron driven by two plane waves, *Phys. Lett. A,* 165: 320 (1992).

4) N. Buric and I.C. Percival, Modular smoothing and KAM tori, accepted in *Physica* D (1993).

SOLITON CHAOS IN ELASTIC CHAINS AND TURBULENCE

Tomasz Kapitaniak

Division of Control and Dynamics
Technical University of Lodz
Stefanowskiego 1/15, 90-924 Lodz, Poland

ABSTRACT

The possibility of purely spatial chaos of loop soliton localization in long elastic strings is considered. Connections between Euler elastica and the trajectory of fluid particle are discussed.

1. INTRODUCTION

In contrast to the evolution of the velocity field of a fluid which is governed by the Navier-Stokes partial differential equations the trajectory of a fluid particle is governed by an ordinary differential equation. Consequently, in a Lagrangian description of a fluid, the results of dynamical systems theory for autonomous ODE's may be used directly. This way the observed randomness of Lagrangian turbulence might be interpreted as deterministic chaos. In the present work we give simple arguments based upon a classical elastic model to confirm the existence of chaotic diffusion-like particle paths in the presence of deterministic wave-like fluctuations. There are numerous analogies between elastomechanical and hydrodynamical problems, such as that holding between the shape of a free fluid surface under tension and the bending of an elastic wire. In what follows, we use another analogy relating to the Euler elastica described in [1-3] and to the lateral displacement of a fluid particle due to the motion of a circular cylindrical solid body in a two dimensional flow described in [2,5].

2. EULER ELASTICA AND PSEUDO-RANDOM WALK OF A FLUID PARTICLE

Consider a circular cylinder of radius a moving in a liquid when its center is at the origin of a fixed Cartesian system (x,y) - Fig. 1. It is easily shown that the curvature of the path of a fluid particle which is displaced laterally by the cylinder is given by [2,5]

$$\kappa = 2/a^2(2\gamma - \nu) \tag{1}$$

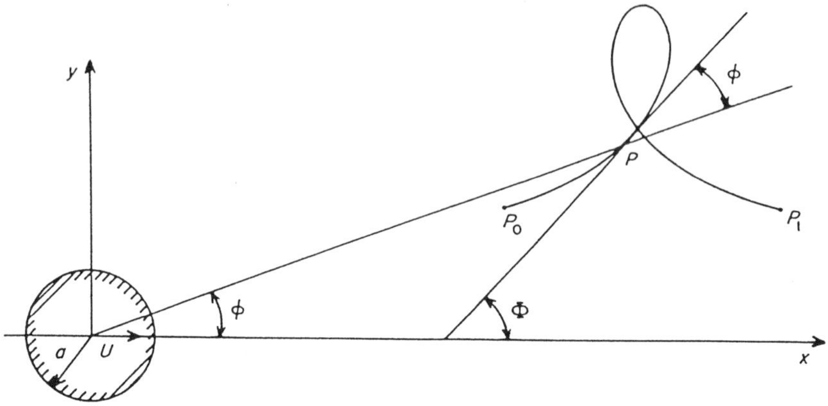

Figure 1. The path of a fluid particle laterally displaced by a rotating cylinder.

where $\eta = y(1 - a^2/r^2)$. Noting that $\kappa = \Phi' = d\Phi/ds$ where s is the arc length of the path and Φ is the slope of the path, we may differentiate the previous equation once and find that

$$\Phi'' = (4/a^2)\frac{dy}{ds} \tag{2}$$

Since $dy/ds = \sin\Phi$, $\lambda = 2/a$ one obtains

$$\Phi'' - \lambda^2 \sin\Phi = 0 \tag{3}$$

Eq. (3) has obviously the same form as classical elastica equation of Euler

$$\Phi'' + \lambda^2 \sin\Phi = 0 \tag{4}$$

One can transpose eq.(3) into eq. (4) by the change of variable $\Phi \rightarrow \pi + \Phi$. The equation of Euler the elastica is related to the oscillations of the pendulum around lower stable equilibrium, while eq. (3) can be related to the oscillations of a pendulum oscillating around upper unstable equilibrium. Both eqs. (3) and (4) have saddle points at $\Phi = k\pi$, $k=0, \pm 1, \pm 2, ..$ and the heteroclinic orbit given by:

$$\begin{aligned} \Phi &= +2\arctan(\sinh s) \\ \Phi' &= +2\,\mathrm{sech}\, s. \end{aligned} \tag{5}$$

Using Melnikov method one can find such small perturbations which when added to eqs. (3) or (4) can lead to the transverse intersection of stable and unstable manifolds [2]. These intersections show the possibility of chaotic behaviour (temporal in pendulum and spatial in problems of elastica or the lateral displacement of a fluid particle due to the motion of the cylinder). Consequently we may state that these looped stereophoid-like paths which were drawn long ago by J. Maxwell and G.I. Taylor [4] are with spatial heteroclinic orbits. In other words they are related to spatial separatrices corresponding in a dynamical analogy to a heteroclinic orbit in phase space [2]. It follows then that the entire chain of reasoning used previously in establishing the possibility for statical loop soliton chaos in the Euler elastica [1-3] may now be carried over to show that deterministic wave-like fluctuations could lead in the appropriate region of parameters and initial conditions to a completely chaotic particle path. The loops themselves may still persist but their spatial distribution will be erratic and will differ essentially from thc classical picture given in text-books. Now, due to the intimate relation between the stability of orbits with irrational winding numbers, damping and random walk, we feel that the preceding discussion might be relevant for the interpretation of diffusion-like process.

3. CHAOTIC VORTICES, CORNU SPIRALS AND THE FLUTTERING ELASTICA

In the previous section we consider the effect of a cylinder moving in a fluid. Here we consider two circular cylinders in a fluid stream [2] . Anyone who has observed the motion of travelling loop solitons in a long flexible wire [1-3] will notice the similarity between them and the vortices observed in fluids [2] which can easily be made visible in an actual experiment. It is this similarity which was the motive for trying to model some of these fluid motions using the elastica. In the case of a Hamiltonian system, the model was relatively straightforward and adding parametrical imperfection we found some interesting spatially chaotic deformations for equation

$$\Phi'' + \lambda^2 \sin\phi = a \sin\omega s \sin\phi \tag{6}$$

with parameters: $\lambda = +0.0272222$, $a = 0.15$ and $\omega = 1$ as shown in Fig. 2.

Some problems arise however in the spatial interpretation of positively and negatively dissipative elastica. In the case of positive dissipation, i.e. damping, this may be interpreted as nonconservative tangential friction forces akin to the so-called follower forces discussed in [2,7]. Negative dissipation is consequently the adjoint system, the so-called flutter set [7]. We may mention that the inclusion of this type of negative damping was motivated by some problems connected to protein deformation.

Eq. (6) have been solved numerically by four-order Runge-Kutta method with integration step $2\pi/200\omega$.

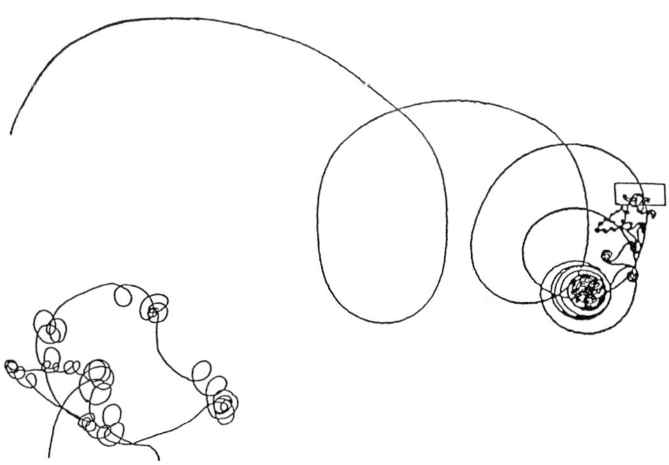

Figure 2. Computer blow-up of the turbulence spirals of the flutter elastica (6) showing the existence of loops on two scales; lower figure: enlargement of the box from the upper figure.

4. CONCLUSIONS

The elastica and in particular the imperfect flutter elastica provides a surprisingly simple model which reflects some fundamental aspects of diffusion and turbulence-like behaviour in fluids. Of course, using our numerical technique we can never show true chaos. Nevertheless, using the pre-entropy related ideas of Rossler and Kahlert [6] we can show asymptotic chaos [2]. When we observed the spiral chaos of the elastica we were initially inclined to regard it as only a numerical instability phenomenon. However, repeated independent calculations using different methods have convinced us that spiral chaos is a true feature of the nonlinear dynamics of our model. The appearance of self-similarity on many scales which these spirals reflect shows that we are dealing with phenomena which may be linked to mixing and diffusion-like process.

REFERENCES

1. M.S. El Naschie and T. Kapitaniak, Soliton chaos models for mechanical and biological elastic chains, Phys. Lett. 147A (1990), 275-281.
2. M.S. El Naschie, S. Al Athel and T. Kapitaniak, A note on elastic turbulence and diffusion, J. Sound Vibr. 155 (1992), 515-522.
3. M.S. El Naschie, On turbulence and complex dynamic in a four-dimensional Peano-Hilbert space, J. Franklin Institute, 330 (1993), 183-198.
4. L. Milne-Thompson, Theoretical hydrodynamics, Macmillian, London 1960.
5. Y. H. Ichikawa, K. Konno and M. Wadati, New nonlinear evolution equation leading to exotic chaos, In: Long-time prediction in dynamics , C. W. Horton (ed.), Wiley-Interscience, New York 1983.
6. C. Kahlert and O.E. Rossler, Chaos in boundary value problems, Z. Naturforschung 39a, (1984) 1200-1204.
7. M.S. El Naschie, Stress, Stability and Chaos, Mc-Graw Hill, London 1990.

STOCHASTIC WEBS WITH FOURFOLD ROTATION SYMMETRY

J.S.W. Lamb

Institute for Theoretical Physics, University of Amsterdam
Valckenierstraat 65, 1018 XE Amsterdam, The Netherlands

1 INTRODUCTION

Weak chaos in nonlinear Hamilton systems has been studied extensively over the last two decades. A lot of attention was paid to the two-dimensional Chirikov-Taylor symplectic (i.e. area-preserving) standard map, modelling (part of) the motion of a charged particle in an inhomogeneous kicking electric field. This map displays weak chaos if it is perturbed from integrability (Chirikov, 1979; Lichtenberg and Lieberman, 1983). In the standard map, the chaotic motion is trapped between KAM-curves if the perturbation from integrability (here the strength of the electric field) is smaller than some threshold value. This causes a separation of chaotic regions in so-called stochastic layers. Only if the perturbation exceeds the threshold value, the chaotic motion is not trapped and the stochastic layers connect to form a two-dimensional infinite net of chaotic saddle connections, a so-called stochastic web, along which chaotic motion proceeds. The latter process is called stochastic diffusion. In the standard map, the threshold for stochastic diffusion is related to the break-up of the last KAM-curve.

This well established mechanism was observed to fail if one considers the same problem but with an additional homogeneous magnetic field (Zaslavsky et al., 1991). In case the frequency of the kicks and the cyclotron frequency of the particle satisfy certain resonance conditions, the threshold for stochastic diffusion in the corresponding two-dimensional symplectic map is observed to disappear, i.e. stochastic diffusion occurs independently of the strength of the electric field (as long it is not zero), the width of the stochastic webs becoming very small for small values of the electric field (Afanasiev et al., 1990). Moreover, the stochastic webs appeared to be extremely symmetric.

The symmetries of these stochastic webs were first thought to be understood on the basis of a Hamiltonian flow approximation (Chernikov et al., 1988; Zaslavsky et al., 1991). However, only recently, in case the resonance satisfies a crystallographic condition, true symmetries of stochastic webs were revealed via an analysis of the reversing symmetry group (of the kth iterate, for some $k \in \mathbb{N}$,) of the symplectic

Hamiltonian Mechanics, Edited by J. Seimenis,
Plenum Press, New York, 1994

map (recently identified as a reversing k-symmetry group (Lamb and Quispel, 1993)). This analysis showed that the stochastic web is left invariant by a crystallographic symmetry group (Lamb, 1993a; Hoveijn, 1992). This crystallographic group was shown to be a true subgroup of the symmetry group of the pattern of saddle connections in the Hamiltonian flow approximation (the so-called web's skeleton (Zaslavsky et al., 1991)).

Furthermore, in the case of a stochastic web with fourfold rotation symmetry, in Lamb (1993a) the disappearance of the threshold for stochastic diffusion was understood on the basis of a symmetry argument.

In this paper we will study the relation between the occurrence of a stochastic web with fourfold rotation symmetry, its threshold for stochastic diffusion and the underlying (reversing) symmetry group possessed by (the kth iterate, for some $k \in \mathbb{N}$, of) a two-dimensional measure preserving map. [1]

First, we will show that reversing symmetry groups possessing a non-reversing mirror symmetry do not allow for the occurrence of a stochastic web. Then, from the observed crystallographic symmetry group of a stochastic web, the determination of possible underlying crystallographic reversing symmetry groups is not very difficult. In fact, this group can be determined precisely by identifying the reversing or non-reversing nature of the various point symmetries of the stochastic web.

Moreover, we conjecture that if (the kth iterate, for some $k \in \mathbb{N}$, of) a two-dimensional measure preserving map possesses a reversing symmetry group that is a crystallographic reversing symmetry group with a fourfold (reversing) rotation, but no non-reversing mirror, this map generically displays a stochastic web implying the absence of a threshold for stochastic diffusion.

In the last section we present some examples area-preserving maps displaying stochastic webs with an underlying fourfold reversing rotation symmetry.

2 SYMMETRIES AND REVERSING SYMMETRIES

In the symmetry analysis of maps, we distinguish between symmetries and reversing symmetries. We say that a map M is a symmetry of a map L if

$$M \circ L \circ M^{-1} = L, \tag{2.1}$$

and that L has a reversing symmetry S, if

$$S \circ L \circ S^{-1} = L^{-1}. \tag{2.2}$$

Note that if the order of the reversing symmetry S, i.e. the smallest integer n such that $S^n = Id$, is odd then automatically $L^2 = Id$.

If L has a reversing symmetry then (historically) the map is called *weakly reversible* in general, and *reversible* if the reversing symmetry is an involution, i.e. its own inverse (Sevryuk, 1986; Roberts and Quispel, 1992).

The symmetries of a map, together with its reversing symmetries form a group under composition.[2] We call this group the *reversing symmetry group* of the map. The composition of a symmetry and a reversing symmetry is a reversing symmetry and

[1] An invertible map L is called measure preserving if there exists a (smooth and continuous) measure μ such that $\mu(S) = \mu(L(S))$ for every subset S of the phase space of L. Symplectic maps are a special class of measure preserving maps, see also Roberts and Quispel (1992).

[2] Here it is assumed that the map and its (reversing) symmetries are invertible.

the composition of two (reversing) symmetries is a symmetry. For a more detailed discussion on the structure of reversing symmetry groups, see Lamb (1992).

In case a map U is not a (reversing) symmetry of L but only of L^k (and not of L^l for some $0 < l < k$), U is called a (reversing) k-symmetry of L. For a more detailed discussion on the concepts and applications of reversing k-symmetries, see Lamb and Quispel (1993), and Quispel and Lamb (1993).

If a map L displays a stochastic web, this web is left invariant by all (reversing) k-symmetries of L for every value of $k \in \mathbb{N}$. In this paper, we discuss the crystallographic symmetry group of a stochastic web possessing a fourfold rotation symmetry. Since in describing the symmetry properties of a stochastic web it is of no use to distinguish between symmetries and reversing symmetries, the symmetry group of the web is induced from the reversing k-symmetry group of the map by neglecting differences between symmetries and reversing symmetries.

3 CRYSTALLOGRAPHIC SYMMETRIES OF STOCHASTIC WEBS WITH FOURFOLD ROTATION SYMMETRY

In a two-dimensional measure preserving map L, weak chaos may arise as a consequence of intersecting stable and unstable manifolds of hyperbolic periodic points (i.e. fixed points of L^n for some value of n). A stochastic web stems from transversal intersections of the stable manifold of one hyperbolic fixed point and the unstable manifold of another hyperbolic fixed point, so-called heteroclinic points. These intersections create a type of saddle connections along which the chaotic motion proceeds. The microscopic structure of these type of saddle connections is very complicated.

The (sudden) ocurrence of a stochastic web, i.e. an infinite net of chaotic saddle connections, seems non-generic unless translation symmetries ensure the existence of heteroclinic connections between saddle points over the entire phase plane. Hence, we are led to investigate stochastic webs in the context of two-dimensional crystallographic space groups.

Crystallographic space groups are generated by translations, spanning a lattice, and linear point transformations (i.e. tranformations leaving at least one point invariant), preserving this lattice. The point transformations in two-dimensional crystallographic space groups are well known to be mirrors and q-fold rotations with $q \in \{2, 3, 4, 6\}$. For a detailed discussion of the possible combinations, see e.g. Loeb (1971).

In dynamical systems we distinguish between reversing symmetries and symmetries, likewise in magnetic crystals spin-reversing symmetries are distinguished from ordinary symmetries (see e.g. Ludwig and Falter (1988)). Because both concepts of reversing symmetries are very similar, we may borrow the Shubnikov-Belov notation from magnetic groups to identify the crystallographic reversing symmetry groups in dynamical systems. In this notation reversing generators are labeled with a prime. It would lead to far to discuss this notation in detail. For a clear exposition, we refer the reader again to Loeb (1971).

In case a two-dimensional crystallographic reversing symmetry group contains a rotation over an angle $\pi/2$, various consequences are seen to follow. First of all, the translation symmetries give rise to a square unit cell. This unit cell can be regarded to be decorated with several additional mirrors and rotocenters. In this particular case, it follows that we have fourfold but also twofold rotocenters in the unit cell. Also mirrors may occur, but three- or sixfold rotocenters cannot arise. A complete analysis leads to 13 different possible crystallographic reversing symmetry groups with a fourfold (reversing) rotation symmetry.

The following proposition reduces this number of possibilities considerably in the case of a stochastic web.

Proposition *Let $L : I\!R^2 \mapsto I\!R^2$ and measure preserving. Then if for some value of $k \in I\!N$, L^k possesses a reversing symmetry group that contains a non-reversing mirror symmetry, then the dynamics of L cannot display a stochastic web.*

Proof.
Let M be a mirror symmetry of L and $W^u(\mathbf{x}_0)$ and $W^s(\mathbf{x}_0)$ be the unstable, resp. stable manifold of a hyperbolic fixed point $\mathbf{x}_0$ of L^n, for some value of n, i.e

$$\begin{aligned} W^s(\mathbf{x}_0) &:= \{\mathbf{x} \in \mathrm{I\!R}^2 | \lim_{m\to\infty} L^{n\cdot m}\mathbf{x} = \mathbf{x}_0\}, \\ W^u(\mathbf{x}_0) &:= \{\mathbf{x} \in \mathrm{I\!R}^2 | \lim_{m\to-\infty} L^{n\cdot m}\mathbf{x} = \mathbf{x}_0\}. \end{aligned}$$

Moreover, let $\mathrm{Fix}(M)$ denote the mirror line. Then it is easily verified that

(i) If $\mathbf{x}_0 \in \mathrm{Fix}(M)$ then either $W^u(\mathbf{x}_0) \subseteq \mathrm{Fix}(M)$ and $W^s(\mathbf{x}_0) \cap \mathrm{Fix}(M) = \emptyset$, or $W^s(\mathbf{x}_0) \subseteq \mathrm{Fix}(M)$ and $W^u(\mathbf{x}_0) \cap \mathrm{Fix}(M) = \emptyset$.

(ii) If $\mathbf{x}_0 \notin \mathrm{Fix}(M)$ then $W^u(\mathbf{x}_0) \cap \mathrm{Fix}(M) = \emptyset$ and $W^s(\mathbf{x}_0) \cap \mathrm{Fix}(M) = \emptyset$.

Hence we find that stable and unstable manifolds cannot (transversally) intersect the mirror line. So if there is a net of saddle connections, there must be hyperbolic fixed points on the mirror. However, these hyperbolic points have a stable or unstable manifold on the mirror line on which no heteroclinic points shared with hyperbolic points on either side of the mirror can be found. This implies that transport through weak chaos stops at a mirror line since saddle points on different sides of the line cannot make a heteroclinic connection. As the mirror line divides the plane, it acts as a blockade for stochastic diffusion. □

Taking the above proposition into account, we find only 6 crystallographic reversing symmetry groups with a fourfold (reversing) rotation symmetry allowing for a stochastic web. In table 1 these groups are presented together with the crystallographic symmetry groups of the stochastic web, as found by neglecting the difference between symmetries and reversing symmetries. For all three (ordinary) crystallographic groups with a fourfold rotation, we find possible underlying crystallographic reversing symmetry groups. In the case of $p4mm$ symmetry we observe that there is only one possible underlying crystallographic reversing symmetry group, whereas in the case of $p4$ and $p4gm$ there are more possibilities. Note that of the 6 crystallographic reversing symmetry groups mentioned in table 1, the groups with a $4'$ possess fourfold rotations only as reversing symmetries, and the groups having a 4 possess fourfold rotations only as non-reversing symmetries. Hence, the twofold rotations in these groups are never reversing symmetries.

To identify from the phase portrait of the map the crystallographic reversing symmetry group of L^k for some value of k, the dynamical implications on the fixed sets of (reversing) point groups, being subgroups of the crystallographic reversing symmetry groups, can be used. In tabel 2 we list the (reversing) point groups occurring as subgroups of the crystallographic symmetry groups with a fourfold (reversing) rotation

Table 1. The crystallographic symmetry group of a stochastic web (SG web) of a two-dimensional measure preserving map L and the various possible underlying crystallographic reversing symmetry groups (RSG) of L^k, for some $k \in \mathbb{N}$. In this table only crystallographic groups with a fourfold rotation are considered.

SG web	RSG
$p4$	$p4$, $p4'$,p'_c4
$p4mm$	$p4m'm'$
$p4gm$	$p4g'm'$, $p4'gm'$

Table 2. Dynamics of the fixed sets of (reversing) point groups occurring as subgroups of the crystallographic reversing symmetry groups, mentioned in table 1.

Point Group	Fixed Set	Dynamics
2	twofold rotocenter	fixed point, generically hyperbolic or elliptic
$2m'$	twofold rotocenter on reversing mirror	fixed point, generically hyperbolic or elliptic
m'	reversing mirror	generically not fixed point (may be hyp. or ellip.)
4	fourfold rotocenter	fixed point, generically elliptic
$4'$	fourfold reversing rotocenter	fixed point, generically hyperbolic

that allow for stochastic diffusion. In particular, fourfold reversing rotocenters can be easily distinguished from fourfold rotocenters, since the former cannot be elliptic and the latter cannot be hyperbolic fixed points. For a discussion of local bifurcations in the presence of a reversing point group symmetry see Lamb and Capel (1993).

To conclude this section, we state the following conjecture

Conjecture *Let $L : \mathbb{R}^2 \mapsto \mathbb{R}^2$ be measure preserving and L^k, for some $k \in \mathbb{N}$, possess a reversing symmetry group that is a crystallographic symmetry group with a fourfold (reversing) rotation symmetry, but no non-reversing mirror, then generically L displays a stochastic web.*

Above conjecture implies the absence of a threshold for stochastic diffusion. It is based on the following observations. Because of the fourfold (reversing) rotocenters, the twofold rotocenters, and the repetition of unit cells, generically, hyperbolic points will occur. Then, because of the fourfold (reversing) symmetry and measure preservation, there will be stable and unstable manifolds running from one unit cell into another. Moreover, transversal intersections with other stable or unstable manifolds seem generically unavoidable. Because of the fourfold rotation and translation symmetries these heteroclinic connections lead to a stochastic web.

Note that integrable maps are nongeneric counterexamples to the above conjecture.

4 EXAMPLES

As an illustration of the above discussion, in this section we will present examples of two-dimensional area-preserving maps possessing a crystallographic reversing crystallographic symmetry group with a fourfold reversing rotation. From the analysis in section 3, it follows that there are only two such crystallographic reversing symmetry groups allowing for a stochastic web, namely $p4'$ and $p4'gm'$. In these cases the stochastic webs have symmetry $p4$ and $p4gm$ respectively. For both groups we will present example maps.

To construct the examples, we make use of a convenient decomposition property, derived in Lamb (1992). Let $R_{\pi/2}$ be a rotation over an angle $\pi/2$ around the origin, i.e. $R_{\pi/2}(x, y) = (y, -x)$. Then, $R_{\pi/2}$ is a reversing symmetry of L if and only if

$$L = R_{\pi/2} \circ A, \tag{4.1}$$

such that $A^2 = R^2_{\pi/2}$ ($\circ$ denotes composition). The construction of such a map A is easy. Just let A be conjugate to $R_{\pi/2}$, i.e. $A = C \circ R_{\pi/2} \circ C^{-1}$, where C commutes with $R^2_{\pi/2}$ but not with $R_{\pi/2}$ (to avoid L being conjugate to $R^2_{\pi/2}$). If we take $C(x, y) = (x + f(y), -y)$ with f being odd, i.e. $f(y) = -f(-y)$, then

$$A : \left\{ \begin{array}{rcl} x' & = & -y - f(y'), \\ y' & = & x + f(y), \end{array} \right. \tag{4.2}$$

satisfies $A^2 = R^2_{\pi/2}$.

With this choice of A, we obtain

$$L : \left\{ \begin{array}{rcl} x' & = & x + f(y), \\ y' & = & y + f(x'). \end{array} \right. \tag{4.3}$$

In case f is not only odd but also periodic, L possesses translation symmetries, and hence a crystallographic reversing symmetry group. The groups $p4'$ and $p4'gm'$ occur with different properties of f:

$$\begin{array}{rcl} p4' & : & f(y) = f(y + a) = -f(-y), \\ p4'gm' & : & f(y) = -f(y + a/2) = -f(-y), \end{array}$$

for some nonzero $a \in \mathbb{R}$. For convenience, in explicit examples we will take $a = 1$.

In figure 4, part of the stochastic webs generated by (4.3) for different choices of f are depicted, one with $p4'$ and one with $p4'gm'$ symmetry. One additional square has been drawn, indicating the size of the unit cell. In the same figure for both cases the symmetry decoration of these square unit cells is given. The reader can easily verify the dynamical implications of the (reversing) point group symmetries, as discussed in the previous section.

In both above examples no threshold for stochastic diffusion is found, i.e. for all nonzero κ and λ a stochastic web is found. This supports the conjecture with respect to the disappearance of the threshold for stochastic diffusion, as stated at the end of the previous section.

Because C was chosen to be an involution, the map L (4.3) can also be written as $L = M^2$ with $M(x, y) = (y, x + f(y))$. In fact, $p4'$ resp. $p4'gm'$ are reversing 2-symmetry groups of M.

a

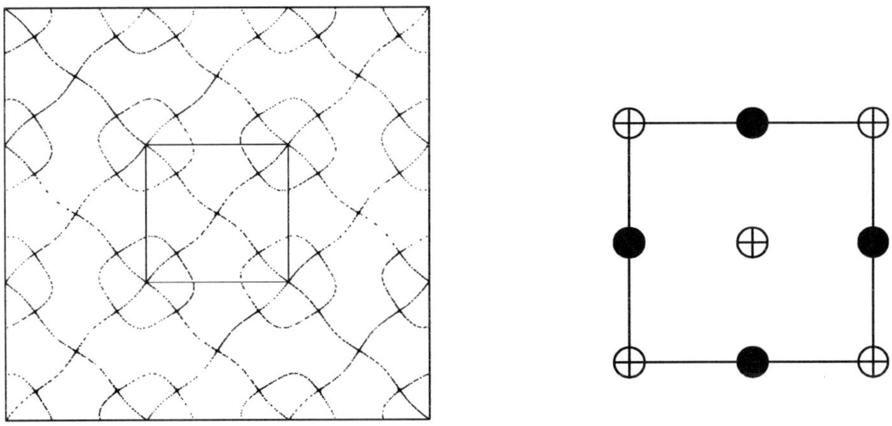

b

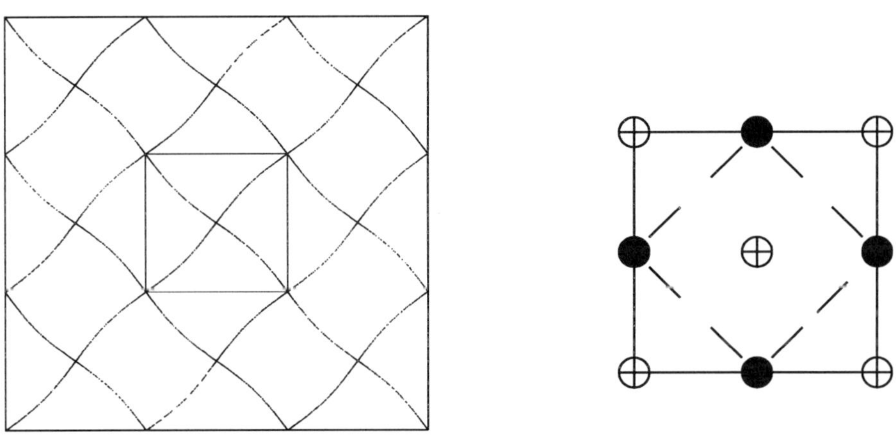

Figure 1. Part of the stochastic web of (4.3) with $f(y) = \kappa \sin(2\pi y) + \lambda \sin(4\pi y)$ on $[-1.5, 1.5] \times [-1.5, 1.5]$ with (a) $\kappa = 0.03$ and $\lambda = 0.06$, and (b) $\kappa = 0.1$ and $\lambda = 0$. The stochastic web in (a) has $p4$ and (b) has $p4gm$ symmetry. Also a square unit cell is drawn. The symmetry decoration of the square unit cells is indicated in accompanying pictures. The crystallographic reversing symmetry groups of the maps are (a) $p4'$, (b) $p4'gm'$. Dashed lines denote reversing mirrors, $\oplus$'s fourfold reversing rotocenters and $\bullet$'s twofold rotocenters.

Besides a crystallographic reversing symmetry group L (4.3) also possesses other (reversing) symmetries. For instance, it possesses the nonlinear reversing mirror symmetry $S(x,y) = (x, -y + f(x))$ (Roberts and Quispel, 1992). This mirror runs through the origin but it does not imply the existence of a non-reversing mirror (preventing stochastic diffusion), as a linear reversing mirror would (because of the fact that the origin is a fourfold reversing rotocenter). This observation illustrates the importance of considering the occurrence of (reversing) symmetries in the light of the structure of the reversing symmetry group.

5 CONCLUDING REMARKS

In this paper we discussed the occurrence of stochastic webs in measure preserving maps possessing a reversing symmetry group that is conjugate to a crystallographic group with a fourfold (reversing) rotation. In particular, we conjecture that in these maps generically stochastic webs occur without a threshold for stochastic diffusion.

In the near future we plan to report in detail on stochastic webs with other crystallographic symmetries as well.

ACKNOWLEDGEMENTS

It is a great pleasure to acknowledge various useful and interesting discussions with G.R.W. Quispel, H.W. Capel, and T. Janssen.

REFERENCES

Afanasiev, V. Chernikov, A.A., Sagdeev, R.Z., and Zaslavsky, G.M., 1990, The width of the stochastic web and particle diffusion along the web, *Phys.Lett. A* 144:229

Chernikov, A.A., Sagdeev, R.Z., and Zaslavsky, G.M., 1988, Stochastic webs, *Physica D* 33:65

Chirikov, B.V., 1979, A universal instability of many-dimensional oscillator systems, *Phys.Rep.* 216:63

Hoveijn, I., 1992, Symplectic reversible maps, tiles and chaos, *Chaos, Solitons and Fractals* 2:81

Lamb, J.S.W., 1992, Reversing symmetries in dynamical systems, *J.Phys.A* 25:925

Lamb, J.S.W., 1993, Crystallographic Symmetries of Stochastic Webs, *J.Phys.A* 26, 2921

Lamb, J.S.W., and Capel, H.W., 1993, Local bifurcations on the plane with reversing point group symmetry, *Chaos, Solitons and Fractals* to appear

Lamb, J.S.W., and Quispel, G.R.W., 1993, Reversing k-Symmetries in Dynamical Systems, preprint.

Loeb, A.L., 1971, "Color and symmetry", Wiley & Sons, New York

Ludwig, W., and Falter, C., 1988, "Symmetries in physics", Springer Series in Solid-State Sciences, Vol. 64, Springer, Berlin

Lichtenberg, A.J., and Lieberman, M.A., 1983, "Regular and stochastic motion", Applied Mathematical Sciences, Vol. 38, Springer, New York

Quispel, G.R.W., and Lamb, J.S.W., 1993, Dynamics and k-symmetries, this volume

Roberts, J.A.G., and Quispel, G.R.W., 1992, Chaos and time-reversal symmetry. Order and chaos in reversible dynamical systems, *Phys.Rep.* 216:63

Sevryuk, M.B., 1986, "Reversible systems", Lecture Notes in Mathematics, Vol. 1211, Springer, Berlin.

Zaslavsky, G.M., Sagdeev, R.Z., Usikov, D.A., and Chernikov, A.A., 1991 "Weak Chaos and Quasiregular Patterns", Cambridge Nonlinear Science Series Vol.1, Cambridge University Press, Cambridge

Zaslavsky, G.M., Zakharov, M.Yu., Sagdeev, R.Z., Usikov, D.A., and Chernikov, A.A., 1986, Generation of ordered structures with a symmetry axis from a Hamiltonian dynamics, *JETP Lett.* 44:451 [*Pis'ma Zh.Eksp.Teor.Fiz.* 44:349],
ibid., 1986, Stochastic web and diffusion of particles in a magnetic field, *Sov.Phys. JETP* 64:294 [*Zh.Eksp.Teor.Fiz.* 91:500]

ANTIBRACKETS AND SUPERSYMMETRIC MECHANICS

Armen Nersessian

Laboratory of Theoretical Physics, JINR, Dubna
Head Post Office, P.O.Box 79, 101000 Moscow, Russia

Abstract

Using the odd symplectic structure constructed over the tangent bundle of the symplectic manifold we construct the simple supergeneralization of an arbitrary Hamiltonian mechanics. In the case where the initial mechanics defines the Killing vector of some Riemannian metric, the corresponding supersymmetric mechanics can be reformulated in terms of the even symplectic structure on the supermanifold.

1 Introduction

It is well-known that on supermanifolds $\mathcal{M}$ the Poisson brackets of two types, even and odd ones, in correspondence with their Grassmannian grading can be defined[1]. These are defined by the expression

$$\{f,g\}_\kappa = \frac{\partial_r f}{\partial z^A}\Omega^{AB}_\kappa(z)\frac{\partial_l g}{\partial z^B}, \tag{1.1}$$

which satisfies the conditions

$$p(\{f,g\}_\kappa) = p(f) + p(g) + 1 \quad \text{(grading condition)},$$
$$\{f,g\}_\kappa = -(-1)^{(p(f)+\kappa)(p(g)+\kappa)}\{g,f\}_\kappa \quad \text{("antisymmetricity")}, \tag{1.2}$$
$$(-1)^{(p(f)+\kappa)(p(h)+\kappa)}\{f,\{g,h\}_1\}_1 + \text{cycl.perm.(f,g,h)} = 0 \quad \text{(Jacobi id.)}, \tag{1.3}$$

where z^A are the local coordinates on $\mathcal{M}$, $\frac{\partial_r}{\partial z^A}$ and $\frac{\partial_l}{\partial z^A}$ denote correspondingly the right and the left derivatives, $\kappa = 0, 1$ denote correspondingly the even and the odd Poisson brackets.

Obviously, the even Poisson brackets can be nondegenerate only if $dim\mathcal{M} = (2N.M)$, and the odd one if $dim\mathcal{M} = (N.N)$.

With nondegenerate Poisson bracket one can associate the symplectic structure

$$\Omega_\kappa = dz^A \Omega_{(\kappa)AB} dz^B, \tag{1.4}$$

where $\Omega_{(\kappa)AB}\Omega^{BC}_\kappa = \delta^C_A$.

The nondegenerate even and odd Poisson brackets can be reduced locally, correspondingly to the forms [2,3]:

$$\{f,g\}^{\mathrm{can}}_0 = \sum_{i=1}^{N}\left(\frac{\partial f}{\partial x^i}\frac{\partial g}{\partial x^{i+N}} - \frac{\partial f}{\partial x^{i+N}}\frac{\partial g}{\partial x^i}\right) + \sum_{\alpha=1}^{M}\epsilon_\alpha \frac{\partial_r f}{\partial\theta_\alpha}\frac{\partial_l g}{\partial\theta_\alpha}, \quad \epsilon_\alpha = \pm 1, \tag{1.5}$$

and

$$\{f,g\}^{\mathrm{can}}_1 = \sum_{i=1}^{N}\left(\frac{\partial_r f}{\partial x^i}\frac{\partial_l g}{\partial\theta_i} - \frac{\partial_r f}{\partial\theta_i}\frac{\partial_l g}{\partial x^i}\right). \tag{1.6}$$

The even Poisson brackets are the straightforward generalization of the ordinary Poisson brackets on the manifolds. They are widely used in physics for the description of the Hamiltonian systems. After quantization such systems described the theories which contain both fermionic and bosonic (real and ghosts) degrees of freedom.

The odd Poisson bracket was introduced in field theory by Batalin and Vilkovisky [4] (they named it antibracket, and following them, we will also use the same term) for the formulation of the covariant (Lagrangian) quantization formalism for the field theories with arbitrary constraints (BV-formalism). BV-formalism and its generalizations [5] are most general methods for the quantization of the gauge theories. A new interest on the subject stimulated recently by the papers [6], which started the developing of the background–independent string field theory on the basis of BV-formalism .

However, the antibrackets can be also used in the direct way – for the formulation of the Hamiltonian mechanics (anti-Hamiltonian mechanics) :

$$\frac{dz^A}{dt} = \{z^A, Q\}_1, \quad p(Q) = 1. \tag{1.7}$$

Leites [2] was the first to point out the possibility of the formulation of such Hamiltonian mechanics.

However, there were a number of obstacles in the quantization of the anti– Hamiltonian mechanics and in the physical interpretation of its grassmannian degrees of freedom: the violated spin–statistics coupling , the necessity of introduction of the odd Planck constant etc.

Possibilities to get over such difficulties were pointed out by D. V. Volkov et al. [7,8]. Particularly, in [8] it was shown that one-dimensional Witten's supersymmetric mechanics can be described using antibrackets, and the role of Hamiltonian in this case is played by one of its supercharges.

Later the Hamiltonian systems, equipped with both even and odd Poisson brackets were studied in more detail in [9].

Nevertheless, the absence of the physical examples, where introduction of the structure of anti–Hamiltonian mechanics would be necessary, or at least successful, gave them the status of non-interesting systems.

Recently a new method for the exact evaluation of the Hamiltonian path integrals was developed[10,11]. It was based on the generalization of the Duistermaat– Heckman localization formula[12] . Using this generalization, one can localize the path integral to the finite-dimensional integral over classical phase space. This forms the basis for the conceptually new approach to the invariant description of supersymmetric theories.

In[13] it was shown that it is convenient to use for the description of this method the odd symplectic structure, constructed on the supermanifold associated with the tangent bundle of symplectic manifold, and the corresponding Hamiltonian dynamic (anti-Hamiltonian dynamic).

Parallelly, this gives the supersymmetrization method for a wide class of the Hamiltonian systems (namely, for the Hamiltonian systems which define Killing vectors of the some Riemannian metrics on the phase space), for which a way to avoid the above mentioned difficulties connected with the quantization exists.

In the present paper we shall study such Hamiltonian systems and their supersymmetrization method more closely.

The paper is organized in the following way:

In *Section 2* we shall construct the odd symplectic structure and the corresponding antibrackets on the supermanifold, associated with the tangent bundle of the initial symplectic manifold. Then we shall define the natural map of the mechanics on the initial manifold in the supersymmetric anti-Hamiltonian mechanics, and interpret this supersymmetry in terms of the basic manifold.

In *Section 3* we shall show that if the initial mechanics define the Killing vector of some Riemannian metric on the phase space, then the corresponding supersymmetric anti-Hamiltonian mechanics can be reformulated with the even Poisson brackets.

2 Anti-Hamiltonian Mechanics and Supersymmetry

Let M be the manifold with the symplectic structure

$$\omega = \frac{1}{2}\omega_{ij} dx^i \wedge dx^j, \tag{2.1}$$

and

$$\{f(x), g(x)\} = \frac{\partial f}{\partial x^i}\omega^{ij}\frac{\partial g}{\partial x^j} \tag{2.2}$$

is the corresponding nondegenerate Poisson bracket on it.

The Hamiltonian $H(x)$ defines on it the Hamiltonian mechanics with the equations of motion:

$$\frac{dx^i}{dt} = \{x^i, H(x)\} \equiv \xi^i. \tag{2.3}$$

It is known that any supermanifold can be associated with some vector bundle [1]. On the supermanifold which is associated with the cotangent bundle of any manifold, one can construct the odd symplectic structure [2] (corresponding Poisson bracket is known in mathematics as Schouten bracket). Indeed, let $\mathcal{M}$ be the supermanifold, associated with M. Then on every chart on $\mathcal{M}$ one can choose the local coordinates (x^i, θ_i) ($p(\theta_i) = p(x^i) + 1 = 1$), which are transformed from chart to chart according to

$$x^i \to \tilde{x}^i = \tilde{x}^i(x), \quad \theta_i \to \tilde{\theta}_i = \sum_{j=1}^{N} \frac{\partial x^j}{\partial \tilde{x}^i} \theta_j, \tag{2.4}$$

i.e. θ_i corresponds to $\frac{\partial}{\partial x^i}$.

Then, obviously, using these coordinates, one can globally define on $\mathcal{M}$ the antibracket (1.6). Let us map the even functions on M onto the odd ones on $\mathcal{M}$:

$$f(x) \to Q_f(z) = \{f(x), F(x,\theta)\}_1, \tag{2.5}$$

where

$$F(z) = \frac{1}{2}\theta_i \omega^{ij} \theta_j = -\frac{1}{2}\theta^i \omega_{(0)ij} \theta^j \tag{2.6}$$

corresponds to the Poisson bracket (2.2) on M.

It is easy to see that

$$\{f(x), g(x)\} = \{f(x), Q_g(x,\theta)\}_1 \quad \text{for} \quad \text{any} \quad f(x), g(x). \tag{2.7}$$

Then, the map (2.5) maps the Hamiltonian mechanics (H, ω, M) to the anti-Hamiltonian mechanics $(Q, \Omega_1, \mathcal{M})$, where

$$Q \equiv Q_H = \{H, F\}_1. \tag{2.8}$$

This mechanics is supersymmetric. Indeed, it is easy to see that the functions H, F, Q form the simplest superalgebra:

$$\begin{aligned} Q &= \{H, F\}_1, \quad & \{H, H\}_1 = 0, \quad \{F, F\}_1 = 0, \\ \{Q, Q\}_1 &= 0, \quad & \{H, Q\}_1 = \{F, Q\}_1 = 0, \end{aligned} \tag{2.9}$$

or, equivalently,

$$\begin{aligned} &\{H \pm F, H \pm F\}_1 = \pm 2Q, \\ &\{H + F, H - F\}_1 = \{H \pm F, Q\}_1 = \{Q, Q\}_1 = 0. \end{aligned} \tag{2.10}$$

The last equation in (2.9) corresponds to the Jacobi identity for (2.2).

For the interpretation of this superalgebra note that in (2.5) Q_f is transformed as df. Correspondingly, one can interpret

$$\theta^i = \{x^i, F\}_1 = \omega^{ij} \theta_j \tag{2.11}$$

as the basic 1-forms dx^i.

Then, one can pass from the description in terms of the coordinates (x^i, θ_i) to that in terms of the coordinates (x^i, θ^i).

Obviously, any function $f(x^i, \theta^i)$ can be interpreted in terms of differential forms on M.

In terms of (x^i, θ^i) antibrackets (1.6) on $\mathcal{M}$ take the form:

$$\{x^i, x^j\}_1 = 0, \quad \{x^i, \theta^j\}_1 = \omega^{ij}, \quad \{\theta^i, \theta^j\}_1 = \frac{\partial \omega^{ij}}{\partial x^k}\theta^k, \tag{2.12}$$

where ω^{ij} is the matrix of the even Poisson bracket (2.2) on M.

The corresponding odd symplectic structure in coordinates $z^A = (x^i, \theta^i)$ takes the form

$$\Omega_1 = \frac{1}{2}\omega_{ij} dx^i \wedge d\theta^j + \frac{1}{2}\omega_{ij,k}\theta^k dx^i \wedge dx^j. \tag{2.13}$$

The equations of motion of the anti-Hamiltonian mechanics $(Q, \Omega_1, \mathcal{M})$ are :

$$\frac{dx^i}{dt} = \{x^i, Q\}_1 = \xi^i \quad \frac{d\theta^i}{dt} = \{\theta^i, Q\}_1 = \frac{\partial \xi^i}{\partial x^j}\theta^j. \tag{2.14}$$

The following correspondences are obvious

$$\begin{aligned}
&\{H, \quad\}_1 = \xi^i \frac{\partial}{\partial \theta^i} \to \imath_H - \text{operator of inner product on } \xi^i; \\
&\{F, \quad\}_1 = \theta^i \frac{\partial}{\partial x^i} \to d - \text{operator of exterior differentiation}; \\
&\{Q, \quad\}_1 = \xi^i \frac{\partial}{\partial x^i} + \xi^i_{,k}\theta^k \frac{\partial}{\partial \theta^i} \to \mathcal{L}_H - \text{Lie derivative along } \xi^i.
\end{aligned} \tag{2.15}$$

Then, using the Jacobi identity (1.3), we show

$$\{H, F\}_1 = Q \to d\imath_H + \imath_H d = \mathcal{L}_H - \text{homotopy formula}. \tag{2.16}$$

As we see, the supercharge $H + F$ which defines the supersymmetry transformation , corresponds to the operator of the equivariant differentiation $d_H = d + \imath_H$.

Therefore the above presented anti-Hamiltonian mechanics is the natural candidate for the description of equivariant localization[13].

However, as we noted in the Introduction, there are many obstacles in the quantization and interpretation of the anti-Hamiltonian mechanics. In the next Section we will show that some additional assumption about the anti-Hamiltonian mechanics (2.14) allows us to reformulate it with the even Poisson brackets and to avoid these obstacles.

3 Reformulation with Even Poisson Brackets

In the previous Section we saw, that defining the map (2.5), we go from the arbitrary Hamiltonian mechanics (2.3) to the supersymmetric one (2.14).

In this Section we shall consider the special case of the mechanics (2.3), when it defines the Killing vector for some Riemannian metric on the phase space.

In this case the corresponding supersymmetrization (2.14) can be reformulated in terms of the *even* symplectic structure on the phase *superspace*.

Let assume that the manifold M is equipped with both the symplectic structure ω_α and the Riemannian one g_{ij}.

Let the local 1-form $A_\alpha = A_{(\alpha)i}dx^i$ defines this symplectic structure on M:

$$dA_\alpha = \omega_\alpha .$$

Then consider on $\mathcal{M}$ the following local 1-form:

$$\mathcal{A}_\alpha = A_{(\alpha)i}dx^i + \theta^i g_{ij} D\theta^j, \tag{3.1}$$

where $D\theta^i = d\theta^i + \Gamma^i_{kl}\theta^k dx^l$ and Γ^i_{kl} – the Cristoffel symbols for metrics g_{ij} on M.

The exterior differential of this 1-form (globally) defines on $\mathcal{M}$ the even symplectic structures:

$$\Omega_\alpha = d\mathcal{A}_\alpha = \frac{1}{2}(\omega_{(\alpha)ij} + R_{ijkl}\theta^k\theta^l)dx^i \wedge dx^j + g_{ij} D\theta^i \wedge D\theta^j, \tag{3.2}$$

where R_{ijkl} is the curvature tensor on M.

The Poisson bracket, which correspond to this structure, is following :

$$\{f(z), g(z)\}_\alpha = \nabla_i f(z)(\omega_{(\alpha)ij} + R_{ijkl}\theta^k\theta^l)^{-1}\nabla_j g(z) + \frac{1}{2}\frac{\partial_r f(z)}{\partial\theta^i} g^{ij} \frac{\partial_l g(z)}{\partial\theta^j}, \tag{3.3}$$

where $g^{ik}g_{kj} = \delta^i_j$,

$$\nabla_i = \frac{\partial}{\partial x^i} - \Gamma^k_{ij}\theta^j \frac{\partial_l}{\partial\theta^k}.$$

Now, let us assume that the Riemannian metric g_{ij} on M Lie-derives with ξ^i (2.3):

$$\mathcal{L}_H g = 0 \Leftrightarrow \xi^k_{,i} g_{kj} + g_{ik}\xi^k_{,j} + g_{ij,k}\xi^k = 0. \tag{3.4}$$

Then the odd function

$$\tilde{Q} = \xi^i g_{ij}\theta^j \tag{3.5}$$

on $\mathcal{M}$ is the motion integral of the anti-Hamiltonian mechanics:

$$\mathcal{L}_H g = 0 \to \{Q, \tilde{Q}\}_1 = 0. \tag{3.6}$$

The functions F and H commute with $\tilde{Q}$ in the following way:

$$\{F, \tilde{Q}\}_1 = -F_2, \quad \{H, \tilde{Q}\}_1 = H_2, \tag{3.7}$$

where

$$H_2 = \xi^i g_{ij}\xi^j, \quad F_2 = \frac{1}{2}\theta^i \omega_{(2)ij}\theta^j, \quad \omega_{(2)ij} = \frac{\partial(g_{ik}\xi^k_H)}{\partial x^j} - \frac{\partial(g_{jk}\xi^k_H)}{\partial x^i}. \tag{3.8}$$

Let us assume that $\det \omega_{(2)ij} \neq 0$.

Then, the mechanics $(H_2, \omega_{(2)ij}dx^i \wedge dx^j, M)$ and (H, ω, M) define the bi-Hamiltonian structure on M (it was first pointed out in[12]):

$$\xi^i = \omega^{ij}\frac{\partial H}{\partial x^j} = \omega^{ij}_{(2)}\frac{\partial H_2}{\partial x^j}. \tag{3.9}$$

It is easy to see that $(\mathcal{H}_\alpha = H_\alpha + F_2, \Omega_\alpha, \mathcal{M})$ and $(Q, \Omega_1, \mathcal{M})$ define the same Hamiltonian dynamic (2.14) on $\mathcal{M}$.

There $\alpha = 0, 2$ and Ω_α is defined by (3.2), $\omega_0 \equiv \omega$, $H_0 \equiv H$.

In other words, we provide the supersymmetric anti-Hamiltonian mechanics (2.14) with the *even Hamiltonian structure* in the case where the initial mechanics (H, ω, M) define the Killing vector for some Riemannian metric on M.

Using this description we can avoid the obstacles connected with the difficulties in the quantization and interpretation of anti-Hamiltonian systems.

However, on the level of classical description the use of antibrackets makes it more simple and clear.

It is obvious that this dynamics has at least two supercharges.

We have got the simple supersymmetrization method for the Hamiltonian systems, which is defined on the symplectic manifolds, equipped with the Riemannian structures, Lie-derives with it.

This class includes the integrable systems on the orbits of the coadjoint representation of semisimple Lie groups and therefore, really all integrable systems of classical mechanics.

Example (1D supersymmetric Witten's mechanics) of dynamics with even and odd Hamiltonian structures was considered at first in[8] (see also[9]).

References

1. F. A. Berezin – Introduction to Superanalysis., D. Reidel, Dordrecht, 1986 ;

   T. Voronov – Geometric Integration Theory on Supermanifolds. Sov. Sci. Rev. C, Math.Phys., **9** (1992), 1 .

2. D. A. Leites – Dokl. Akad. Nauk SSSR, **236** (1977), 804 .

3. V. N. Shander – Dokl. Akad. Nauk. Bulgaria **36** (1983) , 309 .

4. I. A. Batalin , G. A. Vilkovisky – Phys.Lett., **102B** (1981), 27; Phys. Rev., **D28** (1983), 2567 .

5. I. A. Batalin, P. M. Lavrov, I. V. Tyutin –J. Math. Phys., **31** (1990) 1487; *ibid.* **32** (1991), 532; *ibid.* **32** (1991), 2513 ;

   I. A. Batalin, I. V. Tyutin – Int. J. Mod. Phys., **A8** (1993) 2333 .

6. E. Witten– Mod. Phys. Lett., **A5** (1990), 487; Phys. Rev., **D46** (1992), 5446 .

7. D. V. Volkov – JETP Lett., **38** (1983), 508 ;

   D. V. Volkov , V. A. Soroka , V. I. Tkach – Sov. J. Nucl. Phys., **46** (1987) , 110 .

8. D. V. Volkov , V. A. Soroka , A. I. Pashnev , V. I.Tkach – JETP Lett., **44** (1986), 55 .

9. O. M. Khudaverdian , A. P. Nersessian – Preprint YERPHI-1031(81)-1987; J. Math. Phys., **32** (1991), 1938 ; J. Math. Phys., **34** (1993), No. 11 (to appear) ;

   O. M. Khudaverdian – J. Math. Phys., **32** (1991), 1934 ;

   A. P. Nersessian – Theor. Math. Phys.,**96** (1993), 140 .

10. M. Blau, E. Keski- Vakkuri , A. J. Niemi – Phys.Lett., **B246** (1990), 92;

A. Hietamaki, A. Yu. Morozov, A. J. Niemi, K. Palo – Phys. Lett. **B263** (1991), 417 ;

A. Yu. Morozov, A. J. Niemi, K. Palo – Nucl. Phys. **B377** (1992), 295 ;

E. Witten – Preprint IASSNS-HEP-92/15 .

11. A. J. Niemi, O. Tirkkonen – Preprint UU-ITP 3/93 .

12. J. J. Duistermaat , G. J. Heckman – Inv. Math. **69** (1982), 259; *ibid.* **72** (1983), 153 ;

M. F. Atiah , R. Bott – Topology, **23**, No. 1 (1984), 1 .

13. A.P. Nersessian – JETP Lett., **58** (1993), 64 .

INTEGRABLE SYSTEMS AND CONFOCAL QUADRICS

Yuri Fedorov

Department of Mathematics and Mechanics
Moscow State University
Moscow–119899, Russia

INTRODUCTION

A family of confocal quadrics in a projective space and the elliptic coordinates associated with these quadrics are known to be a powerful tool for explicit solving various integrable systems in terms of the Abelian integrals. Using the elliptic coordinates associated with such quadrics K.Jacobi solved the problem on the geodesics on an ellipsoid and K.Neumann [9] did the same for the problem of a mass point motion on a sphere in a force field with a quadratic potential.

The present paper displays a close connection between families of confocal quadrics—a geometrical ground of the Jacobi problem —and some finite-dimensional integrable Hamiltonian systems found during the past twelve years thanks to their known Lax pair representation.

Consider equations of a motion of a mass point on the (n-1)-dimensional ellipsoid in $\mathbf{C}^n$

$$(X, qX) = 1, \tag{1}$$

where q is a $n \times n$ matrix of a nondegenerate quadratic form and $X = (X_1, \ldots, X_n)$ are the Euclidean coordinates. Without loss of generality we may put

$$q = \operatorname{diag}(1/a_1, \ldots, 1/a_n), \quad a_1 < a_2 < \ldots < a_n.$$

The ellipsoid (1) can be included in the family of confocal quadrics

$$Q(s) \; : \; \left\{ \frac{X_1^2}{s - a_1} + \cdots + \frac{X_n^2}{s - a_n} = -1 \right\} \subset \mathbf{C}^n, \quad s \in \mathbf{C}. \tag{2}$$

At $s = a_1, \ldots, a_n$ this family contains n "plane" *focal quadrics* which are (n-2)-dimensional quadratic surfaces lying in the (focal) hyperplanes $\{X_i = 0\}$. On each of the quadrics $Q(s)$, $s \neq a_1, \ldots, a_n$, as on a (n-1)-dimensional manifold, one can define the curvilinear elliptic coordinates $\lambda_1, \ldots, \lambda_{n-1}$ such that

$$X_i^2 = \frac{(a_i - \lambda_1) \ldots (a_i - \lambda_{n-1})(a_i - s)}{\prod_{j \neq i}(a_i - a_j)}. \tag{3}$$

Hamiltonian Mechanics, Edited by J. Seimenis,
Plenum Press, New York, 1994

Using the transformation (3) for $n = 3$ K.Jacobi [6] reduced the geodesic equation to an inversion problem for hyperelliptic integrals of the first and the second kind (so called *Abelian trancendentes*). For the general case of the motion on a quadric $Q(s)$, the latter have the form

$$\int_{\lambda_0}^{\lambda_1} \frac{\lambda^i d\lambda}{\sqrt{R(\lambda)}} + \cdots + \int_{\lambda_0}^{\lambda_{n-1}} \frac{\lambda^i d\lambda}{\sqrt{R(\lambda)}} = \delta_i t + \varepsilon_i, \qquad i = 1, \ldots, n-1$$
$$R(\lambda) = (\lambda - a_1)\ldots(\lambda - a_n)(\lambda - c_1)\ldots(\lambda - c_{n-1}), \quad c_1 = s,$$

where $\delta_{n-1} = 1$ and $\delta_i = 0$ for $1 \le i < n-1$, ε_i are arbitrary constants. These constants $c_2, \ldots, c_{n-1}$ depend on initial conditions only. These can be given a clear geometrical interpretation. Consider a tangent line $l: \ \{X + \tau\dot{X} \mid \tau \in \mathbf{C}\}$ of a geodesic $X(t)$ on $Q(s)$. Then the qualitative behavior of the geodesic is excellently described by the following generalization of the well-known Chasles theorem [2]

Theorem 1. *All the tangent lines $l(t)$ of the geodesic on $Q(s)$ are also simultaneously tangent to the quadrics $Q(c_2), \ldots, Q(c_{n-1})$.*

It follows from this theorem that the common tangent lines of confocal quadrics play an important role in studying the Jacobi problem as well as some other integrable dynamical systems connected with the motion on an ellipsoid (the Neumann system, for example). The nature of this connection was completely investigated by J.Moser [8]. We are going to develop this nice construction and to show that the varieties of common tangent linear spaces of a number of confocal quadrics described in terms of the Plücker coordinates lead exactly to all first integrals of the Clebsch–Perelomov system (the Neumann system is its special case) and the well-known Manakov system dealing with the motion of n-dimensional rigid body.

LINEAR SPACES TANGENT TO QUADRICS

To provide the necessary background we consider k-dimensional linear spaces (k-planes) Λ in n-dimensional projective space $\mathbf{P}^n = (x_0 : x_1 : \ldots : x_n)$, where $x_0, x_1, \ldots, x_n$ are the homogeneous coordinates. The set of all such k-planes is known as the Grassmann variety $G(k+1, n+1)$ of dimension $m = (k+1)(n-k)$, which is also equivalent to the set of all k+1-planes $\overline{\Lambda}$ in $\mathbf{C}^{n+1}$ running through the origin. Any k+1-plane $\overline{\Lambda}$ is represented as a linear span of independent vectors $v_0, \ldots, v_k \in \mathbf{C}^{n+1}$ forming the factorable polyvector ((k+1)-vector)

$$v_0 \wedge \ldots \wedge v_k = \sum G_I e_I,$$

where $I = \{i_0 \cdots i_k\}$ is the multi-index that ranges over all k+1 combinations of $\{0, 1, \ldots, n\}$, $\{e_0, e_1, \ldots, e_n\}$ is a basis in $\mathbf{C}^{n+1}$, $\{e_I = e_{i_0} \wedge \cdots \wedge e_{i_{k+1}}\}$ are basis k+1-vectors in the external product $\bigwedge^{k+1}\mathbf{C}^{n+1}$—the linear space of all (k+1)-vectors in $\mathbf{C}^{n+1}$. The coefficients G_I are called the Plücker coordinates of the k-plane Λ.

Recall that any quadratic surface (a quadric) Q in $\mathbf{P}^n$ can be thought as the set of zeros of the quadratic form

$$\sum_{i,j=0}^{n} \overline{q}_{ij} x_i x_j$$

with a symmetrical $(n+1)\times(n+1)$ matrix $\overline{q}$. In $\mathbf{C}^{n+1}\supset \mathbf{P}^n$ this set is represented by a cone $\overline{Q}$ having its vertex in the origin. In the Euclidean space $\mathbf{C}^n \subset \mathbf{P}^n$ the affine part of Q is given by the equation

$$Q(X) = \sum_{i,j=1}^{n} \overline{q}_{ij} X_i X_j + \cdots + \sum_{i=1}^{n} \overline{q}_{i0} X_i + \overline{q}_{00} = 0.$$

Now consider a set of all k-planes tangent to a non-degenerate quadric Q in $\mathbf{P}^n$. It turns out that this set is also described as a certain quadric in $\bigwedge^{k+1}\mathbf{C}^{n+1}$.

Theorem 2. *The set T^kQ of all k-planes (factorable polyvectors) $\overline{\Lambda} = \sum G_I e_I$ that are tangent to the non-degenerate quadric $\{(x,\overline{q}x) = 0\}$ is the intersection of the Grassmanian $G(k+1,n+1)\subset \bigwedge^{k+1}\mathbf{C}^{n+1}$ with the quadric*

$$\sum_{I,J} \Delta^I_J G_I G_J = 0,$$

where Δ^I_J is the algebraic complement to the minor of $\overline{q}^{-1}$ on the crossings of rows with the indices I and columns with the indices J.

When studying the family of confocal quadrics in the projective space it is useful to consider the family of dual quadrics, because it is a *pencil* of quadrics. For example, the projective closure of quadrics of the family (2)

$$Q(s)\ :\ \left\{\frac{x_1^2}{s-a_1} + \cdots + \frac{x_n^2}{s-a_n} + x_0^2 = 0\right\} \subset \mathbf{P}^n, \qquad s\in \mathbf{C}\cup\infty \tag{4}$$

is dual to the quadrics of the pencil

$$Q^*(s)\ : \{Q^*_0(y) - sQ^*_\infty(y) = 0\} \subset *\mathbf{P}^N,$$
$$Q^*_0(y) = a_1 y_1^2 + \cdots + a_n y_n^2 + y_0^2 = 0, \quad Q^*_\infty(y) = y_1^2 + \cdots + y_n^2 = 0.$$

A general family $Q(r)$ of confocal quadrics (a confocal family) is dual to the general pencil $Q^*(r)\ :\ \{Q^*_0(y) - rQ^*_1(y) = 0\} \subset *\mathbf{P}^n$, Q^*_0, Q^*_1 being any distinct quadrics. Each focal quadric in $Q(r)$ is dual to a degenerate quadric in $Q^*(r)$. A pencil $Q^*(r)$ in $\mathbf{P}^n$ is called non-singular if it contains $n+1$ distinct degenerate quadrics.

It makes sense to consider also general confocal families in $\mathbf{P}^n$. It is easy to show that by an appropriate non-degenerate projective transformation $\mathbf{P}^n \to \mathbf{P}^n$ any non-singular confocal family can be reduced to the *homogeneous* family

$$Q(s)\ ;:\ \left\{\frac{x_1^2}{s-a_1} + \cdots + \frac{x_n^2}{s-a_n} + \frac{x_0^2}{s-a_0} = 0\right\} \quad s\in \mathbf{C}\cup\infty. \tag{5}$$

which is dual to the pencil

$$Q(s)^*\ : \{Q^*_0(y) - s(y,y) = 0\} \subset *\mathbf{P}^n, \quad Q^*_0(y) = a_1 y_1^2 + \cdots + a_n y_n^2 + a_0 y_0^2 = 0.$$

Now we are ready to describe the set of all common tangent linear k-planes of several confocal quadrics in terms of their Plücker coordinates. Consider first the quadrics from the homogeneous confocal family (5). Then, using Theorem 2, we obtain the following tangency condition

$$\mathcal{D}_k(s) = \sum_I (s - a_{l_1}) \cdots (s - a_{l_{n-k}}) G_I^2 = (s - c_1) \cdots (s - c_{n-k}) = 0, \qquad (6)$$

$$\{l_1 \cdots l_{n-k}\} = \{0, 1, \ldots, n\} - I, \#I = k + 1.$$

It is a polynomial of degree $n - k$ in s. Therefore, a k-plane Λ is simultaneously tangent to exactly $n - k$ confocal quadrics in $\mathbf{P}^n$. Of course, the same result is also obtainable by purely geometrical reasons.

Given a fixed k-plane Λ with the Plücker coordinates G_I, the condition (6) can be considered as an algebraic equation whose roots are the parameters $s = c_1, \ldots, s - c_{n-k}$ of the confocal quadrics tangent to Λ. If $\Lambda \subset \mathbf{R}^n$, then these parameters must lie in the different intervals

$$]-\infty; a_1], [a_1; a_2], \ldots, [a_{n-1}; a_n].$$

Conversely, let $n - k$ confocal quadrics for $s = c_1, \ldots, s = c_{n-k}$ be fixed, then (6) gives rise to $n - k$ equations on the coordinates G_I

$$\mathcal{D}_k(s = c_l, G_I) = 0, \quad l = 1, \ldots, n - k.$$

They define cones in $\mathbf{C}^N = \bigwedge^{k+1} \mathbf{C}^{n+1}$ whose joint intersection with the Grassmanian $G(k + 1, n + 1)$ represents the set $\mathcal{C}_k$ of all common tangent k-planes of the quadrics $Q(c_1), \ldots, Q(c_{n-k}) \subset \mathbf{P}^n$.

An analogous approach to the quantitative description of the set $\mathcal{C}_k$ of all common tangent linear k-planes of confocal quadrics from the canonical family (4) leads to the condition

$$\begin{aligned} \mathcal{D}'_k(s) &= \sum_J \frac{\Psi(s)}{(s - D_{j_1}) \cdots (s - D_{j_{k+1}})} G_J^2 + \sum_{J'} \frac{\Psi(s)}{(s - D_{j_1}) \cdots (s - D_{j_k})} G_{J'}^2 \\ &= (s - c_1) \cdots (s - c_{n-k}) = 0, J' = \{0 j_1 \cdots j_k\}, \ 0 < j_1 < \cdots < j_k. \end{aligned} \qquad (7)$$

As shown by M.Reid [11] and R.Donagi [3], the set $\mathcal{C}_1$ is isomorphic to an Abelian torus $\mathbf{T}^{n-1}$ which is the Jacobi variety of the hyperelliptic curve $\{w^2 = (s - a_1) \cdots (s - a_n)(s - c_1) \cdots (s - c_{n-1})\}$. It would be interesting to study the algebraic-geometrical structure of the general sets $\mathcal{C}_k$.

INTEGRABLE SYSTEMS AND THE HYPERELLIPTIC LAX PAIRS

In modern approach to integrable Hamiltonian systems their representation in the Lax form (the Lax pair or L-A pair) plays a key role. Such a representation also makes it possible to construct and solve multi-dimensional integrable generalizations of various problems of dynamics. The most well-known examples are the generalizations of the Euler and the Clebsch classical systems, whose Lax pairs were found by Manakov [17] and Perelomov [10]. These Lax pairs include an additional (spectral) parameter defined on the compactified complex plane or an elliptic curve (Riemann surface of genus one).

In this section for the two mentioned multi-dimensional generalizations we consider their Lax representations with a spectral parameter defined on certain unramified covering of a hyperelliptic curve.

Recall that the Euler-Frahm equations of free motion of an m-dimensional rigid body have the Hamiltonian form

$$\dot{M} = [M, \Omega], \quad \Omega = \frac{\partial H}{\partial M}, \tag{8}$$

where $\Omega \in so(m)$ is the angular velocity, $M \in so^*(m)$ is the angular momentum of the body [6]. As shown by Manakov [17], these equations are integrable provided the operators M and Ω are connected as follows

$$[M, V] = [\Omega, U]. \tag{9}$$

Here U, V are constant diagonal matrices

$$U = \mathrm{diag}(a_1, \ldots, a_m), \qquad V = \mathrm{diag}(b_1, \ldots, b_m),$$

and all the eigenvalues of U and V are different.

This fact follows from the Lax representation of the system (8) with the rational spectral parameter λ

$$\dot{L}(\lambda) = [L(\lambda), A(\lambda)], \quad L(\lambda) = M + \lambda U, \quad A(\lambda) = \Omega + \lambda V. \tag{10}$$

In the sequel we put $V = U^2$. Then, in view of (9), $\Omega = UM + MU$ and equations (8) take the simple form

$$\dot{M}_{ij} = (a_j - a_i) \sum_{k=1}^{m} M_{ik} M_{kj}, \quad i, j = 1, .., m. \tag{11}$$

Apart from this "basic" system, there exists a whole hierarchy of "the higher Manakov systems" commuting with (11):

$$\dot{M} = [M, \{M^l, U^r\}], \quad l = 1, 3, 5, \ldots, r, \tag{12}$$

where $\{M^l, U^r\}$ denotes a homogeneous matrix polynomial in M and U of degree s and r respectively. One can show that the hierarchy (12) is complete, i.e., any system commuting with (11) is defined by a corresponding linear combination of the right-hand sides of (12).

Now turn to another Lax representation of the Euler-Manakov equations (11). It was found in [4] and has the form

$$\dot{L}(s) = [L(s), A(s)] \tag{13}$$

$$L(s)_{ij} = \frac{\sqrt{\Phi(s)}}{\sqrt{(s-a_i)(s-a_j)}} M_{ij}, \quad A(s)_{ij} = \sqrt{(s-a_i)(s-a_j)}\, M_{ij},$$

$$\Phi(s) = (s - a_1)(s - a_2) \ldots (s - a_m), \quad i, j = 1, .., m,$$

where $s \in \mathbf{C}$ is a parameter.

The functions $w_{ij} = \sqrt{(s-a_i)(s-a_j)}$ called *biradicals* are assumed to satisfy the relationships

$$w_{ik} w_{kj} = (s - a_k) w_{ij}, \tag{14}$$

and appear to be single-valued functions on a certain unramified covering of hyperelliptic curve $\Gamma: \ \{w^2 = \Phi(s)\}$. It follows from here that quotients w/w_{ij} in $L(s)$

are also single-valued on the same covering. In this connection we shall refer to the Lax pair (13) and the parameter s as *hyperelliptic*.

Now write the characteristic polynomial

$$|L(s) - zE| = z^m + \sum_{k=2}^{m} z^{m-k}(\Phi(s))^{k/2-1}\tilde{\mathcal{I}}_k(s, M) \quad (k = 2, 4, \ldots), \tag{15}$$

$$\tilde{\mathcal{I}}_k(s, M) = \sum_I \frac{\Phi(s)}{(s - a_{i_1}) \ldots (s - a_{i_k})}|M|_I^I = \sum_{\nu=0}^{m-k} s^\nu H_{k\nu}(M),$$

where $|M|_I^I$ are diagonal minors of order k corresponding to multi-indicies $I = \{i_1 \ldots i_k\} \subset \{1 \ldots m\}$, $1 \leq i_1 < \ldots < i \leq m$.

The polynomials $H_{k\nu}(M)$, $k = 2, 4, \ldots \leq m$, $\nu = 0, 1, \ldots, n$ form a complete set of first integrals in the involution. The major coefficients $H_{k\ m-k}(M)$ are the invariants $\mathcal{I}_k(M) = \sum_I |M|_I^I$ of the algebra $so(m)$. Note that all of these integrals are independent.

The angular momentum M can be associated with the 2-form

$$\omega = \sum_{i,j}^{m} M_{ij}\, dx_i \wedge dx_j, \tag{16}$$

which, in general, has the maximal rank $2[m/2]$. In the special case, when $\mathrm{rank}\omega = k < 2[m/2]$, the invariant $H_{k\ m-k}(M)$ as well as *all* integrals of the same and higher degrees vanishes, and dimension of the corresponding general invariant tori drops.

Another hierarchy of integrable systems arises as a result of a multi- dimensional generalizations of the Clebsch case in the Kirchoff equations describing a rigid body motion in an ideal fluid. These equations systems appears to be equivalent to the equations of the motion of a rigid body with a fixed point in a potential field with a quadratic potential function (Brun's problem [1]). For the case of m dimensions we write this potential function in the form

$$Y(\gamma) = (\gamma, B\gamma)/2, \quad B = \mathrm{diag}(b_1, \ldots, b_n), \quad b_1, \ldots, b_n = \mathrm{const}.$$

Here γ denotes a constant vector in $\mathbf{R^n}$.

Let $K \in so^*(n)$ be the angular momentum and

$$\Omega_{ij} = c_{ij}K_{ij}, \quad c_{ij} = \mathrm{const}, \qquad i, j = 1, .., n.$$

Then the motion equations are Hamiltonian with respect to the Lie-Poisson-Kirillov bracket of the Lie algebra $e(n)$:

$$\begin{aligned} \dot{K} &= [K, \Omega] + [\Gamma, B], \quad \Gamma_{ij} = \gamma_i\gamma_j, \\ \dot{\gamma} &= -\Omega\gamma, \end{aligned} \tag{17}$$

with the Hamiltonian

$$H = -\frac{1}{2}\mathrm{tr}(K\Omega) + Y(\gamma) = \frac{1}{2}\sum_{i<j}^{n} c_{ij}K_{ij}^2 + \frac{1}{2}\sum_{i=1}^{n} b_i\gamma_i^2.$$

Here, as above, Ω is the angular velocity of the body.

As shown by Perelomov [10], the system (17) is completely integrable provided K, Ω are connected as follows

$$[K, B] + [D, \Omega] = 0, \tag{18}$$

where $D = \mathrm{diag}(D_1, \ldots, D_n)$, $D = \mathrm{const}$, and, as a consequence, it admits a representation in the Lax form with a rational spectral parameter

$$(\Gamma/\lambda + K + \lambda D)^{\cdot} = [\Gamma/\lambda + K + \lambda D, \Omega + \lambda B].$$

The hyperelliptic Lax pair for the Clebsch-Perelomov system (17), (18) for the case $B = D^2$ is as follows:

$$\dot{L}(r) = [L(r), A(r)], \quad r \in \mathbf{C},$$
$$L(r) = \begin{pmatrix} W(r) & P(r) \\ -P^t(r) & 0 \end{pmatrix}, \quad A(r) = \begin{pmatrix} \hat{W}_{ij}(r) & \hat{P}(r) \\ -\hat{P}^t(r) & 0 \end{pmatrix},$$
$$W(r)_{ij} = \sqrt{\Psi(r)}/\sqrt{(r-D_i)(r-D_j)}\, K_{ij}, \quad \hat{W}_{ij}(r) = \sqrt{(r-D_i)(r-D_j)}\, K_{ij},$$
$$P(r)_i = \sqrt{\Psi(r)}/\sqrt{(r-D_i)}\, \gamma_i, \quad \hat{P}_i(r) = \sqrt{r-D_1}(r+D_i)\gamma_i,$$
$$\Psi(r) = (r-D_1)(r-D_2)\ldots(r-D_n), \quad i,j = 1,..,n.$$

The functions $\sqrt{\Psi(r)}$, $\sqrt{(r-D_i)(r-D_j)}$ satisfy a condition analogous to (14). Writing out the characteristic polynomial for $L(r)$ we obtain

$$|L(r) - zE| = z^{n+1} + \sum_{k=2}^{n} z^{n+1-k}(\Psi(s))^{k/2-1}\tilde{\mathcal{J}}_k(r,K) \qquad (k = 2, 4, \ldots),$$

$$\tilde{\mathcal{I}}_k = \sum_J \frac{\Psi(r)}{(r-D_{j_1})\ldots(r-D_{j_k})}|K|_J^J + \sum_{J'} \frac{\Psi(r)}{(r-D_{j_1})\ldots(r-D_{j_{k-1}})}\begin{vmatrix} K & \gamma \\ -\gamma^t & 0 \end{vmatrix}_{J'}^{J'}$$
$$= \sum_{\nu=0}^{n+1-k} r^\nu H_{k\nu}(K,\gamma), \quad J = \{j_1 \ldots j_k\} \subset \{1, \ldots, n\}, \tag{19}$$

where, as in (15), $|\cdot|_J^J, |\cdot|_{J'}^{J'}$ are diagonal minors of order k and $J' = \{j_1 \ldots j_{k-1}\ n+1\}$, $1 \le j_1 < \ldots < j_{k-1} \le n$. The major coefficients H_{kn-k} are the invariants $\mathcal{I}_k(K,\gamma)$ of the algebra $e(n)$. As in the polynomial (15) for the Manakov system, all integrals $H_{k\nu}(K,\gamma)$ in (19) are independent and in involution with respect to the Lie–Poisson bracket of the algebra $e(n)$.

For example

$$\tilde{\mathcal{J}}_2(r) = \sum_{j=1}^{n} \overline{(r-D_1)\ldots(r-D_n)}\,\gamma_j^2 + \sum_{j<i}^{n} \overline{(r-D_1)\ldots(r-D_n)}\,K_{ij}^2. \tag{20}$$

(The line above the symbols notifies that in these produces the terms $r - D_i, r - D_j$ are absent). Then putting subsequently into (20) $r = D_j$, $j = 1, \ldots, n$ and dividing by $\prod_{l \neq j}(D_j - D_l)$ yields n independent integrals of the special form

$$F_j(K,\gamma) = \gamma_i^2 + \sum_{i \neq j}^{n} K_{ij}^2/(D_j - D_i).$$

These were obtained by Weber [14] and Ulenbeck [13].

As in (16), the variables K, γ can be associated with the 2-form

$$\omega = \sum_{i<j}^{n} K_{ij}\, dx_i \wedge dx_j + \sum_{i}^{n} \gamma_i\, dx_i \wedge dx_0, \tag{21}$$

which, in general, has the maximal rank $2[(n+1)/2]$.

INTEGRABLE SYSTEMS AND COMMON TANGENT LINEAR SPACES OF CONFOCAL QUADRICS

In order to describe a connection between the sets $\mathcal{C}_k$ and the invariant manifolds of the integrable systems considered in previous section we need to use the following

Proposition. *Let the 2-form ω associated with the angular momentum $M \in so(n+1)$ or the pair $(K, \gamma) \in e(n)$ as in* (16), (21) *have rank $k+1 < n+1$. Then the minimal kernel $\overline{\Lambda}$ of ω such that $\omega \wedge \overline{\Lambda} = 0$ is represented by a factorable (k+1)-vector $\overline{\Lambda}$ which corresponds to k-plane $\Lambda \subset \mathbf{P}^n$ with the Plücker coordinates*

$$G_I = \left(|M|_I^I\right)^{1/2} \text{or, respectively, } G_I = \left(\begin{vmatrix} K & \gamma \\ -\gamma^t & 0 \end{vmatrix}\right)^{1/2}, \tag{22}$$

where the last index $m = n+1$ is replaced by zero.

For the case rank $\omega = 2$ the coordinates G_{ij} are the components of M or the pair (K, γ) themself. And for the case rank $\omega = 4$ we have : $G_{ijlr} = M_{ij}M_{lr} - M_{il}M_{jr} + M_{ir}M_{lj}$.

Now we came to the important

Theorem 3. *Suppose that the 2-form ω defined in (16) or in (21) has rank $k+1 < n+1$, i.e., a degeneration of degree $d = 2[n+1/2] - k - 1$ takes place. Fix all integral constants of the Manakov or the Clebsch–Perelomov systems by putting in the polynomials* (15), (19)

$$H_{k+1\,\nu}(M) = h_{k\nu}, \quad \text{respectively} \quad H_{k+1\,\nu}(K, \gamma) = h_{k+1\,\nu},$$
$$h_{k+1\,\nu} = \text{const}, \nu = 0, 1, \ldots, n-k.$$

Then for these constants to each trajectory $M(t)$ or $(K(t), \gamma(t))$ there corresponds a k-plane $\Lambda(t)$ determined by (22) such that it is simultaneously tangent to $n-k$ quadrics $Q(c_1), \ldots, Q(c_{n-k})$ of the homogeneous confocal family (a_0 replaced by a_m) or, respectively, of the canonical confocal family (a_i replaced by D_i). The parameters $c_1, \ldots, c_{n-k}$ are the roots of the polynomial

$$\sum_{\nu=0}^{n-k} r^{\nu} h_{k+1\,\nu}.$$

The proof is obvious: by virtue of (22), the tangency condition polynomials $\mathcal{D}_k(s), \mathcal{D'}_k(s)$ coincide with the polynomials $\tilde{\mathcal{I}}_{k+1}(s, M)$, respectively $\tilde{\mathcal{J}}_{k+1}(r, K, \gamma)$ (r replaced by s) in (15), (19). Abusing the terminology we may say that *the integrals of the Manakov and the Clebsch–Perelomov systems for the integral constants fixed are the tangency conditions themself.*

References

1. Brun F. Rotation kring fix punkt, *Ofers. Kongl. Svenska Vetensk. Acad. Forhandl. Stockholm* 1893, 7. 455–468.
2. Chasles M. Les lignes géodesiques et les lignes de courbure des surfaces du second degré, *Journ. de Math.* 1846, 11. 5–20.
3. Donagi R. Group law on intersections of two quadrics, *Preprint UCLA* 1978.
4. Fedorov Yu. Lax representations with a spectral parameter defined on coverings of hyperelliptic curves, *Mat. Zametki.* 1993, 6. 34–41 (In Russian).
5. Frahm F. Ueber gewisse Differentialgleichungen, *Mathematiche Annalen* 1874, 8. 35–44
6. Jacobi C. Vorlesungen über Dynamik, *Gesammelte Werke* Berlin, 1884.
7. Manakov S.V. Remarks on the integtals of the Euler eqations of the n-dimensional heavy top, *Functional Anal. Appl.* 1976, 10 93–94.
8. Moser J. Various aspects of Hamiltonian systems, *CIME conference talks* Bressone, 1978.
9. Neumann C. De problemate quodam mechanica, quod ad primam integralium ultraellipticorum classem revocatur, *Journal für die reine und angewandte Mathematik* 1859, 56.
10. Perelomov A. Some remarks on integrability of equations describing a rigid body movement in an ideal liquid, *Functional Anal. Appl.* 1981, 15. 83–85
11. Reid M. The complete intersection of two or more quadrics. *Thesis*, Cambridge (UK) 1972.
12. Schläfli L. *Journal für die reine und angewandte Mathematik* 1852. 43.
13. Ulenbeck K. Minimal 2-spheres and tori in S^k, *preprint*, 1975
14. Weber H. Anwendung der Theta functionen zweir Veränderlicher auf die Theorie der Bewegung eines festen Körpers in einer Flüssigkeit, *Mathematiche Annalen* 1878, 14. 173–206.

AN ELEMENTARY APPROACH TO INTEGRABILITY CONDITION FOR THE EULER EQUATIONS ON LIE ALGEBRA so(4)

Sasho Ivanov Popov[1] and Jean-Marie Strelcyn[2,3]

[1]Institute of Metal Science, Bulgarian Academy of Sciences
67 Shipchenski Prohod Street, 1754 Sofia, Bulgaria

[2] Departement de Mathematiques, Universite de Rouen
76821 Mont Saint Aignan Cedex, France, URA CNRS 1378

[3] Laboratoire Analyse, Geometrie et Applications
URA CNRS 742, Institut Galilee,Departement de Mathematiques
Avenue J.-B.Clement, 93430 Villetaneuse, France

Abstract

The completely elementary derivation of the Manakov integrability condition of the Euler equations on the Lie algebra so(4) is given. Four functionally independent first integrals are written explicitly for all values of parameters. Some simple cases of partial integrability are found.

SETTING OF THE PROBLEM

Given a system of ordinary differential equations depending on parameters, the question arises, how to recognise the values of the parameters for which the equations have first integrals? Except for some simple cases, this problem is very hard and no satisfying methods to solve it are known. To date, the most successful approach is offered by so called Painleve analysis (see [1], [2]), the roots of which can be found in the seminal work of S. Kovalevskaya on the rigid body problem [10], [11]. Unfortunately this method, of high practical value, is not based on a firmly established mathematical ground. Moreover, the Painleve analysis method puts emphasis on complex analytic integrals and is not well adapted to the search of integrals in the real domain.

Already for two degrees of freedom Hamiltonian systems the problem of recognising the values of parameters for which the equations are completely integrable is largely unsolved.

Hamiltonian Mechanics, Edited by J. Seimenis,
Plenum Press, New York, 1994

The same question is also relevant for the so-called Euler equations on Lie algebras ([3], [6]–[9], [12]–[14]). For these equations also, the problem is largely open.

In this note we study one of the simplest example of this kind; the Euler equations on Lie algebra so(4) corresponding to the 'diagonal quadratic hamiltonians'. Because these equations are used here exclusively as an interesting and non trivial example of a multiparameter family of ordinary differential equations we write them down directly without any explanations of their Lie algebraic origin.

Let us fix the real numbers $(\lambda_i)_{1\leq i\leq 6}$ and let us consider the system

$$\left.\begin{aligned}
\frac{dx_1}{dt} &= (\lambda_3-\lambda_2)x_2x_3+(\lambda_6-\lambda_5)x_5x_6,\\
\frac{dx_2}{dt} &= (\lambda_1-\lambda_3)x_1x_3+(\lambda_4-\lambda_6)x_4x_6,\\
\frac{dx_3}{dt} &= (\lambda_2-\lambda_1)x_1x_2+(\lambda_5-\lambda_4)x_4x_5,\\
\frac{dx_4}{dt} &= (\lambda_3-\lambda_5)x_3x_5+(\lambda_6-\lambda_2)x_2x_6,\\
\frac{dx_5}{dt} &= (\lambda_4-\lambda_3)x_3x_4+(\lambda_1-\lambda_6)x_1x_6,\\
\frac{dx_6}{dt} &= (\lambda_2-\lambda_4)x_2x_4+(\lambda_5-\lambda_1)x_1x_5,
\end{aligned}\right\} \tag{1}$$

To simplify the notations we will note in the future $\lambda_{ij}=\lambda_i-\lambda_j$.

The above system is the system of Euler equations on Lie algebra so(4) corresponding to the diagonal quadratic 'Hamiltonian'

$$\frac{1}{2}\sum_{i=1}^{6}\lambda_i x_i^2.$$

It always has three first integrals:

$$H_1 = x_1x_4+x_2x_5+x_3x_6, \qquad H_2=\sum_{i=1}^{6}x_i^2, \qquad H_3=\sum_{i=1}^{6}\lambda_i x_i^2. \tag{2}$$

Unless all the $(\lambda_i)_{1\leq i\leq 6}$ are equal (in which case the system (1) is trivial) these three integrals are functionally independent.

The first integrals H_1 and H_2 are intimately related to the Lie algebra so(4). They represent the 'Casimir functions' of it.

To be integrable (see Sec. 28 of [5]), the system (1) needs to have a supplementary fourth first integral H_4, functionally independent of H_1, H_2, H_3, called shortly *a fourth first integral*. Let us suppose that among the $(\lambda_i)_{1\leq i\leq 6}$, the equality $\lambda_i=\lambda_j$ for $i\neq j$, occurs at most two times. Then the unique known case when such a fourth first integral exists is the so called *Manakov case*, defined by *the Manakov condition*

$$\begin{aligned}M \overset{def}{=} \ &\lambda_1\lambda_4(\lambda_2+\lambda_5-\lambda_3-\lambda_6)+\lambda_2\lambda_5(\lambda_3+\lambda_6-\lambda_1-\lambda_4)\\ &+\lambda_3\lambda_6(\lambda_1+\lambda_4-\lambda_2-\lambda_5)=0\end{aligned} \tag{3}$$

One can prove that, under our assumptions on $(\lambda_i)_{1\leq i\leq 6}$, except for the Manakov case, the system (1) is never algebraically completely integrable; see [3] and [8], [9] where also the precise definition of the algebraically completely integrable system can be found. Although we do not repeat here this definition, let us stress that it is based on the properties of the system, when complex time is considered and go back to S. Kovalevskaya ideas ([10], [11]).

We will apply the following method. To search the integrable cases, we will search the cases when the fourth first integral does not depend on all variables. In fact, in Manakov case, the existence of such fourth first integral follows from the form of fourth first integral reported for example in [3], [8], [9], [13]. In fact in [6] one can find a fourth first integral depending only on three variables (see also [13]). Moreover to study the functional independence of the first integrals we use the method from [18].

For the Euler equations (1) this program can be completely fulfilled and we arrive to Manakov condition of integrability (3). In this way we obtain also a nice expression for the fourth first integral in the generic case, much more direct and simple that those reported in [3], [6]–[9], [13]. This leads also to a coherent, structured and explicit description of first integrals for all values of parameters $(\lambda_i)_{1\leq i\leq 6}$ and not only for a generic one as in above cited papers.

We report also two cases of partial first integrals of system (1). It seems that such cases were unknown up to now.

In fact the method employed is not new. It was already applied in [4] to the successful study of integrable cases of Euler-Poisson equations describing the motion of a heavy rigid body with a fixed point. The first author applied it in [15]–[17] to complete the results from [4].

Although in our knowledge this method was not frequently applied by now, its scope is undoubtedly substantially larger than the two above examples; the system (1) and the Euler-Poisson equations.

It is really astonishing that for those two examples this method provides all integrable cases. We are convinced that this reflects some general fact, not yet understood.

Let us stress that this approach does not require the considerations in the complex domain and we require only the differentiability of the fourth integral.

The main tools used are the simplest facts from linear algebra and the well known following remark. If F is a first integral common to two vector fields X and Y, then F is also a first integral of their Lie bracket $[X,Y]$. Indeed, if $X(F)=Y(F)=0$, then evidently $[X,Y](F)=X[Y(F)]-Y[X(F)]=0$.

To achieve successfully this work, we used the symbolic manipulation language REDUCE. To avoid errors, all statements were verified using it.

THE MAIN RESULT

Theorem *If the system (1) admits, even only locally, a fourth first integral which is independent of the first integrals (2) and does not depend on all the variables, then the Manakov condition (3) is fulfilled.*

Let us verify that this is so if for example the fourth integral is of the form $F = F(x_4, x_5, x_6)$. As

$$\frac{dF}{dt} = \sum_{i=1}^{6} \frac{\partial F}{\partial x_i} = 0,$$

where f_i, $1 \leq i \leq 6$, denotes the right hand side of the ith equation (1), one obtains that

$$\begin{aligned}\frac{dF}{dt} &= x_1\left(\lambda_{16}x_6\frac{\partial F}{\partial x_5}+\lambda_{51}x_5\frac{\partial F}{\partial x_6}\right)+x_2\left(\lambda_{62}x_6\frac{\partial F}{\partial x_4}+\lambda_{24}x_4\frac{\partial F}{\partial x_6}\right)+\\ &\quad x_3\left(\lambda_{35}x_5\frac{\partial F}{\partial x_4}+\lambda_{43}x_4\frac{\partial F}{\partial x_5}\right)= \\ &\quad x_1Y_1(F)+x_2Y_2(F)+x_3Y_3(F)=0.\end{aligned} \tag{4}$$

Since $Y_1(F)$, $Y_2(F)$ and $Y_3(F)$ do not depend on x_1, x_2 and x_3 and since (4) is an identity with respect to all the variables, then

$$Y_1(F) = Y_2(F) = Y_3(F) = 0. \tag{5}$$

Equations (5) give a system of three first order linear homogeneous partial differential equations for determining the function F. This function depends only on three variables x_4, x_5 and x_6 and $\frac{\partial F}{\partial x_i}$, $4 \leq i \leq 6$, do not vanish identically on any open subset of $R^3 = R^3\{x_4, x_5, x_6\}$. Therefore the determinant of the linear system (5) with respect to the unknowns $\frac{\partial F}{\partial x_i}$, $4 \leq i \leq 6$, vanishes identically for all values of x_4, x_5 and x_6. This means that

$$\det \begin{bmatrix} 0 & \lambda_{16}x_6 & \lambda_{51}x_5 \\ \lambda_{62}x_6 & 0 & \lambda_{24}x_4 \\ \lambda_{35}x_5 & \lambda_{43}x_4 & 0 \end{bmatrix} = (\lambda_{43}\lambda_{51}\lambda_{62} + \lambda_{16}\lambda_{24}\lambda_{35})x_4x_5x_6 = 0.$$

and thus

$$\lambda_{43}\lambda_{51}\lambda_{62} + \lambda_{16}\lambda_{24}\lambda_{35} = 0. \tag{6}$$

As it is easy to see, (6) coincides with Manakov condition (3). Thus we proved that Manakov condition is a necessary one for the existence of a non-trivial first integral $F = F(x_4, x_5, x_6)$. Let us stress the incredible simplicity of this derivation of the Manakov condition.

The explicit form of the fourth integral is given in the following table, which cover all the space of parameters $(\lambda_i)_{1\leq i\leq 6}$ satisfying the Manakov condition. The last column of the Table gives the necessary and sufficient conditions for functional independence of first integrals. When many conditions are given, they have to be satisfied simultaneously.

Case	First integrals	Condition
Generic	$H_1, H_2, H_3,$ $\lambda_{16}\lambda_{24}x_4^2 + \lambda_{51}\lambda_{62}x_5^2$ $-\lambda_{16}\lambda_{62}x_6^2$	$\lvert\lambda_{16}\rvert + \lvert\lambda_{26}\rvert > 0,$ $\lvert\lambda_{16}\rvert + \lvert\lambda_{15}\rvert > 0,$ $\lvert\lambda_{24}\rvert + \lvert\lambda_{26}\rvert > 0,$ $\lvert\lambda_{12}\rvert + \lvert\lambda_{13}\rvert > 0$
$\lambda_1 = \lambda_2 = \lambda_6$	$H_1, H_2, H_3, x_3^2 + x_4^2 + x_5^2$	$\lvert\lambda_{34}\rvert + \lvert\lambda_{45}\rvert > 0$
$\lambda_1 = \lambda_2 = \lambda_6, \lambda_3 = \lambda_4 = \lambda_5$	H_1, x_3, x_4, x_5	
$\lambda_1 = \lambda_5 = \lambda_6$	$H_1, H_2, H_3, x_1^2 + x_5^2 + x_6^2$	$\lvert\lambda_{23}\rvert + \lvert\lambda_{34}\rvert > 0$
$\lambda_1 = \lambda_5 = \lambda_6, \lambda_1 = \lambda_3 = \lambda_4$	H_1, x_1, x_5, x_6	
$\lambda_2 = \lambda_4 = \lambda_6$	$H_1, H_2, H_3, x_2^2 + x_4^2 + x_6^2$	$\lvert\lambda_{13}\rvert + \lvert\lambda_{35}\rvert > 0$
$\lambda_2 = \lambda_4 = \lambda_6, \lambda_1 = \lambda_3 = \lambda_5$	H_1, x_2, x_4, x_6	
$\lambda_2 = \lambda_3 = \lambda_4$	$H_1, H_2, H_3, x_1^2 + x_5^2 + x_6^2$	$\lvert\lambda_{15}\rvert + \lvert\lambda_{56}\rvert > 0$
$\lambda_2 = \lambda_3 = \lambda_4, \lambda_1 = \lambda_5 = \lambda_6$	H_1, x_1, x_5, x_6	
$\lambda_1 = \lambda_2 = \lambda_3$	$H_1, H_2, H_3,$ $\lambda_{51}\lambda_{46}x_1^2 + \lambda_{24}\lambda_{65}x_2^2$ $-\lambda_{46}\lambda_{65}x_6^2$	$\lvert\lambda_{14}\rvert + \lvert\lambda_{15}\rvert > 0,$ $\lvert\lambda_{14}\rvert + \lvert\lambda_{16}\rvert > 0,$ $\lvert\lambda_{15}\rvert + \lvert\lambda_{16}\rvert > 0,$ $\lvert\lambda_{45}\rvert + \lvert\lambda_{56}\rvert > 0$
$\lambda_1 = \lambda_2 = \lambda_3, \lambda_1 = \lambda_5 = \lambda_6$	H_1, x_1, x_5, x_6	

Let us note that the results of this table remain valid also when the parameters $(\lambda_i)_{1\leq i\leq 6}$ are complex.

PARTIAL FIRST INTEGRALS

We report here two cases of partial integrability of the system (1). We look for functions P (partial first integral) such that the manifold

$$\{x \in R^6; H_1(x) = P(x) = 0\} \tag{7}$$

is built from the whole orbits of the system (1), i.e. (7) is an invariant manifold.

The cases of partial integrability reported below require some restrictions on $(\lambda_i)_{1\leq i\leq 6}$. Let us note that except the trivial case $\lambda_1 = \lambda_2 = \lambda_3 = \lambda_4 = \lambda_5 = \lambda_6$, the Manakov condition (3) is never satisfied here. As far as we know these partial integrals are new.

Case 1

$$\lambda_1 = -\lambda_4 + \lambda_5 + \lambda_6, \quad \lambda_2 = \lambda_4 - \lambda_5 + \lambda_6, \quad \lambda_3 = \lambda_4 + \lambda_5 - \lambda_6.$$

Suppose that $\lambda_4 > \lambda_5 > \lambda_6$ and let us consider the real functions

$$P_{1\epsilon} = \lambda_{56}x_4^2 + \lambda_{46}x_5^2 + \lambda_{45}x_6^2 + 2\epsilon\sqrt{\lambda_{56}\lambda_{45}}x_4x_6,$$

where $\epsilon = \pm 1$. Then the manifolds (7) with $P = P_{1\epsilon}$ are invariant manifolds for system (1).

Case 2

$$\lambda_3 = \lambda_1 + \lambda_2 - \lambda_6, \quad \lambda_4 = -\lambda_1 + \lambda_2 + \lambda_6, \quad \lambda_5 = \lambda_1 - \lambda_2 + \lambda_6.$$

Suppose that $\lambda_1 > \lambda_2 > \lambda_6$ and let us consider the real function

$$P_{2\epsilon} = \lambda_{26}x_1^2 + \lambda_{16}x_2^2 + \lambda_{12}x_6^2 + 2\epsilon\sqrt{\lambda_{26}\lambda_{12}}x_1x_6,$$

where $\epsilon = \pm 1$. If $P = P_{2\epsilon}$ then (7) are invariant manifolds for system (1).

The detailed description of our results together with the proofs will appear elsewhere.

ACKNOWLEDGEMENTS

Elie Youndje from the Department of Mathematics of the University of Rouen helped us in formal algebra manipulations during the first stage of the preparation of this work.

Alfred Ramani from the Center of Theoretical Physics of the Ecole Polytechnique (Palaiseau, France) and Andrzej J. Maciejewski from Astronomical Institute of Toruń University (Poland) helped us in the final stage of the preparation of this work.

We thank sincerely all of them.

The first author acknowledges the Department of Mathematics of the University of Rouen as well as the URA CNRS 1378 for their hospitality and excellent working conditions during his stay in Rouen. In particular he thanks sincerely Gerard Grancher for his continuous help and attention.

For the first author this work was supported by the fellowship ERB3510 PL922540 from the European Community's Action for Cooperation in Science and Technology with Central and Eastern European Countries.

REFERENCES

1. M.J. Ablowitz, A. Ramani, H. Segur, *Nonlinear Evolution Equations and Ordinary Differential Equations of Painleve type*, Lett. Nuov. Cim., 23, 333–338 (1978).
2. M.J. Ablowitz, A. Ramani, H. Segur, *A Connection Between Nonlinear Evolution Equations and Ordinary Differential Equations of P-type I*, J. Math. Phys., 21, 715–721 (1980).
3. M. Adler and P. van Moerbeke, *The algebraic integrability of geodesic flow on SO(4)*, Invent. Math., 67, 297–331 (1982).
4. Yu.A. Arkhangelskii, *Analytical dynamics of the rigid body* (in Russian), Nauka, Moscow (1977).
5. V.I. Arnold, *Mathematical Methods of Classical Mechanics*, Graduate Texts in Math. 60, 2nd edition, Springer-Verlag, Berlin (1989).
6. O.I. Bogoyavlensky, *Integrable Euler Equations on Six-Dimensional Lie Algebras*, Doklady Akad. Nauk USSR, (in Russian) 268, 11–15 (1983). English transl. Soviet Math. Doklady, 27, N1, 1–5 (1983).
7. A.T. Fomenko, *Integrability and nonintegrability in geometry and mechanics*, Dodrecht: Kluwer Academic Publishers (1988).
8. L. Haine, *Geodesic flow on SO(4) and abelian surfaces*, Math. Ann., 263, 435–472 (1983).
9. L. Haine, *The algebraic complete integrability of geodesic flow on SO(N)*, Commun. Math. Phys., 94, 271–287 (1984).
10. S.V. Kovalevskaya, *Sur le probleme de la rotation d'un corps solide autour d'un point fixe*, Acta Math., 12, 177–232 (1889).
11. S.V. Kovalevskaya, *Sur une propriete du systeme d'equations differentielles qui definit la rotation d'un corps solide autour d'un point fixe*, Acta Math., 14, 81–93 (1880).
12. P.J. Olver, *Applications of Lie groups to differential equations*, Graduate Texts in Math. 107, Springer-Verlag, Berlin (1986).
13. A.M. Perelomov, *Integrable systems of classical mechanics and Lie algebras. The motion of rigid body with fixed point* (in Russian), Preprints of Inst. of Theor. Exper. Phys., Moscow 147 (1983).
14. A.M. Perelomov, *Integrable systems of classical mechanics and Lie algebras.* Vol. I, Birkhauser Verlag, Basel (1990).
15. S.I. Popov, *On the existence of depending on p, q, r, γ first integral of the Euler-Poisson system*, Theoretical and Applied Mechanics. (Bulgarian Academy of Sciences), (in Bulgarian), 2, 28–33 (1981).
16. S.I. Popov, *On the non-existence of a new first integral $F(p,q,r,\gamma,\gamma') = const$ of the problem of a heavy rigid body motion about a fixed point*, Reports of Bulgarian Academy of Sciences, (in Russian), 38 (5), 583–586 (1985).
17. S.I. Popov, *On the non-existence of a new first integral $F(p,q,r,\gamma,\gamma') = const$ of the problem of a heavy rigid body motion about a fixed point*, Theoretical and Applied Mechanics. (Bulgarian Academy of Sciences), (in Russian), 4, 17–23 (1988).
18. S.I.Popov, *On the motion of a heavy rigid body about a fixed point*, Acta Mechanica, 85, 1–11 (1990).

INTEGRABLE HAMILTONIAN SYSTEMS AND POISSON ACTIONS WITH SIMPLE SINGULAR POINTS

Lev M. Lerman[1] * and Yan L. Umanskii[2] **

[1]Research Institute for Applied Mathematics & Cybernetics
10, Ul'yanov str., Nizhny Novgorod 603005, Russia
[2]Department of Mathematics, Agricultural Institute
97, Gagarin ave., Nizhny Novgorod 603078, Russia

INTRODUCTION

The integrable Hamiltonian systems possessing singular points play a specific role in applications to the nonlinear equations of mathematical physics. Especially it concerns separatrix sets of singular points because they relate with a description of soliton solutions of such equations. On the other hand these sets are used in various perturbation methods like the Mel'nikov method for detecting nonintegrability and chaos.

The aim of this paper is to describe a structure of an integrable Hamiltonian vector field (IHVF) in some invariant under flow subsets of a phase space. Each such a subset will contain a singular point p of the vector field and all its orbits for which p belongs to their limit set. In general, a structure of IHVF may be complex enough. So we have to restrict ourselves with some class of "simple" IHVFs which are nevertheless sufficiently generic. In particular, we need in some notion of a generic singular point, periodic orbit, etc. For such IHVFs in the case of two degrees of freedom we shall give a complete description of the orbit structure, shall study a topology of levels of a Hamiltonian. Some results on IHVFs with more than two degrees of freedom will be also presented. Proofs of all results formulated below have been given in our papers [1,2,3].

Let M be a C^∞-smooth $2n$-dimensional symplectic manifold, $H \equiv H_1$ be a C^∞-smooth function on M (Hamiltonian), X_H be a Hamiltonian vector field and $H_2, ..., H_n$ be additional integrals so that a set of functions $H_1, ..., H_n$ is involutive (that is, $\{H_i, H_j\} \equiv 0$, $\{.,.\}$ being a Poisson bracket), and $dH_1, ..., dH_n$ are independent within open dense subset of M. In this case we call the vector field X_H to be IHVF and denote it as $(X_H, H_2, ..., H_n)$.

Any IHVF $(X_H, H_2, ..., H_n)$ generates an Poisson action Φ of the group $\mathbf{R}^n$ on M [4], namely, $\Phi(t_1, ..., t_2; x) = f_1^{t_1} \circ \cdots \circ f_n^{t_n}(x)$ where $f_i^{t_i}$ is a flow of the vector field X_{H_i} (of course, we suppose X_{H_i} to have a complete flow). This action is a Poisson one. It means that a restriction of Φ to any one-parameter subgroup $s \cdot t$ of $\mathbf{R}^n$, $s \in \mathbf{R}$, $t \in \mathbf{R}^n$, gives a Hamiltonian flow on M and their Hamiltonians are linear dependent on t. Each orbit of the action is an invariant

* This research was partially supported by the Committee of Higher Education of Russia under the grant 2-13-01-20

**This research was partially supported by the Committee of Higher Education of Russia under the grant 2-13-01-20 and by the Russian Fund of Fundamental Researches under the grant 93-011-1407 subset of all vector fields X_{H_i} so a study of a foliation into action orbits is very useful for studying IHVF under consideration. The action we call the induced Poisson action for the related IHVF. On the contrary, every Poisson action Φ of the group $\mathbf{R}^n$ on M determines a IHVF if we choose n linear independent vectors in $\mathbf{R}^n$ and consider corresponding commuting Hamiltonian vector fields.

Let p be an isolated singular point of an IHVF X_H, $X_H(p) = 0$. It is an easy task to show that p is a singular point of all vector fields X_{H_i}. Denote $L_1, ..., L_n$ operators of a linearization of vector fields $X_{H_1}, ..., X_{H_n}$ at the point p. Each operator $L_1, ..., L_n$ determines a linear Hamiltonian vector field $\dot{x} = L_i x$ on a tangent space T_pM to the point p. Their flows are commuting and generate linear Poisson action (LPA) of the group $\mathbf{R}^n$ on T_pM. In fact the action itself is defined by a commutative algebra $\mathcal{A}$ coinciding with a linear span of $L_1, ..., L_n$ with real coefficients. Conversely, each commutative algebra of linear Hamiltonian operators $L : T_pM \to T_pM$ generates a LPA on T_pM. So we can define a topology in a space of LPAs inducing it from the space of n-dimensional planes in the space of linear Hamiltonian operators. Now we are able to select a generic class of LPA.

DEFINITION 1. LPA is a simple one if its algebra $\mathcal{A}$ is a n-dimensional and contains an operator L with simple eigenvalues.

The meaning of this definition displays the following proposition that has been proved by us for $n = 2$ and is valid as we think for any n.

PROPOSITION 1. The set of simple LPAs forms an open dense subset in a space of all LPAs.

This result justifies the following definition.

DEFINITION 2. A singular point of IHVF is called to be simple one if the algebra $\mathcal{A}$ generated by linearization operators $L_1, ..., L_n$ is n-dimensional one and the operator L_1 has simple eigenvalues.

Now let us pass to the case $n = 2$. Denote K an additional integral for X_H and let (X_H, K) be an IHVF. Let p be a simple singular point of X_H. Without loss of generality we may assume $H(p) = K(p) = 0$. In this case in some neighborhood of the point p functions H, K in some symplectic coordinate frame (x_1, x_2, y_1, y_2) may be written in the form [5]

$$H = h(\xi, \eta) = \lambda_1\xi + \lambda_2\eta + \cdots, K = k(\xi, \eta) = \nu_1\xi + \nu_2\eta + \cdots \quad (1)$$

where dots mean terms of order greater than two in (x_1, x_2, y_1, y_2) and

$$\xi = (x_1^2 + y_1^2)/2, \ \eta = (x_2^2 + y_2^2)/2 \quad (elliptic\ point)$$

$$\xi = x_1 y_1, \ \eta = (x_2^2 + y_2^2)/2 \quad (saddle-center)$$

$$\xi = x_1 y_1, \ \eta = x_2 y_2 \quad (saddle)$$

$$\xi = x_1 y_1 + x_2 y_2, \ \eta = x_1 y_2 - x_2 y_1 \quad (saddle-focus)$$

It is worth noting that a condition of simplicity of a singular point in this coordinates means an inequality $\Delta = \lambda_1 \cdot \nu_2 - \lambda_2 \cdot \nu_1 \neq 0$ to fulfil.

Let now γ be a periodic orbit of X_H. Such an orbit is called *specific* one (SPO) if vector fields X_H and X_K are collinear along γ. Let us make a reduction of X_H by means of integral H (since $X_H(\gamma) \neq 0$) [6]. Then we obtain a family of Hamiltonian systems with one degree of freedom depending on a parameter h ($H = h$ determines a level of Hamiltonian H). If $h = 0$ the system has a singular point which corresponds to γ. In general case a family of reduced Hamiltonians has the following versal forms:

$$1)\ h + x^2 + y^2, \qquad 2)\ h + x^2 - y^2, \qquad 3)\ hx - y^2 + x^3$$

We call SPO a simple one iff a corresponding versal family is of one of this forms. The SPO will be called elliptic, hyperbolic or parabolic one for cases 1), 2) or 3) respectively. A hyperbolic SPO may be itself orientable or nonorientable depending on topology of its stable or unstable manifolds (cylinders or Möbius strips).

In what follows we consider IHVF possessing the following properties:

1. A closure of any orbit is a compact set.
2. The invariant set V under consideration has only one simple singular point.
3. All SPOs are simple, a set of parabolic SPOs is finite.

To study a structure of IHVF we need in some equivalence relation. The generally used definition of topological equivalence of two vector fields is too rigid for our purposes because it leads to continuous invariants of a type of rotation numbers on Liouville tori. So we have introduced the following definition.

DEFINITION 3. Two IHVFs (X_H, K) and $(X_{H'}, K')$ given on M are isoenergetically equivalent iff there exists such a homeomorphism $g: M \to M$ which transforms each connected component of a level of Hamiltonian H onto that of H' and it transforms each component of joint level of function H and K onto that for H' and K'.

Below we consider a class of Poisson actions which generate IHVFs satisfying conditions 1-3 and use a commonly used definition of topological equivalence of actions (see, for instance [4]).

Let now (X_H, K), $(X_{H'}, K')$ be two IHVFs, p and p' be their singular points of the same type. Throughout below we denote $V_\epsilon = V \cap H = \epsilon$.

1. ELLIPTIC SINGULAR POINT

In this case invariant set V is merely a four dimensional ball small enough containing a singular point because here there are no orbits of IHVF which have a point p in their limit set.

THEOREM 1. There exist invariant neighborhoods V and V' for singular points p, p', respectively, such that (X_H, K) isoenergetically equivalent to $(X_{H'}, K')$ iff $sgn\lambda_1\lambda_2 = sgn\lambda_1'\lambda_2'$ where λ_i, λ_i' are coefficients in (1).

Without loss of generality we may regard $\lambda_1 > 0$, $\lambda_1' > 0$.

PROPOSITION 2. Let $\lambda_1\lambda_2 > 0$ (case a). For $\epsilon < 0$ a set V_ϵ is empty, $V_0 = \{p\}$, V_ϵ is diffeomorphic to a 3-sphere for $\epsilon > 0$. If $\lambda_1\lambda_2 < 0$ (case b) then V_ϵ is diffeomorphic to a solid torus if $\epsilon \neq 0$ and to a cone above torus [1], if $\epsilon = 0$.

The structure of flow X_H on levels V_ϵ is very simple. In case a) each sphere contains two elliptic SPOs linked with the link coefficient 1. Remaining orbits belong to Liouville tori which foliate the 3-sphere. In case b) a solid torus contains one elliptic SPO (the axial circle), the remaining set is foliated with Liouville tori. For $\epsilon = 0$ the set V_0 is obtained from V_ϵ by means of identification of the axial circle in a solid torus to the point p.

THEOREM 2. Two actions Φ_1, Φ_2 with elliptic singular points are always topologically equivalent in some invariant neigbourhoods of its singular points.

These results are generalized to the case $n \geq 3$. Here a simple singular point of IHVF $(X_H, H_2, ..., H_n)$ is called simple elliptic one if additionally to the conditions of definition 2 all eigenvalues of the operator L_1 are pure imaginary. A quadratic part of H (and of other functions $H_2, ..., H_n$) can be written in the form

$$\omega_1(x_1^2 + y_1^2)/2 + \cdots + \omega_n(x_n^2 + y_n^2)/2$$

Let $r = \min(r_1, r_2)$ where r_1 is a number of positive coefficients among $\omega_1, ..., \omega_n$, $r_2 = n - r_1$. We call r the index of a simple elliptic point p.

THEOREM 1′. Two IHVFs are isoenergetically equivalent in some invariant neighborhoods of their simple elliptic singular points p, p', respectively, iff indexes of the points p, p' coincide.

---

[1] It means a set $T^2 \times I$ in which a boundary $T^2 \times \{1\}$ is identified to a point.

To describe a topology of levels V_ϵ we assume without loss of generality that numbers $\omega_1, ..., \omega_r$ are positive, where r being index of the simple elliptic singular point p of an IHVF. Then we have

PROPOSITION 2'. There is a positive constant h_* such that: 1) a manifold $V_\epsilon, -h_* < \epsilon < 0$, is homeomorphic to a $(2n-1)$-dimensional sphere S^{2n-1}, $V_0 = p$, if $r = 0$; 2) for $r > 0$ a manifold V_ϵ, $-h_* < \epsilon < 0$, is homeomorphic to a $D^{2r} \times S^{2(n-r)-1}$, V_0 is homeomorphic to a cone above $S^{2r-1} \times S^{2(n-r)-1}$, and V_ϵ, $h_* > \epsilon > 0$, is homeomorphic to a $S^{2r-1} \times D^{2(n-r)}$, where D^l being a closed l-dimensional ball.

As for Poisson actions with simple elliptic singular points the theorem 2 is also valid for $n > 2$.

2. SADDLE - CENTER

In this case there exist only two one-dimensional orbits containing singular point p in their closure. These orbits are homoclinic to p and form two loops. Since $H(p) = 0$ we derive $H = 0$ on each loop.

PROPOSITION 3. For $|\epsilon|$ small enough there exist two symplectic two-dimensional submanifolds Σ_1, Σ_2 which are transversal to each other at p and contain all SPOs passing near p. The submanifold Σ_2 is a disk foliated with hyperbolic SPOs which lay in levels $H = h$ for $h \geq 0$ only (or $h \leq 0$ only). The submanifold Σ_1 is diffeomorphic to a disk with two disks removed (two holes). Σ_1 contains two loops, each loop surrounds one hole. Two loops divide Σ_2 into two parts corresponding to $H > 0$ and $H < 0$, respectively. One part is diffeomorphic to an annulus foliated with elliptic SPOs ("long" SPOs). Another part is diffeomorphic to disjoined union of two annuli foliated with elliptic SPOs which lie by pairs in each level of H ("short" SPOs).

For definiteness let Hamiltonian H be positive in Σ_2. Then for $|\epsilon|$ small enough we have two cases:

A. The level $H = \epsilon$, $\epsilon > 0$, contains one hyperbolic SPO and one long elliptic SPO.

B. The level $H = \epsilon$, $\epsilon > 0$, contains one hyperbolic and two short elliptic SPOs.

THEOREM 3. There exist invariant neighborhoods V, V' containing their own loops that IHVFs (X_H, K) and $(X_{H'}, K')$ are isoenergetically equivalent iff for each IHVF the same case (A or B) takes place.

PROPOSITION 4. In the case A the sets V_ϵ are diffeomorphic to: 1) *a* connected sum of two solid tori, for $\epsilon > 0$; 2) *a* bouquet of two solid tori glued by their inner points, for $\epsilon = 0$; 3) *a* union of two disjoint solid tori, for $\epsilon < 0$. In the case B the sets V_ϵ are diffeomorphic to: 1) a connected sum of solid torus and $S^2 \times S^1$ (S^n being n-sphere) for $\epsilon > 0$; 2) a solid torus with two inner points identified for $\epsilon = 0$; 3) a solid torus for $\epsilon < 0$. The boundary ∂V of V is homeomorphic to a connected sum of two copies of $S^2 \times S^1$.

THEOREM 4. Two actions Φ_1, Φ_2 with saddle-center singular points are always topologically equivalent in some invariant neighborhoods of their loops.

3. SADDLE

Let p be a saddle singular point. From conditions 1 – 3 for IHVF under consideration it follows that all orbits which belong to the stable manifold of p simultaneously lie in its unstable manifold, i.e. they are homoclinic ones. A closure W of the stable manifold is a two-dimensional cellular complex. Among homoclinic orbits on W there exist exactly four orbits $\gamma_1 - \gamma_4$ such that fields X_H and X_K are collinear along these orbits. Such orbits we call loops. As it is proved in the paper[7] stable and unstable manifolds of p intersect transversally along loops within the three- dimensional level $H = 0$. This allows us to introduce the notion of orientability and nonorientability of a loop.

PROPOSITION 5. There exist exactly two 2-dimensional symplectic manifolds Σ_1, Σ_2 intersecting transversally to each other at p. These Σ_1, Σ_2 contain four loops and all SPOs passing near p. Two of these loops belong to Σ_1 and two others belong to Σ_2. Other orbits in Σ_1, Σ_2 are hyperbolic SPOs and the orbit structure of the field X_H in Σ_1, Σ_2 is the same as for the manifold Σ_1 in Proposition 3.

PROPOSITION 6. Only four cases may occur: 1) all loops are orientable; 2) all loops are nonorientable; 3) three loops are orientable and one loop is nonorientable; 4) two loops in Σ_i are orientable and another two other loops in Σ_j are nonorientable, $i \neq j$.

The set of loops is not a full invariant of isoenergetical equivalency, nevertheless the following assertion is valid.

THEOREM 5. Two Poisson actions Φ_1, Φ_2 with saddle singular points are topologically equivalent in some invariant neighborhoods of W, W', respectively, iff the same case from Proposition 6 takes place for this actions.

For to find conditions of isoenergetical equivalence we need to know more exactly about structure of orbits of X_H in levels $H = \epsilon$. As it follows from Proposition 5 only two possibilities may occur:

A. The level $H = \epsilon \neq 0$ contains four short SPOs, level $H = -\epsilon$ contains two long SPOs.

B. Both levels $H = \epsilon$ and $H = -\epsilon$ contain three SPOs (two short and one long SPOs).

Thus along with Proposition 6 we have 8 cases nA, nB, n=1,...,4.

THEOREM 6. There exist invariant neighborhoods $V \supset W$, $V' \supset W'$ such that IHVFs (X_H, K) and $(X_{H'}, K')$ are isoenergetically equivalent iff for each IHVF the same case (nA or nB) takes place.

To describe topology of V_ϵ and its foliations by the level surfaces of the function K we shall present some construction. Topology of V_ϵ can be depicted by means of the invariant introduced in the paper[8] but for our case we use a somewhat another way.

Let N be a smooth two-dimensional manifold, f be a smooth function on N and Q be a connected compact component of the level line $f = 0$. We denote B connected submanifold of N which is determined by inequality $| f | \leq C$ for C small enough and $Q \subset B$. Additionally we assume that some smooth involution G ($G^2 = id$) acts on B. Let us define a three-dimensional manifold P with a boundary being a bundle with the base S^1 (a circle), the leaf B and the group $\{id, G\}$ acting on B as the structural group of the bundle (if $G = id$ then P is the direct product $S^1 \times B$). This manifold P possesses a foliation generated by the function F being a natural continuation of f to P. Henceforth the manifold P with its foliation we shall call the model manifold.

Let $k = k_i(\epsilon), i = 1, 2$, be smooth functions which define the bifurcation diagram, i.e. the image of Σ_i under the mapping $z \longrightarrow (\epsilon, k) = (H(z), K(z)), z \in V$. Then the curve $k = k_m(\epsilon) = (k_1(\epsilon) + k_2(\epsilon))/2$ in the plane (ϵ, k) is situated between graphs of $k = k_1(\epsilon)$ and $k = k_2(\epsilon)$. It is readily checked that V_ϵ is divided into two pieces V_ϵ^1, V_ϵ^2 by the level $K = k_m(\epsilon)$. For each piece V_ϵ^i we build the model manifold and point out how to glue these manifolds. For $\epsilon \neq 0$ this construction is close to that in the paper.[9] The connected component of V_ϵ^i, $\epsilon \neq 0$, may be represented by means of the following five model manifolds:

$$P_1: \ N = \Re^2, \ f = y^2 - x^2(1 - x^2), \ C = 1/4, \ G = id$$

$$P_2: \ N = \Re^2, \ f = y^2 - (1 - x^2)^2(4 - x^2), \ C = 1, \ G = id$$

$$P_3: \ N = S^2 = \{(x, y, z) \mid x^2 + y^2 + z^2 = 1\}, \ f = yz, \ C = 1/2, \ G = id$$

$$P_4: \ N = \Re^2, \ f = y^2 - x^2(1 - x^2), \ C = 1/4, \ G: (x, y) \rightarrow (-x, -y)$$

$$P_5: \ N = S^2 = \{(x, y, z) \mid x^2 + y^2 + z^2 = 1\}, \ f = yz, \ C = 1/2,$$

$$G: \ (x, y, z) \longrightarrow (x, -y, -z)$$

Each manifold P_i will be symbolically depicted with its graph (see fig.1). The vertices of that graph correspond to boundary tori $F = \pm C$. Two vertices are joined with an edge if there is a trajectory of the vector field -*grad* F passing through corresponding tori. Under

such an agreement P_4 and P_5 would have the same graph. To distinguish them we use two edges for P_5.
Completely the manifold V_ϵ is built by means of gluing of model manifolds along its boundary tori in accordance with fig.2. This gluing is rigidly defined by a behaviour of trajectories of the vector field Y=-*grad* K on V_ϵ. The set V_0 is a submanifold with a singularity in the point p which is also a singular point for the field Y. All the points of loops $\gamma_1 - \gamma_4$ are singular points of Y and stable manifolds of loops intersect each level $K = k > 0$ along some set of smooth closed curves. Under $k \to +0$ these curves stick into $\cup\gamma_i$ and corresponding Liouville tori $K = k, H = 0$ turn into W. For $k < 0$ the similar deformation occurs along unstable manifolds of loops $\gamma_1 - \gamma_4$.

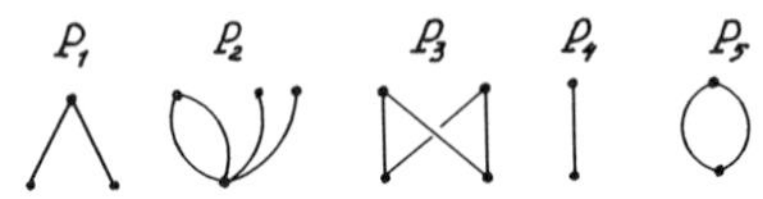

Figure 1. Graphs of model manifolds $P_1 - P_5$

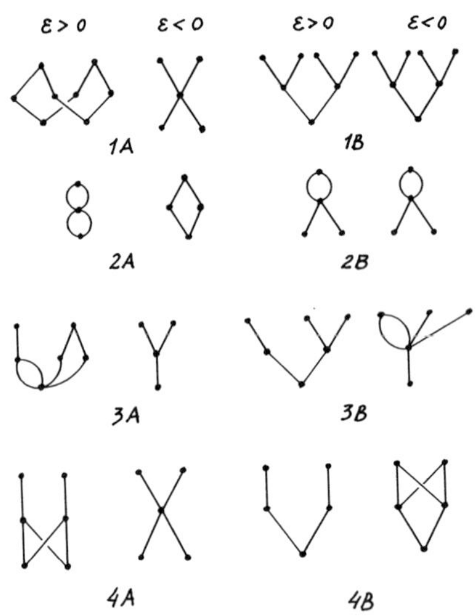

Figure 2. Graphs of complete manifolds V_ϵ for the saddle

4. SADDLE-FOCUS

As follows from (1) there exists such a neighborhood U of p where the fields X_H and X_K are linear independent (excepting point p). Within U all orbits of X_H which contain p in their limit set belong to the stable manifold W^s or the unstable manifold W^u of a point p. Under continuation along orbits the manifold W^u can merge with W^s or with a part of the stable manifold $W^s(\gamma)$ of some SPO γ lying outside of U in the same level of Hamiltonian. Such an SPO must be hyperbolic or parabolic. According with these possibilities we shall consider the following cases:

1. W^u merges with W^s.
2. W^u merges with a component of the set $W^s(\gamma)\backslash\gamma$ of a hyperbolic orientable SPO γ.
3. W^u merges with $W^s(\gamma)$ of a hyperbolic nonorientable SPO γ.
4. W^u merges with $W^s(\gamma)$ of a parabolic SPO γ.

For sake of simplicity we assume further the unstable manifold of γ to merge with W^s in cases 3 and 4. As for case 2 we suppose W^s merges with one component of $W^u(\gamma)\backslash\gamma$ and another component of $W^u(\gamma)\backslash\gamma$ merges with free component of $W^s(\gamma)\backslash\gamma$. Let W be a closure of the set $W^u(\gamma)$ (case 1) or the set $W^u(\gamma) \cup W^s(\gamma) \cup W^u \cup W^s($ cases $2-4)$.

THEOREM 7. There are invariant neighborhoods $V \supset W$, $V' \supset W'$ such that IHVFs (X_H, K) and $(X_{H'}, K')$ are isoenergetically equivalent iff for each IHVF the same case from $1-4$ takes place.

The same assertions is valid for two Poisson actions with saddle-focus singular points.

Topology of sets V_ϵ and its foliation by levels of the function K is described by means of its model manifolds for each case $1-4$ (see construction in sec.3). Denote P/l factor-space of a manifold P in which some smooth closed curve l belonging P is identified to a point. If P is a model manifold and l belongs to some level surface F $=const$ then F is naturally carried to P/l. This function we also denote F.

Case 1. i) $\epsilon = 0$. Here we take the model manifold P_4. Then $V_0 = P_4/l$ where l corresponds to the circle that is determined by the point $(0,0) \in B$. ii) $\epsilon \neq 0$. Here the model manifold is defined by fixing $f(x, y) = x^2 + y^2 - 1$, $c = 1/2$, $G = id$.

Case 2. i). $\epsilon = 0$. Here $f(x, y) = y^2 - x^2(1-x^2)^2(4-x^2)$, $c = 1/10$, $G(x, y) = (-x, -y)$. Thus we obtain the model manifold P. Then $V_0 = P/l$ where l corresponds to the circle that is determined by the point $(0,0) \in B$. ii) $\epsilon \neq 0$. Then V_ϵ is diffeomorphic to the model manifold P_1.

Case 3. i) $\epsilon = 0$. Here the model manifold is P_5. Then $V_0 = P_5/l$ where l is the circle given by the singular point (1,0,0) of the function f. ii) $\epsilon \neq 0$. Here V_ϵ is diffeomorphic to the model manifold P_4.

Case 4. Topology of V_0 and V_ϵ is the same as in the case 1 but the function f is different from that case. Namely $f(x, y) = y^2 - x^2(1-x^2)^3$, $G : (x, y) \rightarrow (-x, -y)$, for $\epsilon = 0$. Let assume for definiteness that hyperbolic and elliptic SPOs emerge from parabolic one under $\epsilon > 0$. Then we can take $f = y^2 - x^2(1-x)(1/2+x)$ for c small enough, $G = id$. For $\epsilon < 0$ we choose $f = x^2 + y^2 - 1$, $c = 1/2$, $G = id$.

REFERENCES

1. L.M.Lerman, and Ya. L.Umanskii, *Classification of four-dimensional integrable Hamiltonian systems and Poisson actions of* $\Re^2$ *within extended neighborhood of simple singular points.I*, Russian Math. Sbornik. 183 (12):141 (1992) (in Russian)
2. L.M.Lerman, and Ya.L.Umanskii, *Classification of four-dimensional integrable Hamiltonian systems and Poisson actions of* $\Re^2$ *within extended neighborhood of simple singular points.II*, Russian Math. Sbornik. 184 (4):105, (1993) (in Russian)
3. L.M.Lerman, and Ya. L.Umanskii, *Isoenergetical classification of integrable Hamiltonian systems within neighborhood of simple elliptic point*, Math.Zametki (in Russian) (to appear)
4. V.I.Arnold. Mathematical Methods of Classical Mechanics. Springer-Verlag: New York, Heidelberg, Berlin (1978)

5. L.H.Elliasson, *Normal forms for Hamiltonian systems with Poisson commuting integrals-elliptic case*, Comment. Math.Helv. 65:4 (1990)
6. J.R.Marsden, and A.Weinstein, *Reduction of symplectic manifolds with symmetry*, Repts Math. Phys. 5:121 (1974)
7. L.M.Lerman, and Ya.L.Umanskii, *Integrable Hamiltonian systems and Poisson actions*, Selecta Math. Sov. 9:59 (1990) (transl. from Russian text of 1984)
8. A.T.Fomenko, and H.Zieschang, *Topological invariant and equivalency criterion of integrable Hamiltonian systems with two degrees of freedom*, USSR Math. Izvestiya. 54:546 (1990) (in Russian)
9. A.T.Fomenko, *Topology of surfaces of constant energy of integrable Hamiltonian systems and obstacles to integrability*, USSR Math.Izvestiya. 50:67 (1986) (in Russian)

DYNAMICS AND BIFURCATIONS IN TWO-PARAMETER UNFOLDING OF A HAMILTONIAN SYSTEM WITH A HOMOCLINIC ORBIT TO A SADDLE-CENTER

O.Yu. Koltsova*, and Lev M. Lerman** [1]

*Department of Mathematics, Institute for Water Transport Engineers, 5, Nesterov str. Nizhny Novgorod, 603600, Russia
**Research Institute for Applied Math.& Cybernetics, 10, Ul'yanov str., Nizhny Novgorod, 603600, Russia

INTRODUCTION

Our goal in this paper is to present some elements of global behavior of Hamiltonian systems with two degrees of freedom in a vicinity of a homoclinic orbit to a saddle-center equilibrium point. Besides of a pure mathematical interest such structure (i.e., a homoclinic orbit to a saddle-center) is surprisingly often encountered in different problems ranging from celestial mechanics (doubly-asymptotic orbits to collinear libration points[1,2]) to modern problems of mathematical physics.[3]

First numerical investigations of homoclinic orbits to a saddle-center were made in connection with the study of solutions doubly asymptotic to the collinear libration point L_2 in the restricted three-body problem. Some questions concerning with periodic orbits near a homoclinic orbit to a saddle-center for Hamiltonian systems containing a one-degree- of-freedom subsystem, were considered in the survey[4]. All of those systems were reversible that one was permitted to study one-parameter families of such systems. The most complete investigation of n-circuited homoclinic orbits to a saddle-center in a reversible Hamiltonian system was accomplished in the paper[5] where chaotic behavior in levels of a Hamiltonian differ from where an equilibrium point is placed had also been established.

For general Hamiltonian systems preliminary classification of orbits that pass through a neighborhood of a saddle-center was made in Ref.[6] but this study was not completed. In the paper[7] the existence of Poincare homoclinic orbits to each hyperbolic periodic orbit of the Lyapunov family have been proved in a presence of a homoclinic orbit to a saddle-center under generic conditions. Besides, four countable families of periodic orbits accumulating to the homoclinic orbit and lying within the level where a saddle-center is placed were discovered. The presence of Poincare homoclinic orbits was also proved in the restricted circular three-body problem near L_2 libration point[8].

Since the paper [7] is not readily available we give here the main statements of it. Besides, we formulate the theorem of the existence of doubling and tripling for homoclinic orbits to a

[1] This research was partially supported by the Russian Fund of Fundamental Researches under grant $N93-011-1787$.

Hamiltonian Mechanics, Edited by J. Seimenis,
Plenum Press, New York, 1994

saddle-center in a general two-parameter unfolding of a Hamiltonian system with a homoclinic orbit (loop) to a saddle-center. Homoclinic doubling was also established by another method in Ref.[9]. The proofs of this theorem and of other results will be published elsewhere.

PRELIMINARIES AND SETTING OF THE PROBLEM

Let us recall some notions. An equilibrium point p of a two-degree-of-freedom Hamiltonian vector field is said to be a saddle-center if eigenvalues of a corresponding linearization operator are $\pm i\omega$, $\pm\lambda$, $\omega\lambda \neq 0$, ω, $\lambda \in \mathbf{R}$. To simplify the exposition we assume in what follows that a symplectic phase space is $\mathbf{R}^4$ and a Hamiltonians is real analytic. Then by Moser theorem[10] with some supplement by Russmann[11] in a neighborhood of such a point there is a symplectic coordinate frame (x_1, x_2, y_1, y_2) where the Hamiltonian H may be written in the form

$$H(x_1, x_2, y_1, y_2) = h(\xi, \eta) = \lambda\xi + \omega\eta + \ldots, \tag{1}$$

$\xi = x_1 y_1$, $\eta = (x_2^2 + y_2^2)/2$. Suppose now that H smoothly depends on some parameters $\mu \in \mathbf{R}^m$ and each Hamiltonian H_μ has a singular point $p(\mu)$ of the same type so $p(\mu)$ is the smooth function. Making use the symplectic change of variables we may regard the form (1) of H_μ to persist for all μ. Without loss of generality we shall assume $\lambda = 1, \omega = \gamma > 0$. In those coordinates differential equations have the form

$$\dot{x}_1 = -Dh_1 x_1, \; \dot{y}_1 = Dh_1 y_1,$$

$$\dot{x}_2 = -Dh_2 y_2, \; \dot{y}_2 = Dh_2 x_2,$$

where Dh_i is a partial derivative of h with respect to i-th variable. Thus the axis x_1 is the local stable manifold and the axis y_1 is the local unstable one. Below we use the local three-dimensional sections $N^s_\pm = \{x_1 = \pm d_1, \eta \leq \eta_0\}$ and $N^u_\pm = \{y_1 = \pm d_2, \eta \leq \eta_0\}$. These sections are foliated by smooth two-dimensional disks $H = c$. We denote these leaves as $N^s_\pm(c)$, $N^u_\pm(c)$. These disks may be represented as graphs of functions $y_1 = \pm a_c(\eta)/d_1$ (in $N^s_\pm$) and $x_1 = \pm a_c(\eta)/d_2$ (in $N^u_\pm(c)$), where $a_c(\eta)$ is the solution of the equation $h_\mu(\xi, \eta) = c$ in some neighborhood of p, $a_c(\eta) = c - \gamma\eta + \ldots$(we often omit the explicit dependence of μ, when it can not lead to a confusion); functions ξ, η are local integrals of the flow.

Let us suppose a Hamiltonian vector field X_H with a Hamiltonian H has a saddle-center equilibrium point p, $H(p) = 0$, and that there is a homoclinic orbit Γ to p (a loop). Since stable and unstable manifolds of p are one-dimensional and lie in the three-dimensional level $V_0 = \{H = 0\}$ the existence of the loop is the codimension 2 phenomenon that should be studied in a two-parameter unfolding. Of course, in special cases the codimension may be reduced to 1 if we deal with reversible Hamiltonian vector fields, for instance. In such situations a two-parameter unfolding allow us to have the curves in a parameter plane which correspond to vector fields with loops.

We shall consider two-parameter family of Hamiltonians $H_\mu, \mu \in \mathbf{R}^2$, and suppose the Hamiltonian H_0 to have a saddle-center p with a loop Γ. All our considerations will be carried out in some small neighborhood U of Γ. Our global aim is to investigate an orbit behavior of H_μ in U for $|\mu|$ small enough. Of course, this aim is very global so we shall concern with the following questions:

1. Integrability and nonintegrability in U, nonintegrability conditions.
2. Twist maps and orbit structure in $U \cap \{H = 0\}$.
3. Periodic orbits in U, their types.
4. Multi-circuited homoclinic orbits when $|\mu| \neq 0$.

All results have been obtained by means of studying the Poincare map f on sections $N^s_\pm$. f is a composition of two maps: local $T : N^s \to N^u$ and global $S : N^u \to N^s$. It is convenient to restrict T and S to an invariant disk $\{H_\mu = c\}$. For definiteness we suppose Γ to coincide locally with the half-axis $y_1 > 0$. It means we may regard N^u_+ as the section to the unstable manifold W^u_{loc}. In dependence on whether Γ returns to p being locally the half-axis $x_1 < 0$ or $x_1 > 0$ we distinguish two cases A and B. These cases have different topology of levels $V_c = \{H = c\}$[12]. We denote below a two-dimensional disk as D^2, a segment as I, a two-dimensional torus as T^2, a circle as S^1.

Proposition 1. In the case A some neighborhood U of $p \cup \Gamma$ in $\mathbf{R}^4$ be: 1) a disjoint union of a solid tori $D^2 \times S^1$ and a solid cylinder $D^2 \times I$, if $c < 0$; 2) the same as in the item 1 only with two inner point in $D^2 \times I$ identified, if $c = 0$; 3) a connected sum of $D^2 \times S^1$and $D^2 \times I$, if $c > 0$. In the case B we have: 1) a solid cylinder $D^2 \times I$, if $c < 0$; 2) the same as in the item 1 only with two inner points in $D^2 \times I$ identified, if $c = 0$; 3) $D^2 \times I$ in which two inner balls have been cut and their boundary spheres have been glued by means of some orientation preserving diffeomorphism.

As follows from this proposition it makes a sense to study an orbit behavior within the level V_0 only for the case A. For $c \neq 0$ we may study both levels $c > 0$ and $c < 0$ for the case A and only levels $c > 0$ for the case B. In other situations we have no a mechanism for a reinjection of orbits. If we have two homoclinic loops to p, all situations make a sense but it is the codimension-four phenomenon.

The local map $T_c : N^s(c) \to N^u(c)$ can be written in the following form[7]:

$$x_2^1 = x_2^0 \cos \Delta_c(\eta^0) - y_2^0 \sin \Delta_c(\eta^0),\ y_2^1 = x_2^0 \sin \Delta_c(\eta^0) + y_2^0 \cos \Delta_c(\eta^0),$$

where the point (x_2^1, y_2^1) belongs to $N^u(c)$ and (x_2^0, y_2^0) belongs to $N^s_+(c)$ or $N^s_-(c)$, $\Delta_c(\eta) = -a_c'(\eta)\ln(\pm d_1 d_2/a_c(\eta))$. Let us remark that the equation $a_c(\eta) = 0$ has the root $\eta^*(c) = c/\gamma + O(c^2)$. It exists, if $c > 0$, and the circle $\eta = \eta^*(c)$ in $N^s_+(c)$ and $N^s_-(c)$ is a track of the stable manifold of the hyperbolic periodic orbit in V_c from the Lyapunov family.

The global map $S_c : N^u(c) \to N^s(c)$ is some analytic symplectic map (here, area preserving map) smoothly depending on parameters μ:

$$x_2^0 = F_\mu(x_2^1, y_2^1, c), \quad y_2^0 = G_\mu(x_2^1, y_2^1, c),$$

where $\det(D(F_\mu, G_\mu)/D(x_2^1, y_2^1)) \equiv 1$, $F_0(0,0,0) = G_0(0,0,0) = 0$ (the loop condition).

It is worth noting that T_0 is discontinuous at $x_2 = y_2 = 0$ since $a_0(0) = 0$, $a_0'(0) = -\gamma \neq 0$. So $f_0 = S_0 \circ T_0$ is also discontinuous at $x_2 = y_2 = 0$. Changing a little d_1 and d_2 (in dependence on μ) we may regard that $D(F_\mu, G_\mu)/D(x_2^1, y_2^1)$ is equal to $\begin{bmatrix} b & 0 \\ 0 & b^{-1} \end{bmatrix}$ at $x_2 = y_2 = c = 0$.

INTEGRABILITY

Let us assume X_H has an additional real analytic integral K, $\{H, K\} \equiv 0$ in U. Then $dK = 0$ at p and the quadratic part K_2 of K has the form $K_2 = k_2(\xi, \eta) = \kappa\xi + \nu\eta$. If $\nu - \gamma\kappa \neq 0$ then Russmann' results[11] become applicable and it yields K to have locally the same form as H has, $K = k(\xi, \eta) = \kappa\xi + \nu\eta + \ldots$ Restrict K to $N^s(c)$. Then $K = k(a_c(\eta), \eta) = g_c(\eta)$, $g_0(\eta) = k(0,0) = 0$, $g_0'(\eta) = -\gamma\kappa + \nu \neq 0$. So $\eta = 0$ is the critical point for any c small enough (maximum or minimum of K). In particular, K is monotonically changed in direction transverse to circles $\eta = const$. Since K is the integral, S_c conserves K and a circle $\overline{\eta} = \eta_0$ is transformed to itself after mapping $T_c \circ S_c$. Then in symplectic polar coordinates $x_2 = \sqrt{2\eta}\cos(\varphi)$, $y_2 = \sqrt{2\eta}\sin(\varphi)$ we have:

$$T_c : \eta^1 = \eta^0,\ \varphi^1 = \varphi^0 + \Delta_c(\eta^0), \quad S_c : \overline{\eta} = \eta^1,\ \overline{\varphi} = \varphi^1 + \Delta_c(\eta^1)$$

$r_c(\eta^1)$ is a smooth function at η^1=0. As a corollary we obtain b=1 or that $D(F_0,G_0)/D(x_2^1,y_2^1)\,|_{(0,0)}$ is a pure rotation. From Propositiion 1 and such the structure of f_c it is easily obtained the full description of a foliation of U by joint levels of two integrals [12].

Remark. Aforementioned considerations show that a presence of a loop to a saddle-center is a common feature for the integrable Hamiltonian systems. More exactly, if X_H has an unstable one-dimensional separatrix that returns to a neighborhood of the local stable manifold then these manifolds are really merged forming a separatrix loop. Unfortunately this simple necessary condition of integrability is tied to the coordinate frame where H takes the form $h(\xi, \eta)$. An invariant form of conditions $b = 1$ and $b \neq 1$ may be done in terms of the system linearized at the loop.

THE LEVEL CONTAINING A SADDLE-CENTER, $\mu = 0$

Consider first the level $H_\mu = 0$ where the equilibrium point is situated. Here we suppose the case A takes place. So when $\mu = 0$ the map f_0 is a symplectic area-preserving diffeomophism from some punctured disk $0 < \eta \leq \eta^*$ to its image. This diffeomophism may be extended to the homeomorphism of the disk $0 \leq \eta \leq \eta^*$ as $\overline{f}_0$, $\overline{f}(0,0) = (0,0)$. The main property of f_0 is the following assertion

Proposition 2. When $\mu = 0$ the area-preserving map f_0 has the twist property on a punctured disk $0 < \eta \leq \eta^0$, if η^0 is small enough.

To utilize the results of the twist map theory[13,14] we need to assume, in addition, that f_0 has some invariant, nonhomotopic to zero, curve in the punctured disk $0 < \eta \leq \eta_0$. It is the case, if we deal with a Hamiltonian close to an integrable one with a loop to a saddle-center. Then existence of an invariant curve follows from KAM theory. Of course, besides one-circuited periodic orbits there are countably many multi-circuited periodic orbits which correspond to periodic points of the map f_0. Moreover, as follows from results[13,14] the map f_0 has a lot of invariant curves or cantori which correspond to irrational rotation numbers. In our case the "twist" tends to infinity when $\eta \to 0$. In general, we do not know if any closed invariant curve exists. Moreover, simulations show that for $| b |$ large enough none invariant curves exist (see also Proposition 4). So the following proposition[7] has a supplementary meaning.

Proposition 3. If $b \neq 1$ there are four countable families of one-circuited periodic orbits accumulating to the loop within the level $H = 0$ (they correspond to the fixed point of the map f_0).

The periodic orbits obtained in Proposition 3 correspond to integral values of rotation numbers. Since "twist" tends to infinity there are countably many such values. Besides, we have found four periodic orbits instead of common two. Types of these orbits in linear approximation may be determined. Namely, four families $p_n^i = (\eta_n^i, \varphi_n^i), i = \overline{1,4}$, can be enumerated such that $0 < \varphi_n^1 < \frac{\pi}{2} < \varphi_n^2 < \pi < \varphi_n^3 < \frac{3\pi}{2} < \varphi_n^4 < 2\pi \ (\mathrm{mod} 2\pi)$[7] (here we assume $b > 1$). The case $0 < b < 1$ is reduced to the case $b > 1$ by means of a change $\varphi \to \varphi + \frac{\pi}{2}$). Then the following assertion is valid.

Proposition 4. If the quantity $s = \gamma(b - b^{-1})$ belongs to the interval (0,2) then p_n^1, p_n^3 be elliptic points and p_n^2, p_n^4 hyperbolic ones. If $s \in (-2,0)$ then p_n^2, p_n^4 be elliptic points and p_n^1, p_n^3 hyperbolic ones. If $| s |> 2$ all point are hyperbolic fixed points.

MULTI-CIRCUITED HOMOCLINIC ORBITS, WHEN $\mu \neq 0$

Now let us consider the case $\mu \neq 0$. We assume H_μ to be a generic family, that is, in our coordinates it means the track of the unstable separatrix on $N^s(0)$ defines a diffeomorghism from some neighborhood of the point *(0,0)* in the parameter plane μ to a neighborhood of the point *(0,0)* in $N^s(0)$. So instead of parameters $\mu = (\mu_1, \mu_2)$ we take coordinates (u_1, u_2) of the track as new parameters.

The assumption of genericity means, in particular, that if $(\mu_1, \mu_2) \neq 0$ the system does not possess one-circuited homoclinic orbits to p. Besides of its own importance the existence of homoclinic orbits has a strong influence to dynamics in U as we have seen before. So it makes a sense to search for multi-circuited homoclinic orbits. We have not obtained so far such complete results as in the paper[5]. Nevertheless the following is valid

Theorem 1. 1) If $b \neq 1$ there is $\epsilon_0 > 0$ such that in the neighborhood $\| \mu \| < \epsilon_0$ there are four countable families of values $\mu_n^i, i = \overline{1,4}$, $\| \mu_n^i \| \to 0$ when $n \to \infty$, such that the Hamiltonian system H_μ, $\mu = \mu_n^i$ has two-circuited homoclinic orbit to p; 2) if $b \neq 1$ and, in addition, $b \neq \sqrt{(3+\sqrt{5})/2}$, there is $\epsilon_1 > 0$ such that in the neighborhood $\| \mu \| < \epsilon_1$ there are eight countable families of values $\mu_n^i, i = \overline{1,4}$, $\| \mu_n^i \| \to 0$ when $n \to \infty$, such that the Hamiltonian system H_μ, $\mu = \mu_n^i$ has triple-circuited homoclinic orbit to p. When $b > (1+\sqrt{5})/2$ there are other families of μ - values which are corresponded to triple-circuited homoclinic orbits to p.

Apparently as in Ref. [5] it may be proved the following assertion.

Hypothesis. For a generic two-parameter Hamiltonian unfolding of a Hamiltonian system with a loop to a saddle-center with $b \neq 1$ and any integer n there are countably many parameter values accumulating to zero for which the corresponding Hamiltonian system has n-circuited homoclinic orbit to a saddle-center.

Such the structure of multi-circuited homoclinic orbits to a saddle-center suggests that the corresponding bifurcation set in the parameter plane are self-limited and possibly self-similar since near every such a point must exist the same set.

The existence of double and triple homoclinic orbits implies the existence of related periodic orbits. For double periodic orbits it have been proved[9] and may be done for triple periodic orbits.

CHAOTIC BEHAVIOR IN LEVELS $c \neq 0$

What can we say about behavior of orbits to the field X_H in the levels differ from $H = 0$? It turns out the existence of homoclinic loop to a saddle-center implies the complicated behavior within nearby levels of Hamiltonian.

Theorem 2[7]. If $b \neq 1$ then there is $h_* > 0$ such that in any level $H = h, 0 < h \leq h_*$, Lyapunov hyperbolic periodic orbit has four transversal one-circuited homoclinic orbits. In particular, there is a positive number ϵ_0 such that for any $\| \mu \| < \epsilon_0$ the vector field X_{H_μ} has not any analytic additional integral.

The presence of a transversal homoclinic orbit to a hyperbolic periodic orbit implies the existence of a countable set of hyperbolic periodic orbits, a continuum of Poisson stable orbits, etc. within a related level of Hamiltonian.

REFERENCES

1. V.Szebehely, "Theory of Orbits", Acad. Press., New York and London (1967).
2. M.L.Lidov, and M.A.Vashkov'yak, Doubly asymptotic symmetric orbits in the plane restricted circular three-body prob- lem., Preprint N 115, Inst. for Appl. Math., USSR Acad. of Sci.(1975).
3. G.L.Alfimov, V.M.Eleonsky, and N.E.Kulagin, Dynamical systems in the theory of solitons in the presence of nonlocal interactions, *Chaos : Interdisc.J.Nonlin.Sci.* 2:565(1992).
4. R.C.Chirchill, G.Recelli, and D.Rod, A survey of the Henon- Heiles Hamiltonian with applications to related examples, *in*: "Stochastic Behavior in Classical and Quantum Hamiltonian systems", Casati et al., eds, Springer Lect. Notes in Phys., v.93, Springer-Verlag, Berlin, Heidelberg, New York, 76:136 (1979).
5. A.Mielke, P.Holmes, and O.O'Reilly, Cascades of Homoclinic orbits to, and chaos near, a Hamiltonian saddle-center, *J.Dyn.Diff.Eq.* 4:95(1992).

6. J.C.Conley, On the ultimate behavior of orbits with respect to an unstable critical point. 1. Oscilating, asymptotic, and capture orbits, *J.Diff.Eq.* 5:136(1969).
7. L.M.Lerman, Hamiltonian systems with a separatrix loop of a saddle-center, *in:* "Methods of Qualitative Theory of Diff.Eq.", L.P.Shil'nikov, ed., Gorky Univ., Gorky, (in Russian) (translated in *SelectaMath.Sov.* 10:297(1991)).
8. J.Llibre, R.Martinez, and C.Simo, Transversality of the invariant manifolds assotiated to the Lyapunov family of periodic orbits near L_2 in the restricted three-body problem, *J.Diff.Eq.* 58:104(1985).
9. O.Yu.Koltsova, On a structure of a bifurcation set in two- parameter family of Hamiltonian systems with a separatrix loop, *in:* "Methods of Qualitative Theory and Bifurcation Theory", L.P.Shil'nikov, ed., Univ. of Nizhny Novgorod, Nizhny Novgorod (1991) (in Russian).
10. J.Moser, On the generalization of a theorem of A.Liapunoff, *Comm.PureAppl.Math.*, 11:257(1958).
11. H.Russmann, Über das Verhalfen analytischer Hamiltonsher Differential gleichungen in der Nahe einer Gleichgewichtslösung, *Math.An.* 154:285(1964).
12. L.M.Lerman, and Ya.L.Umanskii, Classification of four- dimensional integrable Hamiltonian systems and Poisson actions of $\mathbf{R}^2$ within extended neighborhoods of simple singular points.I, *RussianMath.Sbornik* 183(12):141 (1992) (in Russian).
13. J.Mather, Existence of quasi-periodic orbits for twist homeomorphisms, *Topology* 21:457(1982).
14. S.Aubry, and P.Y.Le Daeron, The discrete Frenkel-Kontorova model and its extensions. 1. Exact results for ground- states, *Physica D* 8:381(1983).

SELF-SIMILAR ISOMONODROMY SOLUTIONS OF NONLINEAR SCHRÖDINGER EQUATION

Mehmet Can

Istanbul Technical University
Mathematics Department
Maslak, 80626 Istanbul, Turkiye

INTRODUCTION

Recent works of a group of Japanese mathematicians M. Sato, T. Miwa, M. Jimbo, K. Ueno, E. Date and others [1-2] have attracted major attention of researchers in various areas of mathematics and mathematical physics. One of the interesting gains of the theory of nonlinear equations of mathematical physics is the concept of "isomonodromy" solutions of nonlinear equations admitting a Lax representation. Isomonodromy solutions are a natural generalization of their self similar solutions. Within the framework of the formalism of the matrix Riemann-Hilbert problem finally adopted, isomonodromy solutions are characterized by the constancy in the "spectral parameter" λ of all junction matrices. This leads to the situation that the function corresponding to an isomonodromy solution satisfies, together with the original $U-V$ system, also a third equation in λ . The data of the Riemann-Hilbert problem here acquire the meaning of monodromy data for the λ-equation. The apriori independence of the monodromy data from the dynamical variables x and t means that the original nonlinear equation describes isomonodromy deformations of the coefficients of the λ-equation. This is the motivation for the term *isomonodromy solution.*

Thus, the problem of constructing isomonodromy solutions of NS equation can be formulated as a problem of finding a matrix-valued function Ψ which also satisfies, together with the given $U-V$ system, a third equation, "the equation in λ", where the rational matrix $A(\lambda)$ is determined from given matrices $U(\lambda)$ and $V(\lambda)$.

In this paper, it is shown that when the λ-equation has one irregular singularity at infinity and one regular singular point at the origin, the isomonodromy solutions of the NS equation are self-similar. It is also proved that the similarity equation of the NS equation is equivalent to the Painlevé IV equation.

Hamiltonian Mechanics, Edited by J. Seimenis,
Plenum Press, New York, 1994

THE MATRIX R-H PROBLEM AND THE MANIFOLD OF ISOMONODROMY SOLUTIONS

For reasons explained in the introduction we begin this section with an exposition of basic facts concerning completely integrable evolution equations. Let us consider the overdetermined system of linear differential equations

$$\Psi_x = U\Psi, \quad \Psi_t = V\Psi, \tag{1}$$

where U and V are matrix-valued functions depending rationally on a complex parameter λ. A completely integrable evolution equation is defined as the consistency condition for system (1):

$$U_t - V_x = [V, U] \quad \text{(identically in } \lambda) \tag{2}$$

Equation (2) which is also called the Lax pair or Zakharov-Shabat equation, is equivalent to a system of nonlinear differential equations for the matrix elements of all matrix-valued coefficients contained in the expansions of the functions U and V in their principal parts. The form of this system is completely determined by the matrix dimensions of the matrices U and V, their pole structure as functions of λ, the gauge chosen, and the reduction conditions imposed. Even the simplest completely integrable evolution equations corresponding to 2×2 matrices of polynomial structure in λ are physically interesting evolution systems. Since in the main text of the paper we propose to study the properties of isomonodromy solutions using one of the simplest examples, it appears reasonable to describe them precisely right now.

Let

$$U = \lambda a + b, \quad V = -2\lambda u - iqr\sigma_3 - c \tag{3}$$

where

$$\sigma_3 = \begin{bmatrix} 1 & 0 \\ 0 & -1 \end{bmatrix}, \quad b = \begin{bmatrix} 0 & q \\ -r & 0 \end{bmatrix}, \quad c = \begin{bmatrix} 0 & iq_x \\ ir_x & 0 \end{bmatrix}$$

Then the consistancy condition (2) of the system (1) yields

$$U_t - V_x = [V, U] \Leftrightarrow \begin{cases} iq_t = q_{xx} + 2q^2 r \\ ir_t = -r_{xx} - 2qr^2 \end{cases} \tag{4}$$

Under the reduction constraint $r = \mp\bar{q}$ system (4) becomes the nonlinear Schrödinger (NS) equation

$$iq_t = q_{xx} + 2aq^2\bar{q}, \quad a = \pm 1 \tag{5}$$

Riemann-Hilbert Problem for NS Equation

For NS equation (4) there is one irregular singular point: it coincides with ∞ and is the point of normalization of the problem. The form of the essential singularity of the Ψ function which is responsible for the structure (3) of the matrices U and V is very simple:

$$\Psi(\lambda) = \hat{\Psi}(x, t, \lambda)\, expT(x, t, \lambda)$$

$$\equiv \left(I + \sum_{k=1}^{\infty} \Psi_k(x,t)\lambda^{-k} \right) \exp\left\{2i\lambda^2 t\sigma_3 - i\lambda x\sigma_3\right\} \lambda^{T^0_\infty}. \tag{6}$$

More precisely the following is asserted: substitution of the formal series (6) into system (1) leads to formulas (3) for the functions $U(\lambda)$ and $V(\lambda)$, where

$$q = 2i(\Psi_1)_{12}, \qquad r = 2i(\Psi_1)_{21}. \tag{7}$$

Thus the Riemann-Hilbert problem corresponding to system (4) is characterized by the following collection of data:

$$\begin{array}{ccccccc} \infty, & a_1 & \dots & a_n, & \Gamma_1 & \dots & \Gamma_m \\ T^0_\infty = -\gamma\sigma_3 & T^0_1 & \dots & T^0_n, & G_1(\lambda) & \dots & G_m(\lambda) \\ -ix\sigma_3 & C_1 & \dots & C_n & & & \\ 2it\sigma_3 & & & & & & \\ C_\infty = I & & & & & & \end{array} \tag{8}$$

We are now in a position to turn the main problem of this section, the definition of the concept of an "isomonodromy solution". To isomonodromy solutions correspond their own special type of solvable Riemann problem characterized by the fact that the data outside of the first column of (8) are independent not only of x and t, but also of the variable λ itself. A consequence of this assumption is the rationality in λ of a "third" logarithmic derivative of the function Ψ, the logarithmic derivative with respect to λ: $(\partial\Psi/\partial\lambda)\Psi^{-1}$. In other words, the Ψ function corresponding to an isomonodromy solution together with the system (1) satisfies a "third" linear equation, "the equation in λ":

$$\Psi_\lambda = A\Psi \tag{9}$$

where, in contrast to $U(\lambda)$ and $V(\lambda)$, $A(\lambda)$ is a rational function with poles at all singular points of $\Psi(\lambda)$. (9) has one irregular singularity which is at infinity:

$$A(\lambda) = A_1\lambda + A_0 + \sum \frac{H_j}{\lambda - a_j} \tag{10}$$

Substituting (6) into (9 and (10), we find that

$$A_1 = 4it\sigma_3, \qquad A_0 = 4it[\Psi_1, \sigma_3] - ix\sigma_3. \tag{11}$$

On the other hand, substitution of (6) into the first equation of the system (1)-(3) leads the relation $-i[\Psi_1, \sigma_3] = U(0)$. Hence, the desired expression for the principal part of the matrix $A(\lambda)$ at infinity in terms of the original $U - V$ pair is given by

$$A_1\lambda + A_0 = 4it\sigma_3\lambda - ix\sigma_3 - 4tU_0 = \begin{pmatrix} 4it\lambda - ix & -4tq \\ 4tr & -4it\lambda + ix \end{pmatrix}. \tag{12}$$

Thus, the problem of constructing isomonodromy solutions of system (2) can be formulated as a problem of finding a matrix-valued function $\Psi(\lambda; x, t)$ which also satisfies, together with the given $U - V$ system, a third equation, "the equation in λ" (9), where the rational matrix $A(\lambda)$ is determined from given matrices $U(\lambda)$ and $V(\lambda)$ in a manner indicated above.

EXPLICIT FORMULAS FOR SOME SELF-SIMILAR ISOMONODROMY SOLUTIONS OF THE NS EQUATION

We shall consider the λ-equation (9) in the absence of regular singularities:

$$\frac{\partial \Psi}{\partial \lambda} = (A_1\lambda + A_0)\Psi \tag{13}$$

In order to obtain formulas for the isomonodromy solution itself, it is necessary to compute the Stokes multipliers of the system (13) as functions of the parameters x, t, q and r. The global solution $\Phi(\lambda)$ of (13) is obtained in [3] in terms of Weber-Hermite functions. Actually, one does not even need to compute the Stokes multipliers themselves. In the case being considered the rank of the singularity at ∞ is equal to one; therefore, (13) has solutions Y_+ and Y_- having the standard asymptotics (6) in the open upper and lower half-planes respectively. To construct an isomonodromy solution it suffices to find the matrix $G_0 = Y_- Y_+^{-1}$ as a function of the parameters x, t, q and r.

In [3] the asymptotics of the function $\Phi(\lambda)$ as $\lambda \to \infty$ for $0 < arg\lambda < \pi$ and $-\pi < arg\lambda < 0$ are found. Then from these limiting values, the λ independent matrix G_0 is determined. Its a priori independence of x and t had led to relations on the components of the isomonodromy solution of (2). After translation into our notation here, these relations are

$$-2itqr \equiv \gamma = \text{const.}, \quad \sqrt{t}re^{-ix^2/4t}t^{-\gamma} \equiv -i\gamma/2\mu = \text{const.},$$

from which the expressions $q(x,t)$ and $r(x,t)$ of interest to us are determined:

$$q(x,t) = \mu t^{-\gamma-1/2}e^{-ix^2/4t}, \quad r(x,t) = \frac{-i\gamma}{2\mu}t^{\gamma-1/2}e^{ix^2/4t}. \tag{14}$$

For arbitrary values of the parameters μ and γ, formulas (14) describe solutions of the split equation of the NS system (4). Setting $\gamma = \pm i\delta^2$, $\delta > 0$ and $|\mu| = \delta/\sqrt{2}$, we arrive in (14) at the relations $r = \pm\bar{q}$.

Thus in the absence of regular singularities isomonodromy solutions of NS equation (5) are parametrized by two real parameters δ, $\phi = \arg\mu$ and are described by the exact formula

$$q(x,t;\delta,\phi,a) = \frac{1}{\sqrt{2}}\delta \cdot t^{-ai\delta^2-1/2}e^{-ix^2/4t+i\phi}. \tag{15}$$

A Direct Way to Compute the Isomonodromy Solutions

It is also possible to obtain the above isomonodromy solution of the NS equation, without the usage of a R-H problem. In the previous section, the λ-equation (9) have been being considered in the absence of regular singularities. If we include a regular singularity $a_0 = 0$, we have the λ-equation

$$\frac{\partial \Psi}{\partial \lambda} = (A_1\lambda + A_0 + \frac{1}{\lambda}H_0)\Psi, \tag{16}$$

where

$$A_1 = 4it\sigma_3, \quad A_0 = -ix\sigma_3 - 4tU(0), \quad H_0 = T^0_\infty + xU(0) + 2tV(0) \tag{17}$$

where $T^0_\infty = -\gamma\sigma_3$. Hence the isomonodromy solutions of NS equation obtained in the previous section are characterized by the vanishing of H_0:

$$H_0 = \begin{pmatrix} -\gamma - 2itqr & xq - 2itq_x \\ -xr - 2itr_x & \gamma + 2itqr \end{pmatrix} = 0. \tag{18}$$

This matrix equation is equivalent to three independent scalar equations:

$$-\gamma - 2itqr = 0, \quad xq - 2itq_x = 0, \quad -xr - 2itr_x = 0 \tag{19}$$

Integrating the second of these equations we obtain

$$q(x,t) = C(t)e^{-i\frac{x^2}{4t}} \tag{20}$$

and from the first and third of the equations in (20) we get

$$r(x,t) = \frac{-i\gamma}{2tC(t)} e^{i\frac{x^2}{4t}} \tag{21}$$

Substitution of (20) and (21) into the system in (4) leads to the first order differential equation in $C(t)$:

$$\frac{dC}{dt} + (1/2 + i\gamma)\frac{1}{t}C = 0 \tag{22}$$

whose general solution is $C(t) = \mu t^{-\gamma-\frac{1}{2}}$ with an arbitrary constant μ. from which the expressions in (14) for $q(x,t)$ and $r(x,t)$ of interest to us are reproduced.

The fact that a solution of the NS equation described by (14) is not new. The progres done in [3] consists in the fact that in classifying this solution by means of the Riemann-Hilbert problem as an isomonodromy solution, the Ψ function is explicitly computed. This circumstance had made it possible to construct new exact solutions of the NS equation of the isomonodromy class. Making use of this idea, we have found here the same solution from the condition of absence of regular singularities of the λ-equation.

The λ-equation and the Riemann-Hilbert problem corresponding to these solutions may also have regular singularities. Under the condition $a = -1$ these solutions are smooth and make it possible to find a more general class of self-similar solutions of the NS equation. This possibility is discussed in the next section.

SELF-SIMILAR ISOMONODROMY SOLUTIONS OF THE NS EQUATION AND THE PAINLEVE IV EQUATION

In this section we are going to show that when the λ-equation has a regular singularity at the origin, i.e. $a_1 = 0$, $H_0 \neq 0$ and $H_j = 0$, $j > 0$ in (10), the isomonodromy solutions of the NS equation are self-similar. Then we also prove that in this case the similarity equation of the NS equation is equivalent to the Painlevé IV equation.

Self-Similarity of the Isomonodromy Solutions of the NS Equation

From equations (3) and (16)-(18) one obtains

$$\lambda A(\lambda) = \lambda^2 A_1 + \lambda A_0 + H_0 = xU + 2tV + T^0_\infty \tag{23}$$

One also has from (1) and (9) $\lambda A = \lambda\Psi_\lambda \cdot \Psi^{-1}$, $U = \Psi_x \cdot \Psi^{-1}$, $V = \Psi_t \cdot \Psi^{-1}$. Hence equation (23) leads to a partial differential equation for the common soluton of the equations (1) and (9):

$$\lambda\Psi_\lambda = x\Psi_x + 2t\Psi_t + T^0_\infty\Psi. \tag{24}$$

Equation (24) can readily be solved by characteristics. The solution is:

$$\Psi = t^{-T^0_\infty/2}\phi\left(\lambda t^{\frac{1}{2}},\ xt^{-\frac{1}{2}}\right)\left(\lambda t^{\frac{1}{2}}\right)^{T^0_\infty/2} \tag{25}$$

With characteristic coordinates $z = \lambda t^{\frac{1}{2}}$, $y = xt^{-\frac{1}{2}}$ the function ϕ above has the necessary behavior for $z \to \infty$:

$$\phi(z,y) = \lim_{z\to\infty}\left(I + \frac{\phi_1}{z} + \frac{\phi_2}{z^2} + \dots\right)e^{-(2iz^2 - izy + \gamma)\sigma_3} \tag{26}$$

where $-\gamma\sigma_3 = T^0_\infty$. Comparing (6), (25) and (26), we obtain:

$$\Psi_k(x,t) = t^{-k/2 - T^0_\infty/2}\phi_k(xt^{-1/2})t^{T^0_\infty/2}. \tag{27}$$

From (27) and (7) self similar behaviour of q and r follows:

$$q = t^{-\gamma-1/2}Q(y), \qquad r = t^{\gamma-1/2}R(y). \tag{28}$$

It has been seen that the self similar behaviour in (14) obtained in the case of the absence of the regular singularities is a special case of (28).

Similarity Equation of NS Equation is Equivalent to the Painlevé IV Equation

To obtain the NS similarity equation we substitute (28) into the system (4):

$$-\frac{d}{dy}(Q_y + \frac{i}{2}yQ) = 2Q^2P + i\gamma Q, \quad -\frac{d}{dy}(R_y - \frac{i}{2}yR) = 2R^2Q + i\gamma R. \tag{29}$$

For the class of functions $r = \pm\bar{q}$, which implies $R = \pm\bar{Q}$, the system (29) becomes;

$$-i\gamma Q - \frac{d}{dy}(Q_y + \frac{i}{2}yQ) = 2aQ^2\bar{Q}, \qquad a = \pm 1 \tag{30}$$

Although it is shown in [4] and [5] that the similarity equation for the non- linear Schrödinger equation is equivalent to an equation which is classified by Painlevé as the one which is equivalent to the Painlevé IV equation, and σ-form of the Painlevé IV equation given by Jimbo and Miwa [2] respectively, here the relation between these two equations will be clarified further.

Now it will be shown that for $a = 1$ equation (30) is equivalent to the Painlevé IV equation. Separating Q into a phase θ and an amplitude A one has:

$$Q = Ae^{i\theta}, \quad P = \theta' + \frac{y}{4}, \quad (A^2P)' = -\frac{1}{4}A^2, \quad ' = \frac{d}{dy}. \tag{31}$$

Substitution of (31) into (30) gives;

$$\left(A^2P\right)^2 = A^2\left(AA'' + 2aA^4 + \frac{1}{16}y^2A^2 + i\gamma A^2\right). \tag{32}$$

Defining $A^2P = -4V$ one gets:

$$A^2 = 16V' \text{ and } 16V^2 = A^2\left(AA'' + 2aA^4 + \frac{1}{16}y^2A^2 + i\gamma A^2\right). \tag{33}$$

Eliminating A between the two equations in (33), one obtains a third order and third degree nonlinear ordinary differential equation in V:

$$V'''V' = \frac{1}{2}V''^2 + \frac{1}{8}V^2 - \frac{1}{8}y^2V'^2 - 64aV'^3 - 2i\gamma V'^2. \tag{34}$$

This equation can be integrated once to obtain te second order and second degree nonlinear ordinary differential equation:

$$V''^2 = -\frac{1}{4}\left(yV' - V\right)^2 - 64aV'^3 - CV' - 4i\gamma V'^2 \tag{35}$$

where C is an arbitrary integration constant.

Transforming the independent and dependent variables as;

$$y = 2e^{\frac{1}{4}i\pi}t, \qquad V = \frac{1}{32}e^{-\frac{1}{4}i\pi}\sigma. \tag{36}$$

and defining r_1, r_2, θ_0 and θ_∞ as;

$$r_1 + r_2 = 4\gamma, \quad r_1r_2 = 64C, \quad r_1 = -4\theta_0, \quad r_2 = -2(\theta_0 + \theta_\infty) \tag{37}$$

in (35), we finally obtain for $a = 1$:

$$\sigma''^2 = 4\left(t\sigma' - \sigma\right)^2 - 4\sigma'\left(\sigma' + 4\theta_0\right)\left(\sigma' + 2\left(\theta_0 + \theta_\infty\right)\right). \tag{38}$$

Equation (38) is the σ-form for the Painlevé IV equation given by M. Jimbo and T. Miwa in [2]. Defining

$$\sigma = \frac{2}{y}z^2 - \left(y + 2t + \frac{4\theta_0}{y}\right)z + (\theta_0 + \theta_\infty)y \tag{39}$$

$$\sigma' = -2z, \quad \frac{1}{2}\sigma'' = \frac{2}{y}z^2 + \left(y - \frac{4\theta_0}{y}\right)z - (\theta_0 + \theta_\infty)y \tag{40}$$

the functions y and z satisfies the system of ordinary differential equations

$$y' = -4z + y^2 + 2ty + 4\theta_0, \quad z' = -\frac{2}{y}z^2 + \left(-y + \frac{4\theta_0}{y}\right)z + (\theta_0 + \theta_\infty)y. \tag{41}$$

Elimination of the function $z(t)$ between the two equations of (41) leads to the Painlevé IV equation:

$$y'' = \frac{y'^2}{2y} + \frac{3}{2}y^3 + 4ty^2 + 2(t^2 - \alpha)y + \frac{\beta}{y}, \qquad \alpha = 2\theta_\infty - 1, \quad \beta = -8\theta_0^2. \tag{42}$$

Hence a solution of the nonlinear Schrödinger similarity equation (30), through (36) gives a solution σ of (38), and hence a solution of the Painlevé IV equation. The corresponding solution is obtained by the solution of the Riccati equation:

$$y' - 2ty - y^2 = 2\sigma' + 4\theta_0 \tag{43}$$

or by the second order linear equation

$$u'' - (2\sigma' + 4\theta_0 - 1 + t^2)u = 0, \quad y = -\frac{u'}{u} - t. \tag{44}$$

Conversely any solution of the Painlevé IV equation (42) produces a solution of the NS similarity equation (30) and hence a self similar solution of NS equation.

As an example to the above discussion, for the self similar solution in (14) one obtains the solution $y(t)$ of the Painlevé IV equation (42) in terms of the solutions of the Weber equation:

$$u'' - (4i - 1 + t^2)u = 0, \quad y = -u'/u - t. \tag{45}$$

REFERENCES

1. M. Jimbo and T. Miwa, Monodromy preserving deformation of linear ordinary differential equations with rational coefficients I. General theory and τ-function, Physica 2D, 306(1981).
2. M. Jimbo and T. Miwa, Monodromy preserving deformation of linear ordinary differential equations with rational coefficients II. Physica 2D: 407 (1981).
3. A.R. Its, Isomonodromic solutions of the equations of zero curvature, Math. USSR Izvestia, 26 No.3: 497(1986).
4. M. Boiti and F. Pempinelli, Nonlinear Schrödinger equation, Bäcklund transformations and Painlevé transcendents, Il Nuovo Cimento 59B, 40(1980).
5. M. Can, On the relation between nonlinear Schrödinger equation and Painlevé IV equation, Il Nuovo Cimento 106B N.2, 40(1991).

A METHOD FOR VISUALIZING THE 4-DIMENSIONAL SPACE OF SECTION IN 3-D HAMILTONIAN SYSTEMS

P.A. Patsis[1] and L. Zachilas[2]

[1] ESO, Karl Schwarzschild Str. 2, D-85748, Garching bei München, Germany
[2] Dept. of Chemistry, University of Crete, P.O. Box 1470, GR-71409, Hellas

Abstract. An empirical method that uses colors and rotation of 3D figures is proposed for visualizing the 4D "Poincaré space of section" in 3D Hamiltonian systems. The representation of the 4th dimension as color variation in 3D projections, gives essential information about the areas that are close to each other in the 4D space. This method helps us to reveal existing structures in cases where the consequents in the 3D projections seem to fill densely the space.

1. Introduction

The importance and the necessity for the exploration of the structure of phase space in the vicinity of stable and unstable periodic orbits in Hamiltonian systems with 3 degrees of freedom was recognized already in the very first studies of problems of this kind (Martinet and Magnenat 1981, Contopoulos et al. 1982). However, in 3D, there is an intrinsic difficulty even for visualizing the "Poincaré surfaces of section". For a given value of the Hamiltonian $H(x, \dot{x}, y, \dot{y}, z, \dot{z}) = h$, and for initial conditions $(x_0, \dot{x}_0, y_0, \dot{y}_0, 0, \dot{z}_0)$, we obtain $\dot{z}_0$ from the equation $H = h$, eliminating in this way two variables of the 6-dimensional phase space. By this method we calculate the successive intersections of the phase trajectory with the $z = 0$ plane. The intersections will be distributed in the 4D space $(x, \dot{x}, y, \dot{y})$. Hence, in this case, we speak of a "space" rather than a "surface" of section. Depending on the number of integrals of motion that may exist besides the energy, it is possible to have 2D invariant surfaces in a 3D space (two additional integrals), consequents distributed in a volume of the 4D space, giving surfaces if we take one section more (one additional integral), or consequents which fill densely a 4D region (no additional integral).

One way to study the behavior of non-periodic orbits close to a periodic one, by means of the notion of the "space of section", is to project the obtained 4-tuples of points like $(x, \dot{x}, y)$, $(x, \dot{x}, \dot{y})$, etc in a 3D space. In the past, stereoscopic representations have been used in order to get a better understanding of the shape of the projections (Martinet and Magnenat 1981, Magnenat 1982, Contopoulos et al. 1982, Contopoulos 1984). In this way, different directions for the observer can be chosen and the geometry of the figures can be

Hamiltonian Mechanics, Edited by J. Seimenis,
Plenum Press, New York, 1994

well understood. However, by looking at these sets of consequents in the 3D space, we get no information about their distribution in the fourth dimension. This can be essential for the study of the orbital behavior, especially in the vicinity of unstable periodic orbits.
A big help can be obtained by using special software systems for the analysis of data, as well as plotting and image display packages. A common procedure for visualizing 3D data is to animate the data by rotating it about one or more axes. Also the appropriate use of colors (or shades of a particular color) in plotting the figures offers a lot of additional information. In this paper we use for this purpose the Interactive Data Language package (IDL).
In our applications (§3) we refer to the structure of the 4D spaces of section in the vicinity of periodic orbits belonging to the z-axis family of the potential

$$V(x,y,z) = V_0 + \frac{v_c^2}{2}\ln\left(1 + x^2 + \frac{y^2}{q_a^2} + \frac{z^2}{q_b^2}\right) - \frac{GM_D}{\sqrt{x^2 + \frac{y^2}{q_a^2} + (\alpha + \sqrt{z^2+b^2})^2}} \tag{1}$$

used in a previous paper (Patsis and Zachilas 1990) rotating around the z-axis with angular velocity of the system Ω_s. This potential in its axisymmetric form (q_a=1), has been proposed as a simple disk-halo potential for the Galaxy (May and Binney 1986).

2. The method

2a. The geometry

First we view the distribution of the consequents in the $(x, y, \dot{x})$ space (or in any other 3D space we can obtain by combining the variables of the 4D space $(x, y, \dot{x}, \dot{y})$). In order to have a better qualitative view we scale our axes according to the minimum and maximum values we have in each coordinate. Then we rotate the figure, and by this the geometrical shape of the 3D projection can be cleared up in all details.

2b. The colors

We follow a similar procedure for the colors. First we find the minimum and maximum values of the 4th coordinate ($\dot{y}$). What we do is simply to subdivide the interval [min $\dot{y}$, max $\dot{y}$] in 233 subintervals. This is the maximum resolution offered by IDL. Then each point, plotted in the $(x, y, \dot{x})$ space, is colored according to the subinterval in which the value of its $\dot{y}$ coordinate belongs. Thus we are able to estimate whether our points are scattered or not in the 4th dimension. In the extreme case in which, $\dot{y}$ is constant, our figure would have a unique color. On the other hand, figures with a "multi-colored" appearance indicate that we have a mixing in the 4th dimension. Interesting is also to find out if there are isolated regions with the same color in our figures, or even if chaotic-looking sets of points present some structure when the 4D-color is added.
In many cases the appropriate choice of the color look-up table (the palette defining the succession of the colors or the shades) is important.

3. Examples

A look-up table (LUT) can be defined to give 233 gray shades from black to white. In case the amount of the contribution of the white color in the gray shade, increases in a linear way as we move from 0 to 232, we speak about a linear b/w look-up table. We use this LUT in the following example:
Fig. 1 refers to the orbital behavior in the vicinity of a z-axis periodic orbit in potential (1). As it is known, in 3D Hamiltonian systems we have three different kinds of instabilities

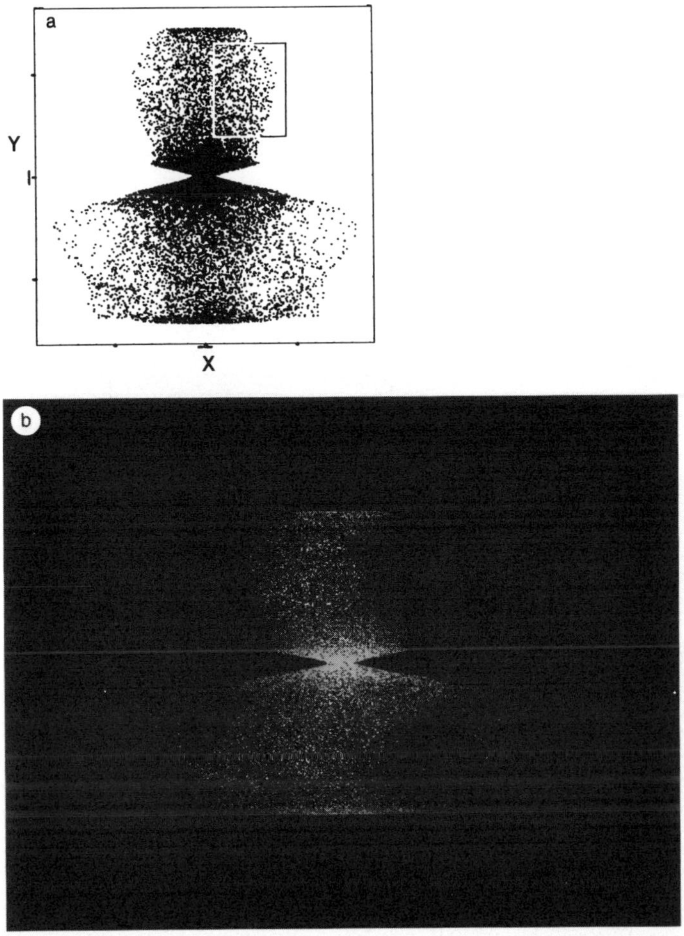

Fig. 1
(a) A $(x, \dot{x}, y)$ projection of the 4D $(x, \dot{x}, y, \dot{y})$ space of section in the vicinity of a z-axis simple unstable periodic orbit of potential (1). In this case we have $\dot{x}$ perpendicular to the plane of the paper (*left*). (b) The same figure using the values of the $\dot{y}$ coordinate to color the consequents with a linear b/w LUT. The obscured areas at the upper-right part of the figure define a set of points, which are close also in the 4-th dimension (*down*)

depending on the number of the eigenvalues of the periodic orbit that are off the unit circle (see e.g. Contopoulos and Magnenat 1985). In the case of Fig. 1 we have given to the parameters V_0, v_c, M_D, α and β in (1), the values proposed by May and Binney (1986), while we have q_a=0.7 and q_b=0.6. For these values of the parameters and for E=-0.32, Ω_s = 0.002 (where E is energy), the z-axis periodic orbit is characterized as Simple Unstable (two eigenvalues are on the real axis and two on the unit circle). By obtaining the 4D space of section $(x, \dot{x}, y, \dot{y})$ for about 17000 consequents as described in the introduction, we consider the $(x, \dot{x}, y)$ projection. In Fig. 1 we have rotated the figure to face the (x, y) side, thus $\dot{x}$ is perpendicular to the plane of the paper. In Fig. 1a we have not taken into account the 4th dimension $(\dot{y})$. If we view the same figure by using the linear b/w LUT to color the consequents according to their values in the $\dot{y}$ direction (Fig. 1b), we realize that parts of the figure on the right side are "missing". This is simply an effect due to the concentration of the color-values in this area close to 0 (black) in combination with the black background. To give this in numbers, we may look at the variation of $\dot{y}$ over the whole spectrum of the 233 shades. This is given in the histogram in Fig. 2a. Although there is a mode value for the "color-value" 110, the whole spectrum is covered. However, if we make again statistics only inside the frame in the upper-right part of Fig. 1a, we realize that almost all values are very close to 0, thus this area is almost obscured (Fig. 2b).

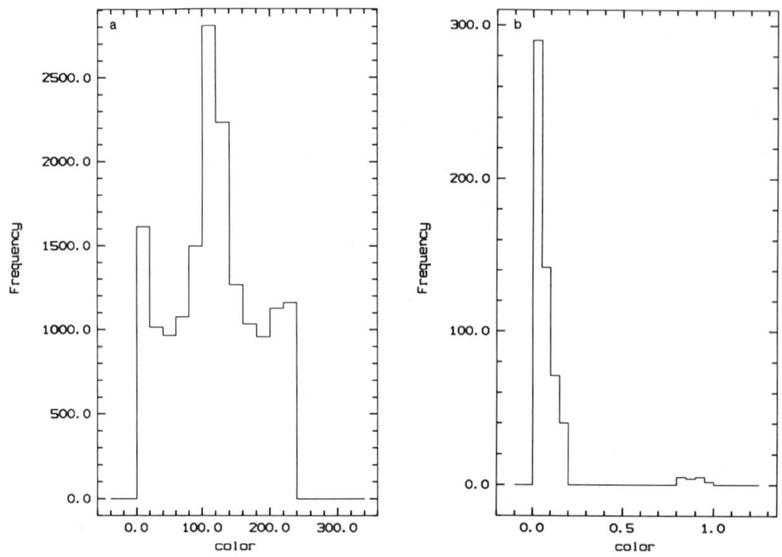

Fig. 2 (a) A histogram giving the distribution of the colors for the 17000 consequents of Fig. 1. (b) A histogram giving the distribution of colors in the part of Fig. 1a inside the drawn frame.

In this way we are able to distinguish the consequents that are close also in the 4th dimension.

In several cases the use of colors is very efficient in revealing existing structures in the forth dimension. In Fig. 3 we observe the (x, y) projection of the $(x, \dot{x}, y, \dot{y})$ space of section in the vicinity of a z-axis periodic orbit for E=-0.31, Ω_s = 0.05. The rest of the parameters in potential (1) are as above. The z-axis family is now complex unstable (all four eigenvalues are complex numbers not on the unit circle). This is a characteristic case, where one obtains additional information on the distribution of the points in the 4D space by using our method. By coloring the consequents according to their $\dot{y}$ values and by using the appropriate LUT it can be seen, that the distribution of the points inside the elliptical area in Fig. 3 presents some structure in the 4D space. Consequents with similar colors (similar $\dot{y}$ values) form "8-like" structures when we face the (x, y) side of the $(x, \dot{x}, y)$ projection. By rotating the figure, we are able to specify the actual shape of these regions.[1]

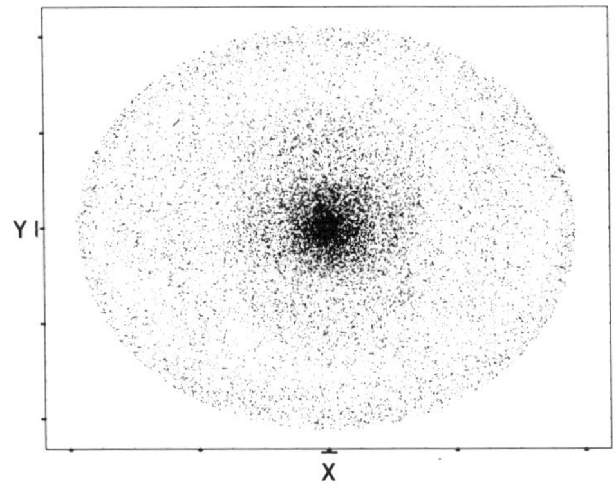

Fig. 3 A (x, y) projection of the 4D $(x, \dot{x}, y, \dot{y})$ space of section close to a z-axis complex unstable periodic orbit of potential (1).

References

Contopoulos G., Magnenat P. and Martinet L.: 1982, Physica D, Vol.6D 123

Contopoulos G. : 1984, in "Nonlinear Phenomena in Physics" (ed.F.Claro), p.238, Springer Verlag

Contopoulos G. and Magnenat P.: 1985, Celest. Mech. 37,387

May A. and Binney J.: 1986, MNRAS 221,857

Magnenat P.: 1982, Cel. Mech. 28,319

Martinet L. & Magnenat P. : 1981, Astron. & Astroph. 96, 68

Patsis P.A. & Zachilas L. : 1990, Astron. & Astroph. 227, 37

Patsis P.A. & Zachilas L. : 1994, "Using colors and rotation for visualizing 4-dimensional Poincare Cross Sections", to appear in Int. J. Bifurcation and Chaos

[1]We have applied this method to study the orbital behavior of the z-axis family in potential (1), as well as of other families of p.o. in the same potential. We have traced a typical orbital behavior in each case of instability. We followed also the changes that occur as we pass from stability to instability or from one to the other kind of instabilities. However, since it is not possible to give here the colored 4D spaces of section we presented in our poster during the conference, these results will be presented elsewhere (Patsis and Zachilas 1993 - in preparation)

NON-ADIABATIC ASPECTS OF TIME-DEPENDENT HAMILTONIAN SYSTEMS

F. Casas,[1] J.A. Oteo,[2] and J. Ros[2]

[1]Departament de Matemàtiques, Universitat Jaume I
12071 Castelló, Spain
[2]Departament de Física Teòrica and IFIC
Universitat de València, 46100 Burjassot, Spain

1. INTRODUCTION

Extreme adiabatic behavior furnishes great simplification in the treatment of linear time-dependent Hamiltonian systems. But the actual time variation of the parameters is only finitely, rather than infinitely, slow. Then one is forced to consider corrections to the adiabatic limit.

In this contribution a practical algorithm for that purpose is proposed. It is based on the Magnus expansion for the classical evolution operator (Magnus, 1954; Oteo and Ros, 1991). In our approach this expansion is carried out after an appropriate coordinate transformation is implemented in order to make it useful in the adiabatic regime. The first order of the resulting expansion is then applied to the evaluation of the change of the so called adiabatic invariant.

2. THE ADIABATIC CLASSICAL MAGNUS EXPANSION

Let us consider a Hamiltonian dynamical system with one degree of freedom and let the two-dimensional vector $\vec{\xi} = (q, p)$ represents its state. Trajectories in phase space can be viewed as the action of the time-dependent evolution operator acting on the initial state: $\vec{\xi}(\tau) = \mathcal{M}(\tau, \tau_0, \vec{\xi}(\tau_0))$. Here we have introduced the new variable $\tau = \epsilon t$, where $1/\epsilon$ sets the time scale and $\epsilon \rightarrow 0$ in the adiabatic limit.

In this formalism a linear system is characterized by the fact that the map $\mathcal{M}$ can be represented by a 2×2 matrix M which satisfies the differential equation

$$\dot{M} = \frac{1}{\epsilon} S M, \qquad M(\tau_0, \tau_0) = I, \tag{1}$$

where S is a 2×2 matrix obtained from the Hamiltonian, I is the identity matrix and the dot stands for derivative with respect to τ.

Hamiltonian Mechanics, Edited by J. Seimenis,
Plenum Press, New York, 1994

Magnus expansion (ME) in the standard form (Magnus, 1954) gives a solution to Eq.(1) of the form $M(\tau,\tau_0) = \exp\Omega(\tau,\tau_0)$ with $\Omega(\tau_0,\tau_0) = 0$. Here Ω is a matrix whose matrix elements are functions of τ and τ_0 but not of the phase space coordinates $\vec{\xi}$. It satisfies its own differential equation which is solved in the form of a series: $\Omega = \sum \Omega_i$. The first term in that expansion is $\Omega_1(\tau,\tau_0) = \frac{1}{\epsilon}\int_{\tau_0}^{\tau} dx S(x)$. Higher order terms can be computed by recursive procedures.

Direct ME as a symplectic integrator for Eq.(1) has been shown (Oteo and Ros, 1991) to work particularly well when H presents sudden time dependence. The situation is different for the adiabatic regime we are now interested in. Here we propose to use still the same scheme but only after some phase space coordinate transformation has been carried out. In the new coordinates the dynamics is exactly solved in the adiabatic limit giving the Adiabatic Approximation (AA). Further corrections which take into account the finite rate at which H varies with time are then obtained via ME.

To see how this procedure works, let us perform a nonsingular time-dependent transformation $\vec{\xi}_R(\tau) = R^{-1}\vec{\xi}(\tau)$. The time evolution of the new coordinates will be governed by

$$\vec{\xi}_R(\tau) = M_R(\tau,\tau_0)\vec{\xi}_R(\tau_0). \tag{2}$$

Here $M_R(\tau,\tau_0) = R^{-1}(\tau)M(\tau,\tau_0)R(\tau_0)$ obeys the equation

$$\dot{M}_R = S_R M_R, \qquad M_R(\tau_0,\tau_0) = I, \tag{3}$$

with

$$S_R = \frac{1}{\epsilon}R^{-1}SR - R^{-1}\dot{R}. \tag{4}$$

We take in the adiabatic regime R so as to instantaneously diagonalize the matrix $S(\tau)$. The diagonal of S_R should then be the dominant term. As we are going to see this procedure leads to an interesting approximation scheme.

The diagonal part $\Delta \equiv \mathrm{diag}(S_R)$ can be easily integrated out by making the factorization

$$M_R = \exp\left(\int_{\tau_0}^{\tau} dx\Delta(x)\right) M_R'. \tag{5}$$

Thus M_R' satisfies

$$\dot{M}_R' = S_R' M_R', \quad S_R' = \exp\left(-\int_{\tau_0}^{\tau} dx\Delta(x)\right)(S_R - \Delta)\exp\left(\int_{\tau_0}^{\tau} dx\Delta(x)\right). \tag{6}$$

The AA amounts to take the simplest approximation $M_R' = I$. In order to improve this zero order step we apply ME to Eq.(6) introducing what we call Adiabatic Classical Magnus Expansion (hereafter referred to as ACME). Then $M_R' = \exp\Omega$, with $\Omega = \Omega_1 + \Omega_2 + \ldots$ In first order

$$\Omega_1 = \int_{\tau_0}^{\tau} dx S_R'(x). \tag{7}$$

As they will be needed in the next section we collect some explicit formulae for the case of the Generalized Harmonic Oscillator (GHO) with Hamiltonian

$$H(q,p,\tau) = \frac{1}{2}[X(\tau)q^2 + 2Y(\tau)qp + Z(\tau)p^2]. \tag{8}$$

The time-dependent functions X, Y, Z will supposed to be gentle (i.e., $C^\infty(-\infty,+\infty)$

with $L^1(-\infty,+\infty)$-derivatives of any order (Wasow, 1973)), and have finite limits as $\tau \to \pm\infty$. In the following we shall assume $X(\tau)Z(\tau) > Y^2(\tau)$ for real τ although formal results can also be obtained in the opposite case. We define $\omega(\tau) = \sqrt{XZ - Y^2}$.

A long but otherwise straightforward calculation allows one to write the matrix M_R to first order in the form:

$$M_R(\tau,\tau_0) = \cosh\eta \begin{pmatrix} e^{+a} & 0 \\ 0 & e^{-a} \end{pmatrix} + \frac{\sinh\eta}{\eta} \begin{pmatrix} 0 & h_1(\tau)e^{+a} \\ h_2(\tau)e^{-a} & 0 \end{pmatrix}. \tag{9}$$

where:

$$\begin{aligned} \eta^2 &= h_1(\tau)h_2(\tau), \\ h_1(\tau) &= \int_{\tau_0}^{\tau} dx\alpha(x)e^{-2a(x,\tau_0)}, \\ h_2(\tau) &= \int_{\tau_0}^{\tau} dx\beta(x)e^{+2a(x,\tau_0)}, \end{aligned} \tag{10}$$

with $\alpha = \gamma[(i\dot{\omega}+\dot{Y})Z-(i\omega+Y)\dot{Z}]$, $\beta = \gamma[(i\dot{\omega}+\dot{Y})X-(i\omega+Y)\dot{X}]$, $\gamma = [2i\omega(i\omega+Y)]^{-1}$, $k(\tau) = [(\omega - iY)X/(\omega + iY)Z]^{1/2}$ and

$$a(\tau,\tau_0) = \frac{1}{2}\log\frac{k(\tau_0)}{k(\tau)} + i\int_{\tau_0}^{\tau} dx\left(\frac{\omega}{\epsilon} + \frac{Y\dot{Z} - Z\dot{Y}}{2\omega Z}\right). \tag{11}$$

This scheme can be proved (Casas et al., 1993) to significantly improve the AA for computing trajectories pretty far from the adiabatic regime. Here, instead, we turn to the analysis of the adiabatic invariant.

3. ADIABATIC INVARIANT AND ITS VARIATION IN THE ACME APPROACH

Under the previous hypothesis for the parameters X, Y, Z of the GHO the action $J(\tau)$ is an adiabatic invariant, i.e. it is constant in the limit $\epsilon \to 0$. Its instantaneous value is $J(\tau) \equiv H(\tau)/\omega(\tau) = -ip_R(\tau)q_R(\tau)$. Furthermore there exist the limit values $J(+\infty)$ and $J(-\infty)$ so that we can define $\Delta J = J(+\infty) - J(-\infty)$ (Arnold et al., 1988). The asymptotic behavior of ΔJ has been profoundly studied over the years (Lenard, 1959; Keller and Mu, 1991). We can consider a finite time interval instead of an infinite one and introduce $\delta J(\tau,\tau_0) = J(\tau) - J(\tau_0)$. We do know that J changes very little over a period of order $1/\epsilon$ and pose the question of how much does it change during a much longer interval. We give here a quantitative answer in the scheme described in the previous section.

Let $m_{ij}(i,j = 1,2)$ be the τ-dependent matrix elements of M_R'. Then the exact expression for δJ is given by

$$\begin{aligned} \delta J(\tau,\tau_0) &= -i\left[m_{11}(\tau)m_{21}(\tau)q_R^2(\tau_0) + m_{22}(\tau)m_{12}(\tau)p_R^2(\tau_0)\right. \\ &\quad \left.+(m_{11}(\tau)m_{22}(\tau) + m_{12}(\tau)m_{21}(\tau) - 1)q_R(\tau_0 p_R(\tau_0)\right]. \end{aligned} \tag{12}$$

We see from this expression that in the AA $m_{11} = m_{22} = 1, m_{12} = m_{21} = 0$ and correspondingly $\delta J = 0$, while any method which goes beyond the AA leads in general to nonvanishing values for δJ.

We want to study this question from the ACME we have just introduced . The previously given formulae allows one to write down the first order approximation for $\delta J(\tau, \tau_0)$, namely

$$\delta J(\tau, \tau_0) = -i \left\{ \frac{\sinh 2\eta}{2\eta} \left[h_2(\tau) q_R^2(\tau_0) + h_1(\tau) p_R^2(\tau_0) \right] + 2(\sinh \eta(\tau))^2 q_R(\tau_0) p_R(\tau_0) \right\} . \tag{13}$$

This expression is a valid approximation provided the integrals involved are convergent. In that case we can either solve them numerically or, in some cases, apply techniques from asymptotic analysis to extract the leading term in their asymptotic expansions.

An important aspect of our scheme worth emphasizing is its possibility to be iterated by evaluating higher order terms in the Magnus expansion for M_R' and the corresponding expressions for δJ. Of course, to rigorously proceed in this way would require a study of the convergence of the algorithm. This remains an open problem for the standard Magnus expansion. Nevertheless, in spite of that, the procedure has been heuristically used in different contexts in Physics and Chemistry.

When an asymptotic determination of the integrals is possible the ACME reproduces the exponentially small character of ΔJ as $\epsilon \to 0$ when $\omega(\tau)$ is an analytic function of τ. This has been discussed in (Casas et al., 1993) where other asymptotic analysis of ΔJ are also carried out.

4. EXAMPLES

To illustrate what has been said so far we present now the comparison between two ways of computing ΔJ: on the one side we do it by numerical integration of the equation of motion, on the other side we apply our first order formula from ACME given in Eq.(13). We treat three different examples. Two of them refer to the GHO while the other concerns the simple time-dependent harmonic oscillator. We think that the time variation of the parameters considered cover a wide spectrum of behavior. For computational reasons we have to take finite values $\tau_\pm$ as the limits in the 'infinite' time interval. But provided their absolute values are large enough ΔJ does not depend of the particular values chosen.

ACME for the Generalized Harmonic Oscillator

We treat a GHO with parameters given by:

$$\begin{aligned} X(\tau) &= \sqrt{a^2 + \frac{s}{\cosh \tau}} + \frac{b}{\cosh \tau}, \\ Y(\tau) &= \sqrt{c^2 - \frac{b^2}{\cosh^2 \tau}}, \\ Z(\tau) &= \sqrt{a^2 + \frac{s}{\cosh \tau}} - \frac{b}{\cosh \tau}, \end{aligned} \tag{14}$$

where a, b, c are constants $a \neq 0, a > c$. The frequency is given as

$$\omega(\tau) = \sqrt{a^2 - c^2 + \frac{s}{\cosh \tau}}. \tag{15}$$

The parameter s serves to distinguish the two cases we are considering: $s = 1$ and $s = 0$.

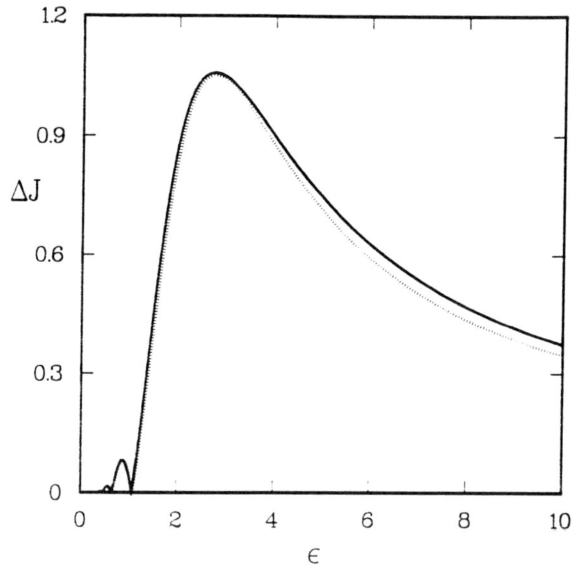

Figure 1. Change in the adiabatic invariant for the example Eq.(14) with $s = 1, a = 3, b = 0.1, c = 2.97$. Solid line corresponds to the exact value and dotted line to first-order ACME.

The asymptotic analysis depends strongly of the transition points of the function $\omega(\tau)$ i.e. of its roots, isolated singular points and branch points.

i) For $s = 1$, $\omega(\tau) > 0$ for real τ and $\omega \to \omega_\pm = \sqrt{a^2 - c^2} > 0$ as $\tau \to \pm\infty$. The function $\omega^2(\tau)$ is analytic in a strip along the real τ-axis, its poles being located at $\tau_p = i(2n+1)\pi/2$, $n = 0, \pm 1, \ldots$. As for its zeros τ_c, if we introduce $\rho = 1/(c^2 - a^2)$, we can distinguish two cases:

- $|\rho| > 1$ then $\tau_c = \log\left(|\rho| \pm \sqrt{\rho^2 - 1}\right) + i(2n+1)\pi$
- $|\rho| \leq 1$ then $\tau_c = \begin{cases} i(\beta_1 + 2n\pi) \\ i(\beta_2 + 2n\pi) \end{cases}$

with $\cos\beta_i = -|\rho|$, $\sin\beta_1 = -\sin\beta_2 = \sqrt{1-\rho^2}$.

Numerical results for this example show that our first order adiabatic Magnus expansion reproduces extremely well the exact behaviour for $|\rho| < 1$. For $|\rho| > 1$ on the average $|\Delta J|$ takes higher values and the agreement worsens, but not dramatically. This can be seen in Figure 1, which corresponds to initial state $(q_R, p_R) = (1, 0)$ and parameters $(a, b, c) = (3, 0.1, 2.97)$. Solid line represents exact values while dotted line (where visible) corresponds to our first order ACME. Furthermore we want to emphasize the fact that our corrections go clearly beyond the exponential asymptotic limit. This is seen by observing the non-exponential character of ΔJ over the whole range of ϵ explored. To the best of our knowledge this is a novel result.

ii) For $s = 0$ the frequency given by Eq.(15) is a constant $\omega = \sqrt{a^2 - c^2} > 0$ on the entire complex τ-plane and so no transition points exist. We have explored a wide range of values for the parameters a, b, c, ϵ and in all cases the exact value and the first order Magnus approximation for ΔJ agree up to the fourth decimal place. An interesting fact to observe here is that notwithstanding the constancy of $\omega(\tau)$, $\Delta J \neq 0$. Methods based on the analysis of singularities in the complex plane cannot cope with this situation. On the contrary, our ACME correction very accurately reproduces the exact behavior.

Asymptotic ACME and the Simple Harmonic Oscillator

Let us consider a simple harmonic oscillator ($X = \omega^2, Y = 0, Z = 1$) in Eq.(8)) with frequency

$$\omega(\tau) = \left[1 + \frac{1}{1 + 2e^{-\tau}}\right]^{1/2}, \tag{16}$$

which was analyzed by Wasow (Wasow, 1974). Now $\omega(\tau) > 0$ for real τ and its limits at $\pm\infty$ are $\omega_- = 1$, $\omega_+ = \sqrt{2}$. ω^2 is a meromorphic function with simple zeroes at $\tau_c = i(2n+1)\pi$ and simple poles at $\tau_p = \log 2 + i(2n+1)\pi$.

Now Eq.(13) simplifies to give

$$\Delta J = \frac{\sinh 2|K|}{2|K|}\left[\omega_- K^* q_R^2(-\infty) - \frac{1}{\omega_-} K p_R^2(-\infty)\right] - 2iq_R(-\infty)p_R(-\infty)\sinh^2|K|, \tag{17}$$

where

$$K \equiv \int_{-\infty}^{+\infty} d\tau \frac{\dot{\omega}(\tau)}{2\omega(\tau)} \exp\left[-\frac{2i}{\epsilon}\Theta(\tau)\right], \quad \Theta(\tau) = \int_{-\infty}^{\tau} \omega(x)dx. \tag{18}$$

Following the asymptotic analysis of the equation of motion by Wasow (1974) and Meyer (1975) we obtain (Casas et al., 1993) that as $\epsilon \to 0$

$$\Delta J \sim \frac{\pi}{6} e^{-2\pi/\epsilon}\left[\varphi p_R^2(-\infty) - \varphi^* q_R^2(-\infty)\right] + O\left(e^{-4\pi/\epsilon}\right). \tag{19}$$

Here φ is given in terms of $\Theta(\tau_c = -i\pi)$. We see from this equation that first order adiabatic Magnus expansion gives correctly the exponentially small character of ΔJ in the limit $\epsilon \to 0$ as well as an approximation to the much more elusive pre-exponential factor.

In Figure 2 we show the function $|\Delta J|(\epsilon)$ calculated a) by numerical integration of the equation of motion (solid line), b) from the Eq.(19) (broken line) and c) by application of ACME, Eq.(17) (dotted line). We take as initial conditions $q_R(-\infty) = 1, p_R(-\infty) = 0$. We conclude that for the values of the parameter ϵ considered the exact and ACME results practically coincide.

5. CONCLUSIONS

A previously proposed scheme for integrating the equations of motion for time-dependent linear Hamiltonian systems has been adapted to work in the adiabatic regime. It is based on the Magnus expansion and as such originates a practical algorithm which can be used as a symplectic integrator to determine the trajectories. Because it is recursive in nature it could be systematically improved by going to higher orders. The first order we have explicitly considered furnishes already extremely good results even far away of the strict adiabatic limit. Nevertheless higher order contributions should be studied to see if these results are confirmed. The scheme has been also applied to the analysis of the adiabatic invariant. As it is well known this magnitude is not strictly speaking a constant and the proposed form of the Magnus expansion reproduces the exponentially small behaviour of its variation. Besides that, however, our scheme gives a way to approximately evaluate ΔJ. We illustrate this point different time-dependences of the parameters of a one-dimensional Generalized Harmonic Oscillator.

Acknowledgments

Two of us, JAO and JR, thank the spanish CICYT for partial financial support under grants number PB91-0119-C02-00 and PB92-0820, respectively.

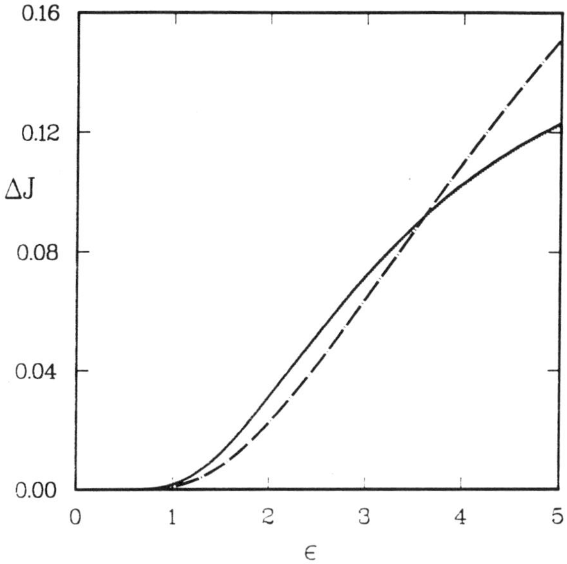

Figure 2. Change in the adiabatic invariant for the simple harmonic oscillator with frequency Eq.(16). Solid line correspond to an exact calculation, broken line to application of Eq.(19) and dotted line to ACME.

REFERENCES

Arnold, V.I., Kozlov, V.V., and Neishtadt, A.I., 1988, Mathematical Aspects of Classical and Celestial Mechanics, *in* "Dynamical Systems III," V.I. Arnold, ed., Springer-Verlag, New York.

Casas, F., Oteo, J.A., and Ros. J., 1993, Linear time-dependent Hamiltonian systems beyond the adiabatic limit, *in preparation.*

Keller, J.B., and Mu, Y., 1991, Changes in adiabatic invariants, *Ann. Phys.* **205**:219.

Lenard, A., 1959, Adiabatic invariance to all orders, *Ann. Phys.* **6**:281.

Magnus, W., 1954, On the exponential solution of differential equations for a linear operator, *Commun. Pure Appl. Math.* **7**:649.

Meyer, R.E., 1975, Gradual reflection of short waves, *SIAM J. Appl. Math.* **29**:481.

Oteo, J.A., and Ros, J., 1991, The Magnus expansion for classical Hamiltonian systems, *J. Phys. A* **24**:5751.

Wasow, W., 1973, Adiabatic invariance of a simple oscillator, *SIAM J. Math. Anal.* **4**:78.

Wasow, W., 1974, Calculation of an adiabatic invariant by turning point theory, *SIAM J. Math. Anal.* **5**:673.

A STUDY OF A FINITE-DIMENSIONAL DYNAMICAL SYSTEM APPROXIMATING THE EVOLUTION OF QUANTUM AVERAGES

S.Yu. SADOV
Moscow, Russia

Introduction

Among the earliest results of quantum mechanics there is P. Ehrenfest's theorem which states that the averages of coordinate and momentum of a quantum particle in the potential $V(x)$ vary as follows :

$$<\dot{x}>=<p>/m,$$

$$<\dot{p}>=-<V(x)>.$$

For the narrow Gauss-shaped initial conditions the JWKB method allows to rewrite the second equation in the form

$$<\dot{p}>=-V(<x>)+O(h),$$

where h is a small parameter. So we have a system of two ordinary differential equations closed modulo $O(h)$.

V.V.Belov developed this idea in [Bel]. Using the JWKB-Maslov technique he derived a closed ODE system modulo $O(h^{3/2})$. Besides the averages of x, p the system contains the second moments

$$\mu_{ij}=<\xi_i\xi_j>-<\xi_i><\xi_j>$$

where ξ is a vector with the components x, p.

In this paper we study the Belov's system with $O(h^{3/2})$ thrown off. The applicability of such results as ours to the source problem is a non-trivial question. Some discussion and further generalization of the "Ehrenfest-Belov system" one shall can find soon in [BBK].

Hamiltonian Mechanics, Edited by J. Seimenis,
Plenum Press, New York, 1994

Two different sides of the system are considered here. First, we ascertain that the system is poissonian, i.e. it has some integrals depending only on the dimension of x-space (the Casimir functions), and the system restricted to the submanifold with fixed Casimirs is hamiltonian.

Second, we study the simplest non-trivial special case— the weakly anharmonic one-dimensional oscillator with the potential

$$V(x) = \frac{x^2}{2} + \epsilon\frac{x^4}{4},$$

where ϵ is small. The method of local nonlinear analysis (reducing to a normal form) [Br 1] allows to clear the behaviour of the solutions for $t \sim O(\epsilon^{-1})$ very explicitly.

Finally, it turns out that in this special case the system has besides two general integrals a third formal integral, i.e. the shortened system up to any order with respect to ϵ is completely integrable.

The author thanks M. F. Kondrat'eva for setting the problem, A. D. Bruno for numerous advices and helpful discussion, and Yu. A. Sadov for the help in numerical research.

1 General case: Casimir Integrals

The system under consideration has a form

$$\dot{\xi} = J\frac{\partial \tilde{H}}{\partial \xi}, \qquad J\dot{\mu} = [H''J, J\mu]. \tag{1}$$

Here $\xi = (x;p)$ is a set of phase variables,

$H(\xi)$ is a symbol of a source quantum problem (we do not describe it more accurately - see [Bel, BBK]),

$H''(\xi)$ is a matrix of second derivatives of H,

$$\tilde{H}(\xi,\mu) = H(\xi) + \frac{1}{2}tr(H''\mu),$$

$J = \begin{pmatrix} 0 & I \\ -I & 0 \end{pmatrix}$ is the standard symplectic matrix.

Let $\psi(\mu)$ be linear operators over symmetric $2n\text{x}2n$ matrices:

$$\psi(\mu)\nu = J\nu\mu - \mu\nu J.$$

Denote

$$\Phi(\mu) = J \oplus \psi(\mu) \tag{2}$$

linear operators on the (ξ, μ)-space.

Theorem

1. A family of operators (2) induces the Poisson bracket on functions depending on (ξ, μ) as below :

$$\{f, g\}(\xi, \mu) =< \Phi(\mu) \bigtriangledown_{\xi,\mu} f, \bigtriangledown_{\xi,\mu} g > \tag{3}$$

2. One can write the system (1) in the form

$$\frac{d}{dt}\begin{pmatrix} \xi \\ \mu \end{pmatrix} = \Phi(\mu) \bigtriangledown_{\xi,\mu} \tilde{H}(\xi, \mu).$$

3. For any λ a function

$$det(J\mu - \lambda I)$$

is the integral of the system.

4. The corank of the Poisson bracket (3) equals to n in any point where all the eigenvalues of matrix $J\mu$ are different. So the system has n functionally independent Casimir integrals. For example, one can take the even coefficients of the characteristic polynomial of $J\mu$ (except the trivial $2n$-th coefficient 1).

The $n+1$ th general integral is $\tilde{H}(\xi, \mu)$, the Hamilton function on symplectic sheets.

2 Special case $H(x,p) = \frac{1}{2}(x^2 + p^2) + \frac{1}{4}\epsilon x^4$

The shortened Normal Form and behaviour of solutions for $t \sim O(\epsilon^{-1})$

If the x-space is one-dimensional the matrix μ has three independent entries

$$\alpha = \mu_{11}, \qquad \beta = \mu_{12} = \mu_{21}, \qquad \gamma = \mu_{22}.$$

The indicated symbol $H(x,p)$ gives the following specific 5-dimensional system

$$\begin{aligned}
\dot{x} &= p, \\
\dot{p} &= -x - \epsilon(x^3 + 3\alpha x), \\
\dot{\alpha} &= 2\beta, \\
\dot{\beta} &= -\alpha + \gamma + \epsilon\, 3\alpha x^2, \\
\dot{\gamma} &= -2\beta - \epsilon\, 6\beta x^2.
\end{aligned} \tag{4}$$

We regard the system with small ϵ. The alternative point of view is that ϵ isn't small but all the variables $x, p, \alpha, \beta, \gamma$ are small. These concepts are bound by a simple scale transformation.

2.1 Reduction to shortened normal form

This calculating procedure includes [Br1]:

a. Diagonalization of linear part of the system and finding the resonance relations;

b. Nonlinear change of variables with a view to get a system with resonance right part up to order ϵ;

c. Power transformation which allows to exclude at least one variable in the shortened normal form.

In our case the diagonalizing variables are

$$\xi = x + ip, \quad \bar{\xi} = x - ip, \quad \eta = \frac{\alpha - \gamma}{2} + i\beta,$$

$$\overline{\eta} = \frac{\alpha - \gamma}{2} - i\beta, \quad \mu = \frac{\alpha + \gamma}{2}.$$

with corresponding eigenvalues

$$\lambda_\xi = -i, \quad \lambda_{\overline{\xi}} = i, \quad \lambda_\eta = -2i, \quad \lambda_{\overline{\eta}} = 2i, \quad \lambda_\mu = 0,$$

so there are four resonance relations

$$\lambda_\xi + \lambda_{\overline{\xi}} = 0, \quad \lambda_\eta - 2\lambda_\xi = \lambda_{\overline{\eta}} - 2\lambda_{\overline{\xi}} = 0, \quad \lambda_\mu = 0. \tag{5}$$

To find the first approximation of normal form one can do without writing the nonlinear normalizing transformation ($\xi_1 = \xi + O(\epsilon), \mu = \mu_1$ and so on) in explicit form [Br1].

Being guided by relations (5), we define new variables

$$R = \xi_1 \overline{\xi}_1, \quad w = \eta_1 \xi_1^{-2}, \quad \overline{w} = \overline{\eta}_1 \overline{\xi}_1^{-2}, \quad u = \mu_1$$

(that is power transformation). It seems more convenient to deal with real variables, so we separate real and imaginary parts of w:

$$w = S + iT$$

Introducing the new time $d\tau = \frac{3}{4}\epsilon dt$, we can write the shortened normal form

$$\begin{aligned} \frac{dR}{d\tau} &= 2R^2 T, \\ \frac{dS}{d\tau} &= (R - 4u - 3RS)T, \\ \frac{dT}{d\tau} &= u + (4u - R)S + 2R(S^2 - T^2). \end{aligned} \tag{6}$$

A study of its solutions yields the information about solutions of (4) for $t \sim O(\epsilon^{-1})$.

2.2 A study of shortened normal form

The Casimir integral and the energy integral of the system (4) beget the integrals

$$E = \frac{1}{2}R + \mu, \qquad \delta = u^2 - R(S^2 + T^2)$$

of (6). In addition the shortened normal form has an integral

$$C_0 = R^2(S - \frac{3}{4}) + 3ER \tag{7}$$

So the shortened normal form is completely integrable. The above formulae yield the equation which concentrates the system's time-dependence, namely

$$\left(\frac{dr}{d\tau}\right)^2 = -64E^2 P_4(r)$$

where:

$$\begin{aligned} r &= R/E, \\ P_4(r) &= \frac{5}{16}r^4 - r^3 + (\frac{3}{8}l + \frac{1}{4}\kappa + \frac{3}{4})r^2 - r - \frac{1}{2}l + \frac{1}{16}l^2 \\ l &= \frac{C_0}{E^2}, \qquad \kappa = \frac{\delta}{E^2} \end{aligned} \tag{8}$$

It is valid if $E \neq 0$. We omit here the formulae concerning to the case $E = 0$.

One can seem that a space of parameters determining the behaviour of solutions of the shortened normal form is a projective plane $(E : \delta : C_0)$. But the sign of E is also of importance because R should be nonnegative. Indeed, the space of (essential) parameters is 2-dimensional sphere.

The motion in the system (6) is everywhere finite. Some sets of parameters do not correspond to any motion, the others in general case correspond to one or two periodic motions of (6) . The areas with different numbers of the motions are separated by branchs of the discriminant curve of the polynomial (8). (See the Figure).

Note that a cycle in the phase space of the system (6) corresponds to a torus in the one of (4).

3 Special case (following)

The formal Third Integral

Theorem. For any given integer $N >= 1$ there exists a function C_N which coincides with (7) modulo $O(\epsilon)$ such that

$$\dot{C}_N = O(\epsilon^{N+2}).$$

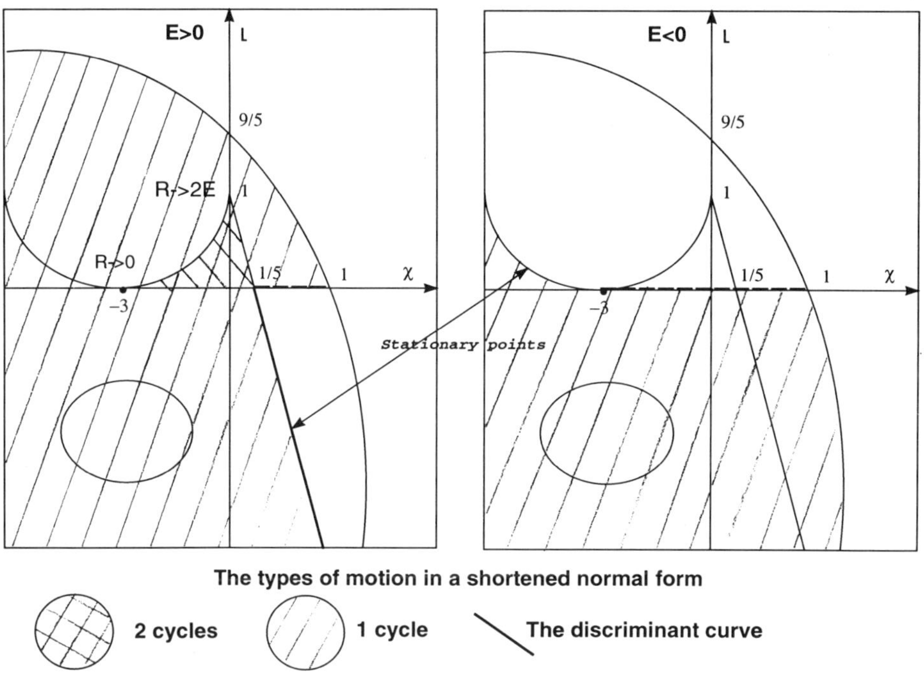
E>0
L
9/5
R->2E
1
R->0
1/5
1
χ
−3
Stationary points
E<0
L
9/5
1
1/5
1
χ
−3
The types of motion in a shortened normal form
2 cycles
1 cycle
The discriminant curve

Figure 1.

Corollary. The trajectories of the normal form of the system (4) shortened up to any finite order with respect to ϵ lie on two-dimensional invariant tori. If the initial data are outside some algebraic set of positive codimension the motion is quasi-periodic.

The author proved the theorem by regarding the iteration process of prolongation of the approximate integral. A simple proof refering to volume-preservingness was proposed by A.D. Bruno. (Unpublished).

The question whether there exists a true third integral (i.e. whether a series defining the formal integral converges) is open.

References

Bel) V.V. Belov, *Quasi-classical trajectory-coherent approximation in the quantum theory*, Thesis (D.Sc.), Moscow,1991, p.130.

BBK) V.G. Bagrov, V.V. Belov, M.F. Kondrat'eva, *Quasi-classical approximation in quantum mechanics.* A new approach, being printed in: Theor.& math.phys.

Br 1) A.D.Bruno, *Local methods in nonlinear differential equations*, Springer-Verlag, Berlin, 1989.

Br 2) — , *Bifurcation of the periodic solutions in the symmetric case of a multiple pair of imaginary eigenvalues*, Selecta Mathematica,v.12, 1, 1993.

INDEX

Adiabatic invariants, 139
Analyzing wavelet, 79-80
Andoyer-Deprit special coordinates, 157-158
Antibracket, 354
Anti-Hamiltonian mechanics, 354, 356-358
Arnold action-angle variables, 91-94
Arnold diffusion, 65, 136, 202
Arnold-Liouville theorem, 91, 205, 214
Artificial families, 82
Asteroids, 73-74, 99
Asymptotic chaos, 344
Asymptotic curves, 64
Averaging theorem, 14

Banach algebras, 107
Band-wavefunctions, 291-292
Bifurcations, 32, 54, 65
Bifurcations subharmonic, 227
Birkhoff-KAM procedure, 234
Birkhoff normal form, 184
Bisection method, 263-264
Bolzano existence criterion, 263

Cantor families, 115
Cantor set, 62
Cat maps, 281
Chaotic layer, 92
Chaotic regions, 51, 83, 89, 164
Chaotic scattering, 62
Chaotic separatrix, 134
Chirikov criterion, 93-94
Cluster analysis, 73
Collisions of bifurcations, 65
Complex instability, 65, 402
Courant-Synder
  transformation, 199-200
  complex coordinates, 199

Eigenfunction-Fourier series, 104-105, 109, 113,118
Elliptic orbit, 46
Escape energy, 62
Escaping orbits, 60
Euler elastica, 341, 343
Extension theorem, 185

Fermi-Ulam model, 297
Floquet theorem, 247
Forcing frequencies, 33
Fourier expansion, 89

Generalized van der Waals Hamiltonian, 190, 241
Gibbs
  calculation, 132
  entropy, 295-296
Gibbsian estimate, 132
Global action, 215
Global angle, 215

Hamiltonian vector field, 20
Henon map, 202-203, 265
Heteroclinic intersection, 52
Heteroclinic orbits, 176, 342
Hierarchical clustering method, 75, 77
Hilbert spaces, 106-107
Hill
  equations, 199
  problem, 44
Homoclinic intersection, 52, 182
Homoclinic lines, 70
Homoclinic orbits, 32, 173, 380, 385, 388
Homoclinic points, 70
Homogeneity dynamical, 84
Homogeneity topological, 85
Homologic equation, 38, 219
Hopf variables, 238-239

Intersection constant, 112
Intersection property, 112
Invariant curves, 59, 62
Invariant manifolds, 23
Invariant torus, 23
Irregular families, 54
Isomonodromy solution, 391
Iterative lemma, 300, 302

KAM
  theorem, 97, 233, 300, 306
  theory, 90, 97, 231
  tori, 123-124, 136, 211, 239, 335-339
  surfaces, 124
Kirchoff equations, 163, 179
Kirkwood gaps, 82, 98
Kowalevski
  exponents, 4, 11
  integrable case, 158

Landau-Teller approximation, 143, 146
Large Hadron Collider, 202
Lax Pair theory, 254, 321, 364
Lax representation, 324, 361

Lobe, 56-60
Locally Hamiltonian, 169
Lyapunov
  exponents, 124, 126-128, 317
  number, 68
  orbit, 60-61
Lyapunov-Schmidt procedure, 105

Magnetic lattice, 198
Manakov
  case, 372
  system, 368
Markov chain modelling, 85
Maupertuis principle, 333
Melnikov
  function, 27, 30, 32
  method, 15, 342, 377
Mexican hat function, 80
Monte Carlo mapping, 83
Multi-valued integral, 169
Multivariate data analysis, 77

Nash-Moser implicit function theorem, 105
Nekhoroshev
  theorem, 220, 299, 304
  theory, 90, 97
Nielsen trace maps, 277-278
Nonlinear Schrodinger equation, 103
Normal form theorem, 301, 303
Normal variational equations, 7, 10

Observer method, 153

Pade approximants, 35, 211
Painleve
  analysis, 371
  property, 3
  IV equation, 396-398
Paul trap, 190
Period doubling bifurcations, 65
Perturbation, 76
  harmonics, 89
  theory, 141, 173, 213
Poincare-Melnikov-Arnold integral, 173, 182-183
Poschel coupling lemma, 120
Pseudo-Anosov mapping, 282

Quantum probability density, 289
Quasiperiodic motion, 68, 74, 213
Quasiperiodic orbits, 125
Quasiperiodic solution, 20, 22

Resonances, 90, 300
  secular, 99
Restricted three-body problem, 43
Reversing symmetries, 308, 311, 346
Riemann-Hilbert problem, 392
Riemann metric, 167
Rotation number, 52

Saddle-center, 385, 388
Semi-numerical methods, 89
Siegel problem, 36, 40
Singular perturbation, 43
Singular point analysis, 2
Singular region, 110-111, 120
Sobolev spaces, 107
Soliton chaos, 341
Spiral helix, 67
Spirals, 64-65
Splitting, 25
Stochastic diffusion, 348-349
Stochastic layer, 134
Stochastic region, 55
Stochastic web, 345, 347
Sturm-Liouville problem, 110
Supersymmetric mechanics, 353-354, 356
Symmetric periodic orbit, 309
Symmetries, 308, 310-311, 346
Symmetry field, 168, 170
Symplectic map, 124, 199
Symplectic structure, 167

Time-periodic solutions, 103
Tisserand criterion, 84
Toda system, 253
Trace maps, 275
Trivial monodromy, 215
Turayev-Shilnikov theorem, 178

Wavefunctions, 287, 291
Wavelet analysis method, 75, 79
Wavepacket, 290
Weak chaos, 347
Whitney extension theorem, 119

Ziglin
  analysis, 6, 7, 10
  theorem, 7